A Dictionary of Chemical Engineering Practice

Sean Moran

Second Edition

Copyright © 2025 Sean Moran. All rights reserved.

No part of this publication may be reproduced or transmitted in any form or by any means, electronic or mechanical, including photocopying, recording, or any information storage and retrieval system, without permission in writing from the author.

This book and the individual contributions contained in it are protected under copyright by the Author (other than as may be noted herein).

Notices
Knowledge and best practice in this field are constantly changing. As new research and experience broaden our understanding, changes in research methods, professional practices, or medical treatment may become necessary.

Practitioners and researchers must always rely on their own experience and knowledge in evaluating and using any information, methods, compounds, or experiments described herein. In using such information or methods they should be mindful of their own safety and the safety of others, including parties for whom they have a professional responsibility.

To the fullest extent of the law, neither the author nor any contributor assumes any liability for any injury and/or damage to persons or property as a matter of products liability, negligence or otherwise, or from any use or operation of any methods, products, instructions, or ideas contained in the material herein.

For Ethel

Preface

The author of any nonfiction work should in my view have a clear definition of the intended scope of that work, and their title should match that scope, otherwise they are just setting the reader up for disappointment. So, this was originally titled Moran's Dictionary of Chemical Engineering Practice, a lexicon of my profession, Chemical Engineering, in which I define the meaning of the terms used professionally by my fellow practitioners of the design and operation of process plants.

However, it was never really just "Moran's Dictionary". That wouldn't be much use, as I already know what I mean! The purpose of the book is to help chemical engineers to communicate unambiguously with each other, and with members of the other professions and trades with whom we interact in professional life. Both my fellow chemical engineers, and those others use the same words to mean different things and different words to mean the same things. I have followed the approach I have developed in writing my previous books, seeking and considering the contributions of hundreds of chemical engineers worldwide on the content, to ensure that it represents a complete, current consensus view.

Even within engineering, there are terms which are commonly used in error, or easily confused for a number of reasons, including the presence of sub-sub-languages, usage differences between industry sectors, and differences between the jargon of chemical engineering and that of the other professions and trades which we work with. Therefore, this dictionary also contains some terms from more general engineering, management, quality etc. Practice does not exist in a vacuum: engineering is a multidisciplinary effort.

Let's also be clear on what this dictionary is not.

Firstly, this is not an encyclopaedia (a set of short articles on a wide range of subjects, summarizing factual knowledge). There are no articles, histories or biographies in this dictionary, though there are a few short discussions of selected terms which merited a more discursive approach.

This is a technical dictionary, rather than a general dictionary such as the OED. As a technical dictionary, this book does not define 'words' as such. The elements of technical vocabulary are terms, or to put it another way, the elements of a sub-language or jargon - the shorthand used within a group. Technical terms may, confusingly, mean different things from the everyday words they resemble.

This is not a dictionary of science. It is a dictionary of engineering, and engineering is not applied science. How can it be when engineering predates it? The engineering sciences which professional engineers utilize, such as thermodynamics, were devised *post hoc* to explain what engineering had already achieved, and the natural sciences are very little to do with engineering practice. Where scientific terms do appear in this dictionary, it is because they are relevant to practice.

By the same token, this is not a dictionary written to accompany the academic study of chemical engineering, though I believe many academics would benefit from using it. In my view (having practised both), academia and Chemical Engineering are entirely different professions. I have therefore left out of this book all that is taught - or held dear - in academia, but which is of no practical utility, as well as the majority of those terms which should be well understood by any graduate of chemical engineering. I have however made an exception for those elements of theoretical knowledge which we all passed exams in, but experience suggests many of us do not remember, or perhaps did not really understand in the first place.

I hope that my fellow practitioners will find this dictionary useful in their professional journey, whether preparing for professional examinations, entering a new sector, progressing in their own, working in a new territory, or collaborating with other disciplines.

Acknowledgements

It took more than three years to compile the first edition of this dictionary, so it was consequently mostly produced under pandemic conditions. A good part of the manuscript was drafted whilst I was working (and locked down) in the Panamanian jungle during the Spring of 2020, on a mining site where facilities where basic but decent Wi-Fi was nonetheless available.

There was no way I could have produced more than 10,000 definitions of terms, which all engineers were going to agree with, on my own. Being as correct as possible involves engagement with the people who use these terms - they mean what we think they mean. So, during a period when we were socially distanced from one another, the extent of the engagement of my fellow engineers in this project was particularly touching.

Some people who would not normally have had time to help found the dictionary a worthwhile distraction from global events. Others who sincerely wanted to help found themselves, understandably, with other priorities. Nevertheless, I would like to thank the great many engineers who assisted with corrections, opinions and necessary additions to the draft text, none more prolifically than Claudio Arato, Martin Armstrong, Laurence Ashley, Warwick Bagnall, Christopher Davis, Peter Elstner, Andy Gelbart, Susan Jaques, Bhavik Mehta, Tristan Reimann, Rick Rhead, Charles Sanderson and Felix Sirovski. Together, their advice and contributions had a significant impact on the content and presentation of the dictionary. Special thanks also to Tom Nicholson at Sulzer, who kindly arranged permission for me to reproduce several illustrations.

In addition, I am grateful to the many other engineers and specialists who took the time to offer comments, including Muhammad Muhammad Aatiq, Daniel Ali, Hassaan Ali, Sayed Ammar, Yacob Banitaha, David Baird, Yinke Bankole, Sanjay Bhagat, Muhammad Butt, Fabio Capezzuoli, VM Chenthil, Mukund Chiplunkar, Marc Clithero, Pukhraj Daga, Harvey Dearden, Roger Freestone, Ronald Frend, David Gladman, Pier-Paolo Greco, Matthew Green, Tim Highfield, Soran Hoseini, Satish Inamdar, Chaitanya Kalipatnapu, Naresh Kandlapalli, Mohan Karmarkar, Myke King, Steve Lancaster, Gert Lubbe, Miriam Abigail Manibo, Mohamed Nadine, Pabba Navya, Mark Nellist, B Nurfarahin, Peter Owoade, Rossiri Pacheco, Nacho Palacios, Dayang Radiah Binti Awang Biak, Alejandro Ramirez De Loza, Alun Rees, Rick Rhead, Ximena Rodriguez, Morgan Rodwell, Venkata Krishnan Sampath, Narinder Singh, Katherine Smith, Achintya Sujan, Jairo Vidal, Jim Warrell and Steven Woolley.

Many others who gave essential assistance are not mentioned - my apologies and thanks in equal measure to all of you.

The nature of the book means that public discussion has also been a key part of the process, as I have been able to draw on many useful social media interactions to clarify the meaning of words, e.g.; around the difference between 'slips' and 'dogs' (thanks due to Tyler Helm et al). I am grateful to everyone who got involved with these discussions, but particularly to Chris Brookes-Mann, Paddy Kitching, Vijay Sarathy, Grant Lukies and Stuart Smart who all engaged regularly and thoughtfully. These discussions were essential to my aims in writing the book, and were particularly useful in understanding the issues around 'risk' (there were many contributors there, but I would like to single out Miguel Piedras, who signposted me to an invaluable article by Chris Paris on the many and varied ISO definitions which proved very helpful).

Sean Moran
Derbyshire, UK
July 2025

Introduction

Commonplace, Generic and Specific Definitions

A technical dictionary such as this, should by definition contain only 'words' with specific meanings within a field of expertise – properly called industry terms. The commonplace meanings of the words are excluded. This is potentially problematic, as misunderstandings may arise when someone uses a word in its commonplace sense, but is understood to be using a technical term, or vice versa.

Several terms are marked in this book as having both 'generic' and 'specific' meanings. 'Generic' meanings are those used professionally in the same way across all industries, rather than being commonplace meanings. I have usually developed a generic meaning from a consensus of my understanding, that of my correspondents, and possibly a number of codes and standards. A generic meaning is intended to convey a succinct view of what I understand to be the core meaning generically given to the term in our profession. 'Specific' meanings are not offered as an illustration of the generic meaning; indeed, they may actually clash with the generic meaning. If in your practice, you come across a term being used in the context (usually industry sector) I reference, then its use could easily be in the specific sense rather than the generic or commonplace ones.

Sector-Specific Terminology

One of the key ways in which terminology differs within our profession is across sectors. I have therefore divided our profession into ten nominal sectors for the purpose of categorizing terms: Hygienic, Biochemical Engineering, Oil and Gas, Bulk Chemicals, Fine Chemicals, Environmental, Products, General Engineering, Nuclear and Safety.

These sector classifications are, to some extent, subjective, and I am aware that engineers in different disciplines (especially oil and gas!) might well disagree with my classifications. However, these are intended simply to align approximately with my understanding of both the professional and linguistic distinctions. Someone in oil refining or petrochemicals might object to a term being lumped in with those involved in primary production, but they understand most of the jargon of that related field in a way that someone involved in pharmaceuticals will not. (I mention oil and gas because a disproportionate amount of the jargon in this dictionary, some of it quite coarse, emanates from this sector!)

We might add an eleventh category, engineering science, a heading under which those few terms from academia which have made their way into this dictionary would tend to fall. Where it has proven necessary to differentiate such academic terminology from professional terminology, I have flagged this up.

In some cases, I may not refer to an industry sector directly, but I signpost a formal definition in an industry-specific standard, such as for example an API standard. API standards are used in many sectors other than oil and gas, especially those influenced by the USA. However, as with my industry classifications, the boundaries of my sectors are linguistic rather than absolute.

It is reasonably frequent that different sectors use a term in different ways. Where I am aware of this, I have flagged this in the definition. For example, chemically combined water means different things to chemists, cement- and ceramic- specialists.

Swearing, Jokes and Insults

I have included some coarse language, as is normal practice in lexicography. Dictionaries (other than those intended for young children) routinely contain terms considered vulgar or offensive so, where these terms are part of engineering practice, I have included them, marking them as either potentially offensive, or offensive.

It is far easier to cause offence nowadays than when I first started in the profession, so my approach will probably not satisfy all readers, some of whom are bound to consider the inclusion of any coarse language at all 'unprofessional'. I would suggest however that swearing and slang can be professional, even when it is obscene. Engineers often use swearing for what research shows it is good for, in the same way as doctors: for stress relief (an effect called lalochezia), to build group solidarity, and for adding emphasis. Like doctors, we may use in-joke acronyms to help us disguise from outsiders that we are swearing; some of these acronyms appear in this dictionary.

These terms however constitute a tiny fraction of one percent of the content. Only people looking for them in the manner of an eleven-year-old boy looking for the naughty words in his first real dictionary are likely to find them, unless they have already heard the term and are wondering what it means.

I have also included some definitions which are 'engineer jokes': a certain flavour of wry commentary which defines words commonly used in engineering in a way which reflects some of the culture of the profession, and some of the nuance of the common understanding of these terms in the minds of professional engineers. I have however marked these as jocular, for avoidance of doubt.

Formal Definitions

Throughout the dictionary, I reference the 'formal definitions' of terms as found in codes and standards, though in almost all cases I have been unable to reproduce them for copyright reasons. This was initially a disappointment, until it became clear that many 'formal definitions' bear the hallmarks of design by committee. Many of them lack succinctness. Some appear to contain errors. Few of them are well written. In many cases the references to a standard and its formal definition are included because the standard is the source of the term, in others it is because there is a different meaning for the term in that context. Terms are not copyright, only definitions.

The point of including reference to the formal definitions is not that they are 'correct', even in their limited sphere of application. The point is that they have a good chance of being the consensus term in that sphere, or at least representing a commonly-held view. This may well prove crucial in the event of a dispute, and being unfamiliar with the definition given in relevant codes and standards is therefore something to be avoided.

Where I have found terms formally defined in different applications, I quote multiple standards. Where I am aware - through my experience or that of my collaborators - of wider applications of a term, I cite those.

For some terms, my references may not always be to the most current edition of a code or standard. Most often this is because the code or standard was updated during the writing process. I reference more than 200 codes and standards in this book, and it took three and a half years to write the first edition. Sometimes, the more recent version no longer defines the term. Occasionally, the version I had on hand was not the most current even on the day I started writing. In a few cases, the original standard was withdrawn some time ago and not replaced, but nevertheless continues to be commonly used.

There are also several cases in the dictionary of clashes between the formal definitions of terms. The variation between them ranges from subtle differences in terminology to complete differences in meaning (see risk for an extreme case). I have flagged where this is the case.

Erroneous, Deprecated and Contested Terms

I have included terms which are commonly used in error, marked as such, as well as a larger number of terms which are best not used at all, in order to avoid confusion; these are marked as deprecated.

However, while compiling this dictionary, I also found several terms, or clusters of interrelated terms, which were so ambiguous that a succinct definition simply could not be written. This was no surprise – in fact I was rather expecting this based on my experience of writing previous books.

My approach has been to avoid relying on the definitions in codes, standards and other dictionaries, especially non-technical dictionaries as being "correct". I have also resisted falling back on what the word/term used to mean when it was first coined, or on what the root version of the word meant in the classical language it was derived from.

Instead, I have taken a descriptivist, rather than prescriptivist, approach reflecting what is, rather than what should be. It does not matter what a term 'should' mean. The raison d'être of this dictionary, after all, is that different groups of engineers use terms to mean different things. I have therefore included more detailed discussion on such terms – though these are in some cases very much my own view on the matter, in some cases expressed in my characteristic occasionally somewhat provocative tone. I did say it was originally called Moran's dictionary!

You are of course free to differ on what you think the correct meaning to be and whether my view is correct, or wide of the mark. My point is that there are a range of opinions - not that mine is the only correct one, but that there is no correct one.

Disclaimers and Apologies

I am all too aware that this book falls short of perfection in ways which I will not be able to address before publication. Codes and standards are updated, popular culture and professional subcultures create new terms, and world events shape our priorities.

The content has been checked thoroughly by myself for completeness, and reviewed by hundreds of other engineers to help ensure correctness and a consensus position. It has been rechecked repeatedly during the course of copyediting and typesetting, and again in producing eth second edition. However, I know that there must be a small residue of errors, and indeed omissions, in a dictionary with over ten thousand entries.

If you find one, please do let me know. I can be contacted through LinkedIn, or directly at sean@independentexpertengineer.co.uk.

Nonalphabetic Terms

%DS	see **percent dry solids**; cf **dry basis**
μ	see **chemical potential**
μmax	see **maximum specific growth rate**
100 % speed	[of an item of equipment, often rotating]; see **rated speed**
12D process	see **botulinum cook**
1400° F (760° C) bypass interlock	An interlock intended to ensure that a combustion chamber is at a temperature above 1400° F (760° C); a term defined formally in NFPA 86: Standard for Ovens and Furnaces, 2019 Edition
1R, 1S	*aka* 2 x 100%. One running, one standby, in the context of equipment redundancy
2 psi regulator system	A liquid petroleum gas vapor regulation system combining a first stage regulator (qv), 2 psi service regulator (qv), and line pressure regulator. Defined formally in NFPA 58: Liquefied Petroleum Gas Code, 2017 Edition
2 psi service regulator	A liquid petroleum gas vapor regulator, delivering LPG at a nominal 2 psig (14 kpag), a term defined formally in NFPA 58: Liquefied Petroleum Gas Code, 2017 Edition
2 x 100%	see **1R, 1S**
2R, 1S	*aka* 3 x 50%. Two running, one standby, in the context of equipment redundancy
3 Rs of waste management	see **waste hierarchy**
3 x 50%	see **2R, 1S**
3-A	*aka* 3-A Sanitary Standards, inc. An organization which produces sanitary standards and accepted practices; used primarily in the USA dairy industry and intended to address the requirements of the USFDA (qv)
3-A Sanitary Standards, inc.	see **3-A**
3Cr12	*aka* stainless steel - ferritic - 1.4003. Arguably the lowest possible grade of stainless steel; it is magnetic, and some do not consider it to be stainless steel at all[1]
3-way valve	see **three way valve**
5 Ss	A Japanese organizational technique for workplace efficiency, based on five words which start with S in Japanese (and can be

---

[1] If you issue a tender invitation specifying only "stainless steel", 3Cr12 might be what you get!

	made to do so in English to some extent): seiri (sort), seiton (set in order), seisō (shine), seiketsu (standardize), and shitsuke (sustain); sometimes made into 6 Ss (qv) by including safety
5 whys analysis	A Japanese root cause analysis for problem-solving, made popular in the 1970's by Toyota, which involves asking 'why' you have a problem. If the answer to that 'why' is another 'why', continue until you reach the root cause or the fifth 'why'
50-year ground snow load	see **ground snow load**
6 sigma	see **six-sigma**
6 Ss	The **5 Ss** (qv) plus safety
8/8ths	*aka* 8/8ths basis, gross operated basis. 100% of the revenues associated with a site (1/8th of revenue being the traditional landowner's royalty for oil exploitation rights)
8/8ths basis	see **8/8ths**
80:20 rule	see **Pareto analysis**
₡	see **center line**
ΔH	Change in enthalpy (qv)
C€	see **CE mark**
ø	*(drawing notation)* Diameter
Φ	Greek character used in various engineering applications, most notably to denote phase; N.B.: this is not the diameter symbol ø (qv)

A

A&E	1. Architect and engineer
	2. Accident and emergency
A/G	see **above ground**
A/G tank	see **aboveground tank**
A/O process	see **anoxic/oxic process**
A0	An ISO paper size similar to ANSI 'E'
A1	An ISO paper size similar to ANSI 'D'
A2	An ISO paper size similar to ANSI 'C'
A2 flask	The shipping container used in the UK to transport spent nuclear fuel from power stations to reprocessing/storage at Sellafield
A3	An ISO paper size similar to US 'Tabloid'/'Ledger' and ANSI 'B'
A4	An ISO paper size similar to US 'Letter' and ANSI 'A'
AACE	Association for the Advancement of Cost Engineering
AAPM	Artificial advanced project management
AB	*(drawing notation)* Anchor bolt
abandonment	1. In the context of pipelines, permanently shutting down the operation of a pipeline or facility where regulatory approval has been received to do so; alternatively, it may refer to the actions to gain such approval
	2. In the context of oil wells, the activities performed to plug a well to prevent fluids from migrating up-hole from a reservoir and contaminating other formations and/or fresh water aquifers
abatement	Reducing a nuisance such as pollution or noise
ABE fermentation	*aka* Weizmann process. ABE (acetone butanol ethanol) fermentation is used to produce acetone, n-butanol, and ethanol from carbohydrates by bacterial fermentation with a *Clostridium* species, most commonly *C.acetobutylicum*
abiotic resources	The non-living parts of an ecosystem
ablation	In an engineering context, the removal of material by erosion, which finds practical application in coatings which protect against biological fouling and thermal effects
ablimation	The opposite of sublimation (qv) i.e. direct transition from vapor to solid phase
abort damper	see **abort gate**

abort gate	*aka* abort damper. A safety device which automatically vents fire, harmful gases and burning material to atmosphere; defined formally in NFPA 654: Standard for the Prevention of Fire and Dust Explosions from the Manufacturing, Processing, and Handling of Combustible Particulate Solids, 2017 Edition. May therefore be described as 'NFPA compliant'
above ground	*aka* A/G. A common designation on a P&ID (qv) or piping layout drawing to indicate if a component is above or below the surface of the Earth; cf under ground (U/G)
aboveground piping	Piping which is located above pieracks or supports according to standards (e.g. ASME Standard B.31.3)
aboveground storage tank	A stationary (often cylindrical) fluid containing vessel, with more than 90% of working volume above local grade
aboveground tank	*aka* A/G tank. A stationary (often cylindrical) fluid containing vessel installed (rather confusingly) above, at, or below grade (without backfill), according to NFPA 30: Flammable and Combustible Liquids Code, 2018 Edition
absolute alcohol	*aka* absolute ethanol. A high purity grade of ethanol
absolute density	1. In the context of ion exchange, the weight in grams of wet resin that displaces a given unit volume; cf relative density and specific gravity 2. (Other than in the context of ion exchange) the ratio of mass to unit volume
absolute error	The difference between a measured or inferred quantity and its true value; see **Box A1.**
absolute filter	A filter which removes 100% of hard spherical particles greater than its rated pore size under constant low pressure in a laboratory; this measure should not however be assumed to predict real world behavior; cf nominal filter
absolute humidity	The water content of air without allowance for temperature, calculated as the mass of water present divided by the total volume (sometimes mass) of air and water
absolute pressure	Fluid pressure relative to absolute vacuum; cf gauge pressure
absolute roughness	(*symbol* ε) A measure of the deviation of a surface from absolutely planar. Commonly used measures of absolute roughness include RMS and Ra; N.B. not the same as hydraulic roughness (qv); cf relative roughness, roughness
absolute temperature	see **Kelvin** or **Rankine**
absolute viscosity	*aka* dynamic viscosity; see **viscosity**
absorbance	see **optical density**

absorbent	*aka* absorbent material. A liquid or solid in which another material is dissolved/dispersed
absorber	1. In nuclear reactors, slightly contested, but may be a material which absorbs alpha particles, neutrons and ionising radiation, or solely a neutron absorber (qv), the more general 'absorber' being referred to as shielding (qv) 2. A unit operation dependent upon absorption, also sometimes used to mean absorbent (qv)
absorption	A mass transfer process in which one fluid component is taken into the bulk of an absorbent material; cf adsorption
absorption tower	*aka* column. Generically, a vertical vessel with a high aspect ratio in which absorption is used to separate a fluid mixture. In natural gas treatment it commonly refers only to acid gas scrubbing towers
ABV	Alcohol by volume; cf proof
AC	see **alternating current**
ACC	American Chemistry Council
accelerant	Generally, a substance which alters a second substance to increase the speed of a chemical process, such as e.g., a catalyst. The most common use of the word is however in the context of fire safety, in which an accelerant accelerates fire development.
acceleration due to gravity	(*symbol* g) the nominal gravitational acceleration of an object in a vacuum near the surface of the Earth. Defined in ISO/IEC 80000 as 9.80665 m/s^2
acceleration phase	In the context of biological growth, the phase of increasing growth in numbers of organisms preceding the log phase (qv)
acceleration transducer	see **accelerometer**
accelerator	see **accelerant**
accelerometer	A transducer which converts an input acceleration to a proportional electrical or electronic output; defined formally in BS ISO 2041:2018 Mechanical vibration, shock and condition monitoring. Vocabulary
acceptance certificate	A certificate formally confirming on behalf of a purchaser that a supplier has successfully met all contractual acceptance criteria
acceptance number	The maximum allowable number of defects in a batch of product for it to pass acceptance criteria
acceptance sampling	The process of testing a sample from a batch for quality assessment purposes
access doors	Doors allowing access for maintenance

access gauge	The space around an item of equipment which must remain unobstructed to enable safe access; defined formally in BS EN ISO 14122 Safety of machinery. Permanent means of access to machinery. Working platforms and walkways
access platform	A platform used for a person to access machinery; defined formally in BS EN ISO 14122 Safety of machinery. Permanent means of access to machinery. Working platforms and walkways
accessible	see **easily or readily accessible**
accessible design product	A design product intended to maximize the number of potential users; defined formally in BS EN 82079-1:2012 Preparation of instructions for use. Structuring, content and presentation. General principles and detailed requirements
accessways	Routes for access
accident	An unexpected occurrence that either leads to death or injury, or has a high probability of causing these; cf incident. Some safety specialists think that there are no true accidents/incidents, as there should be no unexpected occurrences.
Accountable Pipeline Safety and Partnership Act of 1996	A (USA) law intended to reduce the risk to public safety and the environment associated with pipeline transportation of natural gas and hazardous liquids
accumulation	1. The allowed pressure increase over the maximum allowable working pressure (qv) of a vessel during operation of a pressure relief device (qv); defined formally (and very slightly differently) in API RP 520 P1 7th Edition, January 2000 Sizing, Selection, and Installation of Pressure-Relieving Devices in Refineries; Part I - Sizing and Selection, API Standard 521. Pressure-relieving and Depressuring Systems. Sixth Edition | January 2014 and API RP 576 - Inspection of Pressure Relieving Devices; 2. Buildup of a material inside a mass balance envelope
accumulator	1. *aka* pulsation damper. A device which equalizes flow and pressure variation in the output of a reciprocating fluid transfer device 2. A tank in pulp and paper manufacturing used for energy recovery from condensed steam 3. Equipment that stores material in relatively minor quantities to eliminate fluctuation in a continuous process, such as a reflux drum (qv)
accuracy	The closeness of a measured or inferred value to its true value; cf absolute error, precision; see **Box A1**.

Box A1. Accuracy or Precision?

ISO standard 5725-1[2] provides definitions of trueness, accuracy and precision which are founded in statistics, but as my discussion of 'risk' (**Box R2**) shows, that is not the end of the story. This discussion is however informed by the ISO standard, as well as professional usage at varying degrees of formality. It should be noted that engineers might use these terms in any of the following senses, or in the commonplace way, so if in doubt, check!

I define accuracy in this book as "the closeness of a measured or inferred value to its true value", absolute error as "the difference between a measured or inferred quantity and its true value", and precision as "whether an instrument will give the same reading against the same true value the next time it is tested". However, I have written elsewhere[3] about how precision is often defined (by mathematicians) to mean what engineers call resolution, whilst precision is used by engineers to mean all kinds of things, including accuracy.

In engineering, we are using the language of mathematics[4] when we talk about 'implied precision': for example, if we need control of pH to within 0.1 pH units, this is not the same as saying that we need control to within 0.10 pH units, which implies 10 times the (mathematician's) precision. When someone gives us all of the figures on their calculator display without considering how many are significant, the implied precision is spurious, hence we call it 'spurious precision'. The number of significant digits is sometimes called resolution, and is always related to it, but this term is more commonly used for the smallest change in the measured value which an instrument can detect.

Accuracy (also known as, or closely related to trueness) is about the gap between the true value and the value indicated by the instrument. In the example above, I need pH to be controlled within the range around the set point ± 0.05 pH units, therefore my measurement accuracy needs to be reliably at least this good.

Precision in engineering, on the other hand, is not usually the mathematician's precision, nor to do with an engineer's definition of accuracy or resolution. Engineering precision relies upon reproducibility and repeatability, the first encompassing variability over time, and the second being precision under tightly controlled conditions over a short time period.

Related terms
implied precision, limit of detection, limit of quantification, resolution

---

[2] BS ISO 5725-1:1994 Accuracy (trueness and precision) of measurement methods and results. General principles and definitions
[3] Moran, S. (2019) *An Applied Guide to Process and Plant Design*, 2nd Ed. Oxford: Elsevier
[4] By 'the language of mathematics', I am excluding statistics. Statisticians define precision as the reciprocal of variance

acentric factor	A fudge factor (qv) used in thermodynamics to account for the degree to which molecules are not spherical, and therefore do not behave as predicted by the principle of corresponding states (which is based on an assumption of spherical molecules)
acetylene	The simplest alkyne (qv), a gas used along with oxygen for welding and cutting. Explosively unstable as a compressed gas, so normally supplied in an acetone solution under pressure
acetylenes	A generic (non-IUPAC) name for alkynes (qv) still in common use in industry
ACFM	1. Alternating current magnetic flux leakage testing, a plant inspection test method; defined formally in API RP 571 - Damage Mechanism Affecting Fixed Refinery Equipment. 2. Actual cubic feet per minute
ACGIH	see **American Conference of Governmental Industrial Hygienists**
ACI	American Concrete Institute
acicular coke	see **needle coke**
acid	A substance which can act as a proton donor, or less commonly can form a covalent bond with an electron pair
acid egg	An ovoid ceramic vessel used to transport small quantities of acids
acid gas	Strictly, any petroleum gas containing significant quantities of gases which form an acidic solution in water, one such being H_2S, leading to confusion between acid gas and **sour gas** (qv). All sour gases are acid gases but not all acid gases are sour gases; cf **sweet gas**
acid gas respirator	*aka* organic vapor respirator. PPE which provides protection against acidic vapors, chlorine and certain organics as well. Often a limited-life, highly portable item used on refineries and chemical sites handling chlorine
acid number	*aka* acid value, AV, neutralization number, acidity, total acid number. A measure of the amount of carboxylic acidic groups in a substance, measured as the mass of potassium hydroxide (KOH) in milligrams required to neutralize one gram of the substance, especially an oil. In the case of edible oils, it is an indicator of rancidity. In the case of mineral oil, a measure of corrosiveness
acid value	see **acid number**
acidified	Treated with acid, commonly to neutralise alkali present, or to occupy binding sites in ion exchange media or acidified zeolite catalyst

acidity	1. see **acid number**
	2. Ability to donate a proton, the opposite of basicity (qv)
acidogenesis	Generically, a process of acid formation. In environmental engineering, the stage of anaerobic digestion in which fatty acids are produced from more complex substances
ACM	Asbestos-containing material
Aco Channel	A genericized trademark for a sunken covered floor drain; there are a large number of manufacturers and styles; see **Figure A1**.

Figure A1. Aco Channels

acoustic emission testing	*aka* AET. A method for detecting cracks in metallic components, discussed in API RP 571 - Damage Mechanism Affecting Fixed Refinery Equipment
ACS	Automation control systems
ACT	see **automatic custody transfer**
actinides	Elements 90-103 (thorium to lawrencium), named after actinium
actinium	Element 89
activated carbon	An adsorbent material produced from various high-carbon materials such as coal, coconut shells or vegetable matter
activated sludge	see **activated sludge process**
activated sludge process	A long-established process used to treat wastewater by recirculating sludge recovered by settlement after biological treatment to mix with incoming feed
activation energy	(*symbol* E_a) The energy required to initiate a chemical reaction. Catalysts (qv) work by reducing this
active hot standby	Having parallel standby (qv) unit(s) in service, rather than idling
active optoelectronic protective device	Machine safety device based on light sensing; defined formally in BS EN ISO 12100:2010 Safety of machinery. General principles for design. Risk assessment and risk reduction
active pharmaceutical ingredient	*aka* API. The medicinal content in a pharmaceutical product; cf active pharmaceutical intermediate
active pharmaceutical intermediate	*aka* API. A compound forming a step in the chain of syntheses which produce active pharmaceutical ingredients (qv)
active site	The part of a catalyst which binds to a substrate to facilitate a reaction. Almost exclusively used in the context of enzymes

activity	1. In project management, the smallest increment of work; defined more formally in BS EN ISO 9000 Quality management systems Fundamentals and vocabulary 2. The ratio of the change in two related conditions, such as e.g. water activity (qv) 3. In the context of technical documentation, a process, procedure or parts of them; defined formally in ISO 10209:2012 Technical product documentation — Vocabulary — Terms relating to technical drawings, product definition and related documentation
activity coefficient	*aka* fugacity coefficient. A fiddle factor (UK) or fudge factor (USA) used in thermodynamics to account for deviations from ideal behavior in a mixture of chemicals
activity coefficient model	see **Margules' activity model**
activity matrix	A document similar to a project program, placing activities in phases of a product lifecycle; defined formally in ISO 10209:2012 Technical product documentation - Vocabulary - Terms relating to technical drawings, product definition and related documentation
ACTR	see **actuating element**
ACTS	see **Asbestos Contractor Tracking System**
actual cubic feet per minute	*aka* ACFM. A USA customary measure of gas flow defined in a number of ways; cf standard cubic feet per minute
actual discharge area	*aka* actual orifice area. The minimum area that determines the flow through a pressure relief device; defined formally in API RP 520 P1 7th Edition, January 2000 Sizing, Selection, and Installation of Pressure-Relieving Devices in Refineries; Part I - Sizing and Selection
actual flow rate	The flow rate under real working/flowing conditions; defined formally in BS EN ISO 13705:2012/ISO 13705:2012(E) Petroleum, petrochemical and natural gas industries. Fired heaters for general refinery service
actual inside diameter	The inside diameter of a new heat exchanger tube, defined formally in API STD 530 - Calculation of Tube Heater Thickness
actual orifice area	see **actual discharge area**
actuated liquid withdrawal excess flow valve	An adapter operated valve used to withdraw liquid from an LPG container, with integrated excess-flow valve; defined formally in NFPA 58: Liquefied Petroleum Gas Code, 2017 Edition
actuated valve	*aka* AV, CV, motor operated valve. A valve controlled by an actuating element (qv); cf manual valve

actuating element	*aka* ACTR, actuator. An item of instrumentation equipment which receives a signal (pneumatic, hydraulic, or electrical) and adjusts a control element (e.g. valve, motor controller) according to that signal; defined formally in BS1646-1 Symbolic Representation for Process Measurement Control Functions and Instrumentation Part 1: Basic Requirements, ISO 3511/1 (W/D) and BS 1646-3:1984 Symbolic representation for process measurement control functions and instrumentation. Specification for detailed symbols for instrument interconnection diagrams
actuator	see **actuating element**
actuator valve	see **actuated valve**
AD	see **anaerobic digestion**
ADC	see **analogue to digital converter**
additive	1. Generally, a small amount of something added to enhance performance 2. Specifically, in the case of fire protection, foam concentrates, emulsifiers, and so on added to water to enhance firefighting performance; defined formally in NFPA 20: Standard for the Installation of Stationary Pumps for Fire Protection, 2019 Edition
additive pump	1. Generally, a pump used to introduce additives, in which case synonymous with dosing pump 2. In a firefighting context, defined formally in NFPA 20: Standard for the Installation of Stationary Pumps for Fire Protection, 2019 Edition
adequate risk reduction	Risk reduction meeting legal requirements at the very minimum; defined formally in BS EN ISO 12100:2010 Safety of machinery. General principles for design. Risk assessment and risk reduction
adequate ventilation	1. Sufficient ventilation to prevent the accumulation of significant quantities of fuel air mixtures at >20% of their LEL (qv) /LFL (qv); defined formally in API RP 505 2nd Edition, August 2018 Recommended Practice for Classification of Locations for Electrical Installations at Petroleum Facilities Classified as Class I, Zone 0, Zone 1, and Zone 2 2. May also sometimes be used in a broader sense, of ventilation sufficient to maintain a work area as a safe place; cf inadequate ventilation

adiabatic	Strictly, a condition in which heat does not enter or leave the system concerned; defined formally in API RP 2510A - Fire Protection Considerations for the Design and Operation of Liquefied Petroleum Gas (LPG) Storage Facilities
adiabatic efficiency	In the context of compressors and fans, the power theoretically necessary to compress and deliver a gas adiabatically divided by the power supplied at the driveshaft
adiabatic flame temperature	The theoretical flame temperature under adiabatic conditions defined formally in API RP 535 - Burners for Fired Heaters at Refineries
adiabatic flash	see **flash evaporation**
adiabatic saturation temperature	The theoretical temperature of a volume of air if cooled adiabatically to saturation by the evaporation of water. It is approximately equal to the wet bulb temperature (qv)
adjustable guard	A machine safety device which can be adjusted; defined formally in BS EN ISO 12100:2010 Safety of machinery. General principles for design. Risk assessment and risk reduction
adjusted set pressure	Static inlet pressure at which a pressure relief valve is adjusted to open on the test stand; defined formally in API 2000, Venting Atmospheric and Low Pressure Storage Tank, 7th Ed
adjuster	*(jocular)* A hammer used for percussive maintenance (qv); cf Manchester screwdriver
adjutage	A tube or nozzle attached to a pipe or vessel used to control fluid discharge or allow pressure measurement
administrative controls	1. Generally, written procedures rather than (the more reliable, because people are not involved) software or hardware controls 2. In a safety context, essentially procedures to prevent overpressure protection being compromised; defined formally in API Standard 521. Pressure-relieving and Depressuring Systems. Sixth Edition | January 2014
ADR	Accord Européen Relatif au Transport International des Marchandises Dangereuses par Route; United Nations regulations on the transnational carriage of goods including a classification system
adsorbable organically bound halogens	*aka* AOX. A measure of organohalogens in water and soil or sludge used in the environmental field, similar to extractable organically bound halogens (qv)
adsorbate	Something which is adsorbed
adsorbent	*aka* adsorbent material. A solid to the surface of which fluid or solute molecules adhere (usually reversibly)

adsorption	A process in which fluid or solute molecules adhere (usually reversibly) to a solid surface
adsorption isotherm	A graph of the amount of adsorbate on an adsorbent as a function of pressure or concentration at constant temperature, usually normalized by mass of adsorbent. There are quite a few different models, including the Freundlich and Langmuir
adulterant	A contaminant which affects safety or effectiveness of a product, especially one for human consumption. The US FDA was established to prevent the adulteration of food and drugs
advance loss of profit insurance	*aka* ALOP insurance; see **delayed start up insurance**
advanced biofuel	see **second generation biofuel**
advanced gas-cooled reactor	*aka* AGR. The second generation of gas cooled nuclear power stations built in the UK from 1965-1988, which utilise a graphite core as the neutron moderator and carbon dioxide as the reactor coolant
advanced oxidation	A class of water treatment processes involving very strong oxidising agents, such as the hydroxyl free radical
advanced water purification facility	*aka* AWPF. Something of a euphemism for a plant which produces potable water direct from a feed of sewage
adventitious moisture	Surface moisture on coal; cf inherent moisture
adverse influence	In the context of food, something which makes a food less fit for consumption. More formal definition in BS EN 1672-2:2005+A1:2009 Food processing machinery. Basic concepts. Hygiene requirements
AET	see **acoustic emission testing**
aeration	Adding air, usually by mixing air and water. Most commonly achieved by adding fine air bubbles to water, recirculating through a venturi or jet aerator, or using a packed bed
aeration basin	A tank in which aeration of effluent (usually for biological treatment) is carried out
aeration number	*aka* Na, NAe. A dimensionless number used to analyze bioreactor aeration
aerial cooler	*aka* air cooled heat exchanger. Equipment that uses forced air convention to lower the temperature of a gas or a liquid, commonly 'fin-fan' (qv) heat exchangers, and employed in oil refineries as condensers for distillation columns
aerobic	*aka* oxic. In the context of biological effluent treatment, an environment in which elemental oxygen is present; cf anoxic, anaerobic
aerobic digestion	Biological digestion (usually of sludges) under aerobic conditions; cf anaerobic digestion

aerobic process	A process which takes place under aerobic conditions. Most commonly used in a microbiological context
aerobic treatment unit	Unit operations in which aerobic processes (qv) take place
aeroderivative	An electrical generator based on an aircraft jet engine, defined formally in ISO 3977 Gas turbines - Procurement - Part 3: Design requirements
aerogel	A type of very light, strong, fireproof material. They are all thermal (but not necessarily electrical) insulators, and are very expensive to manufacture, and were developed as a result of a bet between two scientists in the 1920s
aerogens	Noble gases (qv)
aerosol	A suspension of liquid droplets or solid particles in gas; defined formally in API RP 535 - Burners for Fired Heaters at Refineries
AET	see **acoustic emission testing**
aethalometer	Device used for monitoring air turbidity as a result of smoke pollution
AFC	1. *(drawing notation)* Approved for construction 2. Air fin cooler
AFD	1. *(drawing notation)* Approved for design 2. Adjustable frequency drive; see **VSD**
AFFF	see **aqueous film-forming foam concentrate**
affination	The first step in sugar refining, in which raw sugar is blended with hot concentrated syrup, and the mixture separated centrifugally
affinity laws	*aka* pump laws, fan laws (qv). A group of equations which allow the prediction of discharge characteristics of a pump or fan under given conditions from characteristics measured at a different speed, impeller diameter, or electrical supply frequency
AFP	Active fire protection (system)
AFPM	American Fuel and Petrochemical Manufacturers
afterburn	Unintended effect of mixing partially combusted gases with tramp air (qv), potentially causing combustion somewhere not intended to contain a combustion process
afterburner	A secondary combustion chamber, or direct thermal oxidizer installed downstream of an item of equipment; defined formally in NFPA 86: Standard for Ovens and Furnaces, 2019 Edition
AG	*(drawing notation)* Above ground or above grade
agglomerate	The product of agglomeration, or its associated verb

agglomeration	1. Generically, collecting fine particles together loosely and reversibly to make larger ones, e.g. flocculation (qv) in water treatment, or the 'sticking together' of two or more sugar crystals during centrifuging and drying operations. There is some contention as to whether the difference between aggregation (qv) and agglomeration is strength of binding between particles, as given here. Whilst this appears to be the consensus view, it might be wise to confirm the meaning being used if the difference is significant 2. By analogy, bringing areas, documents etc. together, whether strongly or weakly
aggregate	The product of aggregation, its associated verb, or graded stones added to concrete
aggregated document	A collected group of individual independent documents; defined formally in BS EN ISO 10209:2012 Technical product documentation — Vocabulary — Terms relating to technical drawings, product definition and related documentation
aggregation	1. Making a stable suspension of fine particles (e.g. a colloid) unstable, as in coagulation in water treatment. Alternatively, binding particles together strongly. There is some contention as to whether the difference between aggregation and agglomeration (qv) is strength of binding between particles as given here. Whilst this appears to be the consensus view, it might be wise to confirm the meaning being used if the difference is significant 2. By analogy, bringing areas, documents etc. together, whether strongly or weakly
agitated vessel	*aka* stirred tank. A tank whose contents are agitated, normally (but not exclusively) mechanically
agitation intensity	A way to define mixing intensity either by reference to turnover time (V/Q) or mixer shaft power
agitator	A blade or propeller mounted on a drive shaft used to stir and mix a vessel's contents
AGO	Atmospheric gas oil; according to API RP 571 - Damage Mechanism Affecting Fixed Refinery Equipment
AGR	see **advanced gas-cooled reactor**
AHERA	Asbestos Hazards Emergency Response Act (US law)
AHJ	see **authority having jurisdiction**
Ahrens-Bode	see **Bode**
AHU	Air handling unit
AI	Active ingredient, analogous in the crop protection industry to active pharmaceutical ingredient (API)

AIChE	American Institute of Chemical Engineers
AIG	Ammonia injection grid; according to API RP 536 - NOx control on Fired Heaters at Refineries
AIHA	American Industrial Hygiene Association
AIM	Architectural and industrial maintenance; in the context of US legislation on coatings
AIME	American Institute of Mining Engineers
air	You know, air!
air admission valve	see **air inlet valve**
air aspirating discharge device	A type of device which makes and discharges firefighting foam; defined formally in NFPA 11: Standard for Low-, Medium-, and High-Expansion Foam, 2016 Edition; cf non air aspirating discharge device
air blinding	A phenomenon in which air bubbles caught between liquid filter media particles impede flow through a filter
air blower	An air delivery device capable of higher pressures than a fan but lower than a compressor. Usually a low pressure gas compressor, such as a Roots blower (qv). However, sometimes used to mean a high pressure (centrifugal) fan
air blowout	*aka* blowout. Blowing debris out of pipes and equipment during commissioning, using compressed air
air compressor	A compressor (qv) for air
air conditioning	A process for controlling indoor air temperature and humidity; cf AHU
air cooled heat exchanger	see **aerial cooler**
air cooler	see **aerial cooler**
air driven diaphragm pump	Pneumatically driven diaphragm pump (qv)
air driven mixer	see **pneumatic mixer**
air fail closed	see **air-to-open**
air fail open	see **air-to-close**
air filter	A filter which removes particulate solids and liquid droplets from air
air fuel gas mixer	A device to combine proportions of air and gaseous fuel prior to combustion; defined formally in NFPA 86: Standard for Ovens and Furnaces, 2019 Edition
air handler	see **air handling unit**
air handling unit	*aka* air handler. Part of a modular HVAC system, usually comprising an air moving device (qv), and ancillary heating, cooling and filtration etc.

air inlet valve	*aka* vacuum breaker, vacuum relief valve, air admission valve. A valve used to let air into a vacuum system; defined formally in NFPA 86: Standard for Ovens and Furnaces, 2019 Edition, or more less formally, and generally, into any system which might develop a vacuum; cf syphon breaker, vent/vac valve
air jet mixer	A type of air / fuel gas mixer; defined formally in NFPA 86: Standard for Ovens and Furnaces, 2019 Edition
air lift	*aka* airlift. A type of simple low head pump, using compressed air to reduce the bulk density of a fluid to cause it to rise up a tube. Used to pump sludge in some types of packaged domestic wastewater treatment plant (qv)
air lift reactor	A reactor, the contents of which are circulated by the air lift (qv) principle
air lock	1. A gas bubble at a high point in a liquid system which restricts liquid flow, which may be an intentional feature or an unintended glitch 2. A device such as a double set of doors which separates an artificial (for example very clean and/or relatively high pressure) environment from the outside world, or other area with different hygiene quality. May be subdivided into material airlock (MAL) and personnel airlock (PAL) in hygienic industries
air-mix mixer	see **pneumatic mixer**
air mixing	In an ion exchange context, the process of mixing two ion exchange materials with different densities in water using air
air moving device	A powered device that moves air; defined formally in NFPA 654: Standard for the Prevention of Fire and Dust Explosions from the Manufacturing, Processing, and Handling of Combustible Particulate Solids, 2017 Edition
air preheater	A device that preheats air going into a fired heater; defined formally (and almost identically) in API STD 530 - Calculation of Tube Heater Thickness, API STD 560 - Fired Heaters for General Refinery Service and BS EN ISO 13705:2012/ISO 13705:2012(E) Petroleum, petrochemical and natural gas industries. Fired heaters for general refinery service
air quality plan	A plan for improvement of ambient air quality with respect to sulfur dioxide (SO_2), nitrogen oxides (NOx), particulate matter (PM10, PM2.5), lead, benzene and carbon monoxide which the EU Ambient Air Quality Directives (AAQDs) require EU member states to establish

air quality standard	In the EU/UK, pollutant concentrations over time considered scientifically acceptable with respect to health and environmental effects
air register	An adjustable vent which controls flow of incoming combustion air; defined formally in API RP 535 - Burners for Fired Heaters at Refineries
air rumbling	The occasional addition of compressed air to the water flowing through a heat exchanger to help prevent fouling
air scour	Using compressed air to enhance solids removal in depth filter backwashing
air seal	*aka* gas seal, purge reduction device. A device preventing backflow of air from flare exit to riser; defined formally in API RP 537 Flare Details for Petroleum, Petrochemical, and Natural Gas Industries
air separation unit	*aka* ASU. Process plant for the separation of air, typically into argon, nitrogen and oxygen, usually through cryogenic distillation
air separator	Equipment which separates entrained air from water
air/fuel ratio	The ratio of flows of combustion air and fuel; defined formally in the context of fired heaters in API RP 535 - Burners for Fired Heaters at Refineries
airlift	see **air lift**
airline BA	Airline breathing apparatus; non self-contained PPE fed by an air line, allowing extended periods of work in an immediately dangerous to life or health atmosphere; cf escape BA, self contained breathing apparatus
air-material separator	A device which separates conveying air from the material conveyed; defined formally in NFPA 654: Standard for the Prevention of Fire and Dust Explosions from the Manufacturing, Processing, and Handling of Combustible Particulate Solids, 2017 Edition
airtight	see **gastight**
air-to-close	*aka* air fail open, fail open, FO. A pneumatically operated control valve which opens without motive air, but requires it to close
air-to-open	*aka* APO, air pressure to open, air fail closed, fail closed, FC. A pneumatically operated control valve which closes without motive air, but requires it to open
air pressure to open	*aka* APO; see **air-to-open**
AIS	Ammonia injection system; defined formally in API RP 536 - NOx control on Fired Heaters at Refineries
AISC	American Institute of Steel Construction

AISI	American Iron and Steel Institute
AIST	Association for Iron and Steel Technology
AIT	see **autoignition temperature**
ALARA	As low as reasonably achievable
alarm	A device which draws the attention of process operators to a defined significant abnormal condition; defined formally in BS1646-1:1979 Symbolic Representation for Process Measurement Control Functions and Instrumentation Part 1: Basic Requirements, ISO 3511/1 (W/D) and BS 1646-3:1984 Symbolic representation for process measurement control functions and instrumentation. Specification for detailed symbols for instrument interconnection diagrams
alarm diagram	A design document which details the alarms on a system; defined formally in BS EN ISO 10209:2012 Technical product documentation — Vocabulary — Terms relating to technical drawings, product definition and related documentation
alarm fatigue	*aka* alert fatigue. The ultimate effect on an operator of alarm flood (qv), desensitization to alarms due to sensory overload
alarm flood	When control room operators are overwhelmed due to the number of simultaneous alarm activations, resulting in the operator being unable to respond appropriately. If this is protracted, it can lead to alarm fatigue (qv)
alarm hierarchy	A method of prioritizing responses to various alarms (e.g. red alarm - immediate action required, yellow alarm - warning, blue alarm - for information only, etc.); can also be used to stop alarm flood (qv) from occurring
alarm management	Design and specification of alarms considering the human factors and human error probability (HEP) in appropriate response; according to EEMUA Publication 191 – Alarm systems: Guide to design, management and procurement
alarm point	The value of a parameter at which an alarm is triggered; defined formally in EN ISO 10437 Petroleum, petrochemical and natural gas industries - Steam turbines - Special-purpose applications
alarm table	The list of all alarms in a plant including their intent, set point and recommended operator action to return the parameter to its normal range
ALARP	As low as reasonably practicable; a legal standard applied in the EU
alc. denat.	*aka* alcohol denat., or denatured alcohol (qv)
alcohol denat.	see **denatured alcohol**

Alclad	Proprietary eponym for corrosion resistant laminated aluminium (aluminum) sheet
alcohol resistant foam concentrate	A firefighting foam concentrate used for to fight fires fuelled by hydrocarbons or materials which destroy other types of foams; as defined in NFPA 11: Standard for Low-, Medium-, and High-Expansion Foam, 2016 Edition
Alconox	Proprietary eponym for industrial detergent
alert fatigue	see **alarm fatigue**
Alfa Laval	Proprietary eponym for plate and frame heat exchanger (qv) (though the company manufactures a wide range of other equipment)
algal bloom	*aka* harmful algal bloom (HAB), red tide. Exponential growth to high numbers of algae (or bluegreen algae (qv)) caused by the discharge of high levels of nitrogen and phosphorus nutrients to a (most commonly warm) water environment. As well as causing environmental damage, harmful algal blooms (HABs) can be dangerous to wildlife and human health and damage process plants using the water for cooling or desalination. Red tide is considered either a synonym or a subset of this phenomenon
algorithm	A logical process or set of rules used in decision making, especially by computers
alignment line	A line drawn parallel to another for the purposes of alignment; defined formally in BS EN ISO 10209:2012 Technical product documentation - Vocabulary - Terms relating to technical drawings, product definition and related documentation
alkali	A water-soluble base, which might be something other than a metal hydroxide; cf base
alkaline stress corrosion cracking	Metal cracking caused by a combination of tensile stress and attack by alkali; defined formally in API RP 579 - Fitness for Service
alkalinity	Theoretically, the simple sum of the amount of hydroxides, carbonates and bicarbonates present in water, but somewhat more complex in practice. Alkalinity is defined by water specialists in terms of the amount of acid used to neutralise a sample to one of two end points provided by the indicators methyl orange and phenolphthalein, expressed as an equivalent amount of carbonate. These two end points yield methyl orange alkalinity (M-Alk) and phenolphthalein alkalinity (P-Alk) respectively
alkane	One of the homologous series of saturated aliphatic hydrocarbons, traditionally known as paraffins

alkene	One of the homologous series of unsaturated aliphatic hydrocarbons containing a C-C double bond, traditionally known as olefins
alkoxylation	Reaction with an epoxide compound (e.g. ethylene oxide)
alkylation	Adding an alkyl group
alkyne	One of the homologous series of unsaturated aliphatic hydrocarbons containing a C-C triple bond, traditionally known as acetylenes
alligator box	Oil drilling term for a very long box in which plastic core liners are stored
alligator teeth	Oil drilling term for rows of mounting pegs for each stand in a birdbath (qv)
allocation	Partitioning system inputs and outputs as part of LCA; defined formally in EN ISO 14040:2006 Environmental management - Life cycle assessment - Principles and framework
allowable nozzle loading	The amount of stress which can safely be exerted on suction and discharge nozzles by piping
allowable operating region	The portion of a pump's range over which it may be allowed to operate to avoid causing excessive vibration; defined formally in EN ISO 13709:2003 Centrifugal pumps for petroleum, petrochemical and natural gas industries; cf preferred operating region
alloy x	Nickel based materials generically referred to as 'alloy', e.g. alloy 20, alloy 50 and alloy 400
alluvium	Geological deposit left by flowing water
ALOP	Advance loss of profit insurance; see **delayed start up insurance**
alpha decay	Spontaneous decay releasing alpha radiation particles (or a helium nucleus)
alpha particle	A positively charged radioactive helium atom nucleus
alteration	Generically, a change in a document or artefact. There are a number of different specific formal definitions in various contexts in API RP 579 - Fitness for Service, API 510 10th Edition, May 2014 Pressure Vessel Inspection Code: In-service Inspection, Rating, Repair, and Alteration, Piping Inspection Code: In-service Inspection, Rating, Repair, and Alteration of Piping Systems, API 570 Fourth Edition, February 2016, API STD 653 5th Edition, November 2014 Tank Inspection, Repair, Alteration, and Reconstruction and NBBI NB 23 (NBIC) 2021 Edition, July 1, 2021 National Board Inspection Code

alternate power	An independent secondary power supply; defined formally in NFPA 20: Standard for the Installation of Stationary Pumps for Fire Protection, 2019 Edition
alternating current	*aka* AC. Electrical supply which reverses polarity at a frequency fixed with respect to time; cf direct current
alternative water supply	An independent secondary firewater supply; defined formally in NFPA 1142: Standard on Water Supplies for Suburban and Rural Fire Fighting, 2017 Edition
alternator	An alternating current generator
alum	A traditional name for a group of acidic aluminium-based chemicals used as coagulants in water treatment, such as aluminium sulfate (papermaker's alum), or more generally for double sulfate salts with potassium or sodium. More generally still, compounds with a similar structure, but with aluminium replaced by other trivalent metals e.g. chromium to give 'chrome alum', or even with sulfur replaced with another group 16 element
alumina	Aluminium oxide used in various forms as a desiccant, adsorbent, and refractory material
aluminium bronze	see **marine bronze**
Amagat's law	*aka* law of partial or additive volumes. A law which states that the volume of a mixture of gases is the sum of the volumes of the pure components. Since it assumes ideality, it is at best only approximate
amalgam	Alloy of mercury
ambient pressure	Local atmospheric pressure
ambient temperature	Local atmospheric temperature
ambient vibration	Local all-encompassing composite vibration; defined formally in BS ISO 2041:2018 Mechanical vibration, shock and condition monitoring. Vocabulary
ambulatory health care occupancy	A term used to describe a building or part of one used for outpatient treatment of individuals who will require assistance from others to escape in an emergency; defined formally in NFPA 30: Flammable and Combustible Liquids Code, 2018 Edition. Ambulatory means 'outpatient', N.B. NFPA 30 appears to assume that outpatients are not the walking wounded
AMD	Air moving device
American Conference of Governmental Industrial Hygienists	*aka* ACGIH. USA based professional association of industrial hygienists
American National Standards Institute	A USA institute that produces standards used to some extent worldwide

American Petroleum Institute	A USA trade organization that produces standards used in the oil and gas industry to some extent worldwide
American screwdriver	(*jocular*) see **Manchester screwdriver**
American Society for Testing and Materials	An obsolete name for what is now known as ASTM International
American Society of Mechanical Engineers	A USA professional society that produces standards used to some extent worldwide
American national standard pipe thread	see **national pipe thread**
amine gas treating process	*aka* amine treating. A process for removing acid gases from flue gas by scrubbing with amines
amine treating	see **amine gas treating process**
ammonia breakthrough	The point at which adding more ammonia to refinery flue gas NOx control systems does not reduce NOx; defined formally in API RP 536 - NOx Control on Fired Heaters at Refineries
ammoniacal nitrogen	*aka* NH3-N, NH4-N. A measure of ammonia concentration in wastewater; commonly used in environmental engineering
ammonia slip	The excess of ammonia in refinery flue gas after NOx removal; defined formally in API RP 536 - NOx Control on Fired Heaters at Refineries
ammonia/NOx ratio	The ratio of ammonia to NOx in a refinery flue gas; defined formally in API RP 536 - NOx control on Fired Heaters at Refineries
ammonia-soda process	see **Solvay process**
ampere	(*symbol* A) The SI unit of electric current
amp hour	see **ampere hour**
ampere hour	(*symbol* Ah) *aka* amp hour. A measure of charging capacity commonly used for batteries
ampere per metre	see **ampere per meter**
ampere per meter	(*units* A/m) *aka* ampere per metre
ampere per square meter	(*units* A/m^2) *aka* ampere per square metre
ampere per square metre	see **ampere per square meter**
amplitude	Size of a quantity; defined formally in BS ISO 2041:2018 Mechanical vibration, shock and condition monitoring. Vocabulary
amplitude distortion	A type of distortion in transducer output; defined formally in BS ISO 2041:2018 Mechanical vibration, shock and condition monitoring. Vocabulary

amplitude of displacement	Dimension of displacement by vibration; defined formally in BS ISO 2041:2018 Mechanical vibration, shock and condition monitoring. Vocabulary
amplitude of velocity	Speed of displacement by vibration; defined formally in BS ISO 2041:2018 Mechanical vibration, shock and condition monitoring. Vocabulary
AMS	see **air-material separator**
AMU	see **atomic mass unit**
anaerobic	In the context of biological effluent treatment, an environment without free oxygen or other common electron acceptor such as nitrate or sulfate; cf anoxic, aerobic
anaerobic digester	1. Generally, a reactor operating under anaerobic conditions. Also used in biochemical engineering in this sense albeit less commonly 2. Most commonly in engineering, a process used for the biological treatment of effluents and sludges with COD >2000 ppm
anaerobic digestion	The complex process which occurs in an anaerobic digester, with multiple stages undertaken by different groups of organisms. In broad outline, complex molecules are broken down into simpler ones; these into fatty acids (qv acidogenesis); and these finally into methane (qv methanogenesis)
analogue signal	A signal which varies continuously, (cf digital signal) usually in practice from 4-20mA
analogue to digital converter	*aka* ADC. A device which converts analogue signals into corresponding digital signals
analysis	As well as the generic meaning, it has a specific one in product development; defined formally in BS EN ISO 10209:2012 Technical product documentation – Vocabulary – Terms relating to technical drawings, product definition and related documentation
analysis of variance	*aka* ANOVA. A statistical method for unpicking influences, when coupled with Duncan's Multiple Range Test. Rapidly becomes impractically complex when more than two or three variables are considered
analysis paralysis	A lack of willingness to act caused by overthinking an issue, often produced by having gathered a mass of irrelevant data
anammox	The natural process of anaerobic ammonium oxidation, and a trademarked process for achieving this in a process plant

anchor	*aka* tieback. In the context of refinery heaters, a retainer of insulating or refractory linings; defined formally in BS EN ISO 13705:2012/ISO 13705:2012(E) Petroleum, petrochemical and natural gas industries. Fired heaters for general refinery service, API STD 530 - Calculation of Tube Heater Thickness and API STD 560 - Fired Heaters for General Refinery Service
anchor bolt	*aka* holding down bolt. A bolt which attaches equipment to its support; defined formally in EN ISO 10437 Petroleum, petrochemical and natural gas industries - Steam turbines - Special-purpose applications
ancillary equipment	1. Secondary, smaller or less critical/significant items of equipment 2. In the context of sewage treatment, pipe connections and internal components that are part of a septic tank; defined formally in EN 12566 - Small wastewater treatment systems for up to 50 PT Part 4: Septic tanks assembled in situ from prefabricated kits. N.B. This definition is rather unusual - pipe connections and internal components would not normally fall under this heading whilst 'equipment' would normally refer to functional assemblies of components
ancillary input	A less significant input of material in life cycle assessment (qv); defined formally in EN ISO 14040:2006 Environmental management - Life cycle assessment - Principles and framework
ancillary system	Generally, a less significant, secondary or supporting system; defined formally in BS EN ISO 10209:2012 Technical product documentation - Vocabulary - Terms relating to technical drawings, product definition and related documentation
anemometer	An instrument which measures windspeed
ANGA	America's Natural Gas Alliance
angle of pitch	In the context of steps, the angle between the pitch line and its horizontal projection; defined formally in BS EN ISO 14122 Safety of machinery. Permanent means of access to machinery. Working platforms and walkways
angle valve	Type of valve which the direction of inlet and outlet flow make an angle (usually 90°), often used when the pressure difference between inlet and outlet is very high.
angström	(*symbol* Å) A non-SI unit of length equal to 0.1 nm
angular acceleration	Acceleration of something with respect to one of its rotational degrees of freedom; defined formally in BS ISO 2041:2018 Mechanical vibration, shock and condition monitoring. Vocabulary

angular chart	A chart which shows the relationship between position and function of a feature; defined formally in BS EN ISO 10209:2012 Technical product documentation — Vocabulary — Terms relating to technical drawings, product definition and related documentation
angular dimension	The angle between features; defined formally in BS EN ISO 10209:2012 Technical product documentation — Vocabulary — Terms relating to technical drawings, product definition and related documentation
angular displacement	Displacement of something with respect to one of its rotational degrees of freedom; defined formally in BS ISO 2041:2018 Mechanical vibration, shock and condition monitoring. Vocabulary
angular frequency	*aka* pulsatance. The product of the frequency and 2π; defined formally in BS ISO 2041:2018 Mechanical vibration, shock and condition monitoring. Vocabulary
angular momentum	The rotational equivalent of linear momentum
angular transducer	A transducer which measures rotational motion; defined formally in BS ISO 2041:2018 Mechanical vibration, shock and condition monitoring. Vocabulary
angular velocity	Velocity of something with respect to one of its rotational degrees of freedom; defined formally in BS ISO 2041:2018 Mechanical vibration, shock and condition monitoring. Vocabulary
anion exchange resin	A positively charged polymeric substance which freely exchanges anions, usually supplied as a 'bead' or spherical particle
anisotropic	Having different properties in different directions, the opposite of isotropic (qv)
annealing	A heat treatment process for metals (usually steel) which reduces hardness and internal stresses; defined formally in ASME BPE (American Society of Mechanical Engineers: Bioprocessing Equipment)
annotated design model	A design model annotated to describe a product; defined formally in BS EN ISO 10209:2012 Technical product documentation — Vocabulary — Terms relating to technical drawings, product definition and related documentation
annotation	Drawing text giving required dimensions, tolerances and notes; defined formally in BS EN ISO 10209:2012 Technical product documentation — Vocabulary — Terms relating to technical drawings, product definition and related documentation

annotation plane	The conceptual plane on which annotations are drawn; described formally in BS EN ISO 10209:2012 Technical product documentation — Vocabulary — Terms relating to technical drawings, product definition and related documentation
annular distributor	1. Generally, anything ring-shaped which distributes a fluid 2. In the context of heat exchangers, a chamber which distributes shell-side fluids more evenly; defined in API 660 - Shell-and-Tube Heat Exchangers and API RP 661 - Heat Exchangers
annular flow	One of the various possible multi-phase flow regimes in a pipe, in which the liquid phase forms a ring in cross section against the pipe wall, and there is a continuous core of gas; cf slug flow, wispy annular flow
annular gas	see **casing gas**
annunciator	A visual indicator of alarm status, intended to draw the attention of the operator
anode	A positive electrode; cf cathode
anodeless riser	A device used to connect plastic and metallic pipes; defined formally in NFPA 58: Liquefied Petroleum Gas Code, 2017 Edition
anodic polishing	see **electropolishing**
anodize	To coat with a protective oxide layer by electrolysis
anolyte	Electrolyte in the vicinity of the anode
anomaly	In the context of bioprocess engineering, an out-of-specification part of a surface; defined formally in ASME BPE (American Society of Mechanical Engineers: Bioprocessing Equipment)
ANOVA	see **analysis of variance**
anoxic	In the context of biological effluent treatment, an environment without free oxygen, but with nitrate or sulfate present as alternative electron acceptors; cf anaerobic, aerobic
anoxic reactor	A reactor (usually a bioreactor) maintained under anoxic (qv) conditions
anoxic/oxic Process	*aka* A/O process. A biological effluent treatment process, a development of the activated sludge process (qv) with an added anaerobic (qv) selector zone (qv), reducing sludge bulking (qv) and promoting biological phosphorus removal (qv)
ANSI	see **American National Standards Institute**
ANSI flanges	Piping class (qv) of flanges from 150 to 2500; defined formally in ASME/ANSI B16.5 Pipe Flanges and Flanged Fittings: NPS 1/2 through NPS 24, Metric/Inch Standard 2020

ANSI pumps	*aka* AVS pumps. Strictly, ANSI pumps are single stage end suction centrifugal pumps built to the dimensional standards of ANSI. However, ASME pumps to ASME B73.1 Specification for Horizontal End Suction Centrifugal Pumps for Chemical Process are considered equivalent, and by extension ASME B73.2 Specification for Vertical In-Line Centrifugal Pumps for Chemical Process and ASME B73.3 Specification for Sealless Horizontal End Suction Centrifugal Pumps for Chemical Process are considered ASME/ANSI pumps. Typically used for less demanding environments compared to API pumps (qv) and are therefore usually cheaper
anthropogenic ground	*aka* anthropogenic soil, made ground, fill. Defined formally in BS 5930:2015 Code of practice for ground investigations
anthropogenic soil	see **anthropogenic ground**
anthropometric data	A data set of typical human body dimensions used to determine the operability and maintenance accessibility of equipment
anti icing system	Heater on an air filter inlet designed to avoid ice formation; defined formally in ISO 3977 Gas turbines - Procurement - Part 3: Design requirements
anti-agglomerants	Hydrate (qv) inhibition agents which keep hydrocarbon hydrates dispersed in the fluid phase, thus preventing them from forming larger agglomerations that could form a plug in a pipeline
anti-solvent crystallization	see **salting out**
antifoam	A chemical used to suppress foam production; cf defoamer
antinode	Place in a system where a wave characteristic peaks; defined formally in BS ISO 2041:2018 Mechanical vibration, shock and condition monitoring. Vocabulary
antirotation device	Part of a seal assembly which prevents rotation of one component relative to the adjacent one; defined formally in API RP 682 - Pump Seals
antithixotropic fluid	*aka* rheopectic fluids. In such fluids, constant shear stress causes an increase in viscosity over time
antithixotropy	see **rheopecty**
Antoine equation	A semi-empirical correlation which describes the relationship between vapor pressure and temperature for pure substance, used in practice to estimate vapor pressure (not always particularly accurately!)
Antonov's rule	An empirical equation which predicts the surface tension between two liquids in equilibrium
AOC	see **area of concern**

AOD	Air operated diaphragm pump
AODD	Air operated double diaphragm pump
AOPD	see **active optoelectronic protective device**
AOX	see **adsorbable organically bound halogens**
aperiodic vibration	A non-periodic vibration (qv); defined formally in BS ISO 2041:2018 Mechanical vibration, shock and condition monitoring. Vocabulary
APH	see **air preheater**
APHA color scale	see **color**
API	1. American Petroleum Institute; a trade association which produces many useful standards and design guides for those working in the sector. These standards are essentially the international standards of the oil and gas industry 2. see **active pharmaceutical ingredient** 3. see **active pharmaceutical intermediate**
API separator	see **API oil separator**
API 421 separator	see **API oil separator**
API ASME container	A container fabricated in accordance with the API/ASME Pressure Vessel Code, a term defined formally in NFPA 58: Liquefied Petroleum Gas Code, 2017 Edition
API ASME tank	see **API ASME container**
API gravity	Usually, an inverse measure of petroleum oil density (higher values of API gravity indicating a lower density) expressed in relation to a nominal water density of 10. There are however other API gravity standards based on materials with a gravity higher than water (e.g. caustic soda)
API oil separator	*aka* API 421 separator, API separator, API oil water separator. Equipment designed to separate oil from water by gravity according to API 421 guidance. Typically used in refinery or natural gas facilities in contaminated drainage systems
API oil water separator	see **API oil separator**
API pumps	Large horizontal single stage centrifugal pumps as described in API 610 and used in the petroleum industry. Typically, they are heavy-duty pumps for demanding processes, compared to ANSI pumps (qv)
APO	Air pressure to open
apparatus list	A list of system components; defined formally in ISO 10209:2012 Technical product documentation — Vocabulary — Terms relating to technical drawings, product definition and related documentation
apparent density	Material mass per unit volume without correction for voids
apparent mass	see **effective mass**

apparent viscosity	Viscosity at constant shear rate. For Newtonian fluids, viscosity is independent of shear rate, but it is not for non-Newtonian fluids
application reference model	An information model which sets out requirements and constraints for an application; defined formally in ISO 10209:2012 Technical product documentation - Vocabulary - Terms relating to technical drawings, product definition and related documentation
approach temperature	1. Generally, the temperature difference between process fluid leaving a heat exchanger and service fluid entering 2. In the case of a heat recovery steam generator (qv), the difference between steam saturation temperature and the economizer water exit temperature; defined formally in API RP 534 - Heat Recovery Steam Generators
approval	Confirmation by relevant authority that something meets specified requirements; defined formally in ISO 10209:2012 Technical product documentation — Vocabulary — Terms relating to technical drawings, product definition and related documentation
approval phase	The stage where a document is checked for approval; defined formally in ISO 10209:2012 Technical product documentation — Vocabulary — Terms relating to technical drawings, product definition and related documentation
approved	Confirmed by relevant authority to meet specified requirements, or to be acceptable; defined formally in API RP 505 2nd Edition, August 2018 Recommended Practice for Classification of Locations for Electrical Installations at Petroleum Facilities Classified as Class I, Zone 0, Zone 1, and Zone 2 and NFPA 11: Standard for Low-, Medium-, and High-Expansion Foam, 2016 Edition
approximation	A nearly - but not exactly - correct estimate of a value, or the process of obtaining such an estimate
appurtenance	1. Generally, an accessory 2. Specifically, an accessory required for a private fire service main's functioning; defined formally in NFPA 24: Standard for the Installation of Private Fire Service Mains and Their Appurtenances, 2019 Edition
APSPA	see **Accountable Pipeline Safety and Partnership Act of 1996**
AQL	A contested term for a quality threshold: held to stand for variously; acceptance quota level, acceptance quality level, acceptable quality limit or acceptable quality level
AQP	Air quality plan

AQS	Air quality standard
aqueous	Watery; in the context of chemical engineering, usually refers to substances dissolved or suspended in water
aqueous film forming foam concentrate	A firefighting foam concentrate of fluorinated surfactants and stabilizers which produces a fluid aqueous film for fighting hydrocarbon vapor fires; defined formally in NFPA 11: Standard for Low-, Medium-, and High-Expansion Foam, 2016 Edition
aquifer	An underground reservoir of water in rock, sand or gravel; defined formally in NFPA 20: Standard for the Installation of Stationary Pumps for Fire Protection, 2019 Edition
aquifer performance analysis	Testing used to determine the minimum spacing of firefighting wells in an aquifer; defined formally in NFPA 20: Standard for the Installation of Stationary Pumps for Fire Protection, 2019 Edition
AR	see **atmospheric residue**
arc	1. In drawing, a curved line with no point of inflection; defined formally in ISO 10209:2012 Technical product documentation — Vocabulary — Terms relating to technical drawings, product definition and related documentation. 2. In welding, an electrically induced plasma discharge
arc gap	In orbital gas tungsten arc welding (qv), this is the nominal distance prior to welding from the electrode tip to the weld surface; defined formally in ASME BPE (American Society of Mechanical Engineers: Bioprocessing Equipment)
arc strike	An imperfection in a weld caused by the passage of an electrical current; defined formally in ASME BPE (American Society of Mechanical Engineers: Bioprocessing Equipment)
arc welding	Welding using an electrical arc to provide heat of fusion as opposed to the chemical energy of fuel gas combustion; cf oxy-acetylene welding
arch	The portion of a fired heater's radiant section opposite the floor, as defined identically in BS EN ISO 13705:2012/ISO 13705:2012(E) Petroleum, petrochemical and natural gas industries. Fired heaters for general refinery service, API STD 530 - Calculation of Tube Heater Thickness and API STD 560 - Fired Heaters for General Refinery Service
arch burner	A downwards firing burner often located in the ceiling of a furnace combustion chamber

arch fired boiler	In US law, "a dry bottom boiler with circular burners, or coal and air pipes, oriented downward and mounted on waterwalls that are at an angle significantly different from the horizontal axis and the vertical axis. This definition shall include only the following units: Holtwood unit 6, and Sunbury units 1A, 1B, 2A, and 2B. This definition shall exclude dry bottom turbo-fired boilers."
Archimedean screw pump	A venerable type of pump comprising a rotating helical screw used to transport water from a low-lying body to an elevated one, still in everyday service at the inlet of sewage treatment works by virtue of its exceptional solids tolerance
architectural drawing	A general arrangement drawing for a building; defined formally in ISO 10209:2012 Technical product documentation – Vocabulary – Terms relating to technical drawings, product definition and related documentation
archive master	A master document produced in a format for long term storage; defined formally in ISO 10209:2012 Technical product documentation – Vocabulary – Terms relating to technical drawings, product definition and related documentation
archiving phase	Project stage where documents are moved from active to archive storage; defined formally in ISO 10209:2012 Technical product documentation – Vocabulary – Terms relating to technical drawings, product definition and related documentation
arcs and sparks	Energy released by flowing current which is a potential source of ignition; defined formally in API RP 2001 - Fire Protection at Refineries
Ardrox	A genericized trademark for surface preparation products commonly used in the aerospace and defence industries
area	Synonymous with location, according to API RP 505 2nd Edition, August 2018 Recommended Practice for Classification of Locations for Electrical Installations at Petroleum Facilities Classified as Class I, Zone 0, Zone 1, and Zone 2
area application of ultra high speed water spray system	In fire protection systems, essentially just what it sounds like; defined formally in NFPA 15 Standard for Water Spray Fixed Systems for Fire Protection

area classification	Classifying areas according to likely concentrations of flammable or explosive air/fuel mixtures. Defined formally in a number of places, including API RP 505 2nd Edition, August 2018 Recommended Practice for Classification of Locations for Electrical Installations at Petroleum Facilities Classified as Class I, Zone 0, Zone 1, and Zone 2, ISO 3977 Gas turbines - Procurement - Part 3: Design requirements, BS EN IEC 60079-10-1:2021 Explosive atmospheres - Classification of areas. Explosive gas atmospheres and BS EN 50281 -3 2002 Electrical apparatus for use in the presence of combustible dust. Classification of areas where combustible dusts are or may be present
area of concern	*aka* area of contamination, AOC. Zone of deleterious environmental conditions resulting from human activities in bodies of water (typically lakes)
area of contamination	see **area of concern**
ARFVTP	Alternative and Renewable Fuel and Vehicle Technology Program, instituted by the California Energy Commission
Argand diagram	A graphical representation of complex numbers
aromatic	A term applied to a substance which is odorous, or has a structure based on a ring or rings with delocalized pi electrons (most commonly a benzene ring). Also used in oil and gas for hydrocarbon mixtures ('aromatic hydrocarbons') with a high proportion of benzene ring-based components
around the pump proportioner	see **pump proportioner**
arrangement 1 seal	Having one seal per cartridge assembly; defined formally in API RP 682 - Pump Seals
arrangement 2 seal	Having two seals per cartridge and a containment seal chamber, defined formally in API RP 682 - Pump Seals
arrangement 3 seal	Having two seals per cartridge assembly with barrier fluid, defined formally in API RP 682 - Pump Seals
Arrhenius equation	An equation which relates temperature to reaction rate for gases
arrival area/exit	The surface which you step on to when exiting a ladder, or step off when accessing it; defined formally in BS EN ISO 14122 Safety of machinery. Permanent means of access to machinery. Working platforms and walkways
ARV	see **automatic recirculation valve**
AS	see **activated sludge**
Asbestos Contractor Tracking System	US EPA database of asbestos removal contractors and sites

as-built drawing	*aka* record drawing. An illustration showing what was actually built; defined formally in ISO 10209:2012 Technical product documentation — Vocabulary — Terms relating to technical drawings, product definition and related documentation and BS 4884-3:1993 Technical manuals. Guide to presentation (Replaced By: BS EN 82079-1:2012)
ASCC	see **alkaline stress corrosion cracking**
ASCE	American Society of Civil Engineers
aseptic	Free of pathogenic organisms (cf sterile); defined formally in ASME BPE (American Society of Mechanical Engineers: Bioprocessing Equipment)
aseptic equipment	According to EHEDG: "Hygienically designed equipment that is sterilizable and is impermeable to microorganisms to maintain its aseptic status"
aseptic process	According to EHEDG: "A process using equipment sterilized before use, and which, in running conditions, is protected against recontamination by microorganisms"
aseptic processing	Processing in a way that prevents contamination; defined formally in ASME BPE (American Society of Mechanical Engineers: Bioprocessing Equipment)
ash content	Mass of solids remaining after ignition (incineration) expressed as a fraction of total sample mass, used most commonly in the context of fuels as a measure of inorganic content; cf dry solids, volatile solids
ASHRAE	American Society of Heating, Refrigeration and Air-Conditioning Engineers
ASME	see **American Society of Mechanical Engineers**
ASME code	The American Society of Mechanical Engineers Boiler and Pressure Vessel Code
ASME container	see **API ASME container** and NFPA 58: Liquefied Petroleum Gas Code, 2017 Edition
ASME F&D	ASME flanged and dished; see **torispherical**
ASME flanged and dished	see **torispherical**
ASME I valve	A type of safety relief valve with two blowdown rings which conforms to the ASME pressure vessel code (Section I)

ASME pumps — In a chemical engineering context, there is some overlap between ASME pumps and ANSI pumps (qv), but ASME pumps are strictly those specified by ASME B73.1 Specification for Horizontal End Suction Centrifugal Pumps for Chemical Process and ASME B73.2 Specification for Vertical In-Line Centrifugal Pumps for Chemical Process and ASME B73.3 Specification for Sealless Horizontal End Suction Centrifugal Pumps for Chemical Process

ASME stamp — Accepted manufacturing standard of pressure vessel

ASME VIII valve — A type of safety relief valve which conforms to the ASME Pressure Vessel Code (Section VIII)

aspect — In a document management context, associated with interrogating a document to get information on a system; defined formally in ISO 10209:2012 Technical product documentation — Vocabulary — Terms relating to technical drawings, product definition and related documentation

aspect ratio — The ratio of the vertical to horizontal dimension

asphalt
1. *aka* bitumen, extra heavy oil. A term for a naturally occurring tar (qv) -like form of crude oil, or a very heavy distillate with a similar appearance. This term tends to be the preferred version in US English for the distillate, bitumen being preferred elsewhere
2. *aka* rolled asphalt, blacktop (USA), pavement (USA), tarmacadam, tarmac (UK). A road surfacing material comprising aggregate and a binding material (historically, but nowadays not necessarily, tar)

asphaltenes — *aka* asphaultines (sic).
1. A generic term essentially meaning the wide range of chemicals which make up asphalt (and are also disproportionately responsible for heat exchanger fouling and other problems in refineries)
2. In the oil and gas industries, material comprising large hydrocarbon chains, solid at room temperature and insoluble in n-heptane; cf wax

aspirator type foam generators — Generators using a stream of foam-aspirating air to produce firefighting foam; defined formally in NFPA 11: Standard for Low-, Medium-, and High-Expansion Foam, 2016 Edition

assembly — A construction of parts which achieves a function; defined formally in ISO 10209:2012 Technical product documentation — Vocabulary — Terms relating to technical drawings, product definition and related documentation

assembly drawing	A drawing illustrating how to assemble parts to achieve a function; defined formally in ISO 10209:2012 Technical product documentation — Vocabulary — Terms relating to technical drawings, product definition and related documentation
assembly instruction	A document illustrating how to assemble parts to achieve a function; defined formally in ISO 10209:2012 Technical product documentation — Vocabulary — Terms relating to technical drawings, product definition and related documentation
assembly model	A model illustrating how to assemble parts to achieve a function; defined formally in ISO 10209:2012 Technical product documentation — Vocabulary — Terms relating to technical drawings, product definition and related documentation
assembly occupancy	A term used to describe a building or part of one used for the assembly of 50 or more people, or the 'special amusement' of any number of people; defined formally in NFPA 30: Flammable and Combustible Liquids Code, 2018 Edition
assist	see **duty/assist**
assist gas	*aka* supplemental gas. A combustible gas added to relief gas to improve its heating value; defined formally in API Standard 521. Pressure-relieving and Depressuring Systems. Sixth Edition | January 2014 and API RP 537 Flare Details for Petroleum, Petrochemical, and Natural Gas Industries
assisted safety valve	A safety valve (qv) with a powered assistance mechanism to lift the valve at less than set pressure, as well as retaining normal functionality unassisted
associated apparatus	Equipment which supports the operation of intrinsically safe equipment (qv), whilst not itself being intrinsically safe; defined formally in API RP 505 2nd Edition, August 2018 Recommended Practice for Classification of Locations for Electrical Installations at Petroleum Facilities Classified as Class I, Zone 0, Zone 1, and Zone 2
associated entities	The things which an annotation refers to; defined formally in ISO 10209:2012 Technical product documentation — Vocabulary — Terms relating to technical drawings, product definition and related documentation
associated equipment	Things which are not themselves machines but are essential to the functioning of a machine; defined formally in BS EN ISO 14159:2008 Safety of machinery. Hygiene requirements for the design of machinery

associated group	User-defined grouping of elements; defined formally in ISO 10209:2012 Technical product documentation — Vocabulary — Terms relating to technical drawings, product definition and related documentation
association	1. In the context of QA and customer satisfaction, an organization; defined formally in BS EN ISO 9000 Quality management systems Fundamentals and vocabulary 2. In many cases, the opposite of dissociation. In the context of chemistry, an interaction between atoms or molecules, or between receptors and ligands
associativity	Established grouping of elements; defined formally in ISO 10209:2012 Technical product documentation — Vocabulary — Terms relating to technical drawings, product definition and related documentation
assumption	Something which is assumed to be true in order to simplify a problem, in the interests of solving it
ASTM	American Society for Testing and Materials; former name for ASTM International
ASTM International	Current name for what was known as the American Society for Testing and Materials
ASU	see **air separation unit**
ATEX	An abbreviation for 'ATmosphere Explosible', and for the European Directive 2014/34/EC concerning the placing on the market of explosion-proof electrical and mechanical equipment, components and protective systems
ATF	Aviation turbine fuel; see **jet fuel**
atline	A field measurement carried out local to the sample collection point; cf online, offline, inline
ATM	see **atmosphere**
atmolysis	Separating mixed gases by differential diffusion through membrane
atmosphere	*aka* ATM, standard atmosphere. A non-SI unit of pressure, equivalent to a standard atmosphere (qv)
atmosphere furnace	A furnace, the purpose of which is the production of a special processing atmosphere; defined formally in NFPA 86: Standard for Ovens and Furnaces, 2019 Edition
atmospheric aerosol particles	see **particulates**
atmospheric bottoms	see **atmospheric residue**
atmospheric	see **atmospheric residue**

atmospheric burner	A burner which requires a secondary air supply; defined formally in NFPA 86: Standard for Ovens and Furnaces, 2019 Edition
atmospheric discharge	Fluid release to atmosphere from pressure relieving devices; defined formally in API Standard 521. Pressure-relieving and Depressuring Systems. Sixth Edition | January 2014
atmospheric distillation	Distillation at around atmospheric pressure, especially crude distillation (qv); cf vacuum distillation
atmospheric distillation unit	see **crude distillation unit**
atmospheric fuel gas system	see **low pressure fuel gas system**
atmospheric gas oil	Gas oil (qv) produced by atmospheric distillation (qv)
atmospheric inspirator mixer	see **gas jet mixer**
atmospheric particulate matter	see **particulates**
atmospheric reduced crude	see **atmospheric residue**
atmospheric resid.	see **atmospheric residue**
atmospheric residue	*aka* atmospheric bottoms, atmospheric reduced crude, atmospheric resid, AR, atmospheric tower bottoms, heavy fuel oil, long residue, reduced crude, residue (petroleum), atmospheric, topped crude. Bottoms (qv) from the atmospheric distillation of crude oil
atmospheric tank	A tank with a headspace at operating pressure from 0.0-1.0 psig; defined formally in NFPA 30: Flammable and Combustible Liquids Code, 2018 Edition; cf low pressure tank
atmospheric tower	Distillation tower operating at around atmospheric pressure to separate components from crude oil usually prior to further processing i.e., carrying out atmospheric distillation (qv)
atmospheric tower bottoms	see **atmospheric residue**
atom balance	A material balance based on moles of elements
atomic mass	The mass of an atom in Daltons (qv)
atomic mass unit	*aka* Dalton (qv)
atomic number	*aka* proton number. The number of protons in the nucleus of an atom of an element
atomic volume	The volume of one gram-atom of an element under standard conditions
atomic weight	see **relative atomic mass**

atomicity	1. The state of being made of atoms 2. The number of atoms in a molecule
atomization	Breaking a liquid into tiny droplets, (usually entrained in a gas). This often precedes combustion of the liquid in chemical engineering; defined formally in API RP 535 - Burners for Fired Heaters at Refineries
atomizer	1. Generally, a device designed to effect atomization 2. In the petrochemical industries, a device which atomizes liquid fuel oil; defined formally in BS EN ISO 13705:2012/ISO 13705:2012(E) Petroleum, petrochemical and natural gas industries. Fired heaters for general refinery service and API STD 530 - Calculation of Tube Heater Thickness, and API STD 560 - Fired Heaters for General Refinery Service
atomizing air	Compressed air used as an atomizing medium; defined formally in ISO 3977 Gas turbines -- Procurement -- Part 3: Design requirements
atomizing burner	A combination burner and atomizer; defined formally in NFPA 86: Standard for Ovens and Furnaces, 2019 Edition
atomizing medium	A fluid (often air or steam for fuel oil) which is used to produce a fine spray of fuel prior to combustion
atomizing medium pressure switch	A switch which shuts down an oil burner if its atomizing medium pressure is too low; defined formally in NFPA 86: Standard for Ovens and Furnaces, 2019 Edition
Atmosphären Überdruck	see **technical atmosphere**
ATS	see **automatic transfer switch**
ATSDR	Agency for Toxic Substances and Disease Registry
attached growth	A description of a biofilm which adheres to a solid support, or systems reliant on such attached biofilms; cf suspended growth
attemporator	see **desuperheater**
attenuation	Reduction of intensity of a flux or signal
atto-	(*symbol* a-) The SI unit prefix denoting a factor of 10^{-18}
attribute	Non-visible but essential notation on a model; defined formally in ISO 10209:2012 Technical product documentation — Vocabulary — Terms relating to technical drawings, product definition and related documentation
attributes sampling	A type of quality sampling in which samples are assessed for the presence (or absence) of a particular attribute; cf variables sampling
attrition	Size reduction and fines (qv) production by breakage and abrasion of bed particles, applied to ion exchange beads, filtration and adsorption media

ATU	see **technical atmosphere**
ATÜ	see **technical atmosphere**
AUBT	Automated ultrasonic backscatter testing, according to API RP 571 - Damage Mechanism Affecting Fixed Refinery Equipment
audio frequency	A frequency normally audible by humans, defined formally in BS ISO 2041:2018 Mechanical vibration, shock and condition monitoring. Vocabulary
audit	A systematic, independent documented onsite investigative process; defined formally (and variously) in BS EN ISO 9000 Quality Management Systems Fundamentals and Vocabulary and ASME BPE (American Society of Mechanical Engineers: Bioprocessing Equipment)
audit client	Entity requesting audit; defined formally in BS EN ISO 9000 Quality Management Systems Fundamentals and Vocabulary
audit conclusion	Audit outcome; defined formally in BS EN ISO 9000 Quality Management Systems Fundamentals and Vocabulary
audit criteria	Documents defining a system against which compliance is checked; defined formally in BS EN ISO 9000 Quality Management Systems Fundamentals and Vocabulary
audit evidence	Evidence relevant to the audit criteria; defined formally in BS EN ISO 9000 Quality Management Systems Fundamentals and Vocabulary
audit findings	The auditors' opinion of their comparison of audit evidence with audit criteria; defined formally in BS EN ISO 9000 Quality Management Systems Fundamentals and Vocabulary
audit plan	A plan of audit activities and arrangements; defined formally in BS EN ISO 9000 Quality Management Systems Fundamentals and Vocabulary
audit programme	A planned, programmed purposeful audit or set of audits; defined formally in BS EN ISO 9000 Quality Management Systems Fundamentals and Vocabulary
audit scope	The planned scope of an audit; defined formally in BS EN ISO 9000 Quality Management Systems Fundamentals and Vocabulary
audit team	Auditors and any supporting technical experts; defined formally in BS EN ISO 9000 Quality Management Systems Fundamentals and Vocabulary
auditee	The entity being audited; defined formally in BS EN ISO 9000 Quality Management Systems Fundamentals and Vocabulary
auditor	An individual responsible for auditing an entity; defined formally in BS EN ISO 9000 Quality Management Systems Fundamentals and Vocabulary

auger	see **screw conveyor**
austenite	A non-magnetic allotrope of iron; cf martensite
austenitic	Iron or steel in the form of austenite (qv); defined formally in API RP 571 - Damage Mechanism Affecting Fixed Refinery Equipment; cf martensitic
austenitic stainless steel	The 300 series stainless steels; defined formally in API RP 571 - Damage Mechanism Affecting Fixed Refinery Equipment
authority having jurisdiction	The entity empowered with the right to enforce a code, standard, or other requirement; defined formally in NFPA 11: Standard for Low-, Medium-, and High-Expansion Foam, 2016 Edition
authorisation	see **authorization**
authorization	*aka* authorisation. Access privileges; defined formally in ISO 10209:2012 Technical product documentation - Vocabulary - Terms relating to technical drawings, product definition and related documentation
autocatalysis	A term which describes a reaction catalyzed (directly or indirectly) by one of its products
autoclave	A device akin to a large pressure cooker, which increases the boiling point of water to enhance temperature dependent thermal sterilisation, or the curing of high-performance composite materials
autogenous fillet weld	A fillet weld (qv) made without a filler rod, and therefore comprised of base metal; defined formally in ASME BPE (American Society of Mechanical Engineers: Bioprocessing Equipment)
autogenous weld	A weld made without a filler rod, and therefore comprised of base metal; defined formally in ASME BPE (American Society of Mechanical Engineers: Bioprocessing Equipment)
autoignition	Temperature-dependent ignition in a normal atmosphere without an external source of ignition; cf selfignition, spontaneous ignition
autoignition temperature	*aka* AIT, kindling point. The temperature at which something ignites in a normal atmosphere without an external source of ignition. Defined slightly differently in API RP 2001 - Fire Protection at Refineries, API RP 535 - Burners for Fired Heaters at Refineries, NFPA 86: Standard for Ovens and Furnaces, 2019 Edition and API RP 505 2nd Edition, August 2018 Recommended Practice for Classification of Locations for Electrical Installations at Petroleum Facilities Classified as Class I, Zone 0, Zone 1, and Zone 2; cf ignition temperature

automated inspection and testing	Remote electronic inspection and testing; defined formally in NFPA 20: Standard for the Installation of Stationary Pumps for Fire Protection, 2019 Edition
automatic aid	A type of plan for immediate mutual aid between fire departments; defined formally in NFPA 1142: Standard on Water Supplies for Suburban and Rural Fire Fighting, 2017 Edition
automatic burners	Burners whose firing rate is automatically controlled; defined formally in BS EN 12952-8:2002 Water-tube boilers and auxiliary installations. Requirements for firing systems for liquid and gaseous fuels for the boiler
automatic changeover regulator	An automated, integrated two-way gas regulator used for changeover on dual cylinder installations; defined formally in NFPA 58: Liquefied Petroleum Gas Code, 2017 Edition
automatic custody transfer	*aka* ACT; see **lease automatic custody transfer**
automatic detection equipment	In a fire safety context, the equipment which detects fire or explosion hazards and initiates fire alarms or protection; defined formally in NFPA 15 Standard for Water Spray Fixed Systems for Fire Protection
automatic drain valve	A valve which automatically drains water from pipes, valves and itself; defined formally in NFPA 24: Standard for the Installation of Private Fire Service Mains and Their Appurtenances, 2019 Edition
automatic drip	see **automatic drain valve**
automatic fire check	A flame arrester and linked gas shutoff valve actuated in the event of a backfire; defined formally in NFPA 86: Standard for Ovens and Furnaces, 2019 Edition
automatic recirculation valve	*aka* ARV. A valve used to maintain a minimum flow through a centrifugal pump by controlling a spillback (qv)
automatic shutdown	A turbine emergency stop; defined formally in ISO 3977 Gas turbines - Procurement - Part 3: Design requirements
automatic transfer switch	Equipment for switching from one power source to another; defined formally in NFPA 20: Standard for the Installation of Stationary Pumps for Fire Protection, 2019 Edition
automatic water spray nozzle	A directional spray nozzle which opens automatically in direct response to local temperature; defined formally in NFPA 15 Standard for Water Spray Fixed Systems for Fire Protection
automatic welding	Using equipment which welds without a need to manually adjust controls; defined in API STD 620 - Low Pressure Storage Tanks and ASME BPE (American Society of Mechanical Engineers: Bioprocessing Equipment); cf machine welding

autoradiolysis	A process in which a chemical compound is decomposed by its own radioactivity
autorefrigeration	A process where the vaporizing of a liquid (such as a hydrocarbon) results in chilling of equipment and/or piping, potentially affecting their mechanical properties; defined formally in API RP 2510A - Fire Protection Considerations for the Design and Operation of Liquefied Petroleum Gas (LPG) Storage Facilities
autothermal	1. Generically, self-heating 2. Commonly used to refer to sludge with sufficient oxidizable matter content for it to burn completely without the need to add additional fuel 3. A method for synthesis gas (qv) production
autotrophic organism	In biology, an organism that does not require organic compounds as a carbon source to make complex organic molecules. Chemoautotrophs power this process with various chemical reactions (chemosynthesis), whilst photoautotrophs use light (photosynthesis); cf heterotrophic organism
auxiliary dimension	A dimension derived from others and given for information only; defined formally in ISO 10209:2012 Technical product documentation — Vocabulary — Terms relating to technical drawings, product definition and related documentation
auxiliary system	A process-supporting system; defined formally in the context of power plant by ISO 10209:2012 Technical product documentation — Vocabulary — Terms relating to technical drawings, product definition and related documentation
AV	1. Acid value; see **acid number** 2. see **actuated valve**
availability	see **Box A2**. A contested term, most commonly the ability of equipment to function at a given time (uptime), expressed as a percentage of total time. This and other definitions however overlap with several definitions of reliability (qv)
available net positive suction head	*aka* NPSHa. The calculated NPSH for a system. A measure of how close the fluid is to flashing, and therefore a measure of how close a pump is to cavitating. It takes into consideration the minimum available static head at the pump suction, headlosses through suction pipework, liquid density at pumping condition, liquid velocity, suction ambient pressure, gravitational acceleration and vapor pressure of pumped fluid at pumping conditions

Box A2. Availability, dependability and reliability

These terms are used most commonly in our profession in the context of reliability engineering. Availability, dependability and reliability have interrelated and somewhat slightly contested meanings in reliability engineering.

I define availability in this book as "most commonly, equipment's ability to function at a given time (uptime), expressed as a percentage of total time. This and other definitions however overlap with several definitions of reliability." Reliability in turn is defined as "most commonly, the ability to perform the required function for a specified time under specified conditions in a given time interval; also used to refer to a probabilistic measure of this ability". I define dependability as the "ability to reliably work on demand".

The use of the phrase 'most commonly' in two of these definitions is intentional: in the case of availability, I have omitted its definition in the context of thermodynamics, whilst in the case of reliability I have sidestepped wider issues which I have discussed in more detail elsewhere[5].

Definitions of availability vary between equipment's 'ability to function at a given time', and the probability that equipment will be able to function, a slightly different concept. I have opted for the former. By contrast, the term reliability conveys the concept of dependability, successful operation/performance and the absence of failures, whilst unreliability (or lack of reliability) conveys the opposite. Since the process of deterioration leading to failure occurs in an uncertain manner, the concept of reliability requires a dynamic and probabilistic framework (see for example Blischke and Murthy, 2000[6]).

Dependability was in turn defined in ISO 9000-4[7] as a "collective term used to describe availability performance and its influencing factors: reliability performance, maintainability performance, and maintenance support performance".

As with all contested terms, the key is to understand the possibility that another person might be using a term to mean something different from your own understanding. In the case of availability and reliability the meaning of these terms can be significant: I have seen complex legal disputes which hinged on whether a given degree of availability had been achieved.

Related Terms
failure, common cause failure, common mode failure, failure mode and effects analysis, failures in time, fault tree analysis, mean time between failures, mean time to failure, mean time to repair, probability of failure on demand, preventive maintenance, reactive maintenance, safety integrity level

---

[5] Moran, S. (2022) Reliability in *Kirk-Othmer Encyclopedia of Chemical Technology*. New York, NY: Wiley
[6] Blischke, W.R. and Murthy, D.N.P. (2000) *Reliability: Modeling, Prediction, and Optimization*, New York, NY: Wiley
[7] ISO (1993) *ISO 9000-4 Quality Management and Quality Assurance Standards - Part 4: Guide to Dependability Programme Management*. Geneva, Switzerland: ISO; N.B. withdrawn

average heat flux density	Heat absorbed per unit area of exposed coil section heating surface; defined formally (and identically) in BS EN ISO 13705:2012/ISO 13705:2012(E) Petroleum, petrochemical and natural gas industries. Fired heaters for general refinery service, API STD 530 - Calculation of Tube Heater Thickness and API STD 560 - Fired Heaters for General Refinery Service
average shaft centreline plot	A plot of average shaft centerline position at a number of frequencies; defined formally in BS ISO 2041:2018 Mechanical vibration, shock and condition monitoring. Vocabulary
average velocity	*aka* mean velocity, superficial velocity. The volumetric flowrate divided by the crossectional area of flow
Avery	Proprietary eponym for scales
Avery Hardoll fitting	see **Hardoll fitting**
aviation turbine fuel	see **jet fuel**
Avogadro's law	One of the gas laws (qv) which states that at a given pressure and temperature, equal volumes of gases contain an equal number of molecules
avoirdupois pound	see **pound**
AVS pumps	Pumps with standard dimensions; also known as ANSI pumps (qv)
AVT	All volatile treatment, in the context of ion exchange
AWG	American wire gauge
AWPF	see **advanced water purification facility**
AWWA	American Water Works Association
axenic culture	A pure culture or monoculture of organisms
axial	In the direction of the axis; cf centrifugal, radial, transverse
axial alignment	see **runout**
axial compressor	A high pressure device in which a gas is compressed in an axial direction, parallel with the drive shaft
axial dispersion	Dispersion along the axis, important in modelling plug flow reactors; cf radial dispersion, transverse dispersion
axial displacement	Movement of a shaft in the direction of the axis
axial flow fan	A low pressure device in which a gas is compressed in an axial direction, parallel with the drive shaft; cf centrifugal fan, radial flow fan
axial flow pump	A low pressure device in which a fluid is moved in an axial direction, parallel with the drive shaft; cf radial flow pump

axially split For pump casings, capable of being split along the line of the drive shaft centerline; defined formally in EN ISO 10437 Petroleum, petrochemical and natural gas industries - Steam turbines - Special-purpose applications and EN ISO 13709:2003 Centrifugal pumps for petroleum, petrochemical and natural gas industries

axis Usually, in engineering, an abbreviation of axis of rotation

axonometric representation A parallel projection on a single plane; defined formally in ISO 10209:2012 Technical product documentation — Vocabulary — Terms relating to technical drawings, product definition and related documentation

azeotrope *aka* constant boiling mixture. A solution that retains its composition when distilled, because the composition of the vapor phase is the same as the liquid phase

azeotropic distillation Purifying a component of a mixture beyond its azeotrope (qv) by distillation, often achieved by the introduction of a third fluid or solvent, or by pressure-swing distillation

azeotropic mixture see **constant boiling mixture**

B

BA set	see **breathing apparatus set**
Babbitt metal	*aka* bearing metal. A soft low-friction alloy (of some combination of antimony, copper, lead, or tin) used in bearing surfaces
Babo's law	*aka* von Babo's law. A law which states that the vapor pressure of a solution decreases proportionally with the concentration of a solute
baboon spanner	*(informal)* see **Stillson**
back blowing	A procedure in which a drain line is blown into the buoyancy seal of a flare; defined formally in API RP 537 Flare Details for Petroleum, Petrochemical, and Natural Gas Industries
back draft damper	1. Generically, the fan equivalent of a nonreturn valve (qv) 2. There is a specific meaning in the case of gas turbines; defined formally in ISO 3977 Gas turbines - Procurement - Part 3: Design requirements
back plate	*aka* backing plate. A plate between some centrifugal pump's chamber and the gland packings of the seal
back pressure	*aka* backpressure. 1. Generically, and slightly informally, closely related to headloss or pressure drop, and often used interchangeably: the pressure at the outlet of a unit operation as a result of downstream static and dynamic pressure differentials 2. There is a specific meaning in the case of PRVs; defined formally in API Standard 521. Pressure-relieving and Depressuring Systems. Sixth Edition | January 2014, API RP 520 P1 7th Edition, January 2000 Sizing, Selection, and Installation of Pressure-Relieving Devices in Refineries; Part I - Sizing and Selection and API RP 576 - Inspection of Pressure Relieving Devices
back pressure valve	An oil industry term for a check valve installed in production well tubing hanger to isolate production tubing, allowing fluids to be pumped in from above, whilst holding pressure from below
back to back configuration	1. In a layout sense, it has the obvious meaning 2. In the case of pump seals, mounting both flexible elements of a dual seal between mating rings; defined formally in API RP 682 - Pump Seals

backfire arrester	A type of flame arrester (qv) with fuel supply shutoff and pressure relief; defined formally in NFPA 86: Standard for Ovens and Furnaces, 2019 Edition; cf automatic fire check
backflow	Reverse or return flow
backflushing	Cleaning by temporary reversal of flow direction; cf backwashing
background radiation	A local environmental level of ionizing radiation which is not attributable to deliberately introduced radiation sources
backing	The material that underlies the joint during a welding operation to ensure full penetration; defined formally in API STD 620 - Low Pressure Storage Tanks
backing pump	see **holding pump**
backmix reactor	see **continuous stirred tank reactor**
backmixing	The mixing of reaction products with reactants: i.e., where reactants or feed materials mix with other materials that were fed in slightly before or after. A continuous stirred tank reactor (cf) (CSTR) is a classic example of a backmixing process. Theoretically, backmixing is 100% in a CSTR, 0% in a plugflow reactor (cf), though neither of these are the case in practice; cf forward mixing
backpressure	see **back pressure**
backpressure regulator	A valve which modulates to maintain a set upstream pressure. The term is commonly used for valves which only reduce pressure under flowing conditions. The formal definition found in NFPA 20: Standard for the Installation of Stationary Pumps for Fire Protection, 2019 Edition however stipulates both static and flowing conditions, so this is not a universally held view; cf pressure reducing valve
backup device	see **standby**
backup layer	A refractory layer behind the hot face layer in a fired heater, defined formally (and identically) in BS EN ISO 13705:2012/ISO 13705:2012(E) Petroleum, petrochemical and natural gas industries. Fired heaters for general refinery service, API STD 530 - Calculation of Tube Heater Thickness and API STD 560 - Fired Heaters for General Refinery Service
backwashing	A special case of backflushing, applied to granular beds in water treatment, pumping water backwards through the filter media to expand or fluidize it, sometimes including intermittent use of compressed air (air scour(qv)) during the process
bacronym	An (often contrived) acronym derived from letters in an existing word, sometimes with humorous intent, or offered as a false etymology such as crud (qv)

BACT	see **best available control technology**
bad oil	see **slop oil**
badger	A cleaning unit for sewage pipes
BAFF	see **biological aerated flooded filter**
baffle	A plate or mechanical device which partitions a vessel, diverts flow, or assists mixing/heat transfer
baffle block	see **stilling baffle**
baffle plate	A plate or mechanical device used frequently in heat exchangers and separators to regulate the flow and reduce flow turbulence, in order to achieve better separation or heat transfer efficiency
bag filter	see **baghouse**
bagacillo	A fine fraction of bagasse (qv) used as a filter aid in sugar refining
bagasse	Crushed sugar cane fiber residue from milling and juice extraction
bagging	A northern UK dialect term used on site for layflat hose (qv)
baghouse	*aka* bag filter, baghouse filter, fabric filter. A sort of fabric filter used to clean air or gases, and/or its housing
BAHX	see **brazed aluminium heat exchanger**
baka yoke	(may be considered impolite) Japanese term meaning 'idiot-proofing', formerly used to describe techniques to avoid human error in the manufacturing industries by preventing, correcting, or drawing attention to human errors as they occur; cf poka yoke
Bakelite	Proprietary eponym for polyoxybenzylmethylenglycolanhydride, the first entirely synthetic plastic
Baker chart	*aka* Baker plot, Baker flow regime map. A graphical representation of flow regimes in two phase flow, popular in the oil and gas industry
balance line	Most commonly, a reduced bore connection between high to low pressure sides of a pump or compressor. Sometimes also expanded in meaning to cover any line that joins two tanks, reservoirs or pipes to maintain equal level or pressure between them
balance of plant	see **offsites**
balanced draught heater	Heater with fans to both supply combustion air and remove flue gases; defined formally (and identically) in BS EN ISO 13705:2012/ISO 13705:2012(E) Petroleum, petrochemical and natural gas industries. Fired heaters for general refinery service, API STD 530 - Calculation of Tube Heater Thickness and API STD 560 - Fired Heaters for General Refinery Service

balanced pressure bladder tank	A foam concentrate tank with a bladder-based mechanism which proportionally controls concentrate injection rate; defined formally in NFPA 11: Standard for Low-, Medium-, and High-Expansion Foam, 2016 Edition
balanced pressure relief valve	*aka* balanced safety relief valve, balanced relief valve, balanced bellows relief valve. A PRV with a balancing mechanism for minimizing the effect of system backpressure on its operation; defined formally in API RP 520 P1 7th Edition, January 2000 Sizing, Selection, and Installation of Pressure-Relieving Devices in Refineries; Part I - Sizing and Selection and API Standard 521. Pressure-relieving and Depressuring Systems. Sixth Edition | January 2014
balanced relief valve	see **balanced pressure relief valve**
balanced safety relief valve	see **balanced pressure relief valve**
balanced seal	A mechanical pump seal with a seal balance ratio (qv) less than or equal to one; defined formally in API RP 682 - Pump Seals
balancing point	see **center of gravity**
ball check valve	A simple, cheap and none too reliable kind of non-return valve with a captive ball seated by line pressure
ball drip	see **automatic drain valve**
ball mill	A rotating cylinder filled with (usually hard steel) balls used to grind (usually primary processed) coarse minerals to fine particles for further processing, and less frequently for blending. In a ball mill, balls are the continuous phase, with interstitial ore; cf SAG mill
ball pass frequency	In rolling element bearings, the number of rolling elements which pass a fixed point per shaft rotation. Note that there are two ball pass frequencies, corresponding with the inside track of the rolling elements and outside of the race; cf fundamental train frequency
ball valve	A valve controlling flow by means of an internal spherical ball with a hole through it, rotating through one quarter of a turn from open to closed. Well suited to potable water shutoff duties in less than 100 mm NB pipes. Not favored in hygienic applications, for larger bore pipes, or for modulating duties; cf plug valve, globe valve
ballast gas	Dry noncondensing gas (such as air) used in rotary vacuum pumps; defined formally in NFPA 86: Standard for Ovens and Furnaces, 2019 Edition
ballast tray	A type of tray used for distillation columns

band screen	A type of fine screen used for initial suspended solids removal in sewage, drinking and cooling water treatment
bank	see **embankment**
banksman	*aka* spotter (US), dogman (Australia). A person responsible for directing large vehicles on site, especially cranes; cf banksman/slinger
banksman/slinger	A UK term for a rigger (qv)
BAP	see **baseline assessment plan**
bar	A commonly used non-SI metric unit of pressure defined as 100 kilopascals
bar gauge	*aka* barg, bar(g). Pressure in bars in excess of ambient pressure
bar rack	see **bar screen**
bar schedule	A schedule of rebar (qv); defined formally in BS EN ISO 10209:2012 Technical product documentation – Vocabulary – Terms relating to technical drawings, product definition and related documentation
bar screen	*aka* bar rack. A type of coarse screen with apertures of 6-200 mm used for initial suspended solids removal in sewage and drinking water treatment
bar stock	A long round, rectangular, square or hexagonal piece (or billet) of steel used to fabricate more complex shapes
bar tube	A tube without fins used in shell and tube heat exchangers (qv)
Bardenpho process	A mulitstage variant of the activated sludge process which achieves biological nitrogen and phosphorus removal
barometer	An instrument which measures atmospheric pressure
barometric leg	An arrangement of piping containing a liquid column used to help create and hold vacuum to provide a liquid seal
barrel	A petroleum industry unit of volume equal to 159L; defined formally in NFPA 30: Flammable and Combustible Liquids Code, 2018 Edition
barrel of oil equivalent	The approximate energy released by burning a barrel of crude oil, which is around 1.7 MWh[8], according to the US IRS
barrel pump	1. *aka* drum pump. Generally, in chemical engineering, a manual or motor-driven self-priming pump used to transfer liquid from a barrel or IBC (qv) 2. In the petrochemical industries, a name for a horizontal double casing centrifugal pump; defined formally in EN ISO 13709:2003 Centrifugal pumps for petroleum, petrochemical and natural gas industries

---

[8] Your mileage may however vary!

barrels per stream day	*aka* BPSD, BSD, barrels per stream per day. A unit of refinery capacity, sometimes used in the oil industry as a unit of pump capacity
barrels per stream per day	see **barrels per stream day**
barrier	Anything used to control, prevent or impede flows; includes engineering (physical, equipment design) and administrative (procedures and work) processes
barrier analysis	A design or investigational methodology involving tracing pathways through which a target is adversely affected by a hazard, and identifying countermeasures that could or should have prevented that from happening
barrier fluid	The pressurized fluid within a pump's dual seal chamber which acts as a barrier between process and environment; defined formally (and slightly differently in each of) EN ISO 13709:2003 Centrifugal pumps for petroleum, petrochemical and natural gas industries, API RP 682 - Pump Seals and ASME BPE (American Society of Mechanical Engineers: Bioprocessing Equipment)
barye	(*symbol* Ba) A non-SI unit of pressure, equal to one dyne (qv) per square centimeter
basal sediment and water	*aka* BS&W. A term used generically to describe many varieties of 'crud' (qv), though strictly a technical specification of crude quality, one of the important product specifications for heavy product streams such as fuel oil
base	1. The chemical opposite of an acid, which accepts protons. Soluble bases may also be alkalis, but the terms are not strictly interchangeable; cf alkali 2. International English equivalent of plinth (qv)
base drawing	A drawing used as the basis for further design; defined formally in BS EN ISO 10209:2012 Technical product documentation — Vocabulary — Terms relating to technical drawings, product definition and related documentation
base exchange	The exchange of cations between a solution and a cationic ion exchange resin
base exchange softening	Softening water by means of base exchange (qv), producing a water with more sodium and fewer calcium and magnesium ions
base metal	1. In welding, the metal being welded; defined formally in one context in API STD 620 - Low Pressure Storage Tanks 2. In electroplating, the lower value metal which is plated with a higher value one

base oils	*aka* base stock(s). Refined oils primarily used to produce lubricating oils and greases, classified by API into five groups (there are other informal groupings); defined formally in API1509 Engine Oil Licensing and Certification System Twentieth Edition, May 2021
base stock(s)	see **base oils**
baseboard	1. Non-technically, US English for what is known in the UK as a 'skirting board' 2. In engineering (perhaps by extension of (1.)), a plate between a working platform and an adjacent construction element; defined formally in BS EN ISO 14122 Safety of machinery. Permanent means of access to machinery. Working platforms and walkways
baseline assessment plan	In the USA, the plan which a pipeline operator must develop to assess the integrity of all of the lines included in its integrity management program. It must show when each line is to be assessed and how; defined formally in Electronic Code of Federal Regulations (e-CFR) Title 49 - Transportation Subtitle B - Other Regulations Relating to Transportation Chapter I - Pipeline and Hazardous Materials Safety Administration, Department of Transportation Subchapter D - Pipeline Safety Part 195 - Transportation of Hazardous Liquids by Pipeline Subpart F - Operation and Maintenance Pipeline Integrity Management § 195.452 Pipeline integrity management in high consequence areas
basement	From a firefighting point of view, any levels of a structure at least half underground, with restricted firefighting access defined formally in NFPA 30: Flammable and Combustible Liquids Code, 2018 Edition, a
baseplate	*aka* base plate, soleplate (qv). The bottom plate of an item of equipment such as a pump, used to bolt it down
basic care areas	see **zoning**
basic design	*aka* conceptual design (qv). There is, however, contention over the names of the stages of design, as discussed in **Box S2**. Defined formally in BS EN ISO 10209:2012 Technical product documentation — Vocabulary — Terms relating to technical drawings, product definition and related documentation.
basic dimension	*aka* dimensional value or simply 'dimension'. Defined formally in BS EN ISO 10209:2012 Technical product documentation — Vocabulary — Terms relating to technical drawings, product definition and related documentation
basic effluent	Waste streams having a pH greater than 7

basic engineering design data	A standard package of information used for early-stage design in the oil and gas industry which sets the basis for design of a plant: atmospheric conditions, wind loads, seismic loads, electrical voltages, steam header levels, etc. Produced between the conceptual design (qv) and FEED (qv) phases; see **Box S2**. The cost variance at this stage is approximately 30%. The design at this stage gives a rough feel for equipment size, layout and mass/energy balance
basic object adapter	*aka* BOA. An interface used by coders
basic oxygen process	*aka* oxygen converter process, BOS, BOP, BOF, OSM, Linz-Donawitz steelmaking. A refinement of the basic Bessemer process (qv) replacing air with pure oxygen
basic process control system	A system which responds to input signals from the process and its associated equipment, other programmable systems, and/or from an operator, and generates output signals causing the process and its associated equipment to operate in the desired manner and within normal production limits. A 'catch-all' term used by process engineers to avoid differentiation between control systems such as DCS (qv), PLC (qv) or SCADA (qv). Typically used to differentiate from a SIS (qv)
basicity	Ability to accept a proton, the opposite of acidity (qv)
basis of calculation	A fixed quantity used as the basis of a calculation, especially a mass, energy or material balance
basis of design	A somewhat contested term (see **Box S2**.). Most commonly refers to preconceptual design definition documents similar to BEDD (qv) or a URS (qv), though sometimes used at later stages of design to describe a more complete set of documents
basis of safety	The primary method of ensuring safe operation in a process e.g., inerting of headspace for a vessel containing flammable material
basket centrifuge	A simple batch centrifuge used to remove solids from liquids with a combination of filtration and sedimentation
BAT	see **best available techniques or best available technology**
BAT reference document	see **BREF**
batch distillation	Distillation carried out in batch mode, normally producing a low volume / high value product
batch furnace	A furnace charged and discharged with work by single discrete operations, defined formally in NFPA 86: Standard for Ovens and Furnaces, 2019 Edition
batch process	A process charged and discharged by single discrete operations; cf continuous process

batch production Production using a batch process; cf campaign manufacture, continuous production
batch reactor A reactor charged and discharged by single discrete operations; cf continuous reactor
bathtub curve In reliability engineering, a description of the shape of the curve of product failure rates over time, with a high but declining early failure rate as defective products are discarded (known as 'infant mortality'), a flat midsection of random 'normal life' failures, and an escalating 'wear-out' failure rate as a result of wear and tear at the end of a product's life. See **Figure B.1**

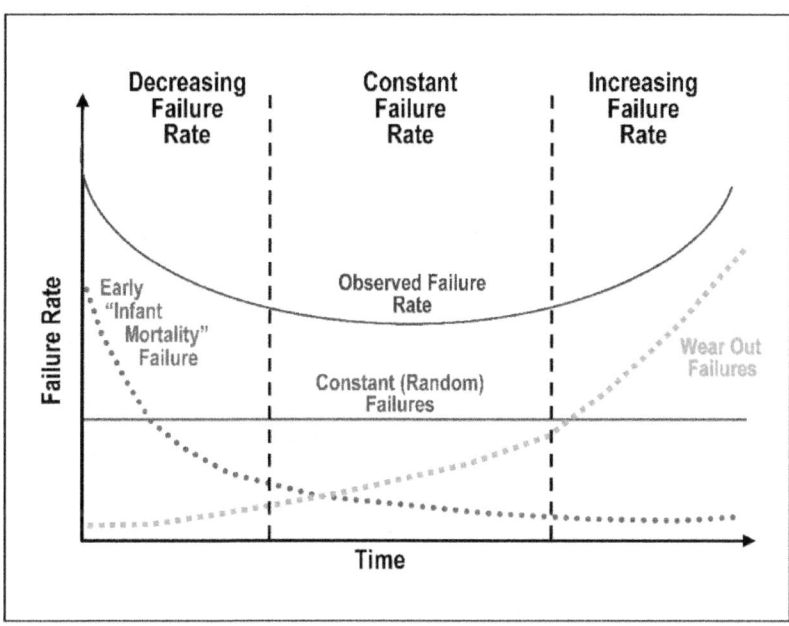

Figure B.1 Bathtub Curve

BATNEEC see **best available technology not entailing excessive costs**
battery limits Geographic boundary which defines the edge of an area from the point of view of design responsibility, may carry over into operational responsibility; cf ISBL, OSBL
Bauer connection see **Bauer coupling**
Bauer coupling *aka* Bauer fitting, Bauer connection. A quick-release hose connection commonly found on tanker hoses
Bauer fitting see **Bauer coupling**
Baur efficiency A fiddle factor (qv) used alongside McCabe-Thiele (qv)

Bayer process	The main process used to refine bauxite to aluminum oxide using caustic soda
Bayesian statistics	Bayesian probability expresses a degree of belief in an event[9]
bayonet adaptor	A 'push and twist' type connector as seen on some old incandescent lightbulbs[10]
bayonet connector	see **bayonet adaptor**
BBE	*(drawing notation)* Bevel both ends
BBL	Abbreviation for barrel[11] (qv)
BC	1. *(drawing notation)* Begin curve
	2. *(drawing notation)* Bolt circle
BCT	see **best conventional pollutant control technology**
Bdr	*(drawing notation)* Bleed ring
BDV	*(drawing notation)* Blowdown valve
BE	1. *(drawing notation)* Bevel end
	2. *(drawing notation)* Buttweld end
bead	1. Small spheroidal (often hard) particles
	2. see **weld bead**
bead blasting	see **media blasting**. A method of removing contaminants and debris or surface preparation of equipment using a stream of beads
bead mill	A device like a very small-scale ball mill used to grind materials to nanoscales
beam pump	see **nodding donkey**
bearing	A mechanical device which constrains the motion of and reduces friction between moving parts
beat frequency	The number of beats per second, equal to the difference in frequency of two oscillations; defined formally in BS ISO 2041:2018 Mechanical vibration, shock and condition monitoring. Vocabulary
beats	see **harmonic excitation**
Becher process	A beneficiation (qv) process which produces rutile (TiO_2) from weathered ilmenite
becquerel	*(symbol* Bq) The SI unit of radioactivity, defined as the activity of a quantity of radioactive material in which one nucleus decays per second
bed	A column of structured, random packed or granular media; qv packed bed, fluidized bed, exchanger bed

---

[9] Invented, incidentally, by a cleric
[10] And certain knives which fit to gun barrels!
[11] Apparently, the abbreviation is BBL rather than BL because the Standard Oil Company painted oil barrels blue, and kerosene barrels red

bed depth	*aka* bed height. The depth of media in a bed
bed expansion	The separation and rise of media in a column during backwashing, or the percentage increase in the bed height during backwashing or fluidization
bed volume	Volume of media contained in a bed at rest
bed volume per hour	Measure of the volumetric flow rate through a bed; cf EBCT
bed warm up	Adding hot dilution water to an ion exchange column immediately prior to regenerant addition
BEDD	see **basic engineering design data**
BEELs	Biological environmental exposure levels
bel	Ten decibels; defined formally in BS ISO 2041:2018 Mechanical vibration, shock and condition monitoring. Vocabulary
bell hole	An excavation made to permit a survey, inspection, maintenance, repair, or replacement of pipe sections, so-called because of its upside-down bell shape, angled to prevent cave-ins
bell metal	An alloy of copper and tin originally used for casting bells. It often contains some zinc and lead
bell nipple	An oil and gas industry term for a section of large diameter pipe fitted to the top of blowout preventers to which the flow line attaches via a side outlet, allowing drilling fluid to flow back over the shale shakers to the mud tanks
Bell-Delaware method	A method of designing a shell and tube heat exchanger (qv) which considers leakage streams as a source of inefficiencies
bellows	A flexible convoluted polymer or metallic fitting in a pipe or duct used to prevent damage from changes in pipe or duct length; not protective against axial distortion
bellows pump	see **reciprocating pump**
bellows safety valve	A type of direct loaded safety valve in which a bellows protects against fluid damage
bellows seal	A mechanical pump seal with flexible bellows providing a secondary seal; defined formally in API RP 682 - Pump Seals
belt conveyor	Solids conveying equipment in which solids are carried by a belt moving on rollers
belt filter press	*aka* belt filter or belt press filter. A device for separating water from sludges and slurries (notably wastewater sludges) in which two filter cloth belts are brought ever closer together by a series of rollers. Also used to express fruit juices[12]
belt weigher	see **weightometer**

---

[12] Not usually on the same machine used to dewater sewage sludge!

benchmark	A standard or reference against which something can be measured or judged
bend	1. Generically, a pipe fitting also known as an elbow (qv). Can be long or short radius; stock versions turn pipe through 90°, 45°, or 22.5°
	2. Sometimes used specifically for a long radius elbow
bending stress	The normal stress induced at a point in a body subjected to load that causes it to bend; defined formally in API RP 579 - Fitness for Service
Benedict Webb Rubin equation of state	*aka* BWR equation of state. An equation of state used in fluid dynamics
beneficiation	*aka* upgradation, ore processing. Improving ore by removing unwanted minerals known as gangue (qv), producing a concentrate product and a tailings (qv) waste stream
Benfield process	The removal of acid gases from petroleum and industrial gases with hot potassium carbonate
bent	see **pipe bent**
Bentley	An engineering design software company
bentonite	A type of clay with a wide range of engineering uses; cf diatomaceous earth
BEP	1. see **best efficiency point**
	2. Basic engineering package: a standard package of information providing process design information in the oil and gas industry. Typically provided by technology supplier/licensor
BER	see **beyond economic repair**
Bergius process	An obsolete coal to liquid fuel process
berl saddle	A type of random packing (qv) used in packed columns or beds.
Bernoulli's equation	see **Bernoulli's theorem**
Bernoulli's principle	see **Bernoulli's theorem**
Bernoulli's theorem	*aka* Bernoulli's principle, Bernoulli's equation. The principle which states that where the velocity of a fluid is high, the pressure is low and vice-versa
Bessemer process	A process used to produce steel from pig iron (qv) by air oxidation. There are basic and acidic variants
best available control technology	One of the pollution control methods covered by the (USA) Clean Air Act
best available techniques	*aka* BAT. The basis for determining (EU/UK) IPPC permit conditions stipulating the minimum required degree of pollution abatement required to permit operation

best available technology	*aka* BAT. The state of the art in abatement technology. Commonly (and incorrectly) used synonymously with best available techniques (qv), but the term is not officially used in an unmoderated form. Rather it is seen in forms which include a consideration of costs such as best available technology economically achievable (qv) or best available technology not entailing excessive costs (qv)
best available technology economically achievable	The USA equivalent of BATNEEC (qv); defined formally in the Clean Water Act section 304(b)(2)
best available technology not entailing excessive costs	*aka* BATNEEC. The most effective commercially available techniques at the appropriate scale with benefits greater than the costs of obtaining them
best conventional pollutant control technology	An approximate equivalent of BAT (qv) used in the USA water industry which addresses conventional pollutants from existing industrial point sources; defined formally at CWA section 304(b)(4)
best efficiency point	*aka* BEP. The flowrate and discharge pressure at which rotodynamic equipment is most efficient; defined formally in EN ISO 13709:2003 Centrifugal pumps for petroleum, petrochemical and natural gas industries
best practicable control technology	*aka* BPCT. An approximate equivalent of BAT (qv); defined at CWA section 304(b)(1)
best practicable environmental option	*aka* BPEO. A UK regulatory measure which "establishes, for a given set of objectives, the option that provides the most benefits or the least damage to the environment, as a whole, at acceptable cost, in the long term as well as the short term."
best practice	The consensus heuristics of practitioners, commercially available technologies; not those considered 'experimental'
BET	Brunauer-Emmett-Teller
BET equation	Brunauer-Emmett-Teller equation: an extension of the Langmuir theory for monolayer molecular adsorption to multilayer adsorption based on a number of assumptions
BET isotherm	Brunauer-Emmett-Teller isotherm: the multilayer equivalent of the Langmuir adsorption isotherm (qv)
beta decay	The spontaneous emission of beta particles (qv) by a heavy radioactive element
beta particle	An electron or positron emitted by a radioactive element
beta radiation	A stream of beta particles
betterment	An insurance term, used to describe the improved state of a damaged item after repair with new parts, compared with its state before damage had occurred

Betts process	An expensive electrochemical method for producing very pure (especially bismuth-free) lead from lead bullion
between bearings pump	A centrifugal pump with one or more impellers on a shaft suspended between bearings; cf overhung pump
beyond economic repair	*aka* BER. A classification of unserviceable equipment, where the diagnosis and repair of faults will not even restore it to a serviceable condition (rather than an as-new condition)
BF	*(drawing notation)* Blind flange
BFG	*(drawing notation)* Bevel for weld[13]
BFH	see **Big Fucking Hammer**
BFO	see **bunker oil**
BFW	1. see **boiler feed water** 2. *(drawing notation)* Bevel for welding
BG	*(drawing notation)* Below ground/grade
Bhopal	A town in India, the site of the worst industrial disaster in history, where a major release of methyl isocyanate resulted in the deaths of thousands of people local to a Union Carbide plant
BHP	Bottom hole pressure, the pressure at the bottom of an oil well
BI	see **business interruption insurance**
bi fuel operation	Running a gas turbine on two (not premixed) dissimilar fuels; defined formally in ISO 3977 Gas turbines - Procurement - Part 3: Design requirements
big bag	see **flexible intermediate bulk container**
big fucking hammer	*aka* BFH. A large hammer used for percussive maintenance (qv)
big Texan pump	see **nodding donkey**
bill of material	*aka* BOM, bill of materials. A list of the raw materials, sub-assemblies, intermediate assemblies, sub-components, parts, and the quantities of each needed to manufacture an end product. Defined formally in the context of drawings in BS EN ISO 10209:2012 Technical product documentation – Vocabulary – Terms relating to technical drawings, product definition and related documentation. Also used in some organizations to list the spare equipment associated with an item of functional equipment for maintenance
bill of materials	see **bill of material**

---

[13] According to some sources, even though weld does not start with a G. Deprecated at best. **BFW** would almost always be clearer

bill of quantities	*aka* BOQ. A descriptive list of quantities of materials, works and so on required for construction works. Commonly used as the basis of pricing by civil engineers, though not always appropriately to process plant[14]; defined formally in BS EN ISO 10209:2012 Technical product documentation —Vocabulary — Terms relating to technical drawings, product definition and related documentation
billet	A small, part-finished piece of metal that is rectangular, circular, or square in shape. Metals such as iron, steel, and plutonium are made into billets for further processing
BIM	Building information modeling; see **building information modeling and management**
bimetallic corrosion	see **dissimilar metals corrosion**
bimetallic joint	A welded joint between two or more dissimilar metals
bimetallic strip	A plate of two metals with different thermal expansion bonded together, which bends on heating, in a way which allows its use as a thermal switch
BIMM	see **building information modeling and management**
bimolecular reaction	A reaction involving two molecules
bin	A short silo
binary	1. Involving two components 2. A base two number system used in computing
binary distillation	Separating two components with different boiling points from each other by distillation; cf fractional distillation
binder	Substance that holds other materials together
Bingham plastic	A material which acts like a solid at low stresses and a viscous fluid at high stresses
bioaerosol	Biological material dispersed in a gas; defined formally in EN ISO 14698-1:2003 Cleanrooms and associated controlled environments — Biocontamination control — Part 1: General principles and methods
bioaugmentation	Adding microorganisms in the hope that this will initiate biological treatment of a contaminant[15]
bioburden	The initial degree of microbiological contamination; defined formally in ASME BPE (American Society of Mechanical Engineers: Bioprocessing Equipment)
biocatalyst	An enzyme or other biologically derived catalyst
biochar	see **char**

---

[14] "How much is a ton of pumps?"
[15] However, in the all-too-common absence of ensuring conditions for the growth of the organisms are met, a worthless practice

biochemical engineering	*aka* bioprocess engineering. The chemical engineering of biologically derived materials
biochemical oxygen demand	*aka* BOD. The result of a test of the amount of biologically oxidizable matter present in a wastewater, expressed as an oxygen concentration. In practice, this is always cBOD (qv) as measured by the 5-day test at 20 °C (BOD5 20); defined formally by a UK Royal Commission on Sewage Disposal in 1912
biocide	A chemical which kills microorganisms; commonly used to prevent biological corrosion of metals, amongst other things
biocontamination	Contamination with biologically viable material; defined formally in EN ISO 14698-1:2003 Cleanrooms and associated controlled environments – Biocontamination control – Part 1: General principles and methods
biodegradable	Capable of being broken down into simpler substances by microorganisms
biodegradable waste	Waste (qv), often but not always of biological origin which is biodegradable (qv)
biodiesel	A fully blendable substitute for diesel fuel, derived from biological materials, usually comprising fatty acid esters. Modern diesel engines need conversion to run on straight vegetable oils, even though these were the original fuel for the diesel engine; cf renewable diesel
bioenergy	All types of energy derived from biomass, including biofuels
bioethanol	Biologically produced ethanol used as a fuel or fuel additive
biofilm	A microbial consortium adhering to a surface, frequently embedded in a polymeric slime which protects against chemical disinfection, making biofilms a source of hard to clear contamination. Defined formally in ASME BPE (American Society of Mechanical Engineers: Bioprocessing Equipment)
biofilter	An attached growth bioreactor used in environmental applications
biofuel	A biologically derived fuel (other than fossil fuels, which are not counted as biologically derived, even though they are), e.g., bioethanol, biodiesel, or biogas
biogas	*aka* biomethane, sludge gas, sewage sludge gas. The mixture of gases produced by the anaerobic digestion of organic material such as sewage sludge or food waste, with variable proportions of carbon dioxide, methane and hydrogen, as well as odorous, reduced compounds such as H_2S and mercaptans
biogenic hydrogen	see **orange hydrogen**

biological aerated flooded filter	A largely superseded attached growth effluent treatment technology with a smaller footprint than the equivalent conventional activated sludge plant
biological dose	see **dose equivalent**
biological oxygen demand	see **biochemical oxygen demand**
biological phosphate removal	The removal of phosphate compounds from wastewater by biological means involving 'luxury phosphorus uptake'
biologics	Therapeutic products produced from or by living organisms[16]; defined formally in ASME BPE (American Society of Mechanical Engineers: Bioprocessing Equipment)
biomass	Biological materials (sometimes 'waste') used as or converted into a fuel
biomass to liquids	*aka* BTL. One of a number of processes which convert biomass into liquid fuels
biomethane	see **biogas**
biomolecule	A loose term for a molecule involved in the growth and development of organisms
biopharmaceuticals	Biologically produced pharmaceuticals; defined formally in ASME BPE (American Society of Mechanical Engineers: Bioprocessing Equipment)
biopolymer	A biologically derived polymer
bioprocess	A process using cells or cell components to produce a product; defined formally in ASME BPE (American Society of Mechanical Engineers: Bioprocessing Equipment)
bioprocess engineering	see **biochemical engineering**
bioprocessing	see **bioprocess**
bioprocessing equipment	Equipment for bioprocessing; defined formally in ASME BPE (American Society of Mechanical Engineers: Bioprocessing Equipment)
bioreactor	A reactor used for bioprocessing
biosafety level	*aka* BSL. One of a numbered series of specific combinations of work practices, safety equipment, and facilities designed to reduce the exposure of workers and the environment to infectious agents to a degree appropriate to the risks
biosynthesis	The production of larger molecules from smaller precursors by bioprocessing
Biot number	(*symbol* Bi) A dimensionless group used in heat transfer calculations

---

[16] Not beer!

biotechnology	A somewhat vague term: "any technological application that uses biological systems, living organisms, or derivatives thereof, to make or modify products or processes for specific use", according to the UN Convention on Biological Diversity
BIPM	International Bureau of Weights and Measures
bird dog	Someone who watches oil-field activities with the intention of gathering information
birdbath	In oil and gas drilling, the storage area for the bottom end of a stand of tubulars (qv) stored vertically in a derrick, which captures any fluids which drain from them
bird's eye perspective	In the context of drawings, a one-point perspective from above; defined formally in BS EN ISO 10209:2012 Technical product documentation – Vocabulary – Terms relating to technical drawings, product definition and related documentation
Birmingham screwdriver	*(jocular)* see **Manchester screwdriver**
Birmingham wire gauge	*aka* BWG, Stubs' iron wire gauge, Stubs' steel wire gauge, Stubs' wire gauge. An old, but not obsolete thickness measurement system, with wider application than wire, commonly used for wall thickness of smaller ID tubes such as exchanger tubes
bitumen	see **asphalt**
BJH analysis	A procedure for calculating pore size distributions from experimental isotherms using the Kelvin model of pore filling, applicable only to the mesopore and small macropore size range
BL	*(drawing notation)* Blower
black area	A separate non-clean room area within a pharmaceutical clean room area in which equipment or machinery is housed for maintenance; cf black utilities
black box	A system considered only at the level of inputs and outputs, with no concern as to internal workings[17]
black hydrogen	see **brown hydrogen**
black liquor	Spent white liquor (qv) from the Kraft papermaking process (which must be used as boiler fuel to render the process economic); defined formally in BS EN 12952-8:2002 Water-tube boilers and auxiliary installations. Requirements for firing systems for liquid and gaseous fuels for the boiler

---

[17] On one level, it's all black boxes in engineering!

black liquor gun	The device which sprays black liquor into a furnace; defined formally in BS EN 12952-8:2002 Water-tube boilers and auxiliary installations. Requirements for firing systems for liquid and gaseous fuels for the boiler
black list	An informal name for the list of dangerous/hazardous aqueous pollutants found at List I of the Annex to Directive 76/464/EEC
black tea	Tea as the term is generally used in the UK and Ireland; defined formally in ISO 3103 Tea - Preparation of liquor for use in sensory tests
black utilities	A term used in the hygienic industries for effluent treatment, or more broadly all non-hygienic utilities (qv) (e.g., process, chilled, cooling and softened water, plant steam and gas) which need to be kept separated from hygienic streams; cf clean utilities, green utilities
black water	In the context of water reuse, water contaminated with fecal material, arising from toilets; cf grey water
blacktop	see **tarmac**
bladder tank	A pressure vessel containing air and liquid separated by a membrane; defined formally in NFPA 22: Standard for Water Tanks for Private Fire Protection, 2018 Edition
Blagden pump	Misspelling of Blagdon pump (qv)
Blagdon pump	*aka* Blagden (sic) pump. A proprietary eponym for an air-operated diaphragm pump; cf Wilden pump
blank	*aka* blank flange, blind flange. A flat plate used to seal the end of a flanged pipe
blank flange	see **blank**
blanketing	*aka* tank blanketing, inerting. Using an inert gas to fill the headspace of a tank; usually to prevent the formation of an explosive mixture of air and vapor, or to prevent reactions between tank contents and air (e.g., oxidation of fats leading to rancidity)
blanking	*aka* blanking off. Sealing a flanged pipe with a flat plate fitted to its terminal flange, or between flanges
blanking off	see **blanking**
Blasius equation	A boundary layer solution over a flat plate for the case in which free stream velocity is constant
blast	see **blast wave**
blast burner	A burner delivering a combustible mixture at 75 kPa or more; defined formally in NFPA 86: Standard for Ovens and Furnaces, 2019 Edition
blast chilling	see **blast freezing**

blast freezing	*aka* flash freezing, shock freezing, blast chilling. Rapid freezing using a blast of cold air, commonly used for foodstuffs
blast furnace	The main reactor in the process of primary iron making: a furnace used to continuously produce impure iron from ores, using a countercurrent flow of pressurized air, oxygen or a mixture of the two and a feed of ore, limestone flux and coke as fuel/carbon source
blast wave	The destructive supersonic wave of pressure arising from an explosion
blasting	Using a blasting agent (qv) or explosive (e.g., in mining)
blasting agent	Any material or mixture of fuel and oxidizer intended for blasting. Blasting agents may not themselves be defined as explosives, even though they contain explosives. The key difference between a blasting agent and an explosive is that blasting agents require a primer, rather than a simple blasting cap to initiate an explosion; cf explosive
blasting plan	A written (or more commonly, drawn) description of the planned use of a blasting agent (qv) or explosive; defined formally in BS EN ISO 10209:2012 Technical product documentation — Vocabulary — Terms relating to technical drawings, product definition and related documentation
BLD	*(drawing notation)* Blind
BLE	*(drawing notation)* Bevel large end
bleach	A substance which removes color, usually by oxidation e.g., hydrogen peroxide or sodium hypochlorite
bleed mass flow	see **extraction mass flow**
bleed valve	A valve used to allow fluid to escape a tank or tube
blender	A name applied to various types of mixers and homogenizers for liquids, powders, or slurries
blending	1. Generally, the process of using a blender (qv) 2. Specifically, in the oil and gas or beverage industries, producing a 'blend' (not necessarily using a blender) which meets a given specification by mixing together disparate components
blendstock	An oil and gas industry term for anything used alone (or more commonly, blended with other substances) to produce a commercial product, such as gasoline (qv)
BLEVE	see **boiling liquid expanding vapor explosion**
blind	see **spade**
blind flange	see **blank**

blind weld	A weld that cannot (by design) be visually inspected; defined formally in ASME BPE (American Society of Mechanical Engineers: Bioprocessing Equipment)
blinding	1. Blockage of a depth filter by solids, liquids or gases leading to increased headloss 2. A synonym of blanking (qv)
blistering	1. Generically, the formation of localized domed imperfections resulting from delamination of a coating 2. Specifically, the formation of localized domed imperfections on a metallic or polymeric surface; defined formally in ASME BPE (American Society of Mechanical Engineers: Bioprocessing Equipment)
block and bleed	see **double block and bleed**
block and bleed valve	According to API 6D, a kind of mixproof valve (qv) with integrated double block and bleed function
block and tackle	see **chain block**
block diagram	see **block flow diagram**; this terminology defined formally in BS EN ISO 10209:2012 Technical product documentation – Vocabulary – Terms relating to technical drawings, product definition and related documentation
block flow diagram	*aka* mass flow diagram, process block diagram. A simplified and highly informal PFD, more popular in academia and management than engineering
block plan	*aka* site plan, site layout plan. A site-wide low resolution general arrangement drawing (qv), which may focus on interconnections; defined formally in BS EN ISO 10209:2012 Technical product documentation – Vocabulary – Terms relating to technical drawings, product definition and related documentation
block valve	An isolation valve: a type giving reliable tight shutoff is preferred
blocked flow or discharge	A standard design case for pressure relief, in which equipment outlets are blocked
blocked in	A standard design case for pressure relief where outlets are blocked off; used especially for heat exchangers, and pressurized hydrocarbon processing equipment
blow back	see **blowback**
blow molding	A type of molding in which compressed air inflates a miniature version of a finished thermoplastic product
blowback	*aka* blow back. Gases travelling in a reverse direction and, by general analogy, other unintended, undesirable outcomes

blowdown	1. Depressurization of plant or equipment; defined formally in API Standard 521. Pressure-relieving and Depressuring Systems. Sixth Edition | January 2014
2. The difference between set and closing pressure of a PRV; defined formally in API RP 576 - Inspection of Pressure Relieving Devices, and API RP 520 P1 7th Edition, January 2000 Sizing, Selection, and Installation of Pressure-Relieving Devices in Refineries; Part I - Sizing and Selection
3. An intentional purge to control impurity levels in steam generating boilers etc.
blowdown drum	1. Usually a knockout drum (qv) protecting a vent to atmosphere of materials arising from depressurization of equipment by a relief device; defined formally in API Standard 521. Pressure-relieving and Depressuring Systems. Sixth Edition | January 2014
2. Sometimes used for a knockout drum on the feed to a flare stack
3. Rarely used to mean a collection vessel for materials arising from depressurization of equipment by a relief device unsuitable for venting to atmosphere or flare system; cf dump tank
blowdown steam	Steam used for furnace cleaning, or in boiler blowdown
blowdown tank	A tank for purging steam or gases via blowdown (qv)
blowdown valve	An automatic valve used for blowdown
blower	Various types of low-pressure gas compressor (qv) (which some dispute are compressors at all). A consensus but necessarily vague definition might be: a piece of equipment for moving gases that delivers higher pressures than a typical fan but lower than typical compressors
blower type foam generators	Foam generators in which foam solution is sprayed onto screens which air is blown through; defined formally in NFPA 11: Standard for Low-, Medium-, and High-Expansion Foam, 2016 Edition
blowing	A term used in distillation for a condition where the rising vapour punches holes through the liquid layer on a tray and carries large drops and slugs of liquid to the next tray
blowoff	Flame being blown off a burner by excessive velocity of fuel-air mixture; defined formally (and differently) in API Standard 521. Pressure-relieving and Depressuring Systems. Sixth Edition | January 2014, API RP 535 - Burners for Fired Heaters at Refineries and API RP 537 Flare Details for Petroleum, Petrochemical, and Natural Gas Industries

blowoff valve	A safety valve used on centrifugal compressors to release pressure to prevent surging; cf **dump valve**
blowout	1. see **air blowout**
	2. A sudden and highly hazardous release of well fluids from an oil well caused by sudden entry of high pressure gas
blowout preventer	A type of valve used to control flow and, if necessary, seal an oil well to prevent blowout
blue diesel	USA term for a type of untaxed dyed diesel (qv) used in government vehicles, the dye in this case being blue; cf **red diesel**
blue hydrogen	'Blue hydrogen' is made from natural gas by steam methane reforming (qv); see **hydrogen colors**
blue water gas	see **water gas**
bluegreen algae	see **cyanobacteria**
blunger	Equipment used for blunging (qv)
blunging	The wetting and mixing of powders, especially clays, to a fluid state
BM	*(drawing notation)* Benchmark
BO	Barrels of oil
BOA	see **basic object adapter**
BOC	*(drawing notation)* Bottom of concrete
BOD	1. see **basis of design**
	2. see **biochemical oxygen demand**
BOD5	*aka* BOD. Biochemical oxygen demand (qv) at 5 days; defined formally in EN 1085:2007 and EN 12566 - Small wastewater treatment systems for up to 50 PT Part 3: Packaged and/or site assembled domestic wastewater treatment plants; usually BOD 520, i.e., cBOD (qv) as measured by the 5 day test at 20 °C
BOD7	Biochemical oxygen demand (qv) at 7 days; defined formally in EN 1085:2007 and EN 12566 - Small wastewater treatment systems for up to 50 PT Part 3: Packaged and/or site assembled domestic wastewater treatment plants
Bode	Proprietary eponym for evaporator (qv), short for Ahrens-Bode
Bode plot	Used in control theory and vibration monitoring; a combination of a Bode magnitude plot, expressing the magnitude (usually in decibels) of the frequency response, and a Bode phase plot, expressing the phase shift; defined formally in BS ISO 2041:2018 Mechanical vibration, shock and condition monitoring. Vocabulary
bodge	see **kludge**
BOE	1. *(drawing notation)* Bevel one end
	2. see **barrel of oil equivalent**

BOEM	Bureau of Ocean Energy Management
BOG	see **boil off gas**
Bogue compounds	*aka* Bogue's compounds, Bogues compounds. Basic oxides, the silicates of which are called Alite, Belite etc., in cement chemist notation (qv)
Bogue's compounds	see **Bogue compounds**
Bogues compounds	see **Bogue compounds**
boil off gas	Vapor generated in cryogenic LNG tanks due to thermal exchange with the environment
boiler	A vessel which raises the temperature of a liquid, but (perhaps confusingly) not necessarily to the point at which it boils or vaporizes. The liquid is often water, and the product often steam, in which case it is a steam boiler
boiler feed water	Highly purified and often chemically treated water suitable for steam boiler feed
boiler horsepower	see **horsepower**
boiler hotwell	A header tank used to collect returned hot condensate for use as boiler feed, often elevated to avoid net positive suction head (qv) problems with feed pumps associated with pumping very hot water
boiler transfer station	see **transfer station**
boiler water	see **boiler feed water**
boiling	Rapid vaporization of a liquid above its boiling point, often accompanied by the formation of bubbles of vapor within the body of the liquid; cf evaporation
boiling liquid expanding vapor explosion	*aka* BLEVE. A catastrophic mode of pressurized tank failure from direct exposure to a fire; defined formally in API RP 2001 - Fire Protection at Refineries, API RP 2510A - Fire Protection Considerations for the Design and Operation of Liquefied Petroleum Gas (LPG) Storage Facilities
boiling point	The temperature at which a liquid's vapor pressure matches ambient pressure, defined formally in API RP 505 2nd Edition, August 2018 Recommended Practice for Classification of Locations for Electrical Installations at Petroleum Facilities Classified as Class I, Zone 0, Zone 1, and Zone 2, and API RP 2001 - Fire Protection at Refineries, NFPA 30: Flammable and Combustible Liquids Code, 2018 Edition
boiling point elevation	An increase in the boiling point of a solvent caused by a non-volatile solute
boiling water reactor	A nuclear reactor whose core is cooled by water which boils in the process; cf pressurized water reactor

boilover	*aka* steam explosion. An escape from containment of a burning liquid as a result of a sudden expansion in volume, commonly as a result of the boiling of water at the bottom of containment. Caused by preferential combustion of lighter fractions in an extended fire in an overlying mixture of hydrocarbons. Defined formally in API RP 2001 - Fire Protection at Refineries and NFPA 30: Flammable and Combustible Liquids Code, 2018 Edition. N.B. This NFPA definition does not seem to require water, only a sudden increase in fire intensity and an escape of burning oil from containment; cf frothover, slopover
boilup	A cleaning technique used in the food and drink industry, comprising filling a vessel with water and boiling it for a time; rather wasteful of water and energy
boilup rate	In a distillation column, the mass or mole rate of vapor generated at the reboiler, which is the same as reflux flowrate at equilibrium
boilup ratio	In a distillation column, the mass or molar ratio of vapor returned to bottom product removed; cf reflux ratio
BOL	*(drawing notation)* Bottom of line
bolt	A threaded rod fitting with a head for tightening
Boltzmann constant	*(symbol* kB or k) A constant which relates the average relative kinetic energy of particles in a gas to the temperature of the gas expressed in units of energy per increment of temperature per particle; cf gas constant
Boltzmann equation	*aka* Boltzmann transport equation, BTE. A statistical treatment of a non-equilibrium thermodynamic system
Boltzmann transport equation	see **Boltzmann equation**
BOM	see **Bill of material**
bond	To glue things together; defined more formally in BS EN ISO 14159:2008 Safety of machinery. Hygiene requirements for the design of machinery; cf earth bonding
Bond number	*(symbol* Bo); see **Eötvös number**, its more common name in Europe
Bond's work index	A semiempirical measure of ore resistance to crushing and grinding, as determined using the Bond grindability test

bonding	*aka* cross bonding (deprecated usage). Joining conductive objects together electrically (not with glue- cf bond) so that they are at the same electrical potential to control static hazards; defined formally in NFPA 30: Flammable and Combustible Liquids Code, 2018 Edition and NFPA 654: Standard for the Prevention of Fire and Dust Explosions from the Manufacturing, Processing, and Handling of Combustible Particulate Solids, 2017 Edition; cf earth bonding
bonnet	The cap on certain types of valves
BOO	see **build own operate**
booster	An auxiliary device amplifying the action of the main device
BOOT	see **build, own, operate, transfer**
BOP	1. see **blow out preventer** 2. see **balance of plant** 3. *(drawing notation)* Battery operated pump 4. *(drawing notation)* Bottom of pipe
BOPD	Barrels of oil per day
BOQ	see **bill of quantities**
bore	Internal hollow part of pipe, commonly used to mean the actual or nominal internal diameter of that part
bore area	*aka* throat area. In the context of pressure relief valve nozzles, the minimum cross-sectional area for flow; defined formally in API RP 520 P1 7th Edition, January 2000 Sizing, Selection, and Installation of Pressure-Relieving Devices in Refineries; Part I - Sizing and Selection
borehole	A deep narrow shaft in the ground often used by investigators to collect samples of underlying ground, or groundwater, or for production of water, oil, gas, geothermal energy etc.; cf well-bore
borescope	*aka* boroscope. An optical or electrooptical device for looking into inaccessible places such as narrow holes for inspection purposes. The industrial equivalent of the medical endoscope. Defined formally in ASME BPE (American Society of Mechanical Engineers: Bioprocessing Equipment)
boroscope	see **borescope**
borosilicate glass	*aka* Pyrex. A type of glass that contains a high level of boron with a low coefficient of thermal expansion and an ability to absorb neutrons
BOT	*(drawing notation)* Bottom
botch	*(informal)*; to bungle; cf bodge
bottleneck	The lowest throughput element of a sequential series of unit operations, which sets the throughput of the series

bottom feeder	*(informal, jocular)*; the last process engineer to leave a project before the lights go out - usually because they are the last to get another job offer
bottom hole assembly	In oil drilling, an assembly comprising the drill bit, drill collars and drilling stabilizers
bottom orifice	A hole in the bottom of a thing
bottom product	see **bottoms**
bottom run-off	The lowest nozzle on a vessel used to fully deinventory (qv) the vessel; often fitted with a bottom run-off valve
bottoms	*aka* bottom product. The high-boiling point product leaving the bottom of a distillation column; cf top product
botulinum cook	*aka* 12D process, twelve D process. Heating and holding food and drink products at 121 °C for 3 minutes, sufficient to deactivate C. botulinum toxin, but not kill its spores
bound moisture	Moisture within the matrix of a solid material which exerts a lower vapor pressure than it would in a free form, very similar to inherent moisture (qv); cf free moisture
boundary layer	In fluid mechanics, the layer adjacent to a system boundary where viscosity becomes significant
Bourdon gauge	The basic process industry standard round-faced mechanical pressure gauge, used to obtain a field pressure reading, containing a Bourdon tube (qv)
Bourdon tube	A curved tube used for fluid pressure measurement
BOV	see **blowoff valve**
bow tie	*aka* bowtie. 1. A diagram used for visualizing the types of preventive and mitigative barriers which can be used to manage risk with the shape of a bow tie 2. A generic P&ID valve symbol to chemical engineers
bowl	The primary spinning element of a centrifuge
bowl centrifuge	*aka* decanter or solid bowl centrifuge. A continuous centrifuge in which solids are sedimented to the surface of a spinning cylindrical 'bowl', with a scroll inside it spinning at a slightly different speed which collects sedimented solids for continuous discharge. Also comes in a three-phase separation version used to separate oil, water and solids, e.g., for olive oil production
bowtie	see **bow tie**
Boyle's law	The law which states that the pressure at constant temperature of a given mass of ideal gas is inversely proportional to its volume. Known in France as Mariotte's law
Boyle-Charles's law	*aka* combined gas law. The law which states that, for the same substance under different conditions: $P_1V_1/T_1 = P_2V_2/T_2$

BPCD	Barrels per calendar day
BPCS	see **basic process control system**
BPD	Barrels per day
BPEO	see **best practicable environmental option**
BPSD	see **barrels per stream day**
BPT	see **best practicable control technology**
Bq	see **becquerel**
braced frame	*aka* rigid frame. In structural engineering, a frame which requires no additional or vertical cross bracing
bracketry	A collective term for the brackets (usually hung from walls or steelwork) which support pipework in the vertical plane. Includes shaped metal pieces that make up the actual bracket, and any other bolts, nuts, rubber etc. needed to complete the mounting of the part
brackets	see **bracketry**
brackish	A term for a mixture of fresh and salt water
brake horsepower	see **horsepower**
branch circuit	The circuit between the final protective overcurrent device and outlets; defined formally in NFPA 20: Standard for the Installation of Stationary Pumps for Fire Protection, 2019 Edition
branches	Subdivisions of a sewer or, more generally, any other larger main pipe. In the case of a sewer, process area drain points feed sublaterals which feed laterals which in turn feed the main sewer, often via a seal
brass	1. An alloy, primarily consisting of copper and zinc 2. Historically the name for a brass medallion with a number or other identifier on it used when entering a plant or construction site - entrants would take it from a board, and then hang it back up on a hook on a board upon exit. Thus, in an emergency, the board could be checked to determine who was in the plant, and potentially in danger
brazed aluminium heat exchanger	aka BAHX. A type of cryogenic heat exchanger, often installed in a cold box (qv)
brazing	Soldering metal using a brass filler material
breaching	*aka* breeching. Flue gas ductwork leading to stack; defined formally in BS EN ISO 13705:2012/ISO 13705:2012(E) Petroleum, petrochemical and natural gas industries. Fired heaters for general refinery service, API STD 530 - Calculation of Tube Heater Thickness and API STD 560 - Fired Heaters for General Refinery Service

break	A discontinuity; defined formally in the context of fittings in ASME BPE (American Society of Mechanical Engineers: Bioprocessing Equipment)
break even point	*aka* breakeven point. The production level at which total expended costs equal revenues gained
break tank	*aka* buffer tank, breakout tank. 1. Generally, a tank with a level which may go up or down to balance incoming supply and outgoing demand, often featuring an air gap 2. In the context of fire protection, it has a specific meaning when used as the supply to a pump with capacity less than the total required fire protection flow volume, as defined formally in NFPA 22: Standard for Water Tanks for Private Fire Protection, 2018 Edition NFPA 20: Standard for the Installation of Stationary Pumps for Fire Protection, 2019 Edition
breakdown maintenance	see **reactive maintenance**
breakeven point	see **break even point**
breaking pin device	A type of pressure-relief device, defined formally in API Standard 521. Pressure-relieving and Depressuring Systems. Sixth Edition | January 2014
breakout tank	A tank used to temporarily store oil to relieve surges in a hazardous liquid pipeline system; defined formally in 49CFR 194.5 and 49CFR 195.2.; cf break tank
breakover	see **slopover**
breakpoint chlorination	The removal of ammonia from water by oxidation with chlorine
breakpot	see **flash drum**
breakthrough	The point, often expressed as a volume of water treated, at which a contaminant appears at a significant concentration in the outflow of an absorption or adsorption vessel
breather valve	see **vent/vac valve**
breathing apparatus set	*aka* BA set. Self-contained PPE providing a personal air supply - either an emergency BA set (qv), or self contained breathing apparatus (qv)
BREEAM	see **Building Research Establishment's Environmental Assessment Method**
breeching	see **breaching**

BREF	*aka* BAT Reference Document, best available techniques reference document. A sector-specific or cross-sector issue-based BAT guidance document developed under the European Union Industrial Emissions Directive (Ied, 2010/75/Eu) and the Integrated Pollution Prevention and Control Directive (2008/1/Ec)
BREF BAT	BAT (qv) as given in a sector level BREF (qv)
Brenner number	(*symbol* Br) The Péclet number (qv) for particulate motion
Brent blend	see **Brent crude**
Brent crude	*aka* Brent blend, London Brent, Brent petroleum. A benchmark classification of sweet light crude oil
Brent petroleum	see **Brent crude**
brewing	Production of alcoholic beverages by fermentation
bridge	see **pipe bridge**
bridge breaker	A powered mechanical device to prevent aggregation (qv) and bridging (qv) of material in the throat of a pump
bridge scraper	*aka* scraper bridge. A scraper blade hanging from a bridge structure used to collect settled sludge in settlement tanks in sewage and effluent treatment. Half bridge scrapers (qv) rotate along with the scraper blade, whilst full bridge scraper bridges do not
bridgewall	The wall separating adjacent fired heater zones; defined formally in BS EN ISO 13705:2012/ISO 13705:2012(E) Petroleum, petrochemical and natural gas industries. Fired heaters for general refinery service, API STD 530 - Calculation of Tube Heater Thickness and API STD 560 - Fired Heaters for General Refinery Service
bridgewall temperature	The temperature of flue gases exiting the radiant section of a fired heater; defined formally in BS EN ISO 13705:2012/ISO 13705:2012(E) Petroleum, petrochemical and natural gas industries. Fired heaters for general refinery service, API STD 530 - Calculation of Tube Heater Thickness and API STD 560 - Fired Heaters for General Refinery Service
bridging	The effective obstruction of an aperture, such as a pump suction, by particles smaller in size than the aperture acting in concert
bridging out	*aka* frigging out (qv). Temporary bypass of safety features or forcing of control signals or actions

brief	*aka* design brief. The initial documentation which specifies the resources of the client and user, and a project's needs, aims, and design requirements, used as the basis of all subsequent design. Defined formally in BS EN ISO 10209:2012 Technical product documentation — Vocabulary — Terms relating to technical drawings, product definition and related documentation
brine	A high concentration solution of sodium chloride in water
Brinell hardness number	*aka* HB. Metal hardness, as measured by an indentation test
British thermal unit	A thermal unit[18] analogous to the calorie (qv). Part of the US Customary Units (qv) system, and may therefore be used in the oil and gas and other US-influenced industries
British thermal unit per hour	see **British thermal unit**
British ton	see **long ton**
British units	*aka* British-American system of units; see **US customary units** and **Box B1**.
British-American system of units	see **British units, Box B1**.
brix	A measure of substances dissolved in water (usually in practice sucrose or other sugars) as determined by hydrometer or refractometer
BRO	see **bottom run-off**
broad weir	see **broadcrested weir**
broadcrested weir	*aka* broad weir. In hydraulics/fluid mechanics, 'broad' means having a significant dimension in the direction of flow. Like so much in hydraulics, it is an approximation; cf thincrested weir
bronze	An alloy, usually mostly copper and tin, though other metals can be used as alloying element as with aluminum bronze (mainly copper and aluminum)
broth	A liquid medium used to grow microorganisms
Brown & Root	1. Genericized trademark - a double pipe heat exchanger (qv) 2. A construction company
brown hydrogen	*aka* black hydrogen. Hydrogen produced from fossil fuels such as coal; cf hydrogen colors
brownfield	In general planning terms, development on previously developed land, or more strictly, any development on property with ground contamination; cf greenfield

---

[18] which has not been British for some considerable time

Box B1. British Units

British units, as a term, is not used much in Britain, nor are the measures which they encompass. 'English units' were superseded by 'British Imperial Units' in the nineteenth century under successive Weights and Measures Acts. These days it would be more accurate to call most of these units 'US customary units' as many Americans do (more formally the 'British-American system of units').

The British, for the most part, haven't used measures such as 'British thermal units' since the 1960s. Liquids are still often sold in pints in the UK, and the Highway Code uses miles, but British professional engineering has considered SI units to be best practice for more than fifty years.

The most well-known engineering problem rooted in this area was the Mars Climate Orbiter, designed by a number of teams, with NASA using 'British Units', and the builders (Lockheed) using metric units (as specified in their contract). The mismatch between NASA's software and Lockheed's caused the failure of the mission on arrival at Mars.

To take a more commonplace scenario, many engineers will have experienced the complexities of specifying pipe sizes using different systems. ISO's metric DN/ID (commonly referred to as nominal bore (NB) in the UK) system of pipe bore sizes increases in multiples of approximately 25 mm to mirror the imperial inch, which is about 25 mm. The USA's alternative nominal pipe size (NPS) is a 'true' imperial pipe size system. The OD of NPS pipe can only be known if you also know the ASME schedule. It should also be noted that NPS and DN are formally dimensionless- they only approximate the pipe bore in inches or mm - and even imperial units are now formally defined in the UK in SI terms.

Added to this is the fact that many engineers appear to use a hybrid system. So, if your project requires 100% imperial systems - not ultimately defined in SI – then you should use the term 'true Imperial', and if you require true DN, you should quote the appropriate ISO standard, otherwise a resulting design may mix and match the two which might have unintended consequences for bore measurement, or more subtly ovality. The ovality of a pipe can matter - you may not be able to weld two pipes together well enough for a hygienic finish if their ovality differs. This is why many pharma designers specify 'true Imperial' pipe sizes. You might specify pipes, valves etc. to ISO 6708 dimensions, which would also be entirely clear, but the problem here is that so few other people do this, so the resulting pipes and fittings are going to be non-standard and therefore expensive.

Related terms
horsepower, long ton, short ton, ton, tonne, British thermal unit

Brownian notion	Erratic random motion of suspended microscopic particles imparted from molecules of the suspension medium
brush pig	A type of pig (qv), specifically a true pipeline scraper (qv)
BRV	*(drawing notation)* Body relief valve
BS	British Standard
BS&W	see **basal sediment and water**
BSD	see **barrels per stream day**
BSE	*(drawing notation)* Bevel small end
BSI	British Standards Institution
BSL	see **biosafety level**
BTE	see **Boltzmann equation**
BTEX	BTX (qv) plus Ethylbenzene
BTFLV	*(drawing notation)* Butterfly valve
BTL	*(drawing notation)* Bottom tangent line
BTL	see **biomass to liquids**
BTU	see **British thermal unit**
BTU/h	see **British thermal unit per hour**
BTX	A term used for mixtures of benzene, toluene, and xylenes in the petrochemical industry
bubble cap	An old-fashioned and expensive type of vapor/liquid contacting device still sometimes used in distillation columns, see **Figure B2**
bubble cap column	A distillation column with bubble cap trays (qv)

 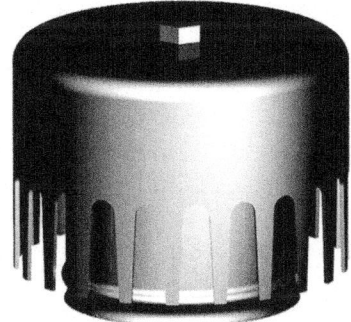

Figure B2 Bubble Caps (Courtesy: Sulzer)

bubble cap tray	An old-fashioned and expensive type of distillation column tray used, based on bubble caps, still used where a positive liquid seal (zero weeping) is required.
bubble column reactor	A multiphase reactor with a high aspect ratio filled with a liquid phase into which gas is sparged at the bottom. The sparged gas serves to agitate the vessel contents, as well as taking part in reaction
bubble flow	The form of two-phase flow in which small bubbles of gas are contained in an otherwise continuous liquid phase
bubble point	The temperature at a given ambient pressure at which the first bubble of vapor is formed in a heated liquid
bubbletight	A description of tightness of sealing of a valve. A closed valve immersed in water and pressurized with air passing no air bubbles is said to be bubbletight
bubbling zone	The part of a fluidized bed in which discrete bubbles coalesce; cf grid zone
bucket	see **rotor blade**
Buckingham π theorem	A theorem which states: "If there are n variables in a problem and these variables contain m primary dimensions, the equation relating all the variables will have (n-m) dimensionless groups"
buckle	A partial collapse of a pipe wall, usually caused by excessive bending or curvature being applied to the pipe
buckling factor	see **in-service margin**
buckling pin device	A type of pressure-relief device; defined formally in API Standard 521. Pressure-relieving and Depressuring Systems. Sixth Edition | January 2014
buffalo	An oil and gas industry term for an amphibious tractor
buffer fluid	A liquid used as a lubricant and/or barrier in a pump seal, held at a lower pressure than pump chamber pressure, so that any cross contamination is outwards from pump chamber; defined formally (and slightly differently) in EN ISO 13709:2003 Centrifugal pumps for petroleum, petrochemical and natural gas industries and API RP 682 - Pump Seals; cf buffer gas
buffer gas	A gas stream in a compressor seal held at a lower pressure than compressor chamber pressure, so that any cross contamination is outwards from compressor chamber. Similar to buffer fluid (qv) but for compressors
buffer store	*aka* decay store. A storage facility for management of nuclear fuel on nuclear power stations used for spent fuel, to reduce by decay heat output prior to dismantling, or to store new nuclear fuel stringers (qv) prior to loading into a reactor
buffer tank	see **break tank**

buffing	A way of smoothing a metal surface; defined formally in ASME BPE (American Society of Mechanical Engineers: Bioprocessing Equipment)
bug blower	An oil and gas industry term for a fan used to disperse noxious gases from a work area
bug wrench	A lineman's term for a speed wrench used to install bolted connectors
Bühler	Proprietary eponym for food processing equipment
build own operate	*aka* BOO. A contractual arrangement in which an EPC company (or more commonly a consortium) builds a plant at their own cost, which they own, to operate on behalf of a customer, often for an agreed unit price of product
build, own, operate, transfer	*aka* BOOT. An arrangement akin to hire purchase in which the client gets to own the plant at the end of the BOO (qv) contract
builder's bag	see **flexible intermediate bulk container**
builders' lime	see **slaked lime**
building	*aka* enclosure. According to firecodes, effectively any enclosed structure intended for any use; defined formally in NFPA 30: Flammable and Combustible Liquids Code, 2018 Edition and NFPA 1142: Standard on Water Supplies for Suburban and Rural Fire Fighting, 2017 Edition
building information modeling	*aka* BIM; see **building information modeling and management**
building information modelling and management	*aka* BIMM, building information and modeling (BIM). Systems commonly used to generate 3D virtual views of buildings; these are becoming a standard feature of architectural design practice. Defined more broadly as "the effective collection and reuse of project data in order to reduce errors and increase focus on design and value" according to the AEC (UK) BIM Standard
building permit drawing	A document used as the basis of a building permit decision; defined formally in BS EN ISO 10209:2012 Technical product documentation – Vocabulary – Terms relating to technical drawings, product definition and related documentation
Building Research Establishment's Environmental Assessment Method	*aka* BREEAM. The UK environmental standard for buildings

built up backpressure	Backpressure on the outlet of a pressure relief device during discharge; defined formally in API Standard 521. Pressure-relieving and Depressuring Systems. Sixth Edition	January 2014, API RP 520 P1 7th Edition, January 2000 Sizing, Selection, and Installation of Pressure-Relieving Devices in Refineries; Part I - Sizing and Selection and API RP 576 - Inspection of Pressure Relieving Devices
bulk chemicals	*aka* heavy chemicals; see **bulk materials**
bulk density	The apparent density of a bulk divided solid material, or of a two-phase flow
bulk materials	*aka* bulk chemicals, commodity chemicals, commodities. Materials produced in large quantities, usually cheaply and sold by weight or volume. Sometimes defined tightly as only dry, divided materials, or sometimes extended to cover things like oil or LPG; cf fine chemicals
bulk plant	A plant with LPG containers of greater than $15.2m^3$ total capacity used as a source for distribution of LPG to other facilities; defined formally in NFPA 58: Liquefied Petroleum Gas Code, 2017 Edition
bulk temperature	Average temperature, often that in the bulk of a pipe or vessel rather than at the wall; cf wall temperature
bulker	*aka* bulk carrier. Most commonly, a ship used for bulk solids transport; sometimes used to mean bulker truck (qv)
bulker truck	*aka* cement bulker. A truck used for bulk solids transport
bull pen	USA lineman's term for where a construction crew gathers before and after work
bull plug	A solid threaded plug used to seal a pipe, especially in oil and gas
bull wheel	A lineman's term for a reel device used to hold tension on a transmission conductor during stringing operations
bulldog	1. *aka* come-along. A lineman's term for a hand operated winch which holds electrical conductors or strands under tension 2. In oil and gas, a fishing tool (qv) used to retrieve items lost in a borehole
bullhead	An oil and gas term for pumping fluids forcibly into a formation, usually formation fluids (qv) that have entered a well-bore (qv) during a well control event (qv)
bump	A highly repetitive shock for test purposes; defined formally in BS ISO 2041:2018 Mechanical vibration, shock and condition monitoring. Vocabulary

bump test	1. see **rotation test**
	2. A quick check that personal gas sensors are responding correctly; cf bump
Buna	Proprietary eponym for polybutadiene synthetic rubber, from which the more modern butadiene styrene rubber (Buna-S, *aka* red rubber) was developed
bunce	*(informal)* A contingency (qv) or allowance of time or money added by estimators
bund	*aka* bunding, bund wall, dyke, dike (US). A wall (often including floor) of containment for vessels "where potentially polluting or flammable substances are handled, processed or stored, for the purposes of containing any unintended escape of material from that area until such time as a remedial action can be taken", according to the US EPA; sized to hold their contents safely in the event of rupture
bund wall	see **bund**
bunding	see **bund**
bunker crude	see **bunker oil**
bunker fuel oil	see **bunker oil**
bunker oil	*aka* bunker crude, bunker fuel oil (BFO), residual fuel oil. A very heavy, often highly polluting grade of oil used as maritime fuel
Bunsen coefficient	The gas solubility coefficient (qv) at 273.15 K and standard pressure
buoyancy	The upwards force on an object immersed in a fluid equal to the weight of the fluid displaced by the object
buoyancy seal	*aka* diffusion seal. A purge gas-sparing type of dry vapor seal in a flare system; defined formally in API Standard 521. Pressure-relieving and Depressuring Systems. Sixth Edition	January 2014 API RP 537 Flare Details for Petroleum, Petrochemical, and Natural Gas Industries
buried	1. Generically, placed below grade
	2. In the context of LPG, those installations in which all of the container except the manway (qv) is below grade; defined formally in NFPA 59: Utility LP-Gas Plant Code, 2018 Edition
burn	1. To combust
	2. An informal term for catalyst regeneration
burn in	The startup procedure of a special atmosphere furnace; defined formally in NFPA 86: Standard for Ovens and Furnaces, 2019 Edition

burn off pilot	A pilot used to ignite special processing atmosphere on exit; defined formally in NFPA 86: Standard for Ovens and Furnaces, 2019 Edition
burn out	The shutdown procedure of a special atmosphere furnace; defined formally in NFPA 86: Standard for Ovens and Furnaces, 2019 Edition
burn pit	A pit used to dispose of materials by burning
burn pit flare	One or more horizontal flares installed in a pit which retains any unburned liquid, allowing multiphase fluids to be handled; defined formally in API Standard 521. Pressure-relieving and Depressuring Systems. Sixth Edition	January 2014 and API RP 537 Flare Details for Petroleum, Petrochemical, and Natural Gas Industries
burn through	A welding error in which the base metal is melted, and a hole results; defined formally in ASME BPE (American Society of Mechanical Engineers: Bioprocessing Equipment)
burnback	Burning within a flare burner (qv); defined formally (and slightly differently) in API Standard 521. Pressure-relieving and Depressuring Systems. Sixth Edition	January 2014 and API RP 537 Flare Details for Petroleum, Petrochemical, and Natural Gas Industries
burner	A device or devices which introduce fuel and, in some cases, oxygen-containing gas into a heater, in such a way as to allow it to burn. Usually fired by oil or gas, and usually located in the radiant section of a fired heaters, heating the fluid in the pipes. More complicated furnaces may include supplementary burners in the convection section. Coal burners are used in various industries in the USA and China, especially in mineral processing. Defined formally and slightly differently in BS EN ISO 13705:2012/ISO 13705:2012(E) Petroleum, petrochemical and natural gas industries. Fired heaters for general refinery service, API STD 530 - Calculation of Tube Heater Thickness, API STD 560 - Fired Heaters for General Refinery Service, NFPA 86: Standard for Ovens and Furnaces, 2019 Edition and API RP 535 - Burners for Fired Heaters at Refineries
burner block	see **tile**
burner group	Burners operated and controlled as a group; defined formally in BS EN 12952-8:2002 Water-tube boilers and auxiliary installations. Requirements for firing systems for liquid and gaseous fuels for the boiler
burner gun	see **atomizing burner**

burner management system	A control system used to assure the safe start-up, operation and shutdown of burners; defined formally and differently in BS EN 12952-8:2002 Water-tube boilers and auxiliary installations. Requirements for firing systems for liquid and gaseous fuels for the boiler and NFPA 86: Standard for Ovens and Furnaces, 2019 Edition
burner system	Burners operated as a group with a common safety shutoff valve; defined formally in NFPA 86: Standard for Ovens and Furnaces, 2019 Edition
burner throat	The narrowest part of the burner air flow path; defined formally in API RP 535 - Burners for Fired Heaters at Refineries
burner tile	see **tile**
burner turndown	Burner fuel input rate turndown; defined formally in NFPA 86: Standard for Ovens and Furnaces, 2019 Edition
burners	Plural of burner (qv); defined formally (and differently from burner!) in BS EN 12952-8:2002 Water-tube boilers and auxiliary installations. Requirements for firing systems for liquid and gaseous fuels for the boiler
burning	see **combustion**
burning rain	see **roman candle effect**
burning velocity	*aka* flame propagation. Flame front speed; defined formally in API Standard 521. Pressure-relieving and Depressuring Systems. Sixth Edition | January 2014 and API RP 537 Flare Details for Petroleum, Petrochemical, and Natural Gas Industries
burnout air	Air introduced to burn away residual flammable atmosphere, soot and carbonaceous material in a furnace; defined formally in NFPA 86: Standard for Ovens and Furnaces, 2019 Edition
burnt lime	see **quicklime**
burr	A rough edge of material produced by welding, tools or molding. Defined formally in a welding context in ASME BPE (American Society of Mechanical Engineers: Bioprocessing Equipment), and more generally in BS EN ISO 10209:2012 Technical product documentation — Vocabulary — Terms relating to technical drawings, product definition and related documentation

burst pressure	In the context of rupture disks, the differential pressure across the disk at which the disk is rated to burst at a specified temperature. Defined formally and slightly differently in API RP 520 P1 7th Edition, January 2000 Sizing, Selection, and Installation of Pressure-Relieving Devices in Refineries; Part I - Sizing and Selection, API Standard 521. Pressure-relieving and Depressuring Systems. Sixth Edition	January 2014 and API RP 576 - Inspection of Pressure Relieving Devices
burst pressure tolerance	The variation around rated burst pressure at which a rupture disk bursts. Defined formally in API RP 520 P1 7th Edition, January 2000 Sizing, Selection, and Installation of Pressure-Relieving Devices in Refineries; Part I - Sizing and Selection and API RP 576 - Inspection of Pressure Relieving Devices
burstel	Device fitted to the non-process side of a burst disk which will send a signal to the control system when the burst disk ruptures
bursting disc	see **rupture disk**
bursting disk	see **rupture disk**
bus bars	see **busbars**
busbars	*aka* bus bars, buss bars. Large, uninsulated common conductors with a rectangular cross section, found in switchgear (qv) etc.
bushel	64 US pints in US customary units (qv); obsolete elsewhere
business interruption insurance	Insurance against loss of income occasioned by events which interrupt a company's ability to trade
business occupancy	A non-mercantile commercial occupancy; defined formally in NFPA 30: Flammable and Combustible Liquids Code, 2018 Edition
buss bars	see **busbars**
BUTEX process	see **PUREX process**
butt joint	A joint between two items placed adjacent to one another, either at right angles or end-to-end; defined formally, in the context of welding, in ASME BPE (American Society of Mechanical Engineers: Bioprocessing Equipment)
butt weld	A butt joint (qv) made by welding
butterfly damper	A centrally pivoted single blade damper similar to a butterfly valve (qv), in ductwork that rotates on its axis to regulate the flow of gases. Defined formally in BS EN ISO 13705:2012/ISO 13705:2012(E) Petroleum, petrochemical and natural gas industries. Fired heaters for general refinery service, API STD 530 - Calculation of Tube Heater Thickness and API STD 560 - Fired Heaters for General Refinery Service

butterfly valve	A centrally pivoted single blade valve commonly used for isolating duties, other than in hygienic service, where its use is controversial. It is cheap, and in its common 'wafer pattern' variant, takes up very little pipeline length. It is however somewhat unreliable, and has poor control characteristics
buttering up	A welding technique used to bridge gaps between two surfaces which require welding together. Used, e.g., when fitting a pipe section into a gap between two fixed pipes, and thus requiring a gap enabling the pipe section to be 'offered up'; further described in ASME VIII
buy-back	A company buying its own shares back
buzzer	An alarm for operator awareness; cf annunciator
BV	*(drawing notation)* Ball valve
BW	see **butt weld**
BWF	*(drawing notation)* Butt weld fitting
BWG	see **Birmingham wire gauge**
BWPD	Barrels of water per day
BWR	1. Boiling water reactor
	2. see **Benedict Webb Rubin equation of state**
BWR equation of state	see **Benedict Webb Rubin equation of state**
bypass	Pipework around valve, heat exchanger, tank, etc.
bypass interceptor	see **interceptor**
bypass pumping	see **overpumping**
byproduct	An (often undesirable, at best of minor revenue value) additional product of a process; cf coproduct

C

C&I	see **control and instrumentation**
C&I engineer	see **control and instrumentation engineer**
C&Q	see **commissioning and qualification**
C2	*aka* C2 hydrocarbons. Short for 'Carbon-2', indicating a hydrocarbon chain length of 2, so a term for ethane, ethylene, acetylene or a mixture of these
C2 hydrocarbons	see **C2**
C3	*aka* C3 hydrocarbons. Short for 'Carbon-3' indicating a hydrocarbon chain length of 3, so a term for propane, propylene, or a mixture of these
C3 hydrocarbons	see **C3**
C4	*aka* C4 hydrocarbons. Short for 'Carbon-4' indicating a hydrocarbon chain length of 4, so a term for butane, butylenes, or a mixture of these
C4 hydrocarbons	see **C4**
CA	1. see **competent authority** 2. see **corrosion allowance**
CAA	see **Clean Air Act**
CABA	Compressed air breathing apparatus, see **breathing apparatus set**
cabinet heater	A type of portable self-contained propane heater; defined formally in NFPA 58: Liquefied Petroleum Gas Code, 2017 Edition
cable avoidance tool	*aka* CAT. Equipment used to locate underground services without excavation
cable diagram	A diagram which identifies the conductors, end locations, routes and function of cables; defined formally in BS EN ISO 10209:2012 Technical product documentation - Vocabulary - Part 1: Terms relating to technical drawings: general and types of drawings
cable run drawing	A cable location drawing; defined formally in BS EN ISO 10209:2012 Technical product documentation - Vocabulary - Part 1: Terms relating to technical drawings: general and types of drawings
cable tray	see **traywork**
CABP	see **cubic average boiling point**
CAD	Computer aided design (qv), computer aided drawing, or computer aided drafting

CAD drawing	A virtual drawing in a computer aided design (qv) model, or a hardcopy of that drawing; defined formally in BS EN ISO 10209:2012 Technical product documentation - Vocabulary - Part 1: Terms relating to technical drawings: general and types of drawings
CAD model	Datafiles which represent various aspects of something being drawn or designed; defined formally in BS EN ISO 10209:2012 Technical product documentation - Vocabulary - Part 1: Terms relating to technical drawings: general and types of drawings
CAD monkey	(*informal*) see **CAD operator**
CAD operator	*aka* CAD monkey. Someone who operates CAD software, with an implication of a lack of engineering knowledge; cf draughtsman
CADD	see **computer aided design and drafting**
CAF	1. see **compressed air foam**
	2. Compressed asbestos fiber; a now-obsolete material for gaskets
	3. Cement aggregate fill, used in stope mining (qv) as a backfill
CAFS	see **compressed air foam system**
cage frequency	see **fundamental train frequency**
cage failing frequency	see **fundamental train frequency**
cake filtration	A filtration mode in which a cake of previously filtered particles forms the filtration medium
caking	The aggregation of a divided solid material into a mass or 'cake'
cal	see **slaked lime**
calandria	A heat exchange device, especially an evaporator's heating element or thermosyphon reboiler, though the term is applied to other heating devices
calcination	see **calcining**
calciner	see **rotary kiln**
calcining	Heating solid materials, often in a low oxygen atmosphere. Used to render materials such as ores friable, or to decompose, evaporate volatile components from, or partially oxidize materials
calcite dissolution capacity	(*symbol* Dc) A measure of potential for corrosion or scaling in carbonate buffered water, having the same value as the calcium carbonate precipitation potential (qv), but the opposite sign
calcium carbonate precipitation potential	A measure of potential for corrosion or scaling in carbonate buffered water; cf Langelier saturation index
calcium hardness	The concentration of the calcium ions present in a sample of water, calcium being one of the ions that produces 'hardness' (qv) in water

calculation pad	An A4 book of blank design calculation sheets, usually marked in a way making it proprietary to the issuing company
calculation sheet	A formal template used for hand calculations, or more generally, hardcopy design calculations; defined formally in BS EN ISO 10209:2012 Technical product documentation - Vocabulary - Part 1: Terms relating to technical drawings: general and types of drawings; cf datasheet
calendering	A method used to thin, coat or smooth a laminar material such as a textile or plastic film by passing it between or over rollers to subject it to heat and pressure
calibration	1. The markings on the scale of a measuring device 2. The act of a comparing a value provided by an instrument being tested with a reliable reference value and correcting any error found, in order to maintain accuracy of readings
calibration factor	The average sensitivity of a transducer under specified conditions; defined formally in BS ISO 2041:2018 Mechanical vibration, shock and condition monitoring. Vocabulary
calibrator	*(jocular) aka* Manchester screwdriver, adjuster. A hammer
caliper pig	*aka* geometry pig. A device to measure internal pipe ovality
calorie	A non-SI, largely obsolete (in engineering) measure of the heat energy required to raise the temperature of 1 gram of water by 1°C
calorific value	The energy yield on complete combustion of a specified quantity of material (qv heat of combustion)
camera ready copy	Materials fit for final reproduction without further revision; defined formally in BS EN 82079-1:2020 Preparation of instructions for use - Structuring, content and presentation. General principles and detailed requirements
Camlok	A genericized trademark for a class of flexible hose connection, or a (usually temporary) electrical connector popular in the USA
campaign	1. *aka* campaign production. A mode of production intermediate between continuous production (qv) and batch production (qv), in which there is a defined period of continuous production followed by cleaning, ready for another campaign of production of either the same or another product 2. Less frequently, the duration of production between planned shutdown/maintenance events
campaign production	see **campaign**

can pump	A vertical multistage centrifugal pump; defined formally in NFPA 20: Standard for the Installation of Stationary Pumps for Fire Protection, 2019 Edition. Also sometimes used in error to mean canned pump (qv)
candela	(*symbol* Cd) The SI base unit of luminous intensity
candela per square meter	(*symbol* Cd/m^2) *aka* candela per square meter. The SI unit of luminance
candela per square metre	see **candela per square meter**
CANDU	CANada Deuterium Uranium reactor; a type of heavy water nuclear reactor
canned motor	A hermetically sealed electric motor
canned pump	A type of centrifugal pump where the impeller and motor are integrated within the same casing, with the motor therefore surrounded by the process liquid. Since the drive shaft of the pump does not pass through the body of the pump (i.e., through the pump bearings and/or a seal), this critical leak point is eliminated, resulting in a high-integrity piece of equipment, often used for very high hazard or air-sensitive liquids; cf magnetic drive pump
cannibalise	see **cannibalize**
cannibalize	*aka* cannibalise. To reuse parts of redundant machinery
canning	A method of preserving food and drink by controlled heating in a hermetically sealed (often metallic) container
CAPA	see **corrective and preventive action**
capability	In a quality management context, the ability of an object to meet requirements; defined formally in BS EN ISO 9000 Quality management systems Fundamentals and vocabulary
capacitive deionisation	see **capacitive deionization**
capacitive deionization	*aka* CDI, capacitive deionisation. A novel electrochemical water deionization technique
capacity	The volume of a vessel or the maximum duty of a process unit, utility or labor supply in suitable units (not always volumetric); defined formally in API RP 520 P1 7th Edition, January 2000 Sizing, Selection, and Installation of Pressure-Relieving Devices in Refineries; Part I - Sizing and Selection
capacity element	see **lag time**
capacity ratio	1. *aka* capacity usage ratio, capacity utilization ratio, labor capacity ratio. In project management, the ratio of hours worked to budgeted hours

	2. In heat exchanger design, the ratio of the product of mass flow and specific heat for each of the fluids undergoing heat exchange
capacity usage ratio	see **capacity ratio**
capacity utilization ratio	see **capacity ratio**
capex	Capital expenditure; cf opex
capillary	A very fine bore channel or tube
capillary bound water	*aka* irreducible water. Water held in capillaries within a material
capital costs	The fixed, one-time cost of a project including purchased land, equipment, engineering, initial inventory and other supplies, debt, insurance, fees, etc.; cf non-capital costs, operating costs
caprolactam	A cyclic amide of caproic acid used to make nylon (qv)
CAR	see **contractors' all risks insurance**
car seal	A metal or plastic cable used to hold (but not lock) a safety-critical valve in an open or closed position, easily broken by cutting or shearing, used only to prevent accidental actuation and to provide a quick and clear visual indication of status; defined formally in the context of PRDs in API RP 576 - Inspection of Pressure Relieving Devices
caramelisation	see **caramelization**
caramelization	*aka* caramelisation. A form of non-enzymic browning resulting from the oxidation of sugars
Carberry reactor	A type of mixed reactor containing a spinning basket filled with catalyst
carbon abatement technology	*aka* CAT. Process plant used to reduce CO_2 emissions, especially from fossil fuel combustion; cf greenwash
carbon black	A material produced by the incomplete combustion of carbonaceous fuel used as a filler in tire production etc.
carbon capture and sequestration	*aka* CCS, carbon capture and storage. Either a synonym of carbon capture and storage (sequestration being a synonym of storage), or a term for non-process plant approaches to permanently preventing anthropogenic CO_2 from reaching the atmosphere; cf carbon sequestration
carbon capture and storage	*aka* CCS, carbon capture and sequestration. Permanently preventing anthropogenic CO_2 from reaching the atmosphere, especially those approaches relying on process plant. Criticized on a number of grounds, most notably for being an experimental technology which promises to allow a 'business as usual' carbon economy to continue. These characteristics however make it something of a grant-magnet for academia despite being uneconomic in reality; cf greenwash, clean coal, green hydrogen, carbon capture and sequestration

carbon credit	A tradeable certificate representing the right to emit a stated mass (tonnes in the EU, tons in the USA) of carbon dioxide equivalent
carbon cycle	The natural processes which recycle the carbon contained in living creatures through the environment
carbon dioxide equivalent	Measure of greenhouse gas emissions as a nominally equivalent quantity of CO_2
carbon fiber(s)	*aka* carbon fibre(s). Carbon filaments and the textiles made from them, used to reinforce composite materials, or such composite materials; cf glass reinforced plastic, FRP
carbon fibre(s)	see **carbon fiber(s)**
carbon footprint	Nominally the total greenhouse gas emissions of an entity, converted to equivalent mass of CO_2; cf greenwash
carbon in leach	*aka* CIL. A process similar to carbon in pulp (qv) except that leaching and adsorption occur simultaneously
carbon in pulp	A simple cheap method of recovering gold obtained by cyanidation using granular activated carbon (qv)
carbon offset	The idea that third party carbon capture and sequestration (qv), or reduction of carbon emissions can offset personal or commercial greenhouse gas emissions, and the commercial offerings based on that idea; cf greenwash
carbon potential	The ability of a heat treatment furnace's atmosphere to increase a steel's carbon content
carbon sequestration	see **carbon capture and storage**
carbon steel	*aka* CS. Unalloyed (or almost unalloyed) steels; defined formally in API RP 571 - Damage Mechanism Affecting Fixed Refinery Equipment
carbon tax	Taxes levied on fossil fuels intended to discourage their use
carbonatation	Clarification process in sugar refining, in which lime and carbon dioxide are added to precipitate calcium carbonate along with colloidal matter present
carbonation	Dissolving carbon dioxide to produce effervescent drinks, or reacting with carbon dioxide in other contexts
carbonization	Pyrolysis (qv) by means of heating in a controlled atmosphere to produce gases and high carbon solids, such as coke (qv) etc.
carboxylic acid	An organic acid with a -COOH functional group
carboy	A container used for small quantities (4-61 liters) of chemicals, especially acids
carburisation	see **carburization**

carburization	*aka* carburisation, carburizing. A form of case hardening (qv) caused by heating low carbon steels in the presence of carbon rich materials. It may be done deliberately as a surface finish on metal, or may be a problem when it has been caused unintentionally in process equipment; cf nitriding
carburizing	see **carburization**
carcinogen	Something known to cause cancer
cargo tank	The tank of a road tanker or seagoing oil tanker; defined formally in the former sense in NFPA 58: Liquefied Petroleum Gas Code, 2017 Edition
Carman-Kozeny equation	*aka* Kozeny-Carman equation. An equation used to calculate head loss through a packed bed
Carnot cycle	A kind of thought experiment used to set the theoretical maximum possible efficiency for heat engines
Carothers equation	*aka* Carothers' equation (but not Carother's equation). An equation which gives the degree of polymerization for a given fractional monomer conversion
carriage paid to	see **Incoterms**
carriage, insurance and freight	Incorrect version of cost, insurance, and freight; see **Incoterms**
carrier gas	An inert gas such as hydrogen[19], nitrogen, helium, argon or carbon dioxide used to carry materials to analysis, or in chromatography
carrier gas special atmosphere	Carrier gas forming the bulk of a special atmosphere; defined formally in NFPA 86: Standard for Ovens and Furnaces, 2019 Edition
carrier water	A loosely controlled flow of water used to dilute and convey precision doses of lime, chlorine and other water treatment chemicals to point of use
carryover leakage	Exhaust leakage across a rotary heat exchanger (qv) as the wheel rotates from the exhaust to supply air stream
cartridge filter	A (usually disposable) filter element
cartridge seal	A completely self-contained pump seal. Defined formally and very slightly differently in API RP 682 - Pump Seals and ASME BPE (American Society of Mechanical Engineers: Bioprocessing Equipment)
cartridge type element	*aka* pump cartridge assembly. All parts of the pump other than the casing; defined formally in EN ISO 13709:2003 Centrifugal pumps for petroleum, petrochemical and natural gas industries

---

[19] 'Inert' is context specific, in case you were wondering why hydrogen was included in a list of inert gases

cascade control	Multiple controllers working together in process control (qv), with the output of one feedback loop controlling the setpoint of a second[20]
cascade plot	*aka* waterfall plot. A diagram which facilitates comparison of frequency analyses; defined formally in BS ISO 2041:2018 Mechanical vibration, shock and condition monitoring. Vocabulary
cascade process	Any process with a number of repeated similar, low efficiency steps, as is common in isotope separation and ore processing
case hardening	The formation of a thin skin (case) of hardened steel on a piece of low carbon steel by carburizing, nitriding etc.
cashflow	The total amount of money being transferred into and out of a business, especially as affecting liquidity
casing	1. The metallic external enclosure of a heater, pump or other machine Defined in the context of heaters in BS EN ISO 13705:2012/ISO 13705:2012(E) Petroleum, petrochemical and natural gas industries. Fired heaters for general refinery service, API STD 530 - Calculation of Tube Heater Thickness, and API STD 560 - Fired Heaters for General Refinery Service 2. A well liner pipe in oil and gas drilling operations
casing bowl and slips	see **casing spider**
casing dogs	see **dogs**
casing gas	*aka* CG, annular gas, casing head gas, casinghead gas. 1. Gas which collects in the space between the casing (qv) and piping in an oil well 2. A synonym of natural gas liquids (qv)
casing head gas	see **casing gas**
casing slips	see **slips**
casing spider	*aka* casing bowl and slips. A rig floor spider (qv) which uses slips or dogs (qv) to secure the casing
casinghead gas	see **casing gas**
Casson's equation	An equation describing the behavior of non-Newtonian fluids

---

[20] Heavily used in academia, perhaps because of all the mathematics involved. Not generally used in practice due to the **KISS principle** (qv) unless unavoidable

cast iron	*aka* CI. Low melting point iron / carbon alloys with >2% carbon content (which is why they are not classed as steel), characterized by high compressive strength and low tensile strength. There are two main types, depending on carbon particle shape: gray iron (qv) has flakes of carbon which promote crack growth, making it very brittle; spheroidal graphite or ductile iron (qv) has rounded nodules of carbon which stop cracks
castable	A shaped refractory made from insulating concrete; defined formally in BS EN ISO 13705:2012/ISO 13705:2012(E) Petroleum, petrochemical and natural gas industries. Fired heaters for general refinery service, API STD 530 - Calculation of Tube Heater Thickness and API STD 560 - Fired Heaters for General Refinery Service
Castell key	A genericized trademark for a trapped key interlock on an enclosure (especially electrical) or spared pressure relief system, providing a mechanical interlock for mechanical, electro-mechanical and electrical systems where exposure to hazard is too frequent for a permit to work (qv) system to be effective
cast-film extrusion	see **chill-roll film extrusion**
Cat	1. catalyst (qv) or catalytic 2. see **Caterpillar truck**
CAT	1. see **carbon abatement technology** 2. Computed tomography, see tomography 3. see **connector actuating tool** 4. see **cable avoidance tool**
CAT and genny	Cable avoidance tool and signal generator; equipment used to locate underground services without excavation
cat cracker	see **catalytic cracker**
cat head	Part of the draw works used to raise/lower an entire drilling string (qv) in oil and gas drilling, which acts like a winch to pull the chain attached to the tongs (large pipe wrenches) used to screw the casing together
cat line	The rope used on the cat head (qv) in oil and gas drilling
catalysis	The action of a catalyst (qv)
catalyst	A substance that promotes a chemical reaction without being consumed
catalyst activity	1. Generically, the ratio of current efficiency of catalyst to that of fresh catalyst 2. In the context of fired heaters, a measurement of the efficiency of catalytic NOx reduction; defined formally in API RP 536 - NOx control on Fired Heaters at Refineries

catalyst dump nozzle	*aka* unloading nozzle, dump nozzle. A nozzle at the bottom of a catalytic reaction vessel used to empty catalyst rapidly under gravity
catalyst handling facility	Lifting beam, containers and hoist used for handling catalyst modules; defined formally in API RP 536 - NOx control on Fired Heaters at Refineries
catalyst matrix	The inert carrier material for a catalytic material; defined formally in API RP 536 - NOx control on Fired Heaters at Refineries
catalyst module	Catalyst assemblies packaged for installation; defined formally in API RP 536 - NOx control on Fired Heaters at Refineries
catalyst poison	A substance which chemically inactivates a catalyst
catalyst poisoning	The reduction of efficiency of a catalyst caused by a catalyst poison; defined formally in the context of flue gas NOx removal in fired heaters in API RP 536 - NOx control on Fired Heaters at Refineries
catalyst space velocity	Volume of flue gas per volume of catalyst per hour; defined formally in API RP 536 - NOx control on Fired Heaters at Refineries
catalyst substrate	see **catalyst matrix**
catalyst support	The structures which support catalyst modules in service; defined formally in API RP 536 - NOx control on Fired Heaters at Refineries
catalyst type	NOx control catalyst typed by active ingredient; defined formally in API RP 536 - NOx control on Fired Heaters at Refineries
catalytic cracker	A unit in which catalytic cracking takes place
catalytic cracking	Breaking long chain hydrocarbons down into shorter ones with the aid of hydrogen and a catalyst (hydrocracking), or temperature and a catalyst (fluid catalytic cracking), making heavy oils into more valuable medium distillates
catalytic distillation	see **reactive distillation**
catalytic hydrocracking	Breaking down long chain less-saturated hydrocarbons into shorter more saturated ones with the aid of hydrogen and a catalyst, making heavy oils into more valuable medium distillates
catalytic oxidizer	see **thermal oxidizer**; defined formally in NFPA 86: Standard for Ovens and Furnaces, 2019 Edition
catalytic reactor	A reactor in which a catalytic reaction takes place
catalytic reforming	*aka* reforming. A process which increases the octane rating of naphthas; the higher-octane products being known as reformates

catalytic rich gas process	*aka* CRG process. A process for producing synthetic natural gas developed by British Gas
catastrophic failure	Sudden, total, unrecoverable failure such as the rapid/instantaneous loss of the entire inventory of a vessel
catch basin	*aka* CB; see **storm drain**
catch can	see **oil separator**
category A welded joint	In the context of heat exchangers, a type of longitudinal welded joint; defined formally in API 660 - Shell-and-Tube Heat Exchangers, API RP 661 - Heat Exchangers and ASME Boiler & Pressure Vessel Code 2017 Edition VIII Rules for Construction of Pressure Vessels, Division 1
category B welded joint	In the context of heat exchangers, welds connecting flanges etc. to other components, defined formally in ASME Boiler & Pressure Vessel Code 2017 Edition VIII Rules for Construction of Pressure Vessels, Division 1, but not in API 660 or 661
category C welded joint	In the context of heat exchangers, welds connecting communicating chambers (qv) or nozzles to other components, defined formally in ASME Boiler & Pressure Vessel Code 2017 Edition VIII Rules for Construction of Pressure Vessels Division 1, but not in API 660 or 661
category D welded joint	In the context of heat exchangers, welds connecting nozzles to shell, defined formally in ASME Boiler & Pressure Vessel Code 2017 Edition VIII Rules for Construction of Pressure Vessels Division 2
category endpoint	Approximately, an environmental, human health or resource issue of concern; defined formally (if rather obliquely) in EN ISO 14040:2006 Environmental management - Life cycle assessment - Principles and framework
category indicator	see **impact category indicator**
catenary bar screen	A type of travelling bar screen used to screen large debris out of wastewater
Caterpillar truck	A dump truck[21]
cathode	1. Generically, the negative electrode; cf **anode**
	2. Specifically, the end product of the electrorefining process for copper
cathodic protection	*aka* galvanic protection, sacrificial protection. Protection against corrosion by electrical connection to a sacrificial metal anode which corrodes preferentially
cation	Positively charged ion

---

[21] Or perhaps more generally 'a generic term for the yellow metal machines on construction sites' - although they're not all yellow any more, are they!

cation exchange resin	Negatively charged crosslinked polymer particles that exchange their bound Na+ or Ca2+ for other cations in solution
CATNIP	*(jocular)* Cheapest available technology not inviting prosecution - a wry joke; cf BATNEEC
catwalk	In oil and gas drilling, the area next to a drill floor where a core (qv) is processed immediately after coming onboard
Cauchy number	*(symbol* Ca) A dimensionless number used to study compressible flows
Cauchy strain	see **strain**
causal factor tree analysis	An investigation and analysis technique used to record and display, in a logical, tree-structured hierarchy, all the actions and conditions that were necessary and sufficient for a given consequence to have occurred
cause and effect	see **cause effect chart/diagram**
cause effect chart/diagram	*aka* cause and effect. In the context of control and instrumentation, a chart/diagram setting out the control logic for trips, interlocks, alarms including instrument, set point, action, etc.; cf fishbone diagram
caustic	1. (adj.) Corrosive to flesh 2. (noun) Commonplace shorthand for caustic soda (qv) 3. Sometimes used more broadly to mean any alkali or basic material
caustic cracking	see **caustic stress corrosion cracking**
caustic lime	see **slaked lime**
caustic soda	*aka* soda lye, lye or caustic. Sodium hydroxide (NaOH)
caustic stress corrosion cracking	*aka* caustic cracking. Cracking of steels under tensile stress exposed to excessive levels of caustic and heat.
causticization	The process of converting carbonates into hydroxides
caution	An instruction that draws attention to a risk; defined formally in BS EN 82079-1:2012 Preparation of instructions for use. Structuring, content and presentation - General principles and detailed requirements
cavitation	Rapid reduction of pressure in a liquid resulting in the formation of vapor bubbles which collapse rapidly when pressure subsequently rises, producing shockwaves which can damage pump impellers and generate noise/vibration; defined formally in ASME BPE (American Society of Mechanical Engineers: Bioprocessing Equipment)
cavity pump	see **progressing cavity pump**
CB	1. Catch basin, see **storm drain** 2. Compact body
CBA	see **cost/benefit analysis**

CBI	Condition based inspection
CBM	Coal bed methane
CBOB	see **conventional blendstock for oxygenate blending**
cBOD	Carbonaceous BOD (qv); i.e., BOD as measured with nitrification inhibitor. This is usually what is meant by BOD
CBR	see **cost/benefit ratio**
C-C	*(drawing notation)* Center to center
CCC	*(drawing notation)* Cold cut and weld
CCF	see **common cause failure**
CCGT	see **combined cycle gas turbine**
CCN	see **cement chemist notation**
CCP	see **critical control point**
CCPP	see **calcium carbonate precipitation potential**
CCPS	Center for Chemical Process Safety, part of the American Institute of Chemical Engineers
CCR	1. see **Conradson carbon residue** 2. Central control room, see control room 3. see **continuous catalytic regeneration**
CCR platforming	Proprietary catalytic reforming (qv) process with a platinum catalyst
CCS	1. see **carbon capture and sequestration** 2. see **carbon capture and storage**
CDC	Centers for Disease Control and Prevention (USA)
CDI	see **capacitive deionization**
CDM	see **Construction, Design and Management Regulations (2015)**
CDTP	see **cold differential test pressure**
CDU	see **crude distillation unit**
CE	Consulting engineer
CE mark	(*symbol* CE) *aka* Conformité Européenne mark. A marking on a product indicating that it is in complete compliance with all relevant EU legislation, fit for intended purpose and not a danger to lives or property
CE marking	Placing a CE mark (qv) on a product
CeeSTAR	see **continuous stirred tank reactor**
ceiling drawing	A drawing which specifies the ceilings of a story of a building, normally in mirrored projection; defined formally in BS EN ISO 10209:2012 Technical product documentation - Vocabulary - Part 1: Terms relating to technical drawings: general and types of drawings

cell	1. A section of a cooling tower (qv), usually consisting of water distributor, fan, stack, and filling
2. see **float cell**
3. see **manufacturing cell**
4. The unit of life; single celled organisms such as bacteria, yeasts and fungi are used in biochemical engineering to produce useful products
cell density	1. A term for the concentration of microorganisms in a broth
2. Hole density in a matrix, defined formally in API RP 536 - NOx control on Fired Heaters at Refineries
cell dry weight	The mass of microorganisms in a broth, usually as grams on a dry weight basis per liter of sample
cell wet weight	The mass of microorganisms in a broth, usually as grams of undried microorganisms per liter of sample
Celsius	see **centigrade scale**
cement	In an engineering context, cement is an inorganic material, which can be mixed with sand, aggregate and water to make concrete. The term is however often misused to refer to the finished product, concrete. If the cement is lime, the resulting concrete is non-hydraulic (cannot be used under water). Hydraulic cements such as Portland cement are made of a mixture of silicates and oxides
cement bulker	see **bulker truck**
cement chemist notation	A simplified notation system, used by those formulating cements, which shortens the names of the complex silicates etc. used in formulation. The most common oxides present are given simple letter notations, and the names of their silicates are based on these letters such as Alite from 'A', Belite from 'B' and so on
cementation	An obsolete steelmaking method
CEMS	Continuous emissions monitoring system
CEng (UK)	A UK Chartered Engineer (qv), commonly seen without the (UK) I have added here. Other countries, however, license chartered engineers, so some non-UK engineering institutions think a following parenthesized 'UK' should be added (a suggestion opposed by UK institutions)
center line	(*symbol* ₵) *aka* CL, centre line, centerline, centreline. Line through the geometric center of a feature; defined formally in BS EN ISO 10209:2012 Technical product documentation - Vocabulary - Part 1: Terms relating to technical drawings: general and types of drawings

center of gravity	*aka* COG, centre of gravity, balancing point. An imaginary point though which gravity is taken to act upon an object; defined formally in BS ISO 2041:2018 Mechanical vibration, shock and condition monitoring. Vocabulary
center of gravity mounting system	*aka* centre of gravity mounting system. A mounting system where the center of gravity (COG) of the mounted equipment coincides with the COG of the thing it is mounted in. Defined formally in BS ISO 2041:2018 Mechanical vibration, shock and condition monitoring. Vocabulary
center of mass	Synonymous with center of gravity (qv) in a uniform gravity field, but defined formally in BS ISO 2041:2018 Mechanical vibration, shock and condition monitoring. Vocabulary
centerline	see **center line**
centi-	A metric unit prefix for x 10^{-2}
centigrade scale	*aka* Celsius. Non-SI unit of temperature, differing from the SI unit kelvin (qv) in that it sets zero as the freezing point of water rather than absolute zero
centipoise	Non-SI unit of dynamic viscosity, one-hundredth of a poise (qv), equivalent to one millipascal-second (mPa·s) in SI units. Frequently used in place of Pa.s due to convenience of scale for commonly used materials
centistoke	Non-SI unit of kinematic viscosity, one-hundredth of a stoke (qv), equivalent to $0.000001 m^2/s$
central control room	see **control room**
central services	Supporting facilities often enclosed within buildings which are neither a direct part of the process reaction train nor utilities, such as telecoms, HVAC, amenities, laboratories, workshops and emergency services
centralized vacuum cleaning system	A system to prevent flammable dust buildup. Defined formally in NFPA 654: Standard for the Prevention of Fire and Dust Explosions from the Manufacturing, Processing, and Handling of Combustible Particulate Solids, 2017 Edition
centrate	The low-solids stream from a decanter centrifuge
centre line	see **center line**
centre of gravity	see **center of gravity**
centre of gravity mounting system	see **center of gravity mounting system**
centreline	see **center line**
centrifugal	A term applied to a rotodynamic device in which a rotating element produces a radial pressure difference; cf axial
centrifugal compressor	A rotodynamic compressor
centrifugal decanter	A bowl centrifuge; see **decanter**

centrifugal fan	*aka* radial flow fan. A low pressure, high volume air mover in which the fan produces a radial pressure difference, moving air centrifugally, rather than axially; cf axial flow fan
centrifugal force	An apparent inertial force acting on all objects viewed in a rotating frame of reference. Opposed by centripetal force (qv)
centrifugal pump	*aka* radial flow pump, rotodynamic pump. Defined formally in NFPA 20: Standard for the Installation of Stationary Pumps for Fire Protection, 2019 Edition; cf axial flow pump, positive displacement pump, rotary pump 1. A pump in which a rotating element generates a radial pressure difference by means of transfer of kinetic energy, rather than by mechanical means 2. *(informal)* may also be used to describe all rotodynamic pumps including axial flow pumps and mixed flow (qv) pumps
centrifugal separator	*aka* separator, cream separator, milk separator. A disk centrifuge used to separate cream from milk
centrifuge	*aka* separator. A machine which separates components of a mixture based on their density through the application of centrifugal forces in a central element spun at high speed
centripetal force	A force towards the center of rotation which keeps a rotating object on a curved path; cf centrifugal force
CEnv	Chartered Environmentalist (UK-only accreditation)
CEO	Chief Executive Officer
CEPCI	see **Chemical Engineering Plant Cost Index**
ceramic	An inorganic non-metallic solid shaped and then hardened by heating to high temperatures, usually hard, corrosion-resistant and brittle, normally differentiated from vitreous materials
ceramic fiber	Fibrous refractory insulation, which confusingly can be vitreous rather than ceramic according to the formal definitions in BS EN ISO 13705:2012/ISO 13705:2012(E) Petroleum, petrochemical and natural gas industries. Fired heaters for general refinery service, API STD 530 - Calculation of Tube Heater Thickness, and API STD 560 - Fired Heaters for General Refinery Service
CERCLA	see **Comprehensive Environmental Response, Compensation, and Liability Act**
certificate	Abbreviation of certificate of authorization (qv) in ASME BPE (American Society of Mechanical Engineers: Bioprocessing Equipment) D170#

certificate holder	In a bioprocessing context, the holder of a certificate of authorization under the ASME BPE Certification Program. Defined formally in ASME BPE (American Society of Mechanical Engineers: Bioprocessing Equipment)
certificate of authorization	A document issued by ASME that allows a certificate holder to market bioprocessing equipment as certified to have been assembled and tested in accordance with the provisions of ASME BPE. Defined formally in ASME BPE (American Society of Mechanical Engineers: Bioprocessing Equipment)
certificate of compliance	*aka* compliance certificate, certificate of conformance. A document, signed by an authorized representative of the manufacturer, certifying that an item or batch of product meets all relevant specifications, regulatory and legal requirements
certificate of conformance	see **certificate of compliance**
certified scrum master	*aka* CSM. A project management certification, less popular than project management professional (qv), associated with the 'agile' school of project management
CET	1. see **critical exposure temperature** 2. see **coil exit temperature**
cetane number	*aka* cetane rating. The equivalent for diesel of the octane number of gasoline; defined formally in ASTM D 613 (52.6)
cetane rating	see **cetane number**
cf	An abbreviation for 'confer/conferatur', meaning 'compare'. A cross reference for comparison to a related entry in the context of this dictionary
CFC	see **chlorofluorocarbon**
CFD	see **computational fluid dynamics**
CFG	see **controlled flow grate**
CFM	Cubic feet per minute
CFPD	Cubic feet per day; US customary units (qv) measure of gas flow
CFPP	see **cold filter plugging point**
CFR	1. Cost and freight, see **Incoterms** 2. see **combined feed ratio**
CFRT	see **covered floating-roof tank**
CG	1. see **casing gas** 2. see **conventional gasoline**
CGA	Compressed Gas Association
cGMPs	Current good manufacturing practices, qv good manufacturing practices
CGPM	General Conference on Weights and Measures
CGR	see **condensate gas ratio**

CGS	An obsolete metric system of units based on centimeter, gram and second base units
chain	1. Colloquially used to refer to a sequence of plants or plant units which produce a single or group of products from the final plant or unit in the 'chain'; cf process train 2. A US customary units (qv) measure equal to 66 yards or 0.1 furlongs
chain block	*aka* block and tackle. A sheave (pulley) and chain used to lift heavy loads by hand
chain dimensioning	Arranging single dimensions in a row; defined formally in BS EN ISO 10209:2012 Technical product documentation - Vocabulary - Part 1: Terms relating to technical drawings: general and types of drawings
chain reaction	A sequence of reactions in which a reaction product or by-product causes additional reactions to take place
chain scission	The degradation of a polymer main chain
chain transfer	A reaction by which the activity of a growing polymer chain is transferred to another molecule
chair	The leader in a HAZOP (qv) or any other form of hazard workshop/study
Chalk River unidentified deposits	A bacronym (qv) for crud (qv) developed during the Manhattan Project
change analysis	An approach which looks systematically for possible risk impacts and appropriate risk management strategies in situations where change is occurring. This includes situations in which system configurations are changed, operating practices or policies are revised, new or different activities will be performed, etc.
change control	Formal methods controlling how, if, and by whom documentation may be changed; related to management of change (MOC) (qv), but not change management (qv). Defined formally in BS EN ISO 9000 Quality management systems Fundamentals and vocabulary
change management	Management preparation, support, and help for organizations undergoing planned organizational change; cf change control
change of phase	*aka* phase change, phase transition. A change from one of the four physical phases (plasma, gas, liquid, solid) to another
change piece	Pipe fitting or adapter connecting different pipe sizes or types
channel	1. A trough which carries water or effluent without piping; cf trench 2. A metal profile such as that used to support ancillaries such as cabling, cable trays and instrumentation

channeling	Non-uniform flow as a result of the formation of low resistance paths in a packed bed, and the process of formation of those paths
chaotic mixing	Turbulent mixing
char	Carbon rich solid residue produced by pyrolysis (qv)
characteristic	Distinguishing feature; defined formally in two different contexts in BS ISO 2041:2018 Mechanical vibration, shock and condition monitoring. Vocabulary and BS EN ISO 9000 Quality management systems Fundamentals and vocabulary
characteristic coefficient	One of the dimensionless coefficients derived from the affinity laws (qv) viz: flow coefficient, ϕ; head coefficient, ψ; performance coefficient, λ; specific speed (qv), n_s; and Strouhal number, S_r
characteristic curves	*aka* Q/H curves, pump curves, pump selection charts. Graphs showing a pump's characteristics across its pumping range. Although the flow vs head curve is most commonly thought of as the pump curve, there are also flow vs power and flow vs efficiency curves. The three of these collectively form the main characteristic curves. There are also operating characteristic curves, constant efficiency, constant head, suction head, and constant discharge curves.
characterization factor	1. A life cycle analysis fiddle factor (qv); defined formally in EN ISO 14040:2006 Environmental management - Life cycle assessment - Principles and framework 2. An abbreviation of Watson characterization factor (qv)
charge	The feed to a batch process
chargehand	The lowest rung of the supervisory ladder in operations, above laborers/operators and below foremen; akin to a roughneck (qv) in the oil and gas industry
Charles's law	A law which states that at constant gas pressure, the volume will be directly proportional to absolute temperature
Charpy impact test	*aka* Charpy vee notch test. Test which determines the toughness (qv) or impact strength of a material
Charpy vee notch energy	see **material impact energy**
Charpy vee notch test	see **Charpy impact test**
Chartered Engineer	Historically, in the UK, Republic of Ireland and some other nations, a near-equivalent of the US's Professional Engineer (PE) (qv)
chase	A vertical space in a wall used to contain or conceal pipes or cables; cf duct

chatter	Undesirable, rapid opening and closing of a pressure relief valve (qv)
CHAZOP	Control HAZOP; a type of HAZOP (qv) undertaken specifically for control systems[22]
cheaters	A lineman's term for channel lock pliers
check valve	see **one way valve**
chelate	A compound containing a ligand (typically organic) bonded to a central metal atom at two or more points
Chem. Eng.	Abbreviation of chemical engineering (qv), especially in an academic context
chemical cleaning	A method for washing and cleaning surfaces and walls of equipment, pipelines, vessels, kettles and heat exchangers of unwanted contaminants using dilute acids and bases
chemical engineer	A practitioner of chemical engineering (qv); see **Box C1.** and **Box E1.**
chemical engineering	1. The well-bounded profession of designing and operating process plants; see **Box C1.** 2. The magazine which publishes the CEPCI (qv)
Chemical Engineering Plant Cost Index	*aka* CEPCI. A composite index published by Chemical Engineering magazine assembled from a set of four sub-indexes: Equipment; Construction Labor; Buildings; and Engineering & Supervision; used for adjusting process plant construction costs with respect to time
chemical equilibrium	see **equilibrium**
chemical hazard	A physical, health or environmental hazard associated with a chemical
chemical indicator	*aka* CI. Chemical which changes color to indicate a pH change, or when effective sterilization has been achieved
Chemical Industries Association	*aka* CIA. UK chemical industry trade association which issues guidance for the chemical process industry (qv), including industry standards for occupied building risk assessment
chemical industry	see **chemical process industry**
chemical oxygen demand	*aka* COD. A standard measure of the amount of oxidizable material in wastewater obtained by chemical oxidation. Far quicker than the BOD test, but less useful in predicting how much of the material will be removed by biological treatment. Defined formally in EN 12566 - Small wastewater treatment systems for up to 50 PT Part 3: Packaged and/or site assembled domestic wastewater treatment plants

---

[22] N.B. used for a different activity in academia

Box C1. Chemical engineer, process engineer

Are process engineers/engineering and chemical engineers/engineering one and the same? The most common view amongst both practitioners and academics nowadays appears to be that chemical engineering is the name of a degree, process engineer is a job title; and neither is the name of a profession. If so, then does the profession of 'Chemical Engineering' which I joined when I became a Chartered Chemical Engineer no longer exist? Despite many publicly available definitions which accord with my understanding of the term, the people who practice my profession no longer have an unambiguous name.

In the 1990s, becoming a Chartered Chemical Engineer required (amongst other things), a degree in chemical engineering plus an absolute minimum of four years' experience designing or supporting the operation of full-scale process plants, applying the principles of process safety in practice. These are no longer the tests to become a Chartered Chemical Engineer, therefore the title no longer carries the same meaning as it did.

More generally, the term chemical engineer has changed its meaning during the time I have practised my now-nameless profession of designing and supporting the operation of process plants. It has not been replaced by the term process engineer – many 'process engineers' have no degree in any kind of engineering, nor any training or experience in applying the principles of process safety. Some of them do not even work with chemical/process plants.

The changes in the meaning of the title have allowed academics and others without practical experience or engineering degrees to call themselves chemical engineers, growing the membership of chemical engineering institutions, but they have also contributed to the loss of association with any profession of the title 'chemical engineer'. At the same time, employers may increasingly hire non-chemical engineers or non-engineers as 'process engineers' because they are cheaper than chemical engineering graduates.

How can we distinguish between them? My fellow professionals could be rather inelegantly defined as 'chemical process engineers with a degree in chemical engineering', but a more succinct solution might involve the use of upper case. Essentially, there are 'Chemical Engineers' (original meaning), and then there are 'chemical engineers', members of the essentially meaningless 'boundaryless profession' of 'chemical engineering'.

Related Terms
Professional Engineer, engineer

chemical plant	A large integrated process plant where substances are produced by or used in chemical reactions. A specific flammable liquids definition is given in NFPA 30: Flammable and Combustible Liquids Code, 2018 Edition
chemical potential	(*symbol* μ) Change in Gibbs free energy (qv) due to a change of the particle number of the given species
chemical process industry	*aka* chemical industry, CPI. A term often used to cover a narrow range of chemical process industries, centered on oil and gas, petrochemicals and bulk organics ('traditional CPI')[23]
chemical recovery boiler start up burner	A burner which initiates black liquor combustion; defined formally in BS EN 12952-8:2002 Water-tube boilers and auxiliary installations. Requirements for firing systems for liquid and gaseous fuels for the boiler
chemical stability	In ion exchange resins, the ability to resist changes in physical properties when exposed to aggressive chemicals
chemical vapor deposition	*aka* CVD. A coating process
chemically combined water	A term which has a number of different context-specific meanings: a chemist would define chemically combined water as the water bound up in things like hydrates, but cement specialists apply the term to all 'non-evaporable water', and ceramicists to water removed during a specific phase of firing
chemically defined medium	see **defined medium**
chemisorption	Adsorption involving a chemical reaction between surface and adsorbate
chemorheology	The study of interactions between chemistry and rheology
chequer plate	see **deckplate**
Chernobyl	The informal name for a power station in modern day Ukraine (properly called the Vladimir Llyich Lenin power station) where the worst 'nuclear accident' ever happened.

---

[23] In my view, the term should encompass all **chemical engineering** (qv) (i.e., design and operation of process plant), but not everything done by chemical engineering graduates (many of whom never become chemical engineers), nor academic activity. There are other views, especially amongst those which my view excludes, though I believe mine represents the consensus amongst those I do include.

Chesterton's fence	The principle which states: "In the matter of reforming things, as distinct from deforming them, there is one plain and simple principle; a principle which will probably be called a paradox. There exists in such a case a certain institution or law; let us say, for the sake of simplicity, a fence or gate erected across a road. The more modern type of reformer goes gaily up to it and says, 'I don't see the use of this; let us clear it away.' To which the more intelligent type of reformer will do well to answer: 'If you don't see the use of it, I certainly won't let you clear it away. Go away and think. Then, when you can come back and tell me that you do see the use of it, I may allow you to destroy it.'" This principle should always be borne in mind when considering improvements on another person's design. Until it is understood why something is there, best leave it be
Chézy formula	An equation which allows the determination of mean velocity for turbulent open channel flow
chicken catcher	A lineman's arm sling
chicken wing	A lineman's term for a steel post insulator standoff for distribution construction
Chiksan	*aka* Chicksan (sic). General term for a marine loading arm, a proprietary eponym for a flexible coupling used in marine loading arms now more generally, especially on fracking rigs
chiller	*aka* cooler, condenser. A device to remove heat for a process or utility stream, typically used when the cold utility temperature must be lower than cooling water temperature (say 10 °C)
chilling	Removing heat from
chill-roll film extrusion	*aka* cast-film extrusion. A plastic film extrusion method in which a sheet of semi-molten plastic is drawn between chilled, highly polished rollers
Chilton–Colburn analogy	*aka* the Chilton–Colburn J factor analogy, modified Reynolds analogy. An analogy between the mathematics of heat, momentum, and mass transfer widely used in chemical engineering
Chilton–Colburn J factor	A dimensionless group used in the Chilton–Colburn analogy (qv)
chimney effect	see **stack effect**
chimney tray	*aka* collector tray. A device used to disengage liquid within a distillation column and redistribute vapor; see **Figure C1**

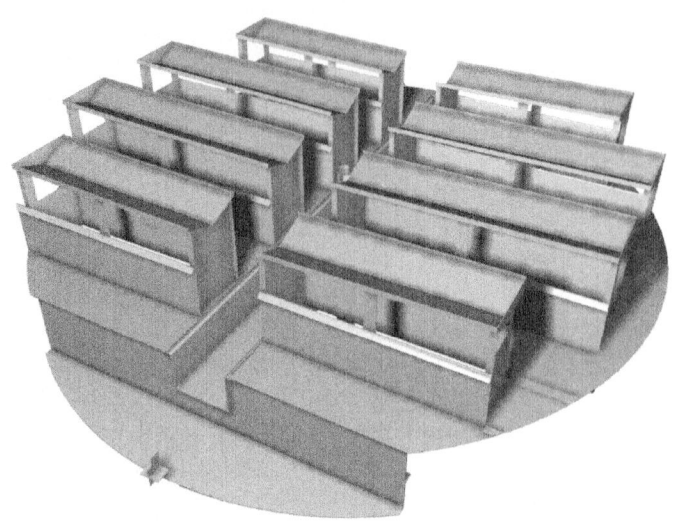

Figure C1 Chimney Tray (Courtesy: Sulzer)

chinesium	*(jocular)* Low grade materials, often metals, as used in the cheapest products, especially when made in China; cf monkey metal
chippy	*(informal)* A joiner or carpenter in UK and Australian site terminology
chlor-alkali process	A process for the industrial scale production of chlorine and NaOH from brine by electrolysis
chloride anion dealkalization	A salt and caustic regenerated anion exchange system which exchanges bound chloride for bicarbonates, carbonates, sulfates, and nitrates in solution
chloride process	A process for industrial scale production of TiO_2. An alternative to the sulfate process (qv)
chloride stress corrosion cracking	Cracking of stainless steels under tensile stress exposed to excessive levels of chloride; defined formally in API RP 932B Corrosion Air Coolers
chlorination	Adding chlorine to a molecule or to a mixture
chlorine covenant	An agreement between the UK government, CIA and HSE on chlorine handling
chlorofluorocarbon	*aka* CFC. A halogenated paraffinic hydrocarbon, formerly known by the proprietary eponym 'Freon'; banned under the Montreal Protocol but still produced
chloroform	Old name for trichloromethane, a trihalomethane (qv)

choke	1. Generically, can be used to refer to any device or obstruction increasing pressure drop/reducing flow 2. Specifically, a control valve on the flowline of a production well 3. *(informal)* A reference to choke point (qv)
choke flow	see **choked flow**
choke point	The maximum stable mass flow above which choked flow (qv) occurs; cf stonewall
choked flow	*aka* choke flow, critical flow. A limitation in the mass flow rate of a fluid caused by a standing shock wave at supersonic velocities. So important in gases that many definitions apply it only to gases, but it has liquid applications in both open channel (qv hydraulic jump) and piped contexts (qv cavitation) see **stonewalling**.
cholesteric phase	A form of the nematic phase of a liquid crystal in which the molecules are spiral
CHOPS	see **cold heavy oil production with sand**
CHP	see **combined heat and power**
CHPS	see **cold high pressure separator**
Christmas tree	*aka* Xmas tree. A wellhead with all its valves in different directions, hence the name; see **Figure C2** for an unusually tidy example

Figure C2 Christmas Tree (courtesy: Armona)[24]

---

[24] Reproduced under Creative Commons Attribution-Share Alike 3.0 Unported License

chromatography	A laboratory separation method based on the different interactions of mixture components with a fluid 'mobile' phase and a solid 'stationary' phase, sometimes scaled up for production of biopharmaceuticals. Defined formally in ASME BPE (American Society of Mechanical Engineers: Bioprocessing Equipment)
chrome-moly	Abbreviation of chrome-molybdenum alloy, a high-grade steel used for corrosive duties
churn	see **no flow**
churn flow	*aka* froth flow, semiannular flow. A multiphase flow regime with large, irregular slugs of gas traveling up the center of the pipe
CI	1. see **cast iron** 2. see **chemical indicator** 3. see **continuous improvement**
CIA	see **Chemical Industry Association**
CIBSE	Chartered Institute of Building Services Engineers
CIBSE guides	Guidance on the best practice and calculation methods for HVAC (qv)
CIF	Cost insurance freight, qv Incoterms
CIL	see **carbon in leach**
CIMAH	Control of industrial major accident hazards; cf COMAH
CIO	Chief Information Officer
CIP	1. see **carbon in pulp** 2. carriage and insurance paid; qv Incoterms 3. see **cleaning in place** 4. see **cold isostatic pressing** 5. Constant injection pressure
CIP circuit	All of the CIP paths (qv) which CIP fluids (qv) must pass to. Defined formally in ASME BPE (American Society of Mechanical Engineers: Bioprocessing Equipment)
CIP cycle	The 'recipe' for cleaning in place (fluid compositions, temperature, flows and durations) of some part of a hygienic plant. Defined formally in ASME BPE (American Society of Mechanical Engineers: Bioprocessing Equipment)
CIP fluids	*aka* CIP solutions. There are four broad types of solutions used for cleaning in place (qv), namely acid, alkali, oxidizing disinfectant, and rinse
CIP kitchen	The area where CIP fluids (qv) are made up, recirculated and recovered to, especially in food and drink factories
CIP path	A point of use of CIP fluids (qv), such as a spray ball (qv). Defined formally in ASME BPE (American Society of Mechanical Engineers: Bioprocessing Equipment)

CIP solutions	see **CIP fluids**
CIP system	see **clean in place system**
CIPM	International Committee for Weights and Measures
circuit diagram	Electrical circuit drawing; defined formally in BS EN ISO 10209:2012 Technical product documentation - Vocabulary - Part 1: Terms relating to technical drawings: general and types of drawings
circular economy	see **Box W1**.
circular runout	see **runout**
circulating load	The material rejected by a separator and returned to the mill for regrinding
circulating pump	A pump which circulates fluid for mixing, heat exchange etc.
circulation relief valve	A type of relief valve used for pump cooling, rather than overpressure protection. Defined formally in NFPA 20: Standard for the Installation of Stationary Pumps for Fire Protection, 2019 Edition
CIS	1. see **close internal survey**
	2. see **Construction Industry Scheme**
Civil Nuclear Constabulary	The heavily armed police force which patrols UK nuclear power stations
CKD	Cement kiln dust
₵	*(drawing notation)* see **center line**
CL	*(drawing notation)* see **center line**
clack valve	see **one way valve**
cladding	A thin outer corrosion-resistant covering such as that found around pipe insulation
claims, arguments, evidence	A common structure for presenting reasoned arguments such as safety cases. High level claims are underpinned by individual arguments, which in turn are underpinned by tangible evidence
clamp	1. A type of support used to fix pipes in a plant
	2. A type of support used to secure pipe connections
Clapeyron equation	see **Clausius–Clapeyron equation**
clarification	Reduction in turbidity, often (but not always) by settlement processes rather than by filtration
clarifier	A name used in the water industry for a settling tank (qv)
clarifying agent	A chemical used to clarify water or alcoholic drinks
Clark's process	see **lime softening**
clash	An error where a design involves two items occupying the same space, usually referring to pipework
class	see **ANSI flanges**
class 1 interceptor	see **interceptor**
class 2 interceptor	see **interceptor**

class A fire	In the USA, UK, Europe and Australia, a fire involving solid materials such as wood, paper or textiles; cf fire classification, **Box F2**.
class A furnace	A furnace with a risk of fire in the materials being heated. Defined formally in NFPA 86: Standard for Ovens and Furnaces, 2019 Edition
class B fire	In the USA, a flammable fluid (gas or liquid) fire, as defined formally in NFPA 11: Standard for Low-, Medium-, and High-Expansion Foam, 2016 Edition, or a flammable liquid fire in the UK, Europe and Australia; cf fire classification, **Box F2**.
class B furnace	A furnace with no risk of fire in the materials being heated. Defined formally in NFPA 86: Standard for Ovens and Furnaces, 2019 Edition
class C fire	In the USA, a fire involving electrical equipment; but in the UK, Europe and Australia, a flammable gas fire; cf fire classification, **Box F2**.
class C furnace	A furnace with a risk of fire in the special atmosphere being used. Defined formally in NFPA 86: Standard for Ovens and Furnaces, 2019 Edition
class D fire	In the USA, UK, Europe and Australia a fire involving metals; cf fire classification, **Box F2**.
class D furnace	Any vacuum furnace; defined formally in NFPA 86: Standard for Ovens and Furnaces, 2019 Edition
class E fire	In Australia, a fire involving live electrical apparatus. Also used in the UK for convenience, in the context of fire extinguishers, as Europe does not have a classification for these fires; cf fire classification, **Box F2**.
class F fire	In the UK, Europe and Australia, a fire involving cooking oils such as in deep-fat fryers; cf fire classification, **Box F2**.
class I liquid	A flammable liquid with a closed cup flash point below 100 °F (37.8 °C) and a Reid vapor pressure (qv) not exceeding 40 psia (2068 mm of mercury) at 100 °F (37.8 °C)
class I, zone 0	The highest classification of likelihood of flammable atmosphere; defined formally in API RP 505 2nd Edition, August 2018 Recommended Practice for Classification of Locations for Electrical Installations at Petroleum Facilities Classified as Class I, Zone 0, Zone 1, and Zone 2, as well as BS EN 50281 -3 2002 Electrical apparatus for use in the presence of combustible dust. Classification of areas where combustible dusts are or may be present

class I, zone 1	The second highest classification of likelihood of flammable atmosphere; defined formally in API RP 505 2nd Edition, August 2018 Recommended Practice for Classification of Locations for Electrical Installations at Petroleum Facilities Classified as Class I, Zone 0, Zone 1, and Zone 2 as well as BS EN 50281 -3 2002 Electrical apparatus for use in the presence of combustible dust. Classification of areas where combustible dusts are or may be present
class I, zone 2	The third highest classification of likelihood of flammable atmosphere; defined formally in API RP 505 2nd Edition, August 2018 Recommended Practice for Classification of Locations for Electrical Installations at Petroleum Facilities Classified as Class I, Zone 0, Zone 1, and Zone 2 as well as BS EN 50281 -3 2002 Electrical apparatus for use in the presence of combustible dust. Classification of areas where combustible dusts are or may be present
class IA liquid	A flammable liquid with a flashpoint lower than 22.8 °C and a boiling point lower than 37.8 °C. Defined formally in NFPA 11: Standard for Low-, Medium-, and High-Expansion Foam, 2016 Edition and NFPA 30: Flammable and Combustible Liquids Code, 2018 Edition
class IB liquid	A flammable liquid with a flashpoint lower than 22.8 °C and a boiling point of 37.8 °C or higher. Defined formally in NFPA 11: Standard for Low-, Medium-, and High-Expansion Foam, 2016 Edition and NFPA 30: Flammable and Combustible Liquids Code, 2018 Edition
class IC liquid	A flammable liquid with a flashpoint of 22.8 °C to 37.8 °C. Defined formally in NFPA 11: Standard for Low-, Medium-, and High-Expansion Foam, 2016 Edition and NFPA 30: Flammable and Combustible Liquids Code, 2018 Edition
class II liquid	A combustible liquid with a flashpoint of 37.8 °C to 60 °C. Defined formally in NFPA 11: Standard for Low-, Medium-, and High-Expansion Foam, 2016 Edition, NFPA 30: Flammable and Combustible Liquids Code, 2018 Edition and API RP 505 2nd Edition August 2018 Recommended Practice for Classification of Locations for Electrical Installations at Petroleum Facilities Classified as Class I, Zone 0, Zone 1, and Zone 2

class IIIA liquid	A combustible liquid with a flashpoint of 60 °C to 93 °C. Defined formally in NFPA 11: Standard for Low-, Medium-, and High-Expansion Foam, 2016 Edition, NFPA 30: Flammable and Combustible Liquids Code, 2018 Edition and API RP 505 2nd Edition, August 2018 Recommended Practice for Classification of Locations for Electrical Installations at Petroleum Facilities Classified as Class I, Zone 0, Zone 1, and Zone 2
class IIIB liquid	A liquid with a flashpoint of at least 93 °C. Defined formally in NFPA 11: Standard for Low-, Medium-, and High-Expansion Foam, 2016 Edition, NFPA 30: Flammable and Combustible Liquids Code, 2018 Edition and API RP 505 2nd Edition, August 2018 Recommended Practice for Classification of Locations for Electrical Installations at Petroleum Facilities Classified as Class I, Zone 0, Zone 1, and Zone 2
class K fire	In the USA, a fire involving cooking oils such as in deep-fat fryers; cf fire classification, **Box F2**.
class location	A regulatory designation for natural gas transmission pipelines that indicates the level of human population within a certain distance on either side of the line under U.S. Code 49cfR 192.5. Defined formally in ASME Gas Transmission and Distribution Piping Systems B31.8 - 2020
classification	1. The sorting of solid particles suspended in water or air using gravity; cf elutriation 2. A method for sorting things according to their characteristics; defined formally in BS EN ISO 10209:2012 Technical product documentation - Vocabulary - Part 1: Terms relating to technical drawings: general and types of drawings
classifier	*aka* elutriator. Equipment which sorts solid particles suspended in water or air using gravity
Claude process	A process for the liquefaction of air
Claus process	A gas sweetening (qv) process
Clausius–Clapeyron equation	*aka* Clapeyron equation. An equation which allows the rate of increase in vapor pressure per unit increase in temperature to be predicted
clean	In practice, 'sufficiently free of contaminants for the intended purpose'; see **Box C2**.
Clean Air Act	1. A 1956 UK Act of Parliament intended to improve air quality 2. A 1963 USA statute intended to improve air quality
Clean Air Act Program	see **prevention of significant deterioration program**

Box C2. Clean/Hygienic/Sanitary

There is a cluster of terms related to the concept of 'clean', a term I define as "sufficiently free of contaminants for the intended purpose". In chemical engineering, making things 'clean enough' has three levels: cleaning, disinfection, and sterilization.

Cleaning is "the removal of objectionable matter", to summarize the formal definitions in codes and standards[25]; it reduces the number of contaminants, thus removing a proportion of organisms present. Cleaning could be as simple as wiping off dirt, but if we are making a product for human consumption, then we need to consider the nature of that 'dirt'. Contaminants is a better word in my view than dirt, evocative of the US FDA's requirements for cleanliness of food and drug processing equipment, which is why I use it in my definition of 'clean'.

Cleaning may be sufficient for the removal of many physical contaminants, such as living or dead organisms, FOGs, etc., and some items are perfectly safe to consume if they are clean. Others, however, are treated with disinfection, which removes most pathogenic organisms; or even sterilization - the killing or removal of all organisms.

Disinfectant is a word used in UK English to mean something close to the US English sanitizer, but the EHEDG defines disinfectant more broadly, bringing in the concepts of food safety, sanitizing and sanitation (which is given as a synonym of hygiene); hygiene, in turn, references food hygiene.

As for the terms hygienic and sanitary, are these synonyms for clean, near synonyms, or is one of them deprecated? In industries effectively regulated by the US FDA there is some evidence that these terms are interchangeable – see for example NSF International's standards for this US view and compare with the EHEDG glossary for the European view, giving sanitization and sanitary as deprecated terms from the USA, preferring disinfection and disinfectant.

In the pharmaceutical industry however, whilst the EHEDG guidance has applicability, the ISPE glossary defines hygienic design more explicitly than EHEDG, evoking the concept of good engineering practice(s) and flagging the key associated design considerations. As for sanitary design, the ISPE glossary states that it is a misnomer for hygienic design, and no longer in use. They think this usage is not just deprecated; it is wrong.

In summary, these terms are not absolute in nature. Engineers in the food and pharma industries in particular should define and where possible quantify such terms in context, so that they can understand and validate compliance with a given specification.

Related Terms
contamination, soil

---

[25] BS EN 1672-2:2005+A1:2009 Food processing machinery. Basic concepts. Hygiene requirements, BS EN ISO 14159:2008 Safety of machinery. Hygiene requirements for the design of machinery, ASME BPE (American Society of Mechanical Engineers: Bioprocessing Equipment)

clean coal	Weasel words: often simply a PR department's greenwash (qv) for coal pollution mitigation (qv), though it can be used to mean anything, from quite literally just washing the coal prior to burning it up, through the commercially unviable and unproven use of CCS (qv) with coal combustion
clean in place	see **cleaning in place**
clean in place system	A system used for cleaning in place (qv). Defined formally in ASME BPE (American Society of Mechanical Engineers: Bioprocessing Equipment)
clean steam	Steam made from pharmaceutical grade water without additives. Defined formally in ASME BPE (American Society of Mechanical Engineers: Bioprocessing Equipment) and 3A Standard 609-02; cf pure steam, culinary steam
clean utilities	Utilities (qv) which may come into contact with a food or pharmaceutical product such as water for injection, clean steam, CIP fluids (qv) etc.; cf black utilities, green utilities
Clean Water Act	The primary federal law in the USA that governs discharges to water
cleanability	*aka* easy cleanability. "Designed and constructed to be cleaned efficiently" according to the EHEDG Glossary
cleanable	"Designed and constructed so that soils are removed by recommended cleaning methods", according to the EHEDG Glossary. Defined formally and similarly in BS EN 1672-2:2005+A1:2009 Food processing machinery. Basic concepts. Hygiene requirements
cleaning	The removal of objectionable matter, to summarize the formal definitions in BS EN 1672-2:2005+A1:2009 Food processing machinery. Basic concepts. Hygiene requirements, BS EN ISO 14159:2008 Safety of machinery. Hygiene requirements for the design of machinery and ASME BPE (American Society of Mechanical Engineers: Bioprocessing Equipment)
cleaning in place	*aka* CIP, clean in place. A fully or partly automated technique used to clean and sanitize process surfaces by circulation of aqueous fluids without dismantling equipment. Defined formally in BS EN ISO 14159:2008 Safety of machinery. Hygiene requirements for the design of machinery and BS EN ISO 14159:2008 Safety of machinery. Hygiene requirements for the design of machinery; cf cleaning out of place

cleaning out of place	*aka* COP, wet cleaning. A "system where equipment is disassembled and cleaned in a tank or in an automatic washer by circulating a cleaning solution and maintaining a minimum temperature throughout the cleaning cycle. (ISO 22000). Note: COP can be done manually or mechanically when the equipment is partially or totally disassembled" according to the EHEDG Glossary; cf wet cleaning
cleanroom	A "room in which the concentration of airborne particles is controlled, and which is constructed and used in a manner to minimize the introduction, generation, and retention of particles inside the room, and in which other relevant parameters e.g., temperature, humidity, and pressure, are controlled as necessary", according to the EHEDG Glossary
clearing factor	see **K-factor**
client	1. Generically, the entity which a contractor works for 2. Specifically, defined in UK Construction Design and Management Regulations 2015 (qv) as an entity carrying certain responsibilities associated with construction and demolition works
climbing film evaporator	see **rising film evaporator**
climbing height	Vertical distance to landing; defined formally in BS EN ISO 14122 Safety of machinery. Permanent means of access to machinery. Working platforms and walkways
climbing height of ladder system	The vertical distance to the top of a ladder system; defined formally in BS EN ISO 14122 Safety of machinery. Permanent means of access to machinery. Working platforms and walkways
clinging nappe	A nappe (qv) with no air between the wall it falls over and its underside; cf detached nappe
close coupled	In the context of pumps, having an impeller mounted directly on an extension of the drive shaft
close interval survey	*aka* CIS. A method of testing corrosion protection systems on pipelines which involves inspection and electrical testing of the corrosion protection system every two to three feet along the pipeline to confirm the status of the protection system and to help identify mechanical damage to the pipeline
closed container	A container sealed so as to prevent the escape of liquid or vapor at ambient temperature
closed cooling system	In the context of flammable and combustible liquid safety, a cooling system which has no observable unrestricted sight drain; defined formally in NFPA 30: Flammable and Combustible Liquids Code, 2018 Edition

closed cycle gas turbine	A gas turbine with energy recovery; cf open cycle gas turbine
closed disposal system	A disposal system which can hold pressure
closed head	An arc welding head containing inert atmosphere, used for making high quality welds in stainless steel, especially in hygienic applications. Defined formally in ASME BPE (American Society of Mechanical Engineers: Bioprocessing Equipment)
closed loop control	An automatic control system which uses feedback based on measurement of a controlled variable to vary a control signal; cf open loop control
closed position indicator switch	A switch which measures and indicates that a valve is 1 mm or less from the fully closed position
closed system	A system in which no material or energy is transferred across the system boundary. A simplifying theoretical assumption; cf open system
closed top diking	A dike or bund (qv) with a cover to exclude precipitation
closing pressure	The pressure at which a pressure relief valve (qv) disk reseats as pressure falls. Defined formally in API RP 520 P1 7th Edition, January 2000 Sizing, Selection, and Installation of Pressure-Relieving Devices in Refineries; Part I - Sizing and Selection and API RP 576 - Inspection of Pressure Relieving Devices
cloud	A cloud shaped boundary on a drawing around a 'Hold' or unresolved design issue; see **Figure C3**

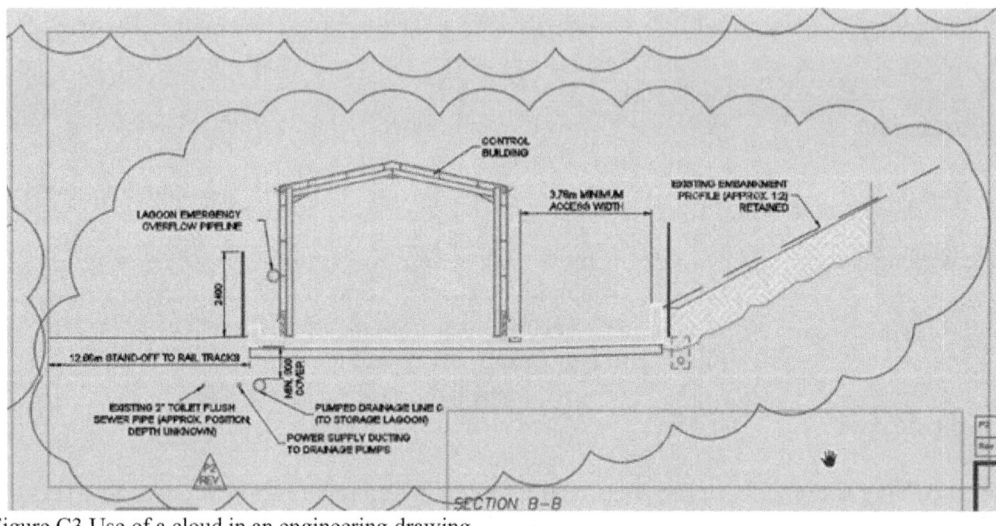

Figure C3 Use of a cloud in an engineering drawing

cloud point	The temperature at which a hydrocarbon liquid starts to appear cloudy due to the precipitation of dissolved waxes
CLPS	see **cold low pressure separator**
clumping	The formation of clumps or agglomerations (qv), especially in a packed bed (qv) due to fouling
cluster of pits	A term describing pitting of stainless steel; defined formally in ASME BPE (American Society of Mechanical Engineers: Bioprocessing Equipment)
cluster porosity	A clumped/clustered form of porosity; defined formally in ASME BPE (American Society of Mechanical Engineers: Bioprocessing Equipment)
CM	see **condition monitoring**
CMF	see **common mode failure**
CMP	Corrugated metal pipe
CNC	see **Civil Nuclear Constabulary**
CNG	see **compressed natural gas**
CO	Carbon monoxide
CO breakthrough	When CO level first increases rapidly as excess air is reduced. Defined formally in API RP 535 - Burners for Fired Heaters at Refineries
CO_2	Carbon dioxide
coagulant	A chemical additive promoting coagulation
coagulation	In water treatment, the process of neutralizing the surface charges which prevent very fine suspended particles and colloidal macromolecules from forming larger particles. Commonly confused with flocculation (qv)
coal ash	see **pulverized flue ash**
coal equivalent	see **tonne of coal equivalent**
coal gas	*aka* manufactured gas. A fuel gas produced by heating coal, toxic due to high CO levels, now superseded by natural gas for domestic purposes in most countries, though still used extensively in South Africa and Hong Kong; cf coke oven gas, town gas, producer gas
coal pollution mitigation	*aka* clean coal (qv). Coal combustion with SOx and NOx reduction, but not (cf) CCS
coal seam gas	Natural gas extracted from within coal seams, to produce a saleable product and enhance mine safety
coal tar	*aka* tar. A thick dark liquid, by-product of the production of coke and coal gas from coal
coal tar acids	*aka* tar acids. Acidic phenolic compounds removed from coal tar by washing with caustic soda

coal tar bases	*aka* tar bases. Basic substances such as pyridine extracted after coal tar acids have been removed
coalesce	To merge into a whole. Defined formally in API RP 535 - Burners for Fired Heaters at Refineries
coalescence	The process of merging into a whole
coalescer	In the context of fired heaters, a device in which aerosol particles are coalesced by filter media. Defined formally in API RP 535 - Burners for Fired Heaters at Refineries
coalescing element	In the context of gas turbines, fibrous coalescing filtration media; defined formally in ISO 3977 Gas turbines - Procurement - Part 3: Design requirements
coalescing media	1. In water treatment, various kinds of packing used to coalesce oil droplets too fine to be removed by gravity; 2. In air treatment, see **coalescing element**
Coanda effect	"The tendency of a jet of fluid emerging from an orifice to follow an adjacent flat or curved surface and to entrain fluid from the surroundings so that a region of lower pressure develops", according to Henri Coanda
Coanda flare	Flare burner which uses the Coanda effect (qv). Defined formally in API RP 537 Flare Details for Petroleum, Petrochemical, and Natural Gas Industries
coarse bar screen	*aka* bar rack. A bar screen with apertures of 40-200 mm used for removal of debris from wastewater as part of preliminary treatment
coating	A "process where a different material is deposited to create a new surface..." according to the EHEDG Glossary
cobra	A pipe cleaning device, a type of snake (qv)
cocurrent flow	*aka* cocurrent operation. Where process streams pass through a unit operation in the same direction; cf countercurrent flow
cocurrent operation	see **cocurrent flow**
COD	1. see **chemical oxygen demand** 2. Cash on delivery; payment terms
coefficient of discharge	*aka* discharge coefficient, efflux coefficient. The ratio of the actual discharge mass flow of a valve, nozzle etc. to the theoretical. Defined formally in API RP 520 P1 7th Edition, January 2000 Sizing, Selection, and Installation of Pressure-Relieving Devices in Refineries; Part I - Sizing and Selection
coefficient of expansion	*aka* expansion coefficient, thermal expansion coefficient. The fractional change in length, area or volume per degree change in temperature at constant pressure

coefficient of performance	*aka* COP. The ratio of useful heating or cooling provided by a heat engine or HVAC (qv) system to work required; the SI equivalent of energy efficiency ratio (qv) (EER). The conversion factor is COP of 1.0 = EER of 3.4.
coefficient of variation	The ratio of the standard deviation to the mean
COG	1. see **center of gravity** 2. see **coke oven gas**
cogen	see **combined heat and power**
cogeneration	see **combined heat and power**
coherence function	Dimensionless number relating two signals. Defined formally in BS ISO 2041:2018 Mechanical vibration, shock and condition monitoring. Vocabulary
coherent units	A set of units derived from a set of base units without the need for conversion factors
coil	Serpentine heat exchanger tubes, or an abbreviation for coil heat exchanger (qv)
coil exit temperature	*aka* CET. The temperature at the outlet of a fired heater or furnace
coil heat exchanger	*aka* coil. A heat exchanger with a serpentine or helical tube (heat exchanger coil), used as part of another item of equipment
coil steam	see **velocity steam**
co-injection molding	*aka* sandwich molding. A polymer injection technology in which different polymers are injected into the same mold
coke	A high carbon solid fuel produced by destructive distillation of coal, or oil (qv petroleum coke), used as a source of low impurity carbon in several processes
coke gas	see **coke oven gas**
coke oven gas	*aka* COG, coke gas, coal gas. Gas produced whilst heating coal to make metallurgical coke, identical in composition to coal gas (qv)
coke strength after reaction	*aka* CSR. A measure used in the coal and steel industries to measure the suitability of coal to produce blast furnace coke; can be estimated from ash's maximum viscosity and trace elements
coking coal	Coal with a high vitrinite (qv) content and key trace elements which will produce coke with a high coke strength after reaction (qv), ideal for blast furnace applications
COL	*(drawing notation)* Column
cold box	A heavily insulated structure in which low temperature process equipment is installed if equipment-specific insulation is impractical, e.g., refrigeration systems or cryogenic distillation columns

cold commissioning	The last stages of commissioning prior to hot commissioning (qv) undertaken without charging with process fluids, off-load and/or at ambient temperatures. It may be split into two stages, dry commissioning (qv) and wet commissioning (qv)
cold differential test pressure	The test-stand set opening pressure of a pressure relief valve (qv); defined formally in API RP 520 P1 7th Edition, January 2000 Sizing, Selection, and Installation of Pressure-Relieving Devices in Refineries; Part I - Sizing and Selection, API Standard 521. Pressure-relieving and Depressuring Systems. Sixth Edition	January 2014 and API RP 576 - Inspection of Pressure Relieving Devices
cold filter plugging point	*aka* CFPP. The lowest temperature at which diesel fuel will pass through a reference filter
cold heavy oil production with sand	*aka* CHOPS. The deliberate collection of large quantities of sand from a well producing heavy oil for enhanced economics
cold high pressure separator	*aka* CHPS. A gas/liquid separator (qv) downstream of a hydrocracking (qv) unit, the bottoms from which are sent to a cold low pressure separator (qv). Defined formally in API RP 932B Corrosion Air Cooler; cf hot high pressure separator, hot low pressure separator
cold isostatic pressing	A process for forming parts by compressing metallic powders
cold low pressure separator	*aka* CLPS. A gas/liquid separator downstream of the cold high pressure separator (qv) which removes some light ends from the bottoms which are sent for fractionation (qv)
cold pasteurization	see **high pressure processing**
cold shot process	Mixing cold fresh feed into the inlet of an exothermic reactor to control temperature, especially in the case of a series of such reactors
cold vent	A relief stream for flammable vapors not routed to a flare or blowdown vessel
cold welding	see **galling**
Colebrook equation	see **Colebrook-White equation**
Colebrook-White equation	*aka* Colebrook equation. An implicit equation which shows the relationship between the friction factor and the Reynolds number (qv), pipe roughness, and inside diameter of pipe. Explicit approximations of the formula exist, allowing direct calculation of desired parameters
coliform	More of a series of test results than a true class of organisms: Gram-negative non-spore forming bacilli which ferment lactose, producing acid and gas when incubated at 35–37°C
collector tray	see **chimney tray**

colligative properties	Properties that depend upon the concentration of solute molecules or ions, but not upon the identity of the solute, e.g., vapor pressure lowering, boiling point elevation, freezing point depression, and osmotic pressure
collision theory	Theory that when suitable particles of the reactant hit each other, only a certain fraction of the collisions, called 'successful collisions', cause any noticeable or significant chemical change
colloid	An evenly dispersed suspension (qv) of extremely small particles (approximately 1 nm to 1 μm) of one material in another, not separable by sedimentation or sand filtration
color	*aka* colour. True color is only seen after particulate matter has been removed. Unfiltered color is called apparent color. In water treatment, true color is measured on a single scale with many names, including the platinum cobalt (Pt Co), APHA and Hazen color scales. Black tea (qv) without milk has a color of about 2500 Hazen. Beer has its own scales - SRM, Lovibond, EBC (qv), which may be mutually interconverted, but are not considered compatible with the Hazen scale. The petroleum industry uses Saybolt color (qv), whilst Gardner color (qv) measures all transparent liquids. What most of these methods have in common is that they measure yellowness. Colors other than yellow are therefore commonly not 'color' in the technical sense
color throw	Color bleed into water from ion exchange resin
colorimeter	A device used to determine the concentration of a solute in a sample by means of a colored reagent
colors of hydrogen	see **hydrogen colors**
colour	see **color**
column	*aka* tower. A vessel with a high aspect ratio, such as a distillation column (qv)
column flooding	The situation where a counter-current packed column, such as a distillation column, changes from having a continuous vapor phase with falling droplets of liquid to having vapor as bubbles in a continuous liquid phase. This increases pressure drop and decreases process efficiency.
column mounting	Baseplate mounting at discrete points; defined formally in ISO 3977 Gas turbines - Procurement - Part 3: Design requirements
COMAH	see **Control of Major Accident Hazards Regulations**
combination burner	see **combination fuel gas and oil burner**

combination fuel gas and oil burner	*aka* combination burner. A burner which runs on fuel gas, oil or both simultaneously. Defined formally in API RP 535 - Burners for Fired Heaters at Refineries and NFPA 86: Standard for Ovens and Furnaces, 2019 Edition
combined audit	A survey of multiple management systems at a single audit; defined formally in BS EN ISO 9000 Quality management systems Fundamentals and vocabulary
combined CO_2	*aka* firmly bound CO_2. In the context of carbonate buffering, a term for CO_2 bound in carbonates; cf free CO_2, semicombined CO_2
combined cycle gas turbine	*aka* CCGT. A power producing machine combining a gas turbine with a steam turbine, delivering very high thermodynamic efficiencies for electrical power generation. Can be installed as part of a combined heat and power plant, but the terms are not synonymous
combined feed ratio	*aka* CFR. The ratio of (fresh feed plus recycle) to fresh feed
combined heat and power	*aka* cogeneration, cogen, CHP. Generating electricity and useful heat simultaneously
combined sewer	A sewer which carries a mix of surface water drainage and effluent
combined system	A common piping system with a single riser and actuation valve for both sprinklers and spray nozzles in a fire area. Defined formally in NFPA 15 Standard for Water Spray Fixed Systems for Fire Protection
combustible	Burnable; sometimes differentiated from flammable (qv). A contested term, see **Box F1**.
combustible dust	A dust which is combustible, or "finely divided solid particles of a substance or mixture that are liable to catch fire or explode on ignition when dispersed in air or other oxidizing media" according to the UN Globally Harmonized System for Classification and Labelling of Chemicals Rev 8. Also see NFPA 654: Standard for the Prevention of Fire and Dust Explosions from the Manufacturing, Processing, and Handling of Combustible Particulate Solids, 2017 Edition and BS EN 50281 -3 2002 Electrical apparatus for use in the presence of combustible dust. Classification of areas where combustible dusts are or may be present

combustible liquid	1. According to the NFPA, any liquid with a flashpoint above 37.8 °C; defined formally in NFPA 30: Flammable and Combustible Liquids Code, 2018 Edition, NFPA 11: Standard for Low-, Medium-, and High-Expansion Foam, 2016 Edition, NFPA 15 Standard for Water Spray Fixed Systems for Fire Protection and API RP 505 2nd Edition, August 2018 Recommended Practice for Classification of Locations for Electrical Installations at Petroleum Facilities Classified as Class I, Zone 0, Zone 1, and Zone 2. Also see UN Globally Harmonized System for Classification and Labelling of Chemicals Rev 8; cf flammable liquid 2. According to Australian Standard AS1940-2017, any liquid with a flash point below its boiling point
combustible particulate solid	Essentially, combustible dust (qv). Defined more formally in NFPA 654: Standard for the Prevention of Fire and Dust Explosions from the Manufacturing, Processing, and Handling of Combustible Particulate Solids, 2017 Edition
Combustibles magazine	A defunct magazine known for the quality of its photography, and its unfortunate associations with fires and explosions
combustion	*aka* burning. Defined formally in API RP 535 - Burners for Fired Heaters at Refineries and API RP 2001 - Fire Protection at Refineries; cf flaming combustion, glowing combustion, smoldering combustion
combustion air	The air used to combust fuel, flare gases etc.; defined formally in API Standard 521. Pressure-relieving and Depressuring Systems. Sixth Edition | January 2014, API RP 537 Flare Details for Petroleum, Petrochemical, and Natural Gas Industries, and NFPA 86: Standard for Ovens and Furnaces, 2019 Edition
combustion air pressure switch	A combustion air pressure-activated safety shutdown switch to stop a burner running with insufficient air pressure. Defined formally in NFPA 86: Standard for Ovens and Furnaces, 2019 Edition
combustion chamber	An enclosed space in which combustion takes place, especially in a fired heater
combustion efficiency	Percentage of burner fuel totally oxidized; defined formally in API RP 537 Flare Details for Petroleum, Petrochemical, and Natural Gas Industries

combustion gases	*aka* flue gases, combustion products. A mixture of gases resulting from combustion. May become, or be synonymous with flue gas (qv) if produced in a system having a flue. May sometimes be used to mean only things produced by combustion, i.e., excluding excess air or even unreacted air components such as nitrogen
combustion products	*aka* flue gases. Defined formally in API RP 535 - Burners for Fired Heaters at Refineries
combustion safeguard	A safety mechanism associated with flame detectors which ensures burner safety. Defined formally in NFPA 86: Standard for Ovens and Furnaces, 2019 Edition
combustion turbine	see **gas turbine**
come-along	USA term for a hand operated ratchet winch such as a bulldog (qv)
Cominco process	1. A process for producing lead alloy strip grid for batteries 2. A lead smelter flue gas treatment process
commercial off the shelf	*aka* COTS. Equipment which is mass produced by OEMs for procurement by industry
commercial sterilization	see **sterilization**
comminution	The reduction of a solid material's average particle size by a process such as chopping. Although comminution, in everyday language, is synonymous with milling (qv), in engineering contexts it has a certain narrow meaning (qv comminutor). N.B. comminutors are never referred to as mills
comminutor	*aka* macerator. A comminution (qv) device. Although comminution, in everyday language, is synonymous with milling (qv), in engineering contexts it has a certain narrow meaning and comminutors are never referred to as mills. One type of comminutor is used to chop coarse debris found in wastewater into fine particles[26], as an alternative to screening (qv)

---

[26] Comminutors of this type come into fashion cyclically on wastewater treatment works (until people remember once again that the chopped material reforms in the system downstream into mats and ropes which clog everything up)

commissioning	The activities/stage between construction and operation which makes a plant or product ready for service. Defined in the context of products in BS EN 82079-1:2012 Preparation of instructions for use. Structuring, content and presentation. General principles and detailed requirements. In the context of process plant, it is a multistage, multidiscipline process, and different disciplines may define it differently. Because of this, some companies no longer routinely refer to 'commissioning', and give names to the specific stages to avoid confusion; see **Box C3**.
commissioning and qualification	*aka* C&Q. Pharmaceutical industry terminology; the commissioning part means what it always does, whilst 'qualification' part is a subset of validation (qv) akin to a performance trial (qv)
commodities	see **bulk materials**
commodity chemicals	see **bulk materials**
commodity release	A euphemism for leak, in the context of pipelines
common cause failure	The failure of multiple components in an electro-mechanical system during a safety incident due to the same cause; defined formally in BS EN ISO 12100:2010 Safety of machinery. General principles for design. Risk assessment and risk reduction
common mode failure	The failure of multiple components in an electro-mechanical system during a safety incident in the same way; defined formally in BS EN ISO 12100:2010 Safety of machinery. General principles for design. Risk assessment and risk reduction
communicating chamber	A separate chamber added on to a heat exchanger's head or shell which is part of its pressure envelope, such as a sump; defined formally in API 660 Shell-and-Tube Heat Exchangers and API RP 661 Heat Exchangers
comonomer	A polymerizable precursor to a copolymer, aside from the principal monomer
comparative assertion	*aka* green claim. A claim that one product is as environmentally good as or better than a competing product. Defined formally in EN ISO 14040:2006 Environmental management - Life cycle assessment - Principles and framework
comparative cleanability	"The cleanability of equipment relative to a reference", according to the EHEDG Glossary

Box C3. Stages Of Commissioning

There is some dispute about the names of various stages of commissioning, though the consensus view is illustrated in **Figure C4**.

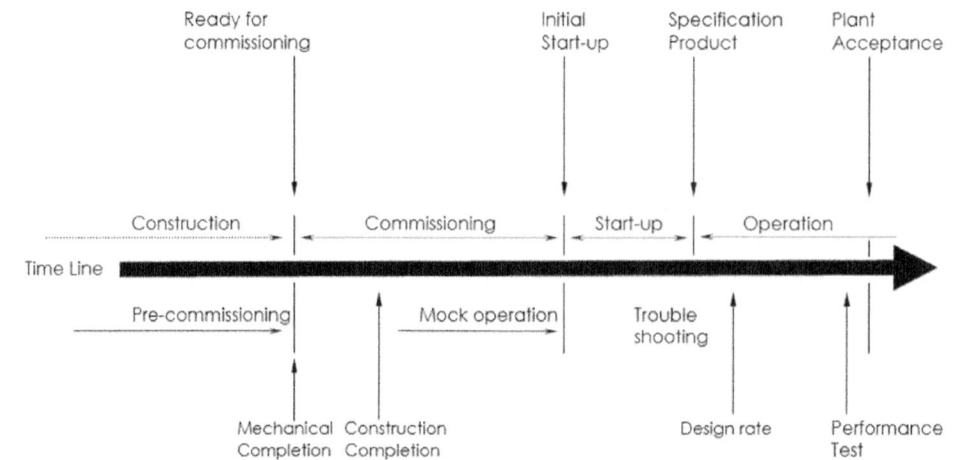

Figure C4 Stages of Commissioning

All of the terms on the graphic are defined in this dictionary. Many of these terms are somewhat contested, and there are additional terms used in specific sectors. The greatest degree of contention may be around what counts as commissioning. Many process engineers see it as shown in Fig. C4, with everything prior to process commissioning being construction, and everything after initial startup being startup or operation. However, performance trials often mark the end of process commissioning, in as much as the process commissioning engineers tend to be present for them. They may therefore occur before startup or before operation, rather than as shown on the diagram, or commissioning may be considered to continue all the way through to plant acceptance. At the maximum, everything shown in the graphic may be considered commissioning of one kind or another. Ultimately, the meanings of all of these terms in practice is that given in contractual documents.

Related Terms
mechanical commissioning, hot commissioning, initial operations, cold commissioning, dry commissioning. wet commissioning, water trials, process commissioning

comparative emission data	Emission data of similar machines collected for comparison; defined formally in BS EN ISO 12100:2010 Safety of machinery. General principles for design. Risk assessment and risk reduction
compartmentation	Controlling the size of a hazard area using barriers to flammable material migration. Defined formally in NFPA 654: Standard for the Prevention of Fire and Dust Explosions from the Manufacturing, Processing, and Handling of Combustible Particulate Solids, 2017 Edition
compendial water	see **purified water**
competence	The ability to achieve desired results by the application of knowledge and skill; defined formally in BS EN ISO 9000 Quality management systems Fundamentals and vocabulary
competence acquisition	The process of acquiring competence; defined formally in BS EN ISO 9000 Quality management systems Fundamentals and vocabulary
competent authority	Generically, the body or bodies responsible for some legal duty. In the EU, a medicines regulatory authority in a European Union Member State, amongst other things. In UK, the combination of the HSE and Environment Agency which regulates compliance with COMAH; see **Box Q1**.
competent person	Someone trained to identifying hazards, authorized to promptly eliminate them; defined formally in NFPA 59: Utility LP-Gas Plant Code, 2018 Edition. see **Box Q1**.
competitive inhibition	Reversible reduction of enzyme activity by something which competes with its substrate for its binding site
complete medium	Biological growth media with more complex nutrients such as amino acids required for growth of cultured organisms; cf minimal medium
completion	see **well completion**
completion of work	The point in commissioning operations when the contractor has erected the plant in accordance with drawings and specifications, completed his specified precommissioning work and completed his final cleanup, painting and thermal insulation work ready for commissioning (qv)
completion certificate	A certificate confirming that work is complete, usually associated with a stage payment; defined formally in BS EN ISO 10209:2012 Technical product documentation - Vocabulary - Part 1: Terms relating to technical drawings: general and types of drawings; cf final certificate, interim certificate

complex	In a plant layout context, a collection of colocated sites (qv) that may or may not be owned by the same business entity
complex device	A device which is sufficiently complicated that each of its components needs its own diagram; defined formally in BS EN ISO 10209:2012 Technical product documentation - Vocabulary - Part 1: Terms relating to technical drawings: general and types of drawings
complex viscosity	A rheological property different from those of the individual components in a material made from two or more constituent materials with significantly different physical or chemical properties. Usually refers to polymeric materials composed of two parts: a polymer matrix and a reinforcement fiber e.g.; composite material (qv)
compliance	1. Most commonly, meeting specifications, legal requirements, etc. 2. The reciprocal of stiffness, defined formally in BS ISO 2041:2018 Mechanical vibration, shock and condition monitoring. Vocabulary
component	1. Part of a mixture of chemicals 2. Part of a machine; defined formally in BS EN 82079-1:2012 Preparation of instructions for use. Structuring, content and presentation. General principles and detailed requirements, BS EN ISO 10209:2012 Technical product documentation - Vocabulary - Part 1: Terms relating to technical drawings: general and types of drawing and NFPA 59: Utility LP-Gas Plant Code, 2018 Edition
component drawing	A drawing of a single machine component; defined formally in BS EN ISO 10209:2012 Technical product documentation - Vocabulary - Part 1: Terms relating to technical drawings: general and types of drawings
component fitting	Fired heater tube fitting; defined formally in API STD 530 - Calculation of Tube Heater Thickness
composite	Something made of disparate parts, or shorthand for composite material
composite material	A material of construction made of two or more materials with properties which differ from each other, e.g., glass reinforced plastic (qv)
composition	A mixture of ingredients, or a description of such a mixture
composting	Aerobic biological decomposition of organic waste
compound	1. Generically, a combination of materials 2. In the plastics industry, mixing additives into a virgin polymer (usually with a mixer-extruder)

compounding	see **compound**
Comprehensive Environmental Response, Compensation and Liability Act	*aka* CERCLA, Superfund. USA environmental legislation
compressed air breathing apparatus	*aka* CABA; see **breathing apparatus set**
compressed air foam	Firefighting foam produced by mixing water, foam concentrate and pressurized gas; defined formally in NFPA 11: Standard for Low-, Medium-, and High-Expansion Foam, 2016 Edition
compressed air foam discharge devices	Discharge of a specified pattern of compressed air foam; defined formally in NFPA 11: Standard for Low-, Medium-, and High-Expansion Foam, 2016 Edition
compressed air foam generating method	A controlled way of making compressed air foam in a mixing chamber; defined formally in NFPA 11: Standard for Low-, Medium-, and High-Expansion Foam, 2016 Edition
compressed air foam system	Compressed air foam delivery system (does not include generation); defined formally in NFPA 11: Standard for Low-, Medium-, and High-Expansion Foam, 2016 Edition
compressed fiber gasket	*aka* compressed fibre gasket. A common replacement for compressed asbestos fiber gaskets, often using polymeric or crystalline fibers
compressed fibre gasket	see **compressed fiber gasket**
compressed natural gas	*aka* CNG. Natural gas which has been compressed in order to be used as fuel, most commonly for internal combustion engines. It differs from liquefied natural gas (qv) in not having been liquefied
compressibility	The relative change in volume of a material in response to a change in pressure
compressible fluid	All fluids are somewhat compressible, but 'compressible fluids', to engineers, are those where any density changes as a result of compression are significant in our process and hydraulic/pneumatic calculations, i.e., gases and vapors. The others (liquids) are 'incompressible' for practical purposes; cf incompressible fluid
compression molding	A molding process that is characterized by the use of a heated mold which is exposed to pressure
compression ratio	The ratio of the discharge pressure to suction pressure in each stage of a compressor

compression set	A permanent deformation of an elastomer after pressure has been applied. There are two types - A and B; formal industry-specific definitions in ASTM D395 and ASME BPE (American Society of Mechanical Engineers: Bioprocessing Equipment)
compressor	A gas pump
compressor inlet drum	see **flash drum**
compressor suction drum	see **flash drum**
computational fluid dynamics	*aka* CFD. Computer modelling of fluid dynamics
computational pipeline monitoring	*aka* CPM. A software method for monitoring the operation of an oil pipeline system for commodity release (qv)
computed tomography	see **tomography**
computer aided design	*aka* CAD. Drawing or design using computer software - the modern norm
computer aided design and drafting	*aka* CADD; see computer aided design
computer aided drafting	*aka* CAD; see computer aided design
computer aided drawing	*aka* CAD; see computer aided design
concarbon	see **Conradson carbon residue**
concavity	*aka* underfill. A depression of the surface of a weld relative to the surface of the workpiece; defined formally in ASME BPE (American Society of Mechanical Engineers: Bioprocessing Equipment)
concentrate	A process to increase concentration, or the product of such a process
concentration	The quantity of a material in a fluid. Usually expressed as a mass per unit volume
concentration polarization	A process in which a membrane filter creates a region of high concentration of the substance being filtered out local to its working surface, thus reducing its performance
concentric	Having a common center (especially applied to circular objects); cf eccentric
conceptual design	1. *aka* conceptual process design. The first stage of process plant design; see **Box S2** 2. In a product design context, defined formally in BS EN ISO 10209:2012 Technical product documentation - Vocabulary - Part 1: Terms relating to technical drawings: general and types of drawings
conceptual process design	see **conceptual design**

concession	1. Generically, a permission 2. In oil, gas and minerals, government permission for a company to explore and exploit underground reserves under government control 3. In quality management, permission to release nonconforming product; defined formally in BS EN ISO 9000 Quality management systems Fundamentals and vocabulary
concrete pad	A poured or placed concrete foundation; defined formally in NFPA 58: Liquefied Petroleum Gas Code, 2017 Edition
concurrent engineering	Coordination of simultaneous activities; defined formally in BS EN ISO 10209:2012 Technical product documentation - Vocabulary - Part 1: Terms relating to technical drawings: general and types of drawings
concurrent operation	Simultaneous or parallel operation. Not the same as cocurrent operation (cf)
condensable gas	1. Generically, a gas or vapor which condenses under process conditions 2. Specifically, (and commonly) used to mean a vapor that might condense in a flare header during flaring. This sense is defined formally in API RP 537 Flare Details for Petroleum, Petrochemical, and Natural Gas Industries; cf noncondensable gas
condensate	*aka* condy. Something which has condensed. Most commonly used to refer to either condensed steam, a very light oil, or natural gas liquids (qv)
condensate gas ratio	*aka* CGR, condensate-gas ratio, condensate to gas ratio. The ratio of gas to condensate in a reservoir, commonly measured in barrels per millions of standard cubic feet (barrels/mmscf)
condensate polisher	An ion exchange column used to purify steam condensate
condensate pump	1. Generally, the pumps which transfer condensate in steam systems 2. Specifically, in compressors, vertical centrifugal pumps mounted in the hot well return condensate from liquefaction in the condenser 3. Also used to refer to natural gas liquids (qv) pumps in the gas industry
condensate recovery unit	*aka* CRU. A packaged plant based on a condensate pump or pumps and feed tank; used to recover hot condensate from a steam system
condensate to gas ratio	see **condensate gas ratio**
condensate-gas ratio	see **condensate gas ratio**

condensation	Phase transition from vapor to liquid (not to solid as sometimes defined in error; cf deposition)
condensation polymerization	*aka* polycondensation or step-growth polymerization. Polymerization by condensation reaction
condensation reaction	Molecules joining together yielding a small molecule, (most commonly water) as a byproduct i.e., the opposite of hydrolysis (qv)
condensation temperature	see **dew point**
condenser	A unit operation which condenses a process stream by transferring heat to a cooler stream, utility or environment
condition monitoring	A process of monitoring a parameter of condition in machinery (vibration, temperature etc.), in order to identify a significant change which is indicative of a developing fault; a major component of predictive maintenance
conditioning tower	*aka* CT, gas conditioning tower. A tower where water is injected to cool process gases
conditions for intended use	"All normal and reasonably anticipated operating conditions, including those of cleaning. These conditions should include limits for variables such as time, temperature and concentration" according to the EHEDG Glossary. Closely related to design envelope (qv)
conductance	An extrinsic property of a component, the inverse of resistance (qv), therefore a measure of the ease of electrical, hydraulic or thermal conduction. The SI unit of conductance is siemens; cf conductivity
conduction	Energy transfer through a material without bulk material movement
conductivity	An intrinsic property of a material, the inverse of resistivity (qv), therefore a measure of the ease of electrical, hydraulic or thermal conduction (qv) of an item. Electrical conductivity is used as a measure of the concentration of dissolved ions in water. The SI unit of conductivity is siemens/meter; cf conductance
condy	see **condensate**
cone and plate viscometer	Equipment found in a quality assurance (qv) laboratory for measuring viscosity
cone roof tank	see **fixed roof tank**

confidence interval	A statistically derived range of plausible values for a population parameter specified as the 'x% Confidence Interval', (and therefore not what most engineers think it means). Defined formally in BS ISO 2041:2018 Mechanical vibration, shock and condition monitoring. Vocabulary
confidence level	The parameter 'x' in the definition of confidence interval (qv); 95% is a common value. This doesn't mean what most people think it means: it is the probability that a population parameter falls inside a range between two values around a mean. Defined formally in BS ISO 2041:2018 Mechanical vibration, shock and condition monitoring. Vocabulary
configurable	Having the option of configuration. Defined in different contexts in BS EN ISO 10209:2012 Technical product documentation - Vocabulary - Part 1: Terms relating to technical drawings: general and types of drawings and BS 1646-4:1984 Symbolic representation for process measurement control functions and instrumentation. Specification for basic symbols for process computer, interface and shared display/control functions
configuration	Setting the functional and physical characteristics of a plant or product, or the sum of those settings/options. Defined in different contexts in BS EN ISO 9000 Quality management systems Fundamentals and vocabulary, and BS EN ISO 10209:2012 Technical product documentation - Vocabulary - Part 1: Terms relating to technical drawings: general and types of drawings
configuration authority	Person or entity assigned the responsibility for choosing a configuration (qv); defined formally in BS EN ISO 9000 Quality management systems Fundamentals and vocabulary
configuration baseline	A standard configuration (qv), akin to 'factory settings'; defined formally in BS EN ISO 9000 Quality management systems Fundamentals and vocabulary
configuration control	Measures for the formal control of configuration (qv); defined formally in BS EN ISO 10209:2012 Technical product documentation - Vocabulary - Part 1: Terms relating to technical drawings: general and types of drawings
configuration control board	see **configuration authority**
configuration management	see **configuration control**

configuration object	A configured element with an end user function. Defined formally in BS EN ISO 9000 Quality management systems Fundamentals and vocabulary
configuration status accounting	The formal recording and reporting of a configuration. Defined formally in BS EN ISO 9000 Quality management systems Fundamentals and vocabulary
confined space	A hazardous space large enough for a person to enter, but at least partially enclosed, with restricted access, and unsuitable for continuous occupancy
conflagration	A widespread destructive fire with low flame front velocity; cf detonation, deflagration
Conformité Européenne mark	see **CE mark**
conformity	Fulfilling requirements, or a way in which something meets requirements; defined formally in BS EN 82079-1:2012 Preparation of instructions for use. Structuring, content and presentation. General principles and detailed requirements and BS EN ISO 9000 Quality management systems, fundamentals and vocabulary
conglomerate	Crystals which grow together during pan boiling stage of sugar refining
conical plate centrifuge	see **disk centrifuge**
coning	A term used in distillation for the condition where rising vapour pushes the liquid back from the top of the hole and passes upward, creating poor vapour – liquid contact
connecting line	The graphical symbol for a functional connection. It might be a wire, a pipe, or a mechanical linkage, depending on drawing type. Defined formally in BS EN ISO 10209:2012 Technical product documentation - Vocabulary - Part 1: Terms relating to technical drawings: general and types of drawings
connection	*aka* nozzle. A joint that connects a port to a pipe; defined formally in API RP 682 - Pump Seals
connection diagram	A diagram of electrical connections; defined formally in BS EN ISO 10209:2012 Technical product documentation - Vocabulary - Part 1: Terms relating to technical drawings: general and types of drawings
connector actuating tool	*aka* CAT. A tool used by a remotely operated vehicle (qv) to connect undersea pipelines
Conradson	A term used in oil and gas as shorthand for the coke forming tendencies of a fluid, as measured by the Conradson carbon residue (qv) test, as in the phrase 'high Conradson feed'

Conradson carbon residue	*aka* microcarbon residue, concarbon, CCR. Carbon residue as determined by ASTM D189-06 (2019) Standard Test Method for Conradson Carbon Residue of Petroleum Products; a measure of the coke-forming tendencies of a hydrocarbon mixture
consequence	A causal outcome. Defined formally in BS EN 82079-1:2012 Preparation of instructions for use. Structuring, content and presentation. General principles and detailed requirements
consequence analysis	An analytical prediction of the expected result from a postulated set of conditions or course of action
consequence modeling	The use of mathematical models to predict consequences from a loss of containment or other process upset
conservation laws	see **laws of conservation**
consistency	1. Can mean 'always the same' 2. A description of liquid 'thickness' ('thin', 'nectar-, honey- or spoon- thick')- related to viscosity, (so you might have a consistent consistency)
consistency check	Verification that a consistent set of assumptions, methods and data have been applied to a life cycle assessment (qv); defined formally in EN ISO 14040:2006 Environmental management - Life cycle assessment - Principles and framework
constant	Something invariant, especially applied to invariant mathematical relationships between variables
constant boiling mixture	*aka* azeotrope, azeotropic mixture. A mixture of liquids, the vapor of which has the same proportions of constituents as the parent mixture, preventing separation of components by distillation
constant molal overflow	see **constant molar overflow**
constant molar overflow	*aka* constant molal overflow. A common assumption in design of distillation columns that the molar flows of liquid and vapor are constant in each section of the column
constant rate drying	The part of a drying cycle in which the water evaporates as from a free water surface, at a constant rate, and at a constant material temperature
construct	1. To build 2. A modeled concept; defined formally in BS EN ISO 10209:2012 Technical product documentation - Vocabulary - Part 1: Terms relating to technical drawings: general and types of drawings
construction	The initial building or erection of a process plant; cf commissioning

construction classification number	A number used in a formula to determine total firefighting water supply requirements. Defined formally in NFPA 1142: Standard on Water Supplies for Suburban and Rural Fire Fighting, 2017 Edition
construction cost	*aka* capital cost. This can be split into direct costs (purchase cost of equipment and site, plus costs of equipment erection, foundations and structural work, piping, electrical, instruments, process buildings and structures, ancillary buildings, storages, utilities, and site preparation) and indirect costs (design and engineering costs, contractors' fees, and contingency allowance)
construction drawing	A drawing which gives information about construction; defined formally in BS EN ISO 10209:2012 Technical product documentation - Vocabulary - Part 1: Terms relating to technical drawings: general and types of drawings; cf working drawing, for construction drawing
Construction Industry Scheme	*aka* CIS. A UK withholding tax regime for contractors and subcontractors working in construction, aimed at preventing tax evasion
construction takeoff	see **material takeoff**
construction work package	*aka* CWP. A project management term for a group of related construction tasks
Construction, Design and Management Regulations 2015	*aka* CDM. Regulations in the UK governing construction and demolition activities
consultant	Ideally, an entity which offers broad advice and guidance, though rarely progresses design beyond conceptual design (qv) stage[27]
consumable	Parts or material consumed through use in a product or plant; also used in a maintenance context to refer to any low value component only used once before disposal. Defined formally in BS EN 82079-1:2012 Preparation of instructions for use. Structuring, content and presentation. General principles and detailed requirements
consumable insert	*aka* insert ring. A metal ring placed between two things to be welded, intended to be consumed as weld filler during welding. Defined formally in ASME BPE (American Society of Mechanical Engineers: Bioprocessing Equipment)

---

[27] All too often, especially with 'management consultants', someone who borrows your watch to tell you the time!

consumer	The end user of a product, human or otherwise; defined formally in BS EN 82079-1:2012 Preparation of instructions for use. Structuring, content and presentation. General principles and detailed requirements and BS EN ISO 14159:2008 Safety of machinery. Hygiene requirements for the design of machinery
consumer product	A product which might be used by consumers; defined formally in BS EN 82079-1:2012 Preparation of instructions for use. Structuring, content and presentation. General principles and detailed requirements
contact angle	(*symbol* θ) The angle (measured through the liquid) where a liquid meets a solid surface; related to surface wettability
contact process	Currently, the most popular process used to make high-strength sulfuric acid in bulk
contact time	The average time that a process fluid spends in contact with something such as an absorption medium, i.e., bed volume/volumetric flowrate
contacting seal	A seal whose design geometry does not prevent faces touching; defined formally in API RP 682 - Pump Seals and ISO 21049:2004 Pumps — Shaft sealing systems for centrifugal and rotary pumps
container	1. Most commonly used to describe an ISO shipping container 2. In oil and gas, either any vessel up to 450 L capacity used to store or transport flammable and combustible liquids, or any vessel used to store and transport LPG (no capacity limit). Both defined formally in NFPA 30: Flammable and Combustible Liquids Code, 2018 Edition and NFPA 58: Liquefied Petroleum Gas Code, 2017 Edition respectively 3. see **enclosure**
container appurtenances	Safety, operating and control devices on a container's openings; defined formally in NFPA 58: Liquefied Petroleum Gas Code, 2017 Edition
container assembly	An LPG container, together with appurtenances and protective housings; defined formally in NFPA 58: Liquefied Petroleum Gas Code, 2017 Edition
containment	Process safety measures which keep hazardous material away from receptors (environmental or human). This includes primary containment (qv), secondary containment (qv) and tertiary containment (qv); cf loss prevention, enclosure

containment building	In the nuclear sector, a reinforced shell surrounding the core of a nuclear power station, ensuring that a failure within the nuclear power station does not cause discharge of radiological material into the environment
containment seal	A seal in which a flexible element, seal and mating rings are chamber mounted; defined formally in API RP 682 - Pump Seals and ISO 21049:2004 Pumps — Shaft sealing systems for centrifugal and rotary pumps
containment seal chamber	A chamber in which a containment seal is installed. Defined formally in API RP 682 - Pump Seals and API RP 682 - Pump Seals and ISO 21049:2004 Pumps — Shaft sealing systems for centrifugal and rotary pumps
contaminants	see **contamination** and **Box C2**.
contamination	Generally, the undesirable presence of something as a result of human activity. Defined formally in two specific contexts in BS 5930:2015 Code of practice for ground investigations and BS EN 1672-2:2005+A1:2009 Food processing machinery. Basic concepts. Hygiene requirements. Other sector-specific definitions exist which fall under the general definition, such as radiological contamination (qv)
continental wash	A simultaneous backwashing and air scour of a depth filter (qv)
contingency	An allowance of time or money added by estimators against the possibility of additional future requirements
continual improvement	An iterative process of performance enhancement; defined formally in BS EN ISO 9000 Quality management systems Fundamentals and vocabulary
continuity equations	Equations based on the conservation laws (qv) which underpin many aspects of chemical engineering, notably mass balance. They describe mathematically a flux of something (mass, momentum or energy to chemical engineers) over an imaginary surface
continuous back mix reactor	see **continuous stirred tank reactor**
continuous catalytic regeneration	*aka* CCR. A refinement of catalytic reforming (qv) in which a portion of the catalyst is continuously removed, regenerated and replaced
continuous furnace	A more or less continuously fed furnace; defined formally in NFPA 86: Standard for Ovens and Furnaces, 2019 Edition

continuous grade of release	A concept in hazardous area classification. A condition in which the release of hazardous material is essentially continuous. Defined formally in BS EN IEC 60079-10-1:2021 Explosive atmospheres. Classification of areas. Explosive gas atmospheres, API RP 505 2nd Edition, August 2018 Recommended Practice for Classification of Locations for Electrical Installations at Petroleum Facilities Classified as Class I, Zone 0, Zone 1, and Zone 2 and IEC 61892-7, Mobile and fixed offshore units – Electrical installations – Part 7: Hazardous areas; cf primary grade of release, secondary grade of release
continuous improvement	*aka* CI; see **continual improvement**
continuous phase	A fluid phase in which particles of solid, droplets of liquid, or bubbles of gas are dispersed
continuous pilot	A pilot burner which burns continuously, irrespective of whether the main burner is burning. Defined formally in NFPA 86: Standard for Ovens and Furnaces, 2019 Edition
continuous process	A more or less continuously fed process; cf batch process
continuous production	Production using a continuous process; cf batch production, campaign manufacture
continuous reactor	A more or less continuously fed reactor. Usually a continuous stirred tank reactor (qv), or plug flow reactor (qv); cf batch reactor
continuous stirred tank reactor	*aka* CeeSTAR, CSTR. A theoretical model of an ideal reactor type, or the real-world item that the model approximates; cf plug flow reactor
continuous vapor concentration controller	A control system for fired heaters which measures, indicates and controls the fuel–air mixture based on the lower flammable limit of the fuel vapor. Defined formally in NFPA 86: Standard for Ovens and Furnaces, 2019 Edition
continuous vapor concentration high limit controller	A safety control for fired heaters which reduces vapor concentrations in the fuel–air mixture if concentration exceeds a set point. Defined formally in NFPA 86: Standard for Ovens and Furnaces, 2019 Edition
contour	see **level contour line**
contour line	see **level contour line**

contract	A binding and (hopefully!) legally enforceable agreement. Defined formally (variously but similarly) in BS EN ISO 9000 Quality management systems Fundamentals and vocabulary, BS EN ISO 10209:2012 Technical product documentation - Vocabulary - Part 1: Terms relating to technical drawings: general and types of drawings, and IEEE 830-1998 - IEEE Recommended Practice for Software Requirements Specifications
contract document	A document which forms part of a contract; defined formally in BS EN ISO 10209:2012 Technical product documentation - Vocabulary - Part 1: Terms relating to technical drawings: general and types of drawings
contracting company	A company which contracts to build plants, and usually does its own detailed design; cf engineering procurement and construction company
contraction flow	see **converging flow**
contractor	Generally speaking, the party to a contract responsible for delivering a service. The contractor might be a contracting company (qv), an individual freelance engineer, or either of the above but with a single client in a manner virtually indistinguishable from an employee/employer relationship. Often used to refer to all non-payrolled workers, particularly in maintenance and construction, including pipe-fitters, laggers and scaffolders among others
contractors' all risks insurance	*aka* CAR. An insurance policy providing coverage for property damage and third-party injury or damage claims, taken out jointly by all parties to a construction project, to make sure that all eventualities are covered (usually other than wear and tear, willful negligence and poor workmanship)
control and instrumentation	*aka* C&I. A process control and measurement engineering specialization, and the equipment associated with this
control and instrumentation engineer	*aka* C&I engineer. A hybrid chemical/software engineer (or sometimes instrument technician) found mainly in petrochemical industry operating companies, involved with plant process control system and instrumentation systems
control area	Any part of a building where the storage, dispensing or use of quantities of flammable liquids up to a maximum allowable quantity is permitted. Defined formally in NFPA 30: Flammable and Combustible Liquids Code, 2018 Edition
control chart	*aka* process behavior chart. A chart such as the Shewhart chart (qv) used in statistical process control to determine if a process is in steady state operation

control element	see **correcting element**
control guard	*aka* interlocking guard with a start function. A machine guard which starts the machine it guards upon closing. Defined formally in BS EN ISO 12100:2010 Safety of machinery. General principles for design. Risk assessment and risk reduction
control logic	The conditions and interrelations expressed within the code of the control system to achieve or prevent certain process conditions. Often expressed diagrammatically or in documents accompanying a new design to allow for interpretation by non-control disciplines
control loop	Physically, the set of instrumentation which, together, automatically measures and adjusts a single process variable to equal a set-point value; or conceptually, the 'loop' of sensor, controller and actuator that defines the scope of this physical kit
control mass	see **control volume**
control measure	"Any action and activity that can be used to prevent or eliminate a food safety hazard or reduce it to an acceptable level" according to the EHEDG Glossary. Might be applicable to other contexts with removal of the word 'food'
control mechanism	1. Generally, a term close in meaning to control loop (qv) 2. Specifically, in the context of steam turbines, the equipment between the speed governor and the valves it controls. Defined formally in EN ISO 10437 Petroleum, petrochemical and natural gas industries - Steam turbines - Special-purpose applications
control of burning	Spraying areas where fire occurs with water to cool and control the fire until it can be extinguished. Defined formally in NFPA 15 Standard for Water Spray Fixed Systems for Fire Protection
Control of Major Accident Hazards Regulations	*aka* COMAH. The UK regulations used to implement the EU 'Seveso' Directives which aim to control major accident hazards involving dangerous substances. Hazard categories include Dangerous to the Environment, Toxic, Self-reactive, Pyrophoric, Explosive, and Oxidizing, amongst others
Control of Substances Hazardous to Health Regulations 2002	*aka* COSHH. UK Regulations requiring assessment of the potential harms associated with use of chemicals
control panel	A deprecated term which might be used (probably by non-engineers) to mean a motor control center (qv), human machine interface (qv), or industrial control panel (qv)

control philosophy	A written description of the process in a plant with an emphasis on how it should be operated and controlled. This document is usually generated by the process engineer and directed to the control systems engineers to assist them in designing their program logic and coding activities; cf functional design specification
control rod	A moveable rod of neutron absorbing, non-fissionable material used to control uranium and plutonium nuclear reactors
control room	*aka* CR, central control room. Generally speaking, a room on a process plant containing whole plant monitoring and control equipment, and capable of continuing operation under all foreseeable loss of containment events (though remote control centers which need not be proofed against loss of containment events might also fall under this name)
control room operator	*aka* control room technician, CRT, panel operator, DCS operator, panel man (irrespective of gender of operator). A process plant operator who works in the control room (qv)
control room technician	see **control room operator**
control valve	A manual or actuated valve controlling the flow of a process fluid or utility
control valve actuator	*aka* actuator. A linear or rotary device which moves a valve to a desired position, most commonly driven by electricity or compressed air
control volume	*aka* control mass. A mathematical abstraction employed in the process of creating mathematical models of physical processes
controlgear	Equipment which is usually known as a motor control center (qv) in Europe and an industrial control panel (qv) in the USA
controlled area	An area defined in relation to hazards, especially radiological hazards
controlled environment	In the hygienic industry sense, see **zoning**
controlled flow grate	A type of clinker cooler technology
controlled safety relief valve	A safety relief valve with additional control device. Defined formally in DIN 3320-1:1984 Safety Valves; Safety Shut-Off Valves; Definitions, Sizing, Marking as a safety relief valve with additional control device
controller gain	see **gain**
convection	Energy transfer through a fluid material by bulk material movement; cf radiation, conduction
convective mass transfer	Mass transfer between a boundary surface and a moving fluid or between two relatively immiscible, moving fluids

conventional blendstock for oxygenate blending	*aka* CBOB, conventional gasoline blendstock for oxygenate blending. A blendstock (qv) which is blended with oxygenates (usually 10% alcohol) to make conventional gasoline (qv), which can still be used in the USA other than in urban areas with poor air quality
conventional gasoline	*aka* CG. A non-reformulated gasoline (qv), which can still be used in the USA other than in areas with poor air quality; cf reformulated gasoline
conventional gasoline blendstock for oxygenate blending	see **conventional blendstock for oxygenate blending**
conventional island	*aka* turbine island. The part of a nuclear power plant which it has in common with a conventional (non-nuclear) power plant, i.e.; the turbine, generator and ancillary systems
conventional pressure relief valve	*aka* conventional safety relief valve. A backpressure susceptible spring-loaded pressure relief valve; defined formally in API RP 520 P1 7th Edition, January 2000 Sizing, Selection, and Installation of Pressure-Relieving Devices in Refineries; Part I - Sizing and Selection and API Standard 521. Pressure-relieving and Depressuring Systems. Sixth Edition | January 2014
conventional safety relief valve	see **conventional pressure relief valve**
converging flow	A fluid flow in which streamlines converge
conversion	How much of a reactant has reacted in a reactor, usually expressed as a fraction or percentage of the original mass of reactant
converter	Usually refers to reactors. e.g., ammonia converter
convexity	*aka* reinforcement. A condition where the surface of a weld is extended relative to the surface of the workpiece. Defined formally in ASME BPE (American Society of Mechanical Engineers: Bioprocessing Equipment)
conveyor	A fixed machine which transports large quantities of solid materials
COO	Chief Operating Officer
coolant	A heat transfer fluid used to reduce temperature
cooler	A device which cools process streams by transfer of heat to a cool utility or the environment
cooling jacket	*aka* jacket. A vessel jacket (qv) used for cooling
cooling tower	A tall air/water direct contact countercurrent heat exchanger in which water loses heat, and to a small extent evaporates

cooling water	*aka* CW. A water utility at ambient temperature, usually raw or relatively low-quality water - even seawater. Whilst some closed loop systems may be treated with oxygen scavengers and corrosion inhibitors, they are never de-oxygenated to the same standard as boiler feed water (qv)
cooling water return	*aka* CWR. Cooling water (qv) which has exchanged heat with the process fluid and is being returned to cooling tower (qv)
cooling water supply	*aka* CWS. Cool water from source or cooling tower (qv)
coolside process	A flue gas desulfurization process
coordinate dimensioning	Dimensioning of a drawing based on either an internal or external coordinate system. Defined formally in BS EN ISO 10209:2012 Technical product documentation - Vocabulary - Part 1: Terms relating to technical drawings: general and types of drawings
COP	1. see **coefficient of performance** 2. see **cleaning out of place** 3. Control of pollution
copolymer	A polymer derived from more than one species of monomer
copolymerization	The polymerization of more than one monomer into copolymers
copperas	*aka* green vitriol. Iron (II) sulfate
COPQ	see **cost of poor quality**
coproduct	Any of two or more products coming from the same unit process or product system. Coproducts can involve similar revenue to main products; defined formally in EN ISO 14040:2006 Environmental management - Life cycle assessment - Principles and framework; cf byproduct
COR	Cell output rate
corbel	In the context of fired heaters, a projection on a refractory surface intended to prevent flue gas bypassing convection section tubes. Defined formally in BS EN ISO 13705:2012/ISO 13705:2012(E) Petroleum, petrochemical and natural gas industries. Fired heaters for general refinery service, API STD 530 - Calculation of Tube Heater Thickness and API STD 560 - Fired Heaters for General Refinery Service
core	A cylindrical sample taken from a borehole
coring tool	*aka* coring bit. An oil and gas industry term for a type of drill bit with a toroidal face used to produce a core (qv)
Coriolis effect	Deflection of an object due to the Coriolis force, a fictitious force rather similar to centrifugal force (qv)

Coriolis flow meter	A mass flow meter based on the Coriolis effect (qv). Often such meters are used where highly accurate measurements are required, or for two-phase mixtures
corrected hydrotest pressure	Hydrotest pressure corrected for temperature. Defined formally in API Standard 521. Pressure-relieving and Depressuring Systems. Sixth Edition | January 2014
correcting element	*aka* control element. The part of a correcting unit (qv) which does the correcting, such as a control valve. Defined formally in BS1646-1 Symbolic Representation for Process Measurement Control Functions and Instrumentation Part 1: Basic Requirements and BS 1646-3:1984 Symbolic representation for process measurement control functions and instrumentation. Specification for detailed symbols for instrument interconnection diagrams
correcting unit	The 'business end' of a control loop, such as an actuated valve. Defined formally in BS1646-1 Symbolic Representation for Process Measurement Control Functions and Instrumentation Part 1: Basic Requirements and BS 1646-3:1984 Symbolic representation for process measurement control functions and instrumentation. Specification for detailed symbols for instrument interconnection diagrams
correction	"Action to eliminate a detected nonconformity. Corrections concern products. A correction may be, for example, reprocessing" according to EHEDG Glossary; also defined formally in BS EN ISO 9000 Quality management systems Fundamentals and vocabulary
corrective action	"Actions implemented to eliminate the cause of a detected nonconformity or other undesirable situation" according to EHEDG Glossary; also defined formally in BS EN ISO 9000 Quality management systems Fundamentals and vocabulary
corrective and preventive action	*aka* CAPA. Part of good manufacturing practices (qv), similar to corrective action (qv)
correlation coefficient	(symbol R) A number between -1 and +1 that represents the strength and direction of any linear relationship between change in two variables
corresponding states	The law of corresponding states is the equation of state for many gases below their critical temperatures, pressures and volumes i.e., their departure from ideality is consistent
corrosion	*aka* CRSN. Deterioration of metal caused by chemical (or electrochemical) means; defined formally in ASME BPE (American Society of Mechanical Engineers: Bioprocessing Equipment) and API RP 579 - Fitness for Service.

corrosion allowance	Additional metal thickness to allow for a component's corrosion in service. Defined formally in BS EN ISO 13705:2012/ISO 13705:2012(E) Petroleum, petrochemical and natural gas industries. Fired heaters for general refinery service, API STD 530 - Calculation of Tube Heater Thickness and API STD 560 - Fired Heaters for General Refinery Service
corrosion circuit	see **corrosion loop**
corrosion loop	*aka* corrosion circuit. A diagrammatic representation of the common process, corrosion environment, and materials of construction of all equipment within a section of plant loop susceptible to the same corrosion/damage mechanisms and rates of deterioration.
corrosion rate	Rate of reduction of material thickness as a result of corrosion. Measurement used is dependent on industry and service, but mm/yr. is the most common. Mutually Identical formal definitions in BS EN ISO 13705:2012/ISO 13705:2012(E) Petroleum, petrochemical and natural gas industries. Fired heaters for general refinery service, API STD 530 - Calculation of Tube Heater Thickness and API STD 560 - Fired Heaters for General Refinery Service
corrosion resistant material	A material of construction with a low corrosion rate under expected service conditions. Three different formal definitions in BS EN 1672-2:2005+A1:2009 Food processing machinery. Basic concepts. Hygiene requirements; BS EN ISO 14159:2008 Safety of machinery. Hygiene requirements for the design of machinery and NFPA 20: Standard for the Installation of Stationary Pumps for Fire Protection, 2019 Edition
corrosion resistant piping	Piping with a low corrosion rate under expected service conditions; defined formally in NFPA 24: Standard for the Installation of Private Fire Service Mains and Their Appurtenances, 2019 Edition
corrosion retarding material	A coating material that makes piping or plant more corrosion-resistant. Defined formally in NFPA 24: Standard for the Installation of Private Fire Service Mains and Their Appurtenances, 2019 Edition
corrosive	A material property; the tendency to damage any materials or living tissue with which it comes into contact; cf irritant
corrugated plate interceptor	*aka* API 421 separator, API separator, API oil water separator, API oil separator. A type of lamella separator with corrugated lamella plates commonly used to contain oil pollution. Most commonly an API oil separator (qv)

COSHH	see **Control of Substances Hazardous to Health Regulations 2002**
cosolvent	A substance added to a primary solvent in small amounts to increase the solubility of a poorly-soluble compound
Cosorb process	A process which produces high purity carbon monoxide by adsorption
cost and freight	see **Incoterms**
cost effective	Providing a good balance between price and function. Most organizations have internal standards based on industry/global metrics such as ROI (qv) and ROCE (qv)
cost index	A ratio of current cost of an item to its cost at an index date, used to allow adjustment for inflation of a historical cost estimate, such as the CEPCI (qv)
cost of poor quality	The quantification of tangible and intangible costs to the supplier of providing poor products or services
cost plus	see **cost reimbursable**
cost reimbursable	*aka* cost plus. A type of contract which compensates a contractor for work done and materials purchased at their cost, plus an agreed margin
cost, insurance, and freight	see **Incoterms**
cost/benefit analysis	An evaluation and comparison of an activity's likely costs and benefits in monetary or monetized units
cost/benefit ratio	The ratio of an activity's likely costs to its benefits in monetary or monetized units
COTS	see **commercial off the shelf**
Couette viscometer	see **cup-and bob-viscometer**
coulomb	(*symbol* C) The SI unit of electric charge
countercurrent flow	A regime where process streams pass through a unit operation in opposite directions. Generally, the optimal flow regime in a heat exchanger for theoretical energy recovery/heat transfer; cf cocurrent flow
countercurrent operation	Operation in countercurrent mode. Not the same as concurrent operation (qv)
coupled water motor driven proportioning pump	A foam concentrate proportioning pump driven by a water motor in the supply line; defined formally in NFPA 11: Standard for Low-, Medium-, and High-Expansion Foam, 2016 Edition
coupling	1. A device used to connect two shafts together at their ends for the purpose of transmitting power 2. A very short length of pipe or tube, with a socket at one or both ends that allows two pipes or tubes to be joined

CoV	see **coefficient of variation**
covalent	In chemistry, usually short for 'non-polar covalent', a form of chemical bonding with equal sharing of electrons which produces a molecule with no electrical charge separation between ends (poles); cf polar
covered floating roof tank	*aka* CFRT. A floating roof tank designed to exclude rainwater; defined formally in API MPMS 19.1 Manual of Petroleum Measurement Standards Chapter 19.1 Evaporative Loss from Fixed-roof Tanks; cf external floating roof tank, internal floating roof tank
cowboying	*(informal)* Acting in a reckless and/or unscrupulous manner
COx	A term for oxides of carbon, less popular than NOx or SOx. In practice, the x in COx is almost always 1 or 2, i.e., carbon monoxide or carbon dioxide, though other oxides are possible
CP	1. see **condensate polishing** 2. see **control panel**
Cp	see **process capability index**
CPA	see **critical path analysis**
CPI	1. see **chemical process industry** 2. consumer price index 3. corrugated plate interceptor
Cpk	A measure of a producer's capability to make a product within customer's range of tolerances; cf process capability index
CPL	see **caprolactam**
CPLG	*(drawing notation)* Coupling
CPM	see **computational pipeline monitoring**
CPQRA	Chemical process quantitative risk assessment/analysis
CPT	Carriage paid to; see **Incoterms**
CPVC	Chlorinated polyvinyl chloride
CR	see **control room**
cracker	A thermal reactor which 'cracks' long chain petrochemicals into short chain ones
cracking	Making long chain petrochemicals into short chain ones
cracking furnaces	*aka* pyrolysis furnaces. Commonly used to produce petrochemicals such as ethylene and vinyl chloride monomer from longer-chain feedstocks using a variety of thermally driven processes
cracking severity	The temperature used for thermal cracking (qv)
cracklike flaw	A flaw that looks like a crack, even if it isn't one; defined formally in API RP 579 - Fitness for Service

cracks	Undesired sharp-edged discontinuities far longer than they are wide in a rigid material, such as those caused by linear ruptures; defined formally in ASME BPE (American Society of Mechanical Engineers: Bioprocessing Equipment); cf gouge
crane	A type of machine, generally equipped with a hoist rope, wire ropes or chains and sheaves, that can be used both to lift and lower materials and to move them horizontally
craneage	Use of a crane
crater	Reduced weld bead end depth; defined formally in ASME BPE (American Society of Mechanical Engineers: Bioprocessing Equipment)
crater crack	Cracks in a crater (qv). Defined formally in ASME BPE (American Society of Mechanical Engineers: Bioprocessing Equipment)
cream separator	see **centrifugal separator**
creation phase	Design documentation stage; defined formally in BS EN ISO 10209:2012 Technical product documentation - Vocabulary - Part 1: Terms relating to technical drawings: general and types of drawings
creative reuse	see **upcycling**
creep	Polycrystalline material deformation as a result of being held for extended periods under stress at high temperature. Potential issue in the steels in very high temperature equipment including reformers, boiler and furnace tubes. Various formal definitions can be found in ASME BPE (American Society of Mechanical Engineers: Bioprocessing Equipment) and API RP 579 - Fitness for Service
creep rupture	A condition where creep leads to gross failure. A concern in steels in very high temperature equipment. Defined formally in API RP 579 - Fitness for Service
Crescent wrench	A proprietary eponym for an adjustable spanner or monkey wrench (qv)
crevice	"Any cavity that can harbor or shelter contaminants such as microbiological cells, resulting from improper hygienic design or from damage of material, such as cracking, corrosion or wear", according to the EHEDG Glossary. There are different formal definitions in BS EN 1672-2:2005+A1:2009 Food processing machinery. Basic concepts. Hygiene requirements and BS EN ISO 14159:2008 Safety of machinery. Hygiene requirements for the design of machinery
crevice corrosion	Major corrosion mechanism associated with crevices within and without process equipment.

CRG process	see **catalytic rich gas process**
cricondenbar	The maximum pressure at which two phases can coexist
cricondentherm	The maximum temperature at which two phases can coexist
criteria pollutant	Air pollutants subject to US Federal air quality standards for maximum allowable concentrations, e.g., carbon monoxide, lead, nitrogen dioxide, ozone, particulate matter and sulfur dioxide
critical bond	Cross bonding (qv) from one pipeline to another allowing their common cathodic protection
critical control	A system or process (procedural or engineered) intended to prevent people from being fatally injured
critical control point	"A step at which control can be applied and is essential to prevent or eliminate a food safety hazard or reduce it to an acceptable level", according to the EHEDG Glossary
critical damping	Just damped enough, (neither overdamped or underdamped); having a damping ratio (qv) of 1
critical defect	In the context of pipelines (and possibly wider), an identified defect requiring immediate attention
critical dilution rate	The maximum dilution rate (qv) which allows steady state in a bioreactor
critical examination	A systematic (some involved might say laborious) design review technique, e.g., HAZOP (qv)
critical exposure temperature	*aka* CET. The lowest reasonably foreseeable design temperature; defined formally in API RP 579 - Fitness for Service
critical flow	see **choked flow**
critical linear damping	see **critical damping**
critical mass	1. The smallest amount of fissile material capable of sustaining fission 2. By analogy, used in management speak (qv) to describe the point at which an organization becomes self-sustaining, or more broadly any other threshold point
critical moisture content	The moisture content at the point in drying when constant rate drying (qv) ends
critical path analysis	A way to identify the chain of events which constrains the end point of a program of activities; used to analyze and optimize scheduling the tasks which form the elements of a project; cf program evaluation and review technique
critical point	Most commonly refers to the 'liquid-vapor critical point', the point on a phase diagram where a liquid and its vapor can coexist. In US math, also used to mean a point of inflection on a curve

critical pressure	The minimum pressure required to liquefy a gas at its critical temperature
critical properties	The properties of substances at their critical point (qv)
critical review	A consistency check, in the context of a life cycle assessment (qv); defined formally in EN ISO 14040:2006 Environmental management-- Life cycle assessment - Principles and framework
critical speed	*aka* resonance speed. 1. Generically, for rotating machinery, the speed at which natural resonant frequency of the assembly becomes equal to the operating speed. Defined formally in different contexts in ISO 3977 Gas turbines - Procurement - Part 3: Design requirements, EN ISO 10437 Petroleum, petrochemical and natural gas industries - Steam turbines - Special-purpose applications and EN ISO 13709:2003 Centrifugal pumps for petroleum, petrochemical and natural gas industries. 2. In the context of ball mills, the speed where the balls do not cascade, but stick to the shell due to centrifugal forces
critical temperature	The temperature above which a gas cannot be liquefied by pressure alone
critical velocity	The velocity above which critical flow (qv) occurs
critical viscous damping	see **critical damping**
critical volume	The volume occupied by a given mass of gas (not necessarily a mole as some define this) at its critical point
criticality	The point at which the number of neutrons released from nuclear fission is sufficient to sustain the nuclear chain reaction
criticality accident	A dangerous, unintended and uncontrolled nuclear fission chain reaction which has two subtypes: 'process accidents', due to poor design; and 'reactor accidents', due to poor operation
CRO	see **control room operator**
crore	*aka* karor, koti. A traditional Indian number multiplier equal to tens of millions, or hundreds of lakhs (qv) (107)
cross bonding	see **bonding**
crossarm	A horizontal support member from which power lines are suspended
crosscurrent separation	An arrangement of multistage solvent extraction processes in which process streams pass through stages in series, whilst fresh solvent passes through them in parallel
crossflow filtration	*aka* tangential flow filtration. The operation of a filter, such as a membrane filter, with far more fluid passing tangentially to the filtration surface than through it, flushing away any cake buildup or concentration polarization (qv)

crossflow plate	Distillation column plates such as the bubble cap, sieve and valve types
crosslinkage	In polymers, crosslinkage is the process of joining polymer chains together, usually to modify their physical properties. Crosslinking tends to make liquids more viscous, and solids stronger and harder
crossover	Process fluid connections between the radiant and convection sections of a furnace or any piping connecting heater-coil sections with each other; defined formally in BS EN ISO 13705:2012/ISO 13705:2012(E) Petroleum, petrochemical and natural gas industries. Fired heaters for general refinery service, API STD 530 - Calculation of Tube Heater Thickness and API STD 560 - Fired Heaters for General Refinery Service
crown	The top of the internal bore of a pipe, directly above the invert (qv)
crown radius	Radius of the spherical cap of a torispherical (qv) head
CRSN	*(drawing notation)*; see **corrosion**
CRT	Control room technician; see **control room operator**
CRU	see **condensate recovery unit**
crud	Unspecified dirt, a polite way of saying crap; qv Chalk River unidentified deposits
crude diet	see **crude slate**
crude distillation	The atmospheric distillation (qv) of crude oil. The first step in oil refining (after the desalter (qv))
crude distillation unit	*aka* CDU, atmospheric distillation unit, crude unit, pipestill. The unit in which a first crude oil distillation takes place at around ambient pressure
crude furnace	Preheater for a crude distillation unit (qv)
crude oil	*aka* petroleum. A liquid fossil fuel, and the starting point for the petrochemicals industry. The NFPA define the pleonasm 'crude petroleum' formally in NFPA 30: Flammable and Combustible Liquids Code, 2018 Edition which includes a flashpoint stipulation
crude slate	*aka* crude diet. The mix of different crude grades that the refinery is running; cf product slate
crude unit	see **crude distillation unit**
crusher	Generically, a machine which reduces large rocks in size. Specifically, in the coal industry, large rocks go into a breaker, the product of breaking into a crusher, and the product of crushing into mills or grinders if required
cryodesiccation	see **freeze drying**

cryogenic distillation	Distillation occurring at very low temperatures, such as the Linde process (qv)
cryogenic fluid	A cold fluid. It might be defined as having a very low boiling point, or just as needing to be produced or stored at low temperatures (formal definitions differ as to what constitutes a low temperature). There are two different formal definitions along these lines in NFPA 30: Flammable and Combustible Liquids Code, 2018 Edition and NFPA 86: Standard for Ovens and Furnaces, 2019 Edition
cryogenic process	A process involving cryogenic fluid (qv)
cryogenic pump	*aka* cryopump. A type of high vacuum pump which condenses gases as solids to remove them
cryogenic service	In flare stack systems, defined as handling materials at 4 °C or less, which represents the 'hot end' of cryogenic. Defined formally in API RP 537 Flare Details for Petroleum, Petrochemical, and Natural Gas Industries
cryophilic bacteria	*aka* psychrophilic bacteria. Bacteria that can grow and reproduce at low temperatures (<10 °C); cf mesophilic bacteria, thermophilic bacteria
cryopump	see **cryogenic pump**
crystal bright	A description of low turbidity in beer
crystallization	Formation of highly molecularly ordered solids (crystals) from solution, by freezing or by deposition from a gas
crystallizer	A machine used for controlled industrial crystallization
crystallizing fluid	A fluid in which crystals might form. Defined formally in API RP 682 - Pump Seals
CS	1. see **carbon steel** 2. Cast steel
CSC	*(drawing notation)* Car seal closed
CSCS	*aka* ticket. A UK site worker certification scheme
CSM	see **certified scrum master**
CSO	*(drawing notation)* Car seal open
CSP	Concentrated solar power
CSR	see **coke strength after reaction**
CSST	Critical steady state temperature
CSTR	see **continuous stirred tank reactor**
CSw	*(drawing notation)* Concentric swage
CT	1. see **computed tomography** 2. see **conditioning tower**
CTR	*(drawing notation)* Center

cubic average boiling point	*aka* CABP. One of five ways used in oil refining to express average boiling point of a mixture of hydrocarbons (the others being molal, weight, volume, and mean)
cubic feet per minute	US customary units (qv) measure of gas flow. Might be actual cubic feet per minute (qv) or standard cubic feet per minute (qv)
cuchasa	see **filter mud**
CUFLA	*(jocular)* Completely unnecessary five letter acronym; cf TLA, ETLA
culinary steam	*aka* clean steam, food steam. Steam which is injected into food products or can find its way into them, so must be food grade, free of additives and other contaminants; defined formally in 3A Standard 609-02
culvert	A tunnel carrying a watercourse or sewer under an obstacle, or the bed of such a tunnel; cf trench
cumene	Isopropyl benzene
cumene process	*aka* cumene-phenol process, Hock process. A process which converts benzene and propylene via cumene into phenol and acetone using atmospheric oxygen
cumene-phenol process	see **cumene process**
cup-and-bob viscometer	Quality assurance (qv) lab equipment for measuring viscosity
curb valve	The underground shutoff valve in a natural gas service line to a building
curing	1. The process of hardening a polymer (by crosslinking) or concrete (by controlling moisture content) 2. A food preservation process
current load	see **environment load**
current-to-pressure transducer	see **I/P transaducer**
curtain area	The area of a discharge opening in a pressure relief valve (qv); defined formally in API RP 520 P1 7th Edition, January 2000 Sizing, Selection, and Installation of Pressure-Relieving Devices in Refineries; Part I - Sizing and Selection
curvature ratio	The ratio of bend curvature radius to pipe diameter
customary units	A relative of UK imperial units (qv) retained by the US, and used in US-influenced industries worldwide; cf US customary units, British units, **Box B1**

customer	The recipient (or potential recipient in some definitions) of a product or service. Some definitions require the customer to pay. Defined formally in BS EN 82079-1:2012 Preparation of instructions for use. Structuring, content and presentation. General principles and detailed requirements, BS EN ISO 9000 Quality management systems Fundamentals and vocabulary and IEEE 830-1998 - IEEE Recommended Practice for Software Requirements Specifications; cf consumer
customer satisfaction	The customer's view of how well its expectations have been met. Defined formally in BS EN ISO 9000 Quality management systems Fundamentals and vocabulary
customer satisfaction code of conduct	A set of written promises to customers by a supplier organization about how they aim to increase customer satisfaction. Defined formally in BS EN ISO 9000 Quality management systems Fundamentals and vocabulary
customer service	Controlled interaction of supplier with customer throughout product/service lifecycle. Defined formally in BS EN ISO 9000 Quality management systems Fundamentals and vocabulary
cut	The set of distillation products yielded between two temperatures, known as cut points (qv)
cut points	A temperature defining the boundary between two cuts in distillation
cut sectional view	A sectional view which also shows outlines of things not in the cutting plane. Defined formally in BS EN ISO 10209:2012 Technical product documentation - Vocabulary - Part 1: Terms relating to technical drawings: general and types of drawings
cut set	In fault tree analysis (qv), a set of basic events whose (simultaneous) occurrence ensures that the top event occurs
cut sheet	see **product specification sheet**
cutaway damper	A furnace damper which lets through a minimal airflow even in the fully closed position. Defined formally in NFPA 86: Standard for Ovens and Furnaces, 2019 Edition
cutoff criteria	Specification of the amount of material or energy flow, or the level of environmental significance associated with unit processes or product system, to be excluded from a study. Defined formally in EN ISO 14040:2006 Environmental management - Life cycle assessment - Principles and framework
cuts	see **cut**
cutting line	Line on a drawing indicating the cutting plane or sectioning axis position. Defined formally in BS EN ISO 10209:2012 Technical product documentation - Vocabulary - Part 1: Terms relating to technical drawings: general and types of drawings

cutting plane	The plane of the imaginary cut through something on a drawing. Defined formally in BS EN ISO 10209:2012 Technical product documentation - Vocabulary - Part 1: Terms relating to technical drawings: general and types of drawings
CV	1. *(drawing notation)* Check valve 2. *(drawing notation)* Control valve 3. see **flow coefficient**
CVD	see **chemical vapor deposition**
CVN	see **Charpy vee notch test**
CW	1. *(drawing notation)* Continuous weld 2. see **cooling water** 3. *(drawing notation)* Chain wheel
CWA	see **Clean Water Act**
CWP	1. Cold water pressure 2. see **construction work package**
CWR	see **cooling water return**
CWS	see **cooling water supply**
CYA	*(jocular, possibly offensive)* Cover your ass
cyanide process	*aka* MacArthur-Forrest process, gold cyanidation. The leaching of gold from low grade ore using cyanide
cyanobacteria	*aka* 'bluegreen algae'. A misnomer, as they are not always blue-green, and never algae; these are photosynthetic bacteria which can cause harmful algal bloom (qv)
cycle time	The time required to pass a volume of process fluid through a batch process, such as a depth filter, plus the time to make it ready for the next batch
cyclic service	A duty characterized by a cyclic nature, in which fatigue can become a significant potential cause of failure. Defined formally in API RP 579 - Fitness for Service and ASME Boiler & Pressure Vessel Code 2017 Edition VIII, Rules for Construction of Pressure Vessels Division 1
cyclone separator	*aka* cyclonic separator. A device in which particles are removed from a fluid stream through vortex separation. A hydrocyclone (qv) is used for liquids; a gas cyclone (qv) for gases
cyclonic separator	see **cyclone separator**
cylinder	In the context of LPG storage, a transportable storage container of less than or equal to 454kg of LPG; defined formally in NFPA 58: Liquefied Petroleum Gas Code, 2017 Edition
Czochralski process	A process used for growing large single crystals, especially those used in the semiconductor industry

D

D value	see **decimal reduction time**
D&D	Deactivation and decommissioning[28]
D&T	*(drawing notation)* Drill and tap (qv)
D&W	*(drawing notation)* Doped and wrapped (pipe)
D/A	see **duty/assist**
D/A/S	see **duty/assist/standby**
D/S	see **duty/standby**
D1160	Shorthand in the oil refining industry for the ASTM D1160-18 Standard Test Method for Distillation of Petroleum Products at Reduced Pressure; commonly used to determine if a product distilled at reduced pressure has the required range of boiling points; cf D86
D2887	Shorthand in the oil refining industry for the ASTM D2887-19AE02 Standard Test Method for Boiling Range Distribution of Petroleum Fractions by Gas Chromatography
d30	see **volume mean diameter**
d50	The diameter below which one will find 50 % w/w (qv) of particles; cf volume mean diameter
D86	Shorthand in the oil refining industry for the ASTM D86-12 Standard Test Method for Distillation of Petroleum Products at Atmospheric Pressure; commonly used to determine if a product distilled at atmospheric pressure has the required range of boiling points; cf D1160
DA	see **deaerator**
Da	1. see **Dalton**
	2. see **Damköhler numbers**
DAC	see **digital to analogue converter**
Dacron	see **Terylene**
DAF	see **dissolved air flotation**
daily discharge	Measurement of effluent stream, during a calendar day (24-hour period) for sampling purposes
Dalton	(*symbol* Da) Non-SI unit of atomic mass, being 1/12th of the mass of a C12 atom
Dalton's law	A law that states that the total pressure of an ideal mixture of gases or vapors is equal to the sum of the partial pressures of its components

---

[28] Not Dungeons and Dragons!

damage	*aka* harm (in certain circumstances). Slightly contentious, but a term for an induced loss of function or quality to environment, machinery or structures rather than to humans, which is always considered harm (qv); see **Box H1**
damage limiting construction	Construction elements intended to limit explosion damage. Defined formally in NFPA 30: Flammable and Combustible Liquids Code, 2018 Edition
damage mechanism	A phenomenon causing harmful changes; defined formally in API RP 579 - Fitness for Service
Damköhler numbers	(*symbol* Da) Dimensionless numbers, which relate rate of reaction to the rate of transport phenomena such as diffusion or convection. Karlovitz numbers (qv) are the inverse of Damköhler numbers
dampening	see **damping**
damper	1. A shock absorber; defined formally in BS ISO 2041:2018 Mechanical vibration, shock and condition monitoring. Vocabulary
2. A butterfly damper is an adjustable plate in a flue similar to a butterfly valve which controls furnace pressure balance or draft, defined formally in BS EN ISO 13705:2012/ISO 13705:2012(E) Petroleum, petrochemical and natural gas industries. Fired heaters for general refinery service and API STD 560 - Fired Heaters for General Refinery Service.
3. A guillotine (qv) damper is akin to a gate valve and is used for isolating duties, as with a fire damper (qv); cf louvre damper
4. see **pulsation damper**
damping	*aka* dampening.
1. Used in the sense of shock absorption; defined formally in BS ISO 2041:2018 Mechanical vibration, shock and condition monitoring. Vocabulary
2. By analogy, suppressing oscillations in signal processing
damping ratio	A dimensionless group used to analyze how rapidly oscillations die down after a disturbance
dangerous substances	Any substances present at work that could, if not properly controlled, cause harm to people as a result of a fire, explosion or metal corrosion. Defined formally in DSEAR (Dangerous Substances and Explosive Atmospheres Regulations) 2002
DAP	Delivered at place, see *Incoterms*
DAQ	Data acquisition system
Darcy friction factor	*aka* Darcy Weisbach friction factor, resistance coefficient. The less commonly used friction factor (qv), four times the Fanning friction factor (qv)

Darcy Weisbach equation	An empirical equation which relates head loss due to friction to average incompressible fluid velocity
Darcy Weisbach friction factor	see **Darcy friction factor**
Darcy's Law	A law which is used to describe the flow of liquids through porous media, which states that flow is proportional to the pressure drop and inversely proportional to the fluid viscosity
DART	see **days away, restricted, or transferred**
dashpot	A type of mechanical damper (qv); defined formally in BS ISO 2041:2018 Mechanical vibration, shock and condition monitoring. Vocabulary
DAT	Delivered at terminal, see **Incoterms**
data	Facts used as the basis of reasoning. Defined in various ways depending on the context in BS ISO 2041:2018 Mechanical vibration, shock and condition monitoring. Vocabulary, BS EN ISO 9000 Quality management systems Fundamentals and vocabulary, and BS EN ISO 10209:2012 Technical product documentation — Vocabulary — Terms relating to technical drawings, product definition and related documentation
data logger	*aka* datalogger, data recorder. A simple, cheap electronic device which logs sensor data against time, and almost certainly nowadays has a facility for remote access to that data by telephone or IP technologies; cf SCADA
data quality	A term related to whether data is fit for purpose, defined formally in an environmental management context in EN ISO 14040:2006 Environmental management - Life cycle assessment - Principles and framework
data quality objectives	*aka* DQO. Statements describing the uncertainty that decision-makers will accept in data. Can be qualitative and/or quantitative
data recorder	see **data logger**
data sheet	see **datasheet**
database	An organized collection of data. Defined formally in BS EN ISO 10209:2012 Technical product documentation — Vocabulary — Terms relating to technical drawings, product definition and related documentation
datalogger	see **data logger**
datasheet	*aka* data sheet, spec sheet. A document summarizing key engineering data/specifications of an item of equipment, especially a pump
date of issue	The official release date of a document; defined formally in BS 8888:2017 Technical product documentation and specification

datum	A reference point, (such as the UK 'ordnance datum' which approximates mean sea level) which acts as a zero point from which other measurements are derived
datum elevation	In the context of net positive suction head (qv), the elevation relative to a datum (qv) point used for NPSH calculations. Defined formally in EN ISO 13709:2003 Centrifugal pumps for petroleum, petrochemical and natural gas industries
daughter	A nuclide formed by the radioactive decay of a different (parent) nuclide
dawai	Urdu for 'medicine'; used on the subcontinent to mean dosed chemicals or additives
daycare occupancy	A building occupancy in which four or more people are cared for by non-family members on a non-residential basis; defined formally in NFPA 30: Flammable and Combustible Liquids Code, 2018 Edition
days away, restricted, or transferred	*aka* DART. USA OSHA term for lost time incident (qv)
DBA	see **design basis accident**
DBB	see **double block and bleed**
DBE	see **design basis event**
DBI	Design basis incident; see **design basis event**
DBP	see **disinfection byproduct**
DC	see **direct current**
DCA	Design code allowable
DCS	see **distributed control system**
DCS operator	see **control room operator**
DDC	Direct digital control
DDP	Delivered duty paid, see **Incoterms**
DE	Dose equivalent
DEA	see **diethanolamine**
deacidification	The process of adding an alkali
Deacon process	A catalytic process which was once used to convert hydrogen chloride gas to chlorine gas; used in the manufacture of bleaching powder
deactivation	1. Generically, a reduction in activity (can be biological, chemical or mechanical-electrical). 2. Used specifically to refer to chemical removal of reactive or corrosive elements, normally oxygen from a fluid
dead area	see **dead space**

dead band	In the context of process control, the range of input values where the output is zero or unchanged i.e., there is no resultant control action. Also used extensively in alarm management to refer to a range around the set point in which the alarm state will not change
dead leg	see **deadleg**
dead load	A load which remains relatively constant over time. In structural engineering, the weight of all structure components including fireproofing; cf live load
dead man's handle	see **dead man's switch**
dead man's switch	*aka* dead man's handle. A hold to run control device (qv) that if released will return to its original (safe) position; cf enabling device
dead space	An "internal section, area or space in equipment wherein a product, ingredient or other extraneous matter may be trapped or retained and wherein the flow rate of agents used for cleaning, disinfection or rinsing is reduced or nil, resulting in accumulation of dirt and inefficacy of cleaning, disinfection and rinsing" according to the EHEDG Glossary. There is a slightly different formal definition in BS EN 1672-2:2005+A1:2009 Food processing machinery. Basic concepts. Hygiene requirements and BS EN ISO 14159:2008 Safety of machinery. Hygiene requirements for the design of machinery; cf crevice, deadleg
dead time	The undesirable delay between an outgoing controller output signal and the first measured change in the controlled variable
dead wood	*(informal)* Underperforming employees, often with long service
deadheading	*aka* churn, shutoff, no flow (qv). Operating a pump against a closed discharge line. In the case of a centrifugal pump (qv), this will give the maximum discharge head of the pump; in the case of a positive displacement pump (qv) this will result in the destruction of the pump or discharge line
deadleg	*aka* dead leg. An area in a vessel or piping which acts as a trap for dirt or fluid. How deep this must be to be of concern is a matter of controversy[29]. Defined formally in ASME BPE (American Society of Mechanical Engineers: Bioprocessing Equipment); cf crevice, dead space

---

[29] At one extreme, the concept is used to justify some terrible design decisions in hygienic engineering to avoid inconsequential 'deadlegs'; at the other extreme, neglect of serious deadlegs has caused serious process incidents in all industries.

deadweight	1. The weight of an unconscious or dead person 2. In shipping, the maximum allowable weight of cargo, crew, fuel, passengers, provisions and water on a vessel 3. A weight used to test pressure gauges 4. *aka* pure mass, lumped mass. A perfectly rigid mass, as defined in the context of vibration in BS ISO 2041:2018 Mechanical vibration, shock and condition monitoring. Vocabulary
deadwood	Things inside a tank which add to or subtract from its working volume; cf dead wood
deaeration	The removal of dissolved gases (not necessarily air) from water
deaerator	*aka* DA, degasifier. A device for deaeration: a common type uses steam to remove O_2 and CO_2 from boiler feedwater. There are also vacuum types
dealkalization	Reduction of water alkalinity by ion exchange
dealkylation	Removal of alkyl groups
Dean number	(*symbol* Dn or D) A dimensionless number to do with flow in curved channels, the product of the Reynolds number (qv) and the square root of the curvature ratio (qv)
death phase	In batch culture, the phase following the stationary phase when the nutrient supply is exhausted and organism numbers start to drop; cf lag phase, log phase, stationary phase
Deborah number	(*symbol* De) A dimensionless number to do with rheology, most useful in working with the time–temperature superposition principle, i.e., using a short experiment at a higher temperature to predict the properties of a material at a lower temperature[30]
debottlenecking	Finding the limiting step in a chain of unit operations by analysis, (the 'bottleneck' which restricts whole-process output), and removing it
debris	Generally, solid wastes from either natural or man-made sources deposited randomly on land and water; cf crud
debutanizer	In the oil and gas industry, a continuous distillation column which removes butane and lighter components
deca-	(*symbol* da-) *aka* deka. SI prefix x 10 (or 10¹)
decaffeination	Removal of caffeine
decanter	1. see **bowl centrifuge** 2. see **liquid-liquid separator**
decanting	Removing supernatant (qv) liquid from a settlement process. The subnatant (qv) may be liquid, solid or a mixture of the two

---

[30] Named after a biblical figure who noted that, on God's timescale, even mountains were a fluid

decationization	Ion exchange of cations for hydrogen ions from a strong acid cationic resin
decay	1. Generically, a change in something over time
2. Most commonly, the spontaneous transformation of one radioactive nuclide into another radionuclide, or into another energy state of the same nuclide with the emission of one or more particles or protons
3. May also refer to biological decomposition (qv), or reduction over time of oscillation (qv)
decay rate	The ratio of activity to the number of radioactive atoms of a particular species
decay ratio	The ratio by which oscillation is reduced during one complete cycle, i.e., ratio of successive peaks above and below the final steady-state value
decay store	see **buffer store**
decay time	The time required for a quantity to fall to $1/e$ times the original value
dechlorination	The removal of chlorine, whether that is removing chlorine atoms from a molecule, or removing chlorine from water
deci-	(*symbol* d-) SI prefix x 1/10 (or x 10^{-1})
decibel	(*symbol* dB) One tenth of a bel (qv)
decimal places	The number of places from the decimal point to the lowest non-zero digit in a number; cf (the too-often confused) significant figures
decimal reduction time	(*symbol* D) *aka* D value. The time required to kill 90% of a population, (a one log reduction) at a specific temperature
decision tree analysis	A management tool used to analyze the likely outcomes of a complex set of decisions in the light of possible alternative scenarios
deck	In the context of storage tanks in oil and gas, the buoyant part of a floating roof; defined formally in API MPMS 19.1 Manual of Petroleum Measurement Standards Chapter 19.1 Evaporative Loss from Fixed-roof Tanks
deck fitting	A device that seals a functional deck (qv) penetration; defined formally in API MPMS 19.1 Manual of Petroleum Measurement Standards Chapter 19.1 Evaporative Loss from Fixed-roof Tanks
deck plate	see **deckplate**
deck seam	The welded, bolted or clamped joint between adjacent sheets or panels in a floating deck; defined formally in API MPMS 19.1 Manual of Petroleum Measurement Standards Chapter 19.1 Evaporative Loss from Fixed-roof Tanks

decking	Floor plates or panels. Commonly made of galvanized mild steel or GRP (qv). Can be open mesh/grate type, or solid deckplate (qv)
deckplate	*aka* chequer plate, tread plate. Non-perforated floor plates, usually galvanized mild steel or GRP (qv); cf Durbar decking
decoking	Cleaning coke build-up inside furnace tubes from the combustion of hydrocarbon fuels, using steam and air
decolorisation	see **decolorization**
decolorization	*aka* decolorisation. Removing color, e.g., in sugar refining: decolorizing filtered raw liquor with an adsorbent such as bone char or GAC
decommissioning	The formal controlled shutdown, dismantling and decontamination of a process plant
decomposition	Breaking down something larger and more complex into smaller, simpler substances
decontamination	Removal of contamination (qv)
decrosslinking	Alteration of a polymer structure by removal of crosslinkage. An aggressive chemical treatment is usually required for this in the case of thermosetting polymers
deep dose equivalent	The dose equivalent (qv) derived from external radiation at a depth of 1 cm in tissue
deep shaft process	An economically non-viable sewage treatment process[31]
deep well pump	A pump which operates beneath the surface of the fluid; cf sump pump
deethanizer	A continuous distillation column in which ethane and other light components are removed overhead, leaving a de-ethanized gas stream
defect	1. Most commonly, a contractual term for non-performing equipment requiring repair/remediation 2. In equipment inspection, a catch-all term for any lower than expected equipment integrity found 3. Imperfections, nonconformities or discontinuities which affect performance. Defined in a quality context in BS EN ISO 9000 Quality management systems Fundamentals and vocabulary, and in the context of hygienic pipes and vessels in ASME BPE (American Society of Mechanical Engineers: Bioprocessing Equipment), where it is referred to as a discontinuity (qv)
defective	Unserviceable due to defects

---

[31] One of several non-viable processes developed by ICI (once one of the world's biggest chemical companies) after it lost its way

defects liability period	*aka* DLP, defect liability period (deprecated). The period after plant handover during which the construction company can be called back to site to fix latent defects, not apparent at the time of handover, at no cost to the client
defects per million opportunities	*aka* DPMO, nonconformities per million opportunities (NPMO). A process performance metric used in process improvement
defined medium	*aka* chemically defined medium, serum free medium, synthetic medium. A microbial growth medium made from defined quantities of pure chemicals, not including ill-defined and variable biological products such as fetal calf serum
deflagration	*aka* flash fire. A rapid fire or subsonic explosion. Slightly variable formal definitions in API Standard 521. Pressure-relieving and Depressuring Systems. Sixth Edition	January 2014, NFPA 15 Standard for Water Spray Fixed Systems for Fire Protection and NFPA 654: Standard for the Prevention of Fire and Dust Explosions from the Manufacturing, Processing, and Handling of Combustible Particulate Solids, 2017 Edition; cf conflagration, detonation
defoamer	Substance that can be used to prevent or control foaming, or destroy foams; cf antifoam
deformation	A shape change
DEFRA	see **Department of Environment, Food and Rural Affairs**
degasifier	A device which removes carbon dioxide from the effluent of hydrogen cation ion exchangers
degassing	The release of dissolved, absorbed, or adsorbed gases from a liquid or solid
degradation	Chemical decomposition (qv) or a reduction in process performance or safety measures
degradation factor	In a process safety context, a situation, condition, defect or error that compromises the function of a main pathway barrier, through either defeating it or reducing its effectiveness
degree	A non-SI unit of measurement of angle equivalent to $\pi/180$ radians
degree Celsius	(*symbol* °C) The SI unit of temperature
degree Fahrenheit	(*symbol* °F) Non-SI US Customary Unit (qv) of temperature
degree of hydration	see **hydration**
degrees Hazen	Units of color (qv) of a water sample
degrees Lovibond	Units of color (qv) used in the brewing industry

degrees of freedom	The minimum number of dimensions across which a system may vary, or to put it another way, the number of values necessary to define its state; defined formally in the context of vibration monitoring in BS ISO 2041:2018 Mechanical vibration, shock and condition monitoring. Vocabulary; cf statistical degrees of freedom
degrees SRM	Standard reference measure degrees; a color (qv) measurement scale used in the brewing industry
degrees Twaddell	A measure of the specific gravity of a solution with an 19th century proprietary hydrometer, for some reason still current as a measure of the strength of caustic soda solutions. 0 °Tw is 1000 kg/m^3, 2 °Tw is 2000 kg/m^3
degritting	A preliminary stage in sewage treatment; the removal of small, dense (usually mineral) particles
dehumidification	Decreasing the humidity (water content) of a gas; part of air conditioning
dehydration	1. A drying process 2. A chemical reaction involving loss of water from a molecule 3. A dangerous state in living creatures resulting from lack of water
dehydrochlorination	A chemical reaction involving removal of HCl molecule from a low molecular weight chemical compound or a polymer
dehydrogenation	A chemical reaction involving loss of hydrogen from a molecule
deinventory	A fancy way of saying 'to empty' an item of equipment
deionisation	see **deionization**
deionised water	see **deionized water**
deionization	*aka* deionisation. Removal of dissolved ionizable substances and silica from a solution, usually by ion exchange
deionized water	*aka* deionised water. Water which has been subjected to deionization; defined formally in a hygienic context in ASME BPE (American Society of Mechanical Engineers: Bioprocessing Equipment)
del factor	In a sterilization process, the ratio of the number of viable contaminating microorganisms to the initial number
delamination	A mode of failure of materials involving separation into layers; defined formally in ASME BPE (American Society of Mechanical Engineers: Bioprocessing Equipment)
delay	In contracts, infractions of the determined schedule for mechanical completion and initial acceptance on capital projects (i.e., being late); cf dead time

delayed coking	The product of thermally cracking resid. (qv) by heating it in a furnace, then moving it to a large coke drum where cracking continues. Cracked vapors come out the top of the drum to be fractionated downstream, and petcoke (qv) is left in the drum as a byproduct
delayed neutrons	Neutrons released by fission products up to several seconds after fission which enable control of fission in a nuclear reactor
delayed start up insurance	*aka* DSU, advance loss of profit (ALOP) insurance. A type of insurance intended to ensure that debt may be serviced and anticipated profit received, even if accidental physical damage to plant interferes with the insured's business operations
deliquescence	The process by which a hygroscopic substance absorbs moisture from the atmosphere, and dissolves in the absorbed water to form a solution
deliquoring	*aka* dewatering (qv). 1. Generally, removal of water without changing its phase 2. Specifically, removing water from filter cake, especially in filter dryers
deliverables	Items delivered under a contract; in a plant design context, mainly drawings
delivered at place	see **Incoterms**
delivered at terminal	see **Incoterms**
delivered duty paid	see **Incoterms**
delivery pressure	The outlet pressure of a pump, blower or fan
delta	(*symbol* δ) 1. A differential value as in for example δP, pressure difference 2. A triangular pattern of three phase electrical connection, as in wye delta (qv)
deluge system	A system which rapidly deluges an area with large volumes of water; used widely in fire protection of various types of process equipment
deluge valve	An actuated valve which supplies water to all spray nozzles in a water spray fire protection system; defined formally in NFPA 15 Standard for Water Spray Fixed Systems for Fire Protection
demand	A process event which (by design) will cause a safety instrumented function (qv) to activate.
demarcation	In the context of hygienic design, a distinct area with a defined boundary; defined formally in **ASME BPE** (American Society of Mechanical Engineers: Bioprocessing Equipment)

demethanizer	A continuous cryogenic high pressure distillation column in which methane is removed overhead, leaving ethane and heavier hydrocarbons as a bottom product called natural gas liquids (qv)
demin.	Demineralization or demineralized (water)
demineralisation	see **deionization**
demineralization	see **deionization**
demister	see **mist eliminator**
demulsifier	see **emulsion breaker**
demurrage	Cost associated with delivery vehicle/vessel waiting time
denaturation	The process of irreversibly changing the structure of natural polymeric materials such as proteins so that they are no longer functional. NB denatured alcohol (qv) is not denatured in this sense, merely contaminated to discourage drinking; cf renaturation
denatured alcohol	Ethanol with poisonous, bad-tasting or nauseating additives intended to make it non-potable
dendrite	Characteristic tree-like structure of crystals in metals
denitrification	1. The conversion of nitrates and nitrites produced by nitrification (qv) into gaseous nitrogen 2. In oil refining, the removal of nitrogen compounds in oil products, chiefly by hydrotreating (qv)
dense gas	A gas with a specific gravity greater than air: 'heavier than air'
dense media	A fluid suspension of superfine particles (usually magnetite for recoverability) with a density controlled through concentration; used in minerals processing to separate ore from gangue (qv) by dense medium separation (qv)
dense media separation	see **dense medium separation**
dense medium bath	*aka* DMB. A static dense medium separation (qv) device
dense medium cyclone	*aka* DMC. A cyclonic dense medium separation (qv) device
dense medium separation	*aka* DMS, dense media separation, heavy medium separation, sink and float separation. A process using a suspension of finely divided minerals in water to separate ore into valuable mineral particles and gangue (qv) based on their different specific gravities
dense non-aqueous phase liquid	*aka* DNAPL. An environmental engineering term for a liquid denser than, immiscible with and insoluble in water
dense phase	A fourth state of matter (or fifth if you count plasma). A dense, highly compressible fluid that demonstrates properties of both liquid and gas formed above critical temperature and pressure

dense phase conveying	A method of pneumatic solids conveying where the solids are kept in a homogenous 'slug' to reduce the velocity of the solids and erosion on the conveying pipe; cf lean phase conveying
Denseveyor	Proprietary eponym for a dense phase conveying (qv) system
density	1. Generally, the mass per unit volume 2. Specifically, in a fire protection context, the volumetric flow of water application per unit of area; defined formally in NFPA 15 Standard for Water Spray Fixed Systems for Fire Protection; cf gravity
Denso tape	Proprietary eponym for a petrolatum/wax impregnated tape used for underground pipe protection
Densowrap	see **Denso tape**
dent	A large smooth-bottomed depression in a pipe or vessel wall which does not reduce wall thickness, usually caused by pressure; defined formally in ASME BPE (American Society of Mechanical Engineers: Bioprocessing Equipment)
denudation	1. Physical, chemical, and/or natural factors that drastically reduce animal and plant life 2. Erosion or wear of the earth's surface
deoxygenation	The natural biochemical, or artificial chemical removal of dissolved oxygen from a liquid
Department of Environment, Food and Rural Affairs	*aka* DEFRA. The UK government department responsible for the natural environment, food and farming industries
departure area/entrance	In the context of working platforms and walkways, the place from where a ladder or stairs are accessed (ground level or intermediate platform); defined formally in BS EN ISO 14122 Safety of machinery. Permanent means of access to machinery. Working platforms and walkways
dependability	Ability to reliably work on demand; defined formally in BS EN ISO 9000 Quality management systems Fundamentals and vocabulary; see **Box A2**.
DEPG	Dimethyl ether polyethylene glycols
dephlegmator	A device which partially condenses a multicomponent vapor stream, allowing a better separation than a simple reflux condenser (qv)
depletion	*aka* resource depletion. The removal of non-renewable or renewable resources from the environment faster than they can be replenished
deposition	A phase change direct from vapor to solid; cf condensation

deprecated definition	A definition of a term which is less preferred or obsolete[32]
depression of freezing point	Decrease of the freezing point of a solvent caused by the addition of a non-volatile solute, such as common salt in water
DePriester charts	Nomograms of the vapor-liquid equilibrium ratios of a range of substances, (notably hydrocarbons) at different conditions of pressure and temperature
depropanizer	A continuous distillation column fed by the deethanizer (qv) bottom product in which propane is removed overhead, leaving a depropanized liquid stream
depth filter	*aka* granular bed filter. A nominal filter (qv) comprising a relatively deep bed of particles such as a sand filter; cf absolute filter, surface filter
depth of blanket	Level of sludge in the bottom of a secondary clarifier
depth of fusion	*aka* penetration. In welding, the depth from the surface to which fusion extends; defined formally in API STD 620 - Low Pressure Storage Tanks
depth of step	Distance from front to back of a step; defined formally in BS EN ISO 14122 Safety of machinery. Permanent means of access to machinery. Working platforms and walkways
derivative	1. A compound that can be imagined to arise from a parent compound by replacement of one atom with another atom or group of atoms 2. The product of differential calculus
derivative action control	In the context of process control, using controller output proportional to the derivative (qv) of the difference between set point and measured value. The 'D' of PID control (qv), with an anticipatory action
derived units	Strictly called SI derived units, these are units derived from the SI base units (second, meter, kilogram, ampere, kelvin, mole, and candela). There are of course other units derived from different base units, but they are not usually referred to by this shorthand term
dermal toxicity	Adverse health effects resulting from skin exposure to a substance
derrick	A type of lifting device (similar to a crane), or the steel trussed tower supporting the drilling apparatus in an oil rig
derrick floor	see **drill floor**

---

[32] Some of these are however still in common use. Terms mean what people think they mean, and not everyone has 'read the memo' telling them to use another term

derrick support	A steel trussed tower support for tall flare stack risers or drilling apparatus; defined formally in API RP 537 Flare Details for Petroleum, Petrochemical, and Natural Gas Industries
derrickhand	see **derrickman**
derrickman	*aka* derrickhand. In oil drilling, the person (of any gender) responsible for guiding the drill stand at the top of a derrick (qv)
DERV	Diesel Engined Road Vehicle, also a UK synonym for diesel (qv) fuel itself
DES	see **double extra strong**
desal.	Desalinated water
desalination	The removal of dissolved salts from a solution, especially the removal of salt from brackish or sea-water
desalter	1. Most commonly, a device which removes any salt water and sediment from crude oil, usually the first stage of refining 2. Used occasionally to describe brackish water desalination plant
descaling	*aka* scaling. Removing scaling (qv); defined formally in ASME BPE (American Society of Mechanical Engineers: Bioprocessing Equipment)
desiccant wheel	see **rotary heat exchanger**
desiccation	Extreme drying; cf freeze drying
design	The narrowest definition of the term is perhaps that defined in EN ISO 10437 Petroleum, petrochemical and natural gas industries - Steam turbines - Special-purpose applications, but see **Box D1.** and **Box S2.**
design and development	Processes that turn a statement of requirements into a detailed description of a product or plant; defined formally in a quality management context in BS EN ISO 9000 Quality management systems Fundamentals and vocabulary
design basis	1. Generally, information compiled to allow design at any stage 2. Strictly, a short document produced early in design which defines the broad limits of the FEED (qv) study, including such things as operating and environmental conditions, feedstock and product qualities, and the acceptable range of technologies. Often used as a living document through the design process to document the assumptions, inputs and requirements that form part of the design output; cf basic engineering design data for a sector-specific exception.
design basis accident	*aka* DBA; see **design basis event**
design basis event	*aka* DBE, maximum credible accident, design basis incident, design basis accident. A term for various worst-case scenarios used in nuclear reactor design

design basis incident	*aka* DBI; see **design basis event**
design brief	see **brief**
design certification	Independent third-party confirmation that a design complies with requirements; defined formally in NFPA 58: Liquefied Petroleum Gas Code, 2017 Edition

Box D1. Design

The ability to design is a natural human ability: designers imagine an improvement on reality as it is, think of a number of ways to achieve the improvement, select one of them, and then communicate it with those who will realize the plan.

Design is in essence the same process, whether we are designing a process plant, a vacuum cleaner, or a wedding cake. Process engineering design is simply a more specialized version of this ability but all types of design share some common characteristics.

All designers need to make sure that they are answering the right question and that their solution will thus be fit for the purpose for which it is intended. They need to consider the resource implications of their choices and whether their designs will be safe even if it not used exactly as intended. In engineering, those design choices can have life and death implications, and almost always involve significant financial commitments.

Like all designers, the engineer's potential solutions will include approaches to similar or analogous problems which they have seen to work, so the application of experience and professional judgement is crucial. Process plant design is an art and whilst its practitioners use science and mathematics, models and simulations, drawings and spreadsheets, these play a supporting role to professional judgement. Equally, if not more important are the heuristics, approximations, probabilities and workable approaches used by all experienced professional engineers.

Process plant design is system level design, and drawings are its best expression. However, the documents with which we transmit design intentions are just a means to the ultimate end of design - the improvement on reality itself.

I discuss this area in more detail in my book An Applied Guide to Process and Plant Design[33].

Related Terms
heuristics, systems engineering, task, problem, total design, partial design, process design

---

[33] Moran, S. (2019) *An Applied Guide to Process and Plan Design*, 2nd ed. Oxford: Elsevier

design codes	*aka* design standards. In some jurisdictions, such as the USA, the design of certain items must meet published codified minimum standards, *aka* design codes. There is controversy over whether such an approach produces better designs than one which allows more scope for exercising professional judgement
design and development	Processes that turn a statement of requirements into a detailed description of a product or plant; defined formally in a quality management context in BS EN ISO 9000 Quality management systems Fundamentals and vocabulary
design draughtsman	see **draughtsman**
design envelope	The definition of the full range of expected operating conditions, including transient and unsteady state conditions. The operating envelope (qv) and safe operating envelope (qv) are best specified to lie comfortably within the design envelope
design failure mode and effects analysis	*aka* DFMEA. A subtype of failure mode and effects analysis (qv), as applied to design
design flare capacity	Maximum flare design flow; defined formally in API RP 537 Flare Details for Petroleum, Petrochemical, and Natural Gas Industries
design for manufacturing	A type of design for x (qv): designing products in such a way that they are easy to manufacture
design for manufacturing and assembly	A type of design for x (qv): design for manufacturing (qv) plus consideration of ease of assembly
design for sustainability	A type of design for x (qv): designing products with sustainability (qv) in mind
design for the environment	A type of design for x (qv): design based on life cycle assessment (qv), which may have substantial overlap with design for sustainability (qv)
design for x	Where x is manufacturing, assembly, disassembly, sustainability, reuse, recycling, repairability etc. Design with some aim (x) in mind as well as/other than the usual cost, robustness, and safety metrics
design freeze	A quality assurance (QA) procedure in which no further modification is allowed to any of (or any specified part of) a design - which does not really mean in practice that a design cannot be changed under any circumstances - a design may change at any stage with approval by the project manager
Design Institute for Emergency Relief Systems	USA organization which develops and publishes guidance for the design and operation of pressure relief systems

design life	The time of operation to failure or before replacement is required which is assumed in the design basis (qv); defined formally in the context of heater tubes in API STD 530 - Calculation of Tube Heater Thickness
design meeting minutes	A record of decisions made in design meetings; defined formally in BS EN ISO 10209:2012 Technical product documentation — Vocabulary — Terms relating to technical drawings, product definition and related documentation
design metal temperature	Design temperature in the context of heater tubes; defined formally in API STD 530 - Calculation of Tube Heater Thickness; cf minimum design metal temperature
design of experiment(s)	*aka* DOE. A systematic method used to design investigations of the relationship between factors affecting a process and the output of that process; qv experimental design
design operating windows	aka DOW. Limits within which equipment can be safely operated without the risk of complete failure.
design philosophy	A written account of decisions made on how a number of common design problems and issues will be handled during a design. Best generated early in the design process, since stating the selection made at the start of the project prevents expensive redesign on another basis later (qv Chesterton's fence)
design pressure	The maximum (and arguably also minimum) pressure at design temperature (or temperatures) used in design calculations. This is usually significantly greater than maximum operating/working pressure. There are various formal definitions, not all of which accord with the definition here; found in API RP 520 P1 7th Edition, January 2000 Sizing, Selection, and Installation of Pressure-Relieving Devices in Refineries; Part I - Sizing and Selection, API Standard 521. Pressure-relieving and Depressuring Systems. Sixth Edition | January 2014, API RP 576 - Inspection of Pressure Relieving Devices, API STD 620 - Low Pressure Storage Tanks and NFPA 58: Liquefied Petroleum Gas Code, 2017 Edition
design qualification	see **V model**
design reference temperature	(*symbol* Θr) A characteristic design temperature, taking into consideration multiple factors; defined and tabulated in PD 5500:2018+A1:2018 Specification for unfired fusion welded pressure vessels
design reference thickness	(*symbol* E) A characteristic design thickness, taking into consideration multiple factors; defined and tabulated in PD 5500:2018+A1:2018 Specification for unfired fusion welded pressure vessels

design review	Ideally, an independent, impartial multidisciplinary process for evaluating the quality of a design
design standards	Can be synonymous with design codes (qv), or it may reflect documents intended as guidance for a design approach based in professional judgement
designer	Someone who designs a plant, and is therefore to some extent legally responsible for its quality. Defined formally defined in the UK within the Construction, Design and Management Regulations (2015) (qv)
desilt	Removal of sediment from water storage or soil conservation sites
desilting	Removal of silt from water bodies
desorption	The opposite of sorption (qv) (ab- or ad-)
destruction efficiency	In the context of flares, the fraction of the mass of incoming combustible fluid oxidized. Slightly different formal definitions in API Standard 521. Pressure-relieving and Depressuring Systems. Sixth Edition | January 2014 and API RP 537 Flare Details for Petroleum, Petrochemical, and Natural Gas Industries
destruction of microorganisms	"Irreversible physical or chemical damage to microorganisms to prevent them from surviving and multiplying", according to EHEDG Hygienic design of valves for food processing
desublimation	*aka* deposition. Changing directly from gas to solid without a liquid stage; the reverse of sublimation (qv)
Desulphovibrio desulfuricans	A common sulfate-reducing bacterium causing accelerated corrosion of pipes
desulfurization	Generally, the removal of sulfur. Important examples are the production of elemental sulfur from hydrogen sulfide in natural gas by the Claus process (qv), removal of sulfur as hydrogen sulfide from liquid hydrocarbon fuels by hydrotreating (qv), and removal of SOx in flue gas desulfurization (qv)
desuperheater	1. Generically, a device installed upstream of a steam-condensing heat exchanger to reduce steam temperature to its saturation point to optimize heat exchanger efficiency 2. Specifically, a heat exchanger associated with a heat recovery steam generator to control steam exit temperature; defined formally in API RP 534 – Heat Recovery Steam Generators; cf superheater
detached nappe	A nappe (qv) with air between the wall it falls over and its underside; cf clinging nappe

detached stable flame	A stable flame detached from the flare burner; defined formally in API RP 537 Flare Details for Petroleum, Petrochemical, and Natural Gas Industries
detachment	Hazard management by spatial separation; defined formally in NFPA 654: Standard for the Prevention of Fire and Dust Explosions from the Manufacturing, Processing, and Handling of Combustible Particulate Solids, 2017 Edition; cf segregation
detail design	Design of details by mechanical/structural engineers, rather than designing in detail led by process engineers. A subcategory of detailed design (qv), readily confused with it
detailed design	The third stage of process plant design development; defined formally in the context of product design development in BS EN ISO 10209:2012 Technical product documentation – Vocabulary – Terms relating to technical drawings, product definition and related documentation; see **Box S2**.
detailed engineering	see **detailed design**
detection limit	see **limit of detection**
detection systems	Arrangements of equipment which automatically identify the presence of material or a change in environmental conditions such as pressure, temperature or composition
detention and correctional occupancy	A term used to describe a building or part of one used for secure housing of detainees who will require assistance from others to escape in an emergency, defined formally in NFPA 30: Flammable and Combustible Liquids Code, 2018 Edition
determinand	A term used in water quality analysis for measures of pollutant species
determination	Conclusive investigation of the values of product characteristics in a quality management context; defined formally in BS EN ISO 9000 Quality management systems Fundamentals and vocabulary
determine	1. To establish conclusively, e.g., the concentration of a determinand (qv) in a sample 2. In contracts, to decide how much money is due after contract termination
detonation	Supersonic explosion; defined formally in API Standard 521, Pressure-relieving and Depressuring Systems. Sixth Edition | January 2014 and NFPA 15 Standard for Water Spray Fixed Systems for Fire Protection; cf conflagration, deflagration
detoxification	The removal of toxic materials
detraying booths	Enclosed areas where toxic product may be removed from drying trays without contaminating operators
Detritor	Proprietary eponym for a particular kind of grit trap (qv)

detritus	1. Generically, biological organic debris
	2. In sewage treatment, debris capable of being transported by moving water despite being denser than water
	3. Debris from the weathering of rocks
deuterium	(*symbol* D) A stable, heavy isotope of hydrogen; cf protium
developer	1. An applicant to the competent authority (qv) for authorization to develop (build on/alter) a site
	2. A chemical used in dye penetrant testing of weld quality
development consent	Consent from the competent authority (qv) to develop a site
development plan	A document covering planned development, often used as part of an application by a developer (qv); defined formally in BS EN ISO 10209:2012 Technical product documentation — Vocabulary — Terms relating to technical drawings, product definition and related documentation
development well	A well drilled in or adjacent to a proven part of a pool to optimize petroleum production; cf wildcat
deviation	1. In process control, the difference between measured value and setpoint. If sustained, this is called offset (qv)
	2. In quality management, a departure from specification
deviation permit	In quality management, permission to accept a deviation; defined formally in BS EN ISO 9000 Quality management systems Fundamentals and vocabulary
device	A component assembly which performs a duty; defined formally in BS EN ISO 10209:2012 Technical product documentation — Vocabulary — Terms relating to technical drawings, product definition and related documentation
devitrification	The crystallization of glass
dew point	*aka* condensation temperature. The temperature at which air is completely saturated with water vapor, below which drops of condensate form
dewar	see **vacuum flask**
dewar flask	see **vacuum flask**
dewar vessel	see **vacuum flask**
dewatering	1. Generally, the removal of bulk liquid from a slurry, differentiated from drying which removes liquid/moisture ab-/ad-sorbed by the solid phase, or removal of liquid (especially water) without changing its phase
	2. In water treatment, the next level of water removal after thickening (qv) in sludge treatment, and preceding drying (qv)
dewaxing	*aka* winterizing. Removing wax from base or vegetable oils to improve low temperature viscosity
DFAH	see **direct fired air heater**

DfE	see **design for the environment**
DfM	see **design for manufacturing**
DfMA	see **design for manufacturing and assembly**
DFMEA	see **design failure mode and effects analysis**
DfS	see **design for sustainability**
Dfx	see **design for x**
Dh	see **mean hydraulic diameter**
DHA	see **dust hazards analysis**
DHPT	see **down hole pressure and temperature transducer**
DHSV	see **downhole safety valve**
DI	see **ductile iron**
diafiltration	A type of ultrafiltration membrane separation process using a semi-permeable membrane to remove salts and microsolutes from a solution
diagram	A symbolic engineering drawing in single-plane form showing the interconnections between components; defined formally in BS 1553-1:1977 Specification for graphical symbols for general engineering. Piping systems and plant and BS EN ISO 10209:2012 Technical product documentation – Vocabulary – Terms relating to technical drawings, product definition and related documentation; cf drawing
dialysis	Separating dissolved molecules by the difference in their diffusion rates through a semipermeable membrane
diaphragm	1. Generally, a relatively thin sheet of material 2. In a hygienic context "A thin sheet of material forming a non-porous partition between the product and a measuring sensor or an actuator" according to the EHEDG Glossary
diaphragm pump	*aka* Blagdon pump (qv). A type of reciprocating positive displacement pump
diaphragm safety valve	Safety valve with a diaphragm to protect internal elements from relieved fluid
diaphragm valve	*aka* Saunders valve (qv). A valve sealed by a moveable diaphragm which has good hygienic qualities
diatomaceous earth	*aka* diatomite, kieselguhr, kieselgur. Fossilized diatoms used in water treatment as a filter aid, filter medium and oil absorbent. Used as an inert filler, abrasive etc. in other sectors; cf bentonite
diatomic	Composed of two atoms, of the same or different elements
diatomite	see **diatomaceous earth**
diauxic growth	Microbiological cell growth in two phases
DIB	1. Deisobutanizer 2. *(drawing notation)* Double isolation and bleed
die	An extrusion nozzle

dielectric heating system	A high frequency (at least 3 Mhz) induction heater; defined formally in NFPA 86: Standard for Ovens and Furnaces, 2019 Edition
DIERS	Design Institute for Emergency Relief Systems
diesel	*aka* DERV, distillate, gas oil. Fuel for diesel engines, made from fractions in oil refining, specified in Europe by EN 590:2013+A1:2017 Automotive fuels - Diesel - Requirements and test methods
diesel index	*aka* cetane number. A measure of ignition properties, the diesel equivalent of (cf) octane number
diethanolamine	*aka* DEA. A chemical used in the amine gas treating process (qv) to remove acid gases (qv) from hydrocarbons
differential distillation	Single stage batch distillation with constant rate heating and constant vapor removal and condensation
differential flow switch	A flow switch based on differential pressure; defined formally in NFPA 86: Standard for Ovens and Furnaces, 2019 Edition
differential pressure	The difference in pressure between two points, used to monitor things such as filter headloss; cf net pressure
differential scanning calorimetry	*aka* DSC. A thermal technique used for the analysis of substances in which heat is electrically added or removed to change the temperature, thereby allowing enthalpy changes due to thermal decomposition to be accurately studied
diffuser	Strictly, a device for increasing the pressure of a confined moving fluid by reducing its velocity. In practice, something which disperses a gas. Can be a ductwork component, or a perforated membrane used in effluent treatment aeration systems
diffusion	The movement of substances from areas of high concentration to areas of lower concentration as a net result of the random movement of atoms; cf effusion
diffusion coefficient	(*symbol* D) *aka* diffusion constant, mass diffusivity. Proportionality constant between the molar flux and the concentration gradient for a given pair of substances
diffusion constant	see **diffusion coefficient**
diffusion flame	A flame in which the fuel and oxidizer are not premixed. The oxidizer consequently combines with the fuel by molecular and turbulent diffusion, and flame speed is limited by these rates of diffusion
diffusion pump	A type of vacuum pump; defined formally in NFPA 86: Standard for Ovens and Furnaces, 2019 Edition
diffusion seal	see **buoyancy seal**
diffusivity	see **diffusion coefficient**

digester	Can refer, depending on context, to any kind of reactor (especially a bioreactor), but usually one of several disparate types which break something down. The term is used for the pressure cooker in the Kraft process (qv), and is most commonly applied to the bioreactor used in anaerobic (and less frequently aerobic) digestion of effluent and sludge
digital document management system	see **electronic document management system**
digital signal	A signal comprising a sequence of discrete values from a finite range; cf analogue signal
digital to analogue converter	*aka* DAC. A device which produces analogue signals corresponding with a digital input
digitization	Converting information from physical to digital format
dike	US English for bund (qv)
dilatant	*aka* shear thickening. A fluid property of apparent viscosity increasing with rate of shear (a time-independent non-Newtonian behavior)
dilatation	see **dilation**
dilation	Increased volume, or more broadly dimension
dilbit	Diluted bitumen
diluent	*aka* dilutant, dilution medium, filler or thinner. A substance used to dilute, perhaps to decrease viscosity
dilutant	see **diluent**
dilution	Making more dilute, by adding more dilution medium (qv)
dilution medium	Solvent, dispersant or carrier medium. In the case of flue gas treatment, the fluid dispersing the contaminant in the flue gas; defined formally in API RP 536 - NOx control on Fired Heaters at Refineries
dilution rate	(*symbol* D) In bioreactors, the incoming volumetric flowrate divided by culture volume; cf hydraulic retention time
dimension ratio	*aka* DR. The ratio of pipe OD (qv) to wall thickness, commonly used in the USA as a pressure rating or class for PVC pipe; cf standard dimension ratio
dimensional analysis	A form of algebra where the base units of the parameters are analyzed to check expected units of outputs and sense check calculations; more specifically, used in the application of Buckingham pi theorem (qv) to analyze scaleup (qv)
dimensional drawing	see **dimensioned drawing**
dimensional tolerance	see **tolerance of dimension**
dimensional value	see **basic dimension**

dimensioned drawing	A drawing marked with dimensions of the real-world counterparts of illustrated items; not guaranteed to be a scale drawing. Defined formally in BS EN ISO 10209:2012 Technical product documentation — Vocabulary — Terms relating to technical drawings, product definition and related documentation
dimensionless group	A combination of quantities having zero overall dimension, which in coherent units, can consequently be represented by a simple number; used in the scaleup of experimental data
dimensionless number	see **dimensionless group**
dimer	A molecule formed by combination of two smaller identical molecules
DIN	Deutsches Institut für Normung; German National Standards Institution. A German body producing metric based standards in common use in Europe; as opposed to the USA's ANSI/ASME US customary units (qv) based standards
dinosaur pump	see **nodding donkey**
dioxins	Highly toxic chemicals which may be produced unintentionally from industrial activities, especially the incineration of chlorinated waste
DIP	*(drawing notation)* Ductile iron pipe
direct acting controller	A controller with output proportional to input; the opposite of a reverse acting controller (qv)
direct acting solenoid valve	see **solenoid valve**
direct air preheater	*aka* direct APH. A heat exchanger which preheats combustion air directly with hot flue gases; defined formally in API STD 530 - Calculation of Tube Heater Thickness, API STD 560 - Fired Heaters for General Refinery Service and BS EN ISO 13705:2012/ISO 13705:2012(E) Petroleum, petrochemical and natural gas industries. Fired heaters for general refinery service
direct APH	see **direct air preheater**
direct catalytic oxidizer	A combustion system which heats VOCs to destruction temperature, then passes them through a catalyst without heating incoming VOCs with exhaust gases either by regeneration or recuperation. Defined formally in NFPA 86: Standard for Ovens and Furnaces, 2019 Edition; cf regenerative catalytic oxidizer, recuperative catalytic oxidizer
direct condenser	Barometric condenser used with vacuum systems
direct current	*aka* DC. An electrical supply with constant polarity; cf alternating current

direct discharger	A producer of industrial effluent who discharges effluent directly to environment (almost always after onsite treatment); cf indirect discharger
direct dryer	Dryers that expose solids to hot gas
direct fired air heater	*aka* DFAH. Heater type most commonly used for startup of FCC units
direct fired air makeup unit	Ambient pressure Class B fuel-fired heat utilization unit, which heats replacement process air; defined formally in NFPA 86: Standard for Ovens and Furnaces, 2019 Edition
direct fired external heating system	A heating system with separate combustion and heating chambers connected by a circulating fan; defined formally in NFPA 86: Standard for Ovens and Furnaces, 2019 Edition
direct fired heating system	Any heating system where combustion products pass through the heating chamber; defined formally in NFPA 86: Standard for Ovens and Furnaces, 2019 Edition
direct fired internal heating system	A heating system with a combined combustion and heating chamber; defined formally in NFPA 86: Standard for Ovens and Furnaces, 2019 Edition
direct fired vaporizer	A vaporizer in which the LPG vaporization heat exchanger is directly fired (exposed to flame); defined formally in NFPA 58: Liquefied Petroleum Gas Code, 2017 Edition
direct gas fired tank heater	A direct tank heater (qv) in which the LPG container surface is directly fired (exposed to flame); defined formally in NFPA 86: Standard for Ovens and Furnaces, 2019 Edition
direct ignition	Flare ignition by a high-energy source (rather than a pilot); defined formally in API RP 537 Flare Details for Petroleum, Petrochemical, and Natural Gas Industries
direct immersion electric vaporizer	An LPG vaporizer with an electric heating element directly immersed in the LPG; defined formally in NFPA 58: Liquefied Petroleum Gas Code, 2017 Edition
direct injection	Specifically in the context of LPG: the delivery of LPG fuel directly into a combustion chamber at high pressure through a fuel injector; defined formally in NFPA 58: Liquefied Petroleum Gas Code, 2017 Edition
direct injection variable output proportioning pump	A foam concentrate proportioning pump controlled by flowmeters for foam concentrate and water; defined formally in NFPA 11: Standard for Low-, Medium-, and High-Expansion Foam, 2016 Edition
direct loaded safety valve	A safety valve closed against fluid pressure solely by a weight or spring-based device. Defined formally in BS EN ISO 4126-1:2013+A2:2019 Safety devices for protection against excessive pressure. Safety valves

direct orthographic projection	*aka* orthogonal projection. One of the ways of representing a 3D object on a 2D drawing by projection. Defined formally in BS EN ISO 10209:2012 Technical product documentation — Vocabulary — Terms relating to technical drawings, product definition and related documentation; cf oblique projection
direct steam stripper	A stripping column with direct steam injection as the heat source instead of a reboiler
direct tank heater	A device which heats LPG in a container by heating part of the container directly; defined formally in NFPA 58: Liquefied Petroleum Gas Code, 2017 Edition; cf indirect tank heater
direct thermal oxidizer	*aka* afterburner. A combustion system which heats VOCs to destruction temperature without preheating incoming VOCs with exhaust gases; defined formally in NFPA 86: Standard for Ovens and Furnaces, 2019 Edition; cf regenerative thermal oxidizer, recuperative thermal oxidizer
dirty	Contaminated, or the opposite of clean (qv) (whatever that might mean! - see **Box C2**); defined formally in ASME BPE (American Society of Mechanical Engineers: Bioprocessing Equipment)
dirty heat transfer coefficients	Heat transfer coefficients in a fouled condition
disc and doughnut	see **disk and donut**
disc centrifuge	see **disk centrifuge**
disc filter	see **disk filter**
discharge coefficient	see **coefficient of discharge**
discharge devices	Foam water solution delivery devices; defined formally in NFPA 11: Standard for Low-, Medium-, and High-Expansion Foam, 2016 Edition
discharge head	Pump or compressor discharge pressure, expressed as an additional head of pumped fluid
discharge pressure	The total pressure available at a pump or compressor discharge point under given conditions; defined formally for firepumps in NFPA 20: Standard for the Installation of Stationary Pumps for Fire Protection, 2019 Edition
discharging	Releasing (e.g., effluent), usually to the environment
discoloration	1. In the context of welding, an undesirable change in surface color; defined formally in ASME BPE (American Society of Mechanical Engineers: Bioprocessing Equipment) 2. Also used for color change when nitrogen compounds in light distillates (e.g., naphtha, kerosene) get exposed to oxygen and sunlight

disconnecting means	Mechanisms for disconnecting an electrical circuit from power supply; defined formally in NFPA 20: Standard for the Installation of Stationary Pumps for Fire Protection, 2019 Edition
discontinuity	An interruption or heterogeneity in weldment structure (not always a defect); defined formally in ASME BPE (American Society of Mechanical Engineers: Bioprocessing Equipment)
discontinuous phase	The dispersed phase of a dispersion, such as an emulsion
discovery of a condition	Recognizing a condition for the first time
discrete settling	The removal of suspended solids in dilute solutions at their terminal settling velocities; cf hindered settling
disinfectant	"A chemical used after cleaning in order to reduce the population of viable microorganisms remaining on a surface", according to the EHEDG Glossary; cf sanitizer
disinfection	"The reduction, by means of chemical agents and/or physical methods, of the number of microorganisms in the environment, to a level that does not compromise food safety or suitability. Disinfection reduces microorganism population to a level acceptable for a defined purpose e.g., a level which is harmful neither to health nor to the quality of food" according to the EHEDG Glossary. Other formal definitions can be found in BS EN 1672-2:2005+A1:2009 Food processing machinery. Basic concepts. Hygiene requirements and BS EN ISO 14159:2008 Safety of machinery. Hygiene requirements for the design of machinery; see also **Box C2**
disinfection byproduct	*aka* DBP. Undesirable substances formed by the action of disinfectants on water such as THMs or bromates
disk and donut	*aka* disc and doughnut, disk and doughnut. A low efficiency mass transfer plate configuration used for highly fouling or viscous materials
disk and doughnut	see **disk and donut**
disk bowl centrifuge	see **disk centrifuge**
disk centrifuge	*aka* disc centrifuge, conical plate centrifuge, disk bowl centrifuge, disk stack separator. A centrifuge containing a stack of disks which give a short sedimentation path; used predominantly in hygienic applications with low solids feeds
disk filter	*aka* disc filter. A dewatering filter using concentric discs of cloth which rotate through a bath and then eject the cake with a puff of air
disk stack separator	see **disk centrifuge**
dislocation	A direct defect in a crystalline array of atoms

dispenser	A device which transfers LPG into mobile containers of various kinds; defined formally in NFPA 58: Liquefied Petroleum Gas Code, 2017 Edition
dispensing system	A dispenser (qv) and associated LPG storage containers; defined formally in NFPA 58: Liquefied Petroleum Gas Code, 2017 Edition. This term might also be used in any sector where relatively small, accurately measured quantities are discharged, either as a feed to the next stage of a process, or to a customer
dispersed gas flotation	see **induced gas flotation**
dispersed phase	The solute-like discontinuous phase in a colloid or other dispersion; cf dispersing medium
disperser	1. A device used to disperse material 2. In smelting, a modified burner used to react metal sulfides with oxygen to form oxides
dispersing medium	The solvent-like continuous phase in a colloid or other dispersion
dispersion	1. As a verb, combining immiscible substances so that one is scattered through the other reasonably evenly as discrete particles, drops or bubbles; or as a noun, the product of this process 2. Alternatively, atmospheric dilution of gaseous combustion products after release, as in the formal definitions in API RP 537 Flare Details for Petroleum, Petrochemical, and Natural Gas Industries, and API Standard 521. Pressure-relieving and Depressuring Systems. Sixth Edition | January 2014
dispersion modelling	A mathematical modeling technique which is employed to assess the impact of loss of containment on people, community and environment
dispersion number	The inverse of the Péclet Number (qv) used in studies of vessel mixing
displacement ton	see **long ton**
displacement transducer	A transducer which produces an electrical output proportional to spatial displacement, used in vibration monitoring; defined formally in BS ISO 2041:2018 Mechanical vibration, shock and condition monitoring. Vocabulary
display	The part of a device which is used to show system states and also often to interact with the device. Defined formally in BS EN 82079-1:2012 Preparation of instructions for use. Structuring, content and presentation. General principles and detailed requirements; cf human machine interface
disposal	see **waste disposal**
dispositioning authority	see **configuration authority**

dispute	1. Generally, disagreement about a contractual matter 2. In quality management, a disagreement arising from a consumer complaint, as defined formally in BS EN ISO 9000 Quality management systems Fundamentals and vocabulary
dispute resolution process provider	*aka* DRP provider. In quality management, an entity providing external dispute resolution service. Defined formally in BS EN ISO 9000 Quality management systems Fundamentals and vocabulary
dispute resolver	In quality management, an individual appointed by a DRP provider (qv) to assist dispute resolution. Defined formally in BS EN ISO 9000 Quality management systems Fundamentals and vocabulary
dissimilar metals corrosion	*aka* galvanic or bimetallic corrosion. The preferential corrosion of one of two metals in electrical contact when submerged in an electrolyte. May be accidental and undesirable, or intentional and desirable as in cathodic protection (qv)
dissociation	The reversible (and usually partial) splitting of ionic compounds into ions when dissolved in water, or complexes into components; cf ionization
dissociation constant	*aka* KD. A type of equilibrium constant (qv) in which the reaction is dissociation (qv)
dissolution	see **solvation**
dissolve	Incorporate into a liquid to form a solution
dissolved air flotation	*aka* DAF. A process used in water treatment for removing fine solids from water by attaching them to microbubbles of air, produced by dissolving air in water, then rapidly depressurizing; cf froth flotation, induced air flotation
dissolved oxygen	Usually refers to oxygen dissolved in water, measured in ppm
dissolved silica	see **monomeric silica**
dissolved solids	see **total dissolved solids**
distal cause	see **ultimate cause**
distance monitoring	*aka* remote monitoring. Remote (qv) rather than local (qv) monitoring. Defined formally in NFPA 20: Standard for the Installation of Stationary Pumps for Fire Protection, 2019 Edition
distance/velocity lag	*aka* transportation or transport lag. Process control time delay associated with the controlled medium and/or control signal moving from place to place
distillate	1. Generally, any liquid condensed from vapor in a distillation process 2. Specifically in the refining industry (and more widely in Australia), diesel (qv) fuel

distillate of agricultural origin	see **neutral spirit**
distillation	Separation of substances on the basis of their differences in volatility/boiling points
distillation column	*aka* fractionating column, tower. A vessel with a high aspect ratio used for continuous distillation or fractionation
distilled water	Water purified by distillation
distillery	A plant where the alcohol from fermented liquids is concentrated by distillation; defined formally in NFPA 30: Flammable and Combustible Liquids Code, 2018 Edition
distributed control system	*aka* DCS. Strictly, a process plant control system that uses multiple electronic controllers, typically located in a dispersed way throughout a process plant and connected by high-speed communication buses. They perform a similar job to SCADA (qv) and, outside of control/instrument engineering, PLC (qv), BPCS (qv), SCADA and DCS are used interchangeably to refer to the control system for a plant. Defined formally in BS EN ISO 10209:2012 Technical product documentation — Vocabulary — Terms relating to technical drawings, product definition and related documentation and BS 1646-4:1984 Symbolic representation for process measurement control functions and instrumentation. Specification for basic symbols for process computer, interface and shared display/control functions
distributed flush system	An arrangement designed to distribute flush fluid in a pump seal; defined formally in API RP 682 - Pump Seals
distribution	Moving products or substances outwards from a point of origin
distribution block	A connection between engine fuel line and rail, defined formally in NFPA 58: Liquefied Petroleum Gas Code, 2017 Edition
distribution box	see **possum belly**
distribution coefficient	*aka* distribution ratio. The ratio of concentrations of a substance in two immiscible solvents in contact at equilibrium; useful in solvent extraction. To a chemist, may be distinct from partition coefficient (qv) in as much as it takes into consideration ionized and unionized forms of a molecule and is therefore pH-dependent
distribution ratio	see **distribution coefficient**
distribution system	In a hygienic context, a system which delivers fluids from point of supply, such as a CIP kitchen (qv), to point of use (qv); defined formally in ASME BPE (American Society of Mechanical Engineers: Bioprocessing Equipment)

distributor	A device which distributes a flow of fluid in a controlled manner (commonly evenly), especially over the surface of a packed bed
distributor zone	see **grid zone**
distrometer	Equipment used to measure the size distribution and speed of raindrops
disturbance	In a process control context, an unwanted input outside the control of the system which affects process outputs
Dittus–Boelter correlation	An empirical correlation used to predict the Nusselt number (qv)
diurnal	Daily
diurnal variation	Recurrent variation in some parameter over the course of a day, such as for example the flow of sewage
diverter	see **rotary valve**
divider	A device used in process modelling to split components in an optimal fashion, which may well have no real-world counterpart
dividing wall column	A type of distillation column which can separate a mixture into three or more high-purity streams with lower footprint, capital investment, and energy requirements than a conventional column
divinylbenzene	*aka* DVB. A difunctional monomer used to crosslink polymers
DLP	see **defects liability period**
DLP + 1	*(jocular)* A wry term for a plant design life of the defects liability period (qv) plus one day
DM	Dry matter; see **dry solids**
DM water	Demineralized water
DMB	see **dense medium bath**
DMC	see **dense medium cyclone**
DMS	see **dense media separation**
DN	Diamètre nominal/nominal diameter/durchmesser nach norm; see **NB**
DNAPL	see **dense non-aqueous phase liquid**
DNB	1. Departure from nucleate boiling 2. Dinitrobenzene
DO	see **dissolved oxygen**
doctor sweet	In the context of gasoline and kerosene, very 'sweet' (as opposed to sour), as measured by the doctor test (qv)
doctor test	A test comprising shaking with sodium plumbite solutions, and adding powdered sulfur. If there is no dark precipitate of lead sulfide in the test, a distillate may be called doctor sweet (qv); defined formally in UOP 41-07 Doctor Test for Petroleum Distillates, a standard published by UOP LLC

document	The medium information is transmitted in; defined formally in BS EN 82079-1:2012 Preparation of instructions for use. Structuring, content and presentation. General principles and detailed requirements, BS EN ISO 10209:2012 Technical product documentation — Vocabulary — Terms relating to technical drawings, product definition and related documentation and BS EN ISO 9000 Quality management systems Fundamentals and vocabulary
document issue	A document version usually characterized by a date and or revision number; defined formally in BS EN ISO 10209:2012 Technical product documentation — Vocabulary — Terms relating to technical drawings, product definition and related documentation
document list	A formal inventory of documents; defined formally in BS EN ISO 10209:2012 Technical product documentation — Vocabulary — Terms relating to technical drawings, product definition and related documentation
document replica	A copy of a document which is sufficiently accurate for the desired purpose; defined formally in BS EN ISO 10209:2012 Technical product documentation — Vocabulary — Terms relating to technical drawings, product definition and related documentation
document revision	An approved document issue (qv); defined formally in BS EN ISO 10209:2012 Technical product documentation — Vocabulary — Terms relating to technical drawings, product definition and related documentation
documentation	A systematic collection of related documents; defined formally in BS EN 82079-1:2012 Preparation of instructions for use. Structuring, content and presentation. General principles and detailed requirements and BS EN ISO 10209:2012 Technical product documentation — Vocabulary — Terms relating to technical drawings, product definition and related documentation
DOE	see **design of experiment(s)**
DOF	see **degrees of freedom**
Dogbone	Proprietary eponym for a type of powerline vibration damper
dog collar	A clamp placed tightly around a drill collar (qv) suspended in the rotary table by drill collar slips (qv)
doghouse	*(informal)* An enclosure on the drill floor (qv) where the driller works and/or the drill crew take breaks

dogleg	In the oil and gas industry, a particularly crooked place in a well-bore (qv); commonly refers to a section of the hole that changes direction faster than desired, or more generally any deviation in pipework
dogman	see **banksman**
dogs	Mainly in oil and gas drilling, a term for a large number of things, most of which are devices which grip something. Although slips (qv) are sometimes known as dogs, the term dogs has a wider range of meanings than slips, including: 1. Piston-like devices driven into an engineered profile to lock components together 2. *aka* casing dogs. A lesser-used synonym of slips 3. *aka* casing dog. A fishing tool (qv), such as a bulldog (qv) 4. Short for dog collar (qv) 5. Short for dog shifts (night shifts) 6. *aka* pipe dogs. Devices used to hold and lift pipe, quite different from slips 7. *aka* rachet dog. A chain tensioner
domestic water	*aka* potable water, town mains. Municipal water supply
dominant dead time process	In a process control context, a process where dead time (qv) is greater than lag time (qv)
dominant frequency	The peak in a vibration frequency vs amplitude chart; defined formally in BS ISO 2041:2018 Mechanical vibration, shock and condition monitoring. Vocabulary
domino effect	1. Generally, any propagating or knock-on effect, or chain reaction (in an informal sense) 2. In process safety, a domino effect accident is one in which a chain of secondary effects is propagated after the failure of an item of equipment
domino groups	Groups of equipment with the potential to affect each other via the domino effect (qv); defined formally by COMAH Regulation 24
donkey pumper	see **nodding donkey**
donkey's dick	*(informal, potentially offensive)* 1. Generally used for a telescopic chute 2. In the water industry, an inclined pipe with a filter nozzle on the end, used to remove clarified water from a DAF unit 3. In the oil and gas industry it can mean a long tubular bladder in a pressure sensor, or a flexible guide on the end of logging tools
dope	1. In the context of textiles, a solution for spinning 2. In piping, a liquid thread sealant/lubricant

dose	1. *aka* exposure. A quantity of bioactive material or electromagnetic radiation to which a worker is exposed 2. A discrete quantity of process active material added to a process
dosing	*aka* metering. Reliably and accurately delivering a discrete quantity of process active material; cf dawai
dose equivalent	(*symbol* H) *aka* biological dose, equivalent dose. A measure of damage to living tissue from radiation exposure. The SI unit of dose equivalent is the sievert (qv)
dosing pump	*aka* additive pump, metering pump. A pump which reliably delivers a precisely measured quantity of process active material
dosing system	Dosing pump(s) (qv) and their associated instrumentation
DOT	USA Department of Transportation
double acting	A motor which is driven both forth and back (as opposed to spring return), especially in the case of valve actuators
double block and bleed	Usually an arrangement of two isolation (block) valves in series, with a bleed valve connected to the pipe between the two block valves, common in water and hygienic applications. However, API 6D defines this term as a single valve having the same functionality, operated with both sides under pressure (like a mixproof valve (qv)). Defined formally in API Specification 6D Specification for Pipeline and Piping Valves Twenty-fourth Edition |August 2014; cf **Box M2**.
double casing	In the context of pumps, having a casing separate from pumping elements; defined formally in EN ISO 13709:2003 Centrifugal pumps for petroleum, petrochemical and natural gas industries
double extra heavy	see **double extra strong**
double extra strong	*aka* DES, double extra heavy, extra extra strong (XXS), extra extra heavy (XXH). The heaviest grade of ANSI schedule (qv) steel pipe
double isolation and bleed	API 6D defines this identically to its definition of a double block and bleed (qv) valve, except that pressure is from a single source, rather than both ends. Defined formally in API Specification 6D Specification for Pipeline and Piping Valves Twenty-fourth Edition |August 2014
double jeopardy	A term used in HAZOP for circumstances where two things failing simultaneously is thought too unlikely to be worthy of consideration. Some say this 'get-out' is identified rather too frequently; all agree that it should never be applied to common cause failures (qv)

double layer	Normally refers to an electric double layer, where interactions at the interface between suspended particles or an electrode and electrolyte lead to oppositely charged particles becoming aligned
double mechanical seal	see **dual mechanical seal**
double pipe heat exchanger	A simple heat exchanger of concentric pipes, usually with cooling water in the annulus between pipes
double skinned tank/pipe	A tank or pipe having a twin wall to prevent escape of harmful contents
double submerged arc welded	*aka* DSAW. In the context of welded seam pipe, having a seam produced by at least one pass each on the inside and outside of the pipe. A method required by some pipe design standards, such as ASTM A-381 and API 5L
DOW	see **design operating windows**
Dow process	*aka* Dow's process. There are a number of Dow processes. Most notably, there is one to recover bromine from seawater electrolytically, and another to obtain magnesium chloride from seawater
down hole pressure and temperature transducer	*aka* DHPT, permanent downhole gauge (PDG). A device permanently installed in an oil or gas well to measure pressure and/or temperature
downcomer	1. Generically, a pipe which transports water or gas downwards from the top of a process unit; defined formally in API RP 534 – Heat Recovery Steam Generators; cf riser 2. More specifically, the part of a fractionation column that redirects liquid to the tray below
downcycling	*aka* recycling. Turning 'waste' into a product of lower quality/value; cf upcycling
downdraft	Downflow (qv), but only applied to gases; cf updraft
downflow	In a vertical column, flow downwards, from top to bottom
downflow velocity	Superficial downward velocity of fluid in a filter etc.; cf surface loading
downhole safety valve	*aka* DHSV, subsurface safety valve (SSSV or SCSSV). A device installed within an oil or gas well to isolate wellbore pressure and fluids in the event of an emergency or catastrophic failure at the surface
Downs process	A way of obtaining magnesium (and chlorine byproduct) by electrolysis of molten magnesium chloride, or sodium and chlorine from a molten sodium/calcium chloride mixture

downslope	In the context of automatic orbital welding, the part of the process during which current is reduced; and the product of this part of the process - a weld bead tapering towards minimal penetration; defined formally in ASME BPE (American Society of Mechanical Engineers: Bioprocessing Equipment)
downstream	1. Generically, later in the process 2. Used specifically in the oil and gas and biotech industries to refer to the refining, packaging and marketing steps; cf upstream, midstream
downstream processing	see **downstream**
downtime	A period of unavailability of equipment or personnel
Dow's process	see **Dow process**
Dowtherm	Proprietary eponym for heat transfer fluids
DP cell	A differential pressure cell/transducer
DPD method	A standard analytical method for determining free chlorine residual using the reagent DPD (n-diethyl-p-phenylenediamine)
DPMO	see **defects per million opportunities**
DQ	see **design qualification**
DQO	see **data quality objectives**
DR	see **dimension ratio**
Draeger	Proprietary eponym for gas testing tube
draffie	A draughtsman (qv)/draughtswoman, or sometimes a CAD operator (qv)
draft	*aka* draught. 1. The negative pressure which draws combustion air and flue gases through a heater; defined formally in API RP 535 - Burners for Fired Heaters at Refineries, BS EN ISO 13705:2012/ISO 13705:2012(E) Petroleum, petrochemical and natural gas industries. Fired heaters for general refinery service, API STD 530 - Calculation of Tube Heater Thickness and API STD 560 - Fired Heaters for General Refinery Service 2. A preliminary version of a drawing, sketch, or less frequently in our context written document 3. To produce a preliminary drawing, sketch, or less frequently in our context, written document
draft drawing	A preliminary drawing; defined formally in BS EN ISO 10209:2012 Technical product documentation — Vocabulary — Terms relating to technical drawings, product definition and related documentation

draft loss	*aka* draught loss. Pressure drop/headloss in the context of a fired heater. Defined formally in API RP 535 - Burners for Fired Heaters at Refineries, BS EN ISO 13705:2012/ISO 13705:2012(E) Petroleum, petrochemical and natural gas industries. Fired heaters for general refinery service, API STD 530 - Calculation of Tube Heater Thickness and API STD 560 - Fired Heaters for General Refinery Service
draftsman	see **draughtsman**
drag chain	see **drag link**
drag coefficient	(*symbol* Cd, Cx, Cw) A dimensionless number based on drag force, fluid density and flow speed which can be used to predict drag on an object moving in a fluid
drag link	*aka* drag chain. A conveyor with an endless belt moving in a closed trough with crossmembers which drag solids along
drain valve	A valve at a low point in process equipment used to empty them of liquid
drainage	Pipes and channels, usually carrying surface water
drainage cleanout	see **rodding point**
drainage drawing	A drawing of drainage provision at a site; defined formally in BS EN ISO 10209:2012 Technical product documentation - Vocabulary - Terms relating to technical drawings, product definition and related documentation
drainage field	An arrangement of perforated pipes and sufficiently permeable ground which offers passive tertiary biological treatment of effluent, rather than just flow buffering, like a soakaway. There are those who claim that a drainage field is a subtype of soakaway, but this interpretation is not supported by British Standards. It is not the same as a soakaway, though the terms are commonly confused in error
draught	see **draft**
draught loss	see **draft loss**
draughtsman	*aka* draftsman, draughtswoman, draughtsperson. A highly skilled (especially if called a 'design draughtsman') producer of engineering drawings, with a level of engineering knowledge sufficient to add or improve upon detailed design features; cf CAD operator
draughtsperson	see **draughtsman**
draughtswoman	see **draughtsman**
drawdown	Height difference between pumping and static water levels; defined formally in NFPA 20: Standard for the Installation of Stationary Pumps for Fire Protection, 2019 Edition

drawing	A general term for technical information presented in a graphical format; defined formally in BS 1553-1:1977 Specification for graphical symbols for general engineering. Piping systems and plant and BS EN ISO 10209:2012 Technical product documentation — Vocabulary — Terms relating to technical drawings, product definition and related documentation; cf diagram
drawoff	see **outlet**
dredge	To remove sediments from the bed of a body of water
drift	In the context of process control, a change in the average value of a controlled variable over time
drift loss	*aka* windage (loss). Losses of water through entrainment in air flow through a cooling tower
drill bit	The 'business end' of a drilling string
drill collar	An oil and gas industry term for a length of heavy pipe above a drill bit (qv)
drill collar slips	see **slips**
drill floor	*aka* derrick floor, rig floor. The (very hazardous) place where drilling string (qv) components are added and taken away by an oil and gas drilling crew, and their doghouse (qv) is located
drill pipe	An oil and gas industry term for the pipe which connects the transition pipe (qv) to the surface. The drill pipe will constitute the great majority of the length of a drilling string (qv).
drill slips	see **slips**
drill string	see **drilling string**
driller	1. Generically, someone who operates a drilling rig of some type 2. In the oil and gas industry, the supervisor of drilling operations
drilling areas	Areas in which drilling is taking place. A more convoluted formal definition is given in API RP 505 2nd Edition, August 2018 Recommended Practice for Classification of Locations for Electrical Installations at Petroleum Facilities Classified as Class I, Zone 0, Zone 1, and Zone 2
drilling fluid	see **drilling mud**
drilling mud	*aka* drilling fluid. Sometimes complex mixtures of solids and fluids (water, non-water and gases) used to control pressure, cool, lubricate and clean a drill bit, and carry spoil away from a drilling operation
drilling slips	see **slips**
drilling stabilizer	A device which stabilizes the bottom hole assembly (qv) in a borehole

drilling string	*aka* drill string. A collective term used in the oil and gas industry for the bottom hole assembly (qv), transition pipe (qv) and drill pipe (qv) used to transmit fluids and torque between the surface and drill bit
drilling superintendent	*aka* driller. In oil and gas, the leader of a drilling crew
drillship	A type of oil and gas production platform used in very deep waters
drinking water	*aka* DW; see **potable water**
drip gas	see **natural gas liquids**
dripproof guarded motor	A dripproof motor with guarded ventilation openings; defined formally in NFPA 20: Standard for the Installation of Stationary Pumps for Fire Protection, 2019 Edition
dripproof motor	A motor with ventilation openings constructed such that it can operate in an environment of falling liquid or solid particles; defined formally in NFPA 20: Standard for the Installation of Stationary Pumps for Fire Protection, 2019 Edition
drive	Used to mean either a motor, the complete electrical system which powers and controls it, or just the motor starter (qv)
drive collar	1. In oil and gas drilling operations, a pipe drivehead 2. An annular component transmitting torque from pump drive shaft to seal sleeve, defined formally in API RP 682 - Pump Seals
drive schedule	A list of all prime movers on a plant, with their kW rating, any required starter type etc. Prime movers may be driven by electricity, steam, hydraulic fluid or compressed gas
drive shaft	A mechanical connection for transmitting torque and rotation
drive starter	see **starter**
drive train component	A constituent part of the system which drives a pump; defined formally in EN ISO 13709:2003 Centrifugal pumps for petroleum, petrochemical and natural gas industries
drivehead	A fitting which allows a mechanical part to be efficiently driven
driven unit	Plant components driven by a prime mover. Defined formally in the context of gas turbines in ISO 3977 Gas turbines - Procurement - Part 3: Design requirements
driving force	1. Generically, any impetus which makes things happen 2. In chemical engineering, usually refers to the gradient of values of a property which drives a flux of mass or energy
dropleg	*aka* downcomer. This term is usually applied to air containing downcomers in effluent aeration systems
dropwise condensation	Condensation by discrete drops on an otherwise unwetted cold surface; cf film condensation

dross	1. A floating solid impurity on molten metal in smelting; cf slag, gangue 2. *(informal)* Also, by analogy, inferior quality work or staff members; cf dead wood
drowning out	see **salting out**
DRP provider	see dispute resolution process provider
drug delivery systems	Engineered technologies for the targeted delivery and/or controlled release of therapeutic agents
drum filter	*aka* rotary drum filter. A type of gravity driven water filter with the filtration medium laid on a rotating drum. May use filter aid and vacuum to enhance filtration performance, though this is expensive. Obsolete for many of its former duties
drum pump	see **barrel pump**
dry barrel hydrant	*aka* frostproof hydrant. A fire hydrant for use where there is a danger of freezing, the barrel of which is dry since the main control valve is mounted below the frost line; defined formally in NFPA 24: Standard for the Installation of Private Fire Service Mains and Their Appurtenances, 2019 Edition; cf wet barrel hydrant
dry basis	A basis of calculation ignoring the presence of water and other solvents; cf dry solids
dry bottom boiler	A boiler with a bottom temperature below ash melting point, from which ash is removed in a solid form
dry bulb temperature	True thermodynamic temperature; cf wet bulb temperature
dry cleaning	The cleaning of equipment and the workplace to prevent or reduce the build-up of objectionable matter, such as residues of aged or modified product, without the use of water; mostly done manually using brushes and/or vacuum cleaners - based on the EHEDG Glossary
dry commissioning	The first part of cold commissioning (qv) undertaken without fluids in the system
dry critical speed	Rotor critical speed in the absence of liquid effects; defined formally in EN ISO 13709:2003 Centrifugal pumps for petroleum, petrochemical and natural gas industries; cf wet critical speed
dry gas	Gas almost free of condensables, either a naturally occurring type of almost pure methane or natural gas processed to remove liquid hydrocarbons and impurities; cf wet gas
dry hydrant	*aka* dry riser (UK). An unpressurized piped firewater system intended to supply a fire department pump; defined formally in NFPA 1142: Standard on Water Supplies for Suburban and Rural Fire Fighting, 2017 Edition; cf dry barrel hydrant

dry ice	Solid carbon dioxide
dry matter	*aka* DM; see **dry solids**
dry matter content	see **dry solids**
dry riser	see **dry hydrant**
dry scrubbing	A unit operation which removes acid or odorous gases from exhaust gases using finely divided dry reagents
dry solids	*aka* dry solids content, dry matter (DM), dry matter content. That part of a mixture of solvent (usually water) and dissolved and suspended solids which remains after drying at 105 °C, i.e., the sum of dissolved solids (qv) and suspended solids (qv). Often expressed as percent dry solids (qv)
dry solids content	see **dry solids**
dry steam	Steam without suspended liquid water, usually defined as less than 0.5% moisture. Commonly used as synonymous with saturated steam (qv), though superheated steam (qv) is at least as dry
dry weight capacity	*aka* DWC. The adsorption capacity of a unit weight of a sorbent, such as ion exchange resin, expressed in units such as mEq/gram or mEq/ml of dry resin
dry well	A sewage pumping station comprising two adjacent chambers, one of which acts as a sump and the second used as the pump mounting chamber, avoiding the use of submersible pumps; cf wet well
dryer	A device which reduces the moisture/volatile component content of a material using some combination of temperature and pressure change. Defined formally in NFPA 654: Standard for the Prevention of Fire and Dust Explosions from the Manufacturing, Processing, and Handling of Combustible Particulate Solids, 2017 Edition
drying	Reducing the moisture/volatile component content of a material using some combination of temperature and pressure change; cf dewatering, thickening, desiccation
drying rate	Speed of drying
dryness	1. Generally, the degree of absence of water 2. In the context of steam, the degree of absence of liquid water; cf steam wetness
dryness fraction	*aka* steam dryness fraction. The mass of steam as a proportion of the total mass of steam and entrained liquid water
DSAW	see **double submerged arc welded**
DSC	see **differential scanning calorimetry**
DSEAR	Dangerous Substances and Explosive Atmospheres Regulations 2002: UK Safety Legislation; qv dangerous substances

DSU	see **delayed startup insurance**
dual fuel burner	A burner which can burn either fuel gas or oil separately; defined formally in NFPA 86: Standard for Ovens and Furnaces, 2019 Edition; cf multifuel burner
dual fuel system	A system which allows a gas turbine to burn two dissimilar fuels separately. Defined formally in ISO 3977 Gas turbines - Procurement - Part 3: Design requirements
dual mechanical seal	*aka* double mechanical seal. Two independent seals, separated by a flushed chamber. An API RP 682 Arrangement 2 or 3 seal, defined formally in API RP 682 - Pump Seals
Dubai crude	A price benchmark oil; cf West Texas Intermediate, Brent Crude
Duck tape	*aka* duct tape, gaffer tape (disputed). Proprietary eponym for a plasticized pressure sensitive textile tape, originally used for sealing ammunition cans against water ingress[34]. Subsequently used for a multitude of purposes including bringing Apollo 13 back to Earth.
duct	A conduit for gases, or suspensions of solids in gas. Defined formally in BS EN ISO 13705:2012/ISO 13705:2012(E) Petroleum, petrochemical and natural gas industries. Fired heaters for general refinery service, API STD 530 - Calculation of Tube Heater Thickness, API STD 560 - Fired Heaters for General Refinery Service and NFPA 654: Standard for the Prevention of Fire and Dust Explosions from the Manufacturing, Processing, and Handling of Combustible Particulate Solids, 2017 Edition; cf chase
duct tape	see **Duck tape**
ductile	Practically, the opposite of brittle. Strictly, the ability to be drawn in into a wire without breaking; cf malleable
ductile iron	*aka* spheroidal graphite (SG) iron. A non-brittle form of cast iron (qv), spun when molten, in which the graphite inclusions are rounded, used extensively in water and sewage applications to form pipes and fittings e.g., manhole covers, valves and penstocks; cf gray iron
ductility	The degree to which something is ductile. Defined formally in API RP 579 - Fitness for Service
ductwork	A collective term most commonly referring to a system of ducts which carry air and other gases
due diligence	Generating sufficient certainty in your opinions, considering the potential downside if you are wrong

---

[34] Also, half of the universal toolkit (qv)

Duhem's theorem	A theorem which states that, for any closed system formed initially from given masses of prescribed chemical species, the equilibrium state is completely determined when any two independent variables are fixed
Duhring rule	An empirical rule which states that the boiling point of a given solution is a linear function of the boiling point of pure water at same pressure
Dulong–Petit law	An empirical rule which states that, for any solid element, the heat capacity per mole is about 25 J/K
dump nozzle	see **catalyst dump nozzle**
dump tank	Collection vessel for dumped materials, such as a blowdown drum (qv)
dump valve	A type of automatic pressure relief valve downstream of a positive displacement compressor or pump; defined formally for pumps in NFPA 20: Standard for the Installation of Stationary Pumps for Fire Protection, 2019 Edition; cf blowoff valve
dumping	The loss of all liquid from trays in a distillation column via a domino effect (qv) caused by a low vapour flow rate, requiring the column to be restarted
dunder	*aka* vinasse. Waste water from fermentation of molasses to ethanol
duplex process	1. Generically, any steelmaking process involving two sequential melting/refining furnaces 2. Specifically, the now-obsolete Bessemer/open hearth steelmaking process
duplex pump	A pump with two linked pumping heads. Most commonly a positive displacement pump, either piston or diaphragm type
duplex stainless steel	Strong, corrosion-resistant stainless steel with a roughly equal mix of austenite and ferrite; defined formally in ASME BPE (American Society of Mechanical Engineers: Bioprocessing Equipment), and API RP 571 - Damage Mechanism Affecting Fixed Refinery Equipment
durability	Hardwearingness
durable	Hard-wearing; defined formally in BS EN 1672-2:2005+A1:2009 Food processing machinery. Basic concepts. Hygiene requirements
Duralumin	*aka* duraluminum, dural. Proprietary eponym for a copper-aluminum alloy, differing from aluminum bronze (qv) in having a far higher aluminum content
duraluminum	see **Duralumin**
duraluminium	see **Duralumin**

dural	see **Duralumin**
Durbar decking	Proprietary eponym for a diamond-patterned plate used as deckplate (qv). Durbar refers to a particular pattern of shapes, see **Figure D1**
durometer	*aka* Shore durometer. A device for measuring the resistance to surface penetration of non-metallic materials, measured on a Shore A or D scale. Defined formally in ASME BPE (American Society of Mechanical Engineers: Bioprocessing Equipment)
dust	Small solid particles. Definitions of how small particles have to be to count as dust vary quite widely, but ISO quote 75 micrometers or less in ISO 4225
dust cloud explosion	see **dust explosion**
dust collection system	Equipment which contains and removes dust from a process area; defined formally in NFPA 654: Standard for the Prevention of Fire and Dust Explosions from the Manufacturing, Processing, and Handling of Combustible Particulate Solids, 2017 Edition
dust deflagration hazard	The condition in which there is a sufficient quantity of a dust/oxidizing medium mixture to support a harmful dust deflagration; defined formally in NFPA 654: Standard for the Prevention of Fire and Dust Explosions from the Manufacturing, Processing, and Handling of Combustible Particulate Solids, 2017 Edition

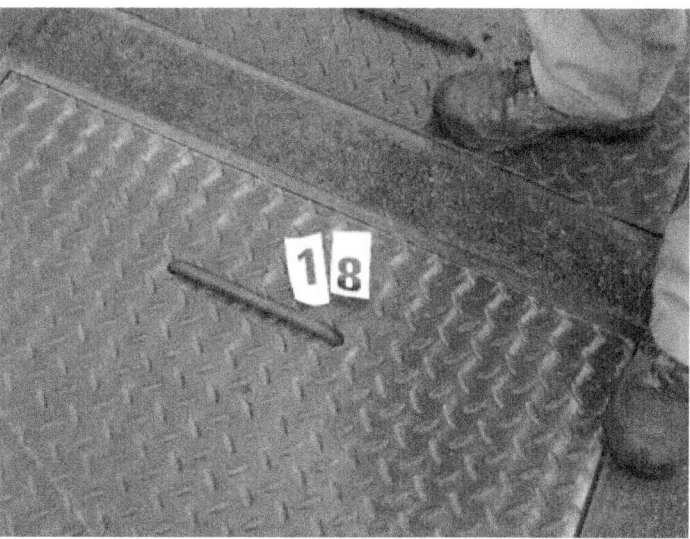

Figure D1 Durbar decking

dust explosion	The rapid combustion of an enclosed dust/oxidizing medium mixture; can be highly energetic; cf primary explosion, secondary explosion
dust explosion hazard	An enclosed dust deflagration hazard (qv) capable of generating sufficient pressure to rupture its enclosure. Defined formally in NFPA 654: Standard for the Prevention of Fire and Dust Explosions from the Manufacturing, Processing, and Handling of Combustible Particulate Solids, 2017 Edition
dust explosion hazard area	Part of a building where a foreseeable dust explosion might cause structural damage; defined formally in NFPA 654: Standard for the Prevention of Fire and Dust Explosions from the Manufacturing, Processing, and Handling of Combustible Particulate Solids, 2017 Edition
dust flash fire hazard area	An area where there is a combustible dust cloud (qv primary explosion), or where combustible dust has accumulated on surfaces, creating the possibility of a dust cloud being created when accumulated dust is disturbed (qv secondary explosion); defined formally in NFPA 654: Standard for the Prevention of Fire and Dust Explosions from the Manufacturing, Processing, and Handling of Combustible Particulate Solids, 2017 Edition
dust hazards analysis	A systematic study of the potential fire and explosion hazards of combustible dusts at a process facility. Defined formally in NFPA 654: Standard for the Prevention of Fire and Dust Explosions from the Manufacturing, Processing, and Handling of Combustible Particulate Solids, 2017 Edition
dust ignition proof motor	A motor with an enclosure which prevents dust getting in; and potential sources of ignition getting out. Defined formally in NFPA 20: Standard for the Installation of Stationary Pumps for Fire Protection, 2019 Edition
duty	1. The specified quantity of material handled, power used, heat transferred, etc. for an item of equipment 2. The item of equipment available for service in a redundant or parallel system (qv duty/standby, duty/assist) 3. A type of import tax
duty point	The intersection of a given pump's Q/H curve (qv) and the system curve (qv); cf characteristic curves
duty/assist	In parallel systems, one item of equipment is (usually temporarily) designated the duty unit, and is brought into service on demand. An additional unit or units will be held in reserve (assist) to be brought into parallel service in the case of insufficiency of the duty unit. This is commonly combined with redundancy, in duty/assist/standby (qv) systems

duty/assist/standby	The combination of parallel duty/assist (qv) units and redundant standby
duty/standby	In redundant systems, one item of equipment is (usually temporarily) designated the duty unit, and is brought into service on demand. An additional unit or units will be held in reserve (standby) to be brought into service in the case of unavailability of the duty unit
DVB	see **divinylbenzene**
DW	1. see **deionized water** 2. Drinking water, see **potable water**
DWC	see **dry weight capacity**
dwelling	A detached building containing no more than two dwelling units; defined formally in NFPA 1142: Standard on Water Supplies for Suburban and Rural Fire Fighting, 2017 Edition
dwelling unit	A set of rooms with facilities for cooking, eating, sleeping and sanitation. Defined formally in NFPA 30: Flammable and Combustible Liquids Code, 2018 Edition and NFPA 1142: Standard on Water Supplies for Suburban and Rural Fire Fighting, 2017 Edition
dxf	File extension for 'drawing exchange format': a file format developed by Autodesk (developers of AutoCAD) which allows a degree of file sharing with other CAD software programs
dye penetrant inspection	see **liquid penetrant indication**
dyed diesel	Diesel which has been dyed red or blue to denote its untaxed status; used in specified heating and vehicle fueling applications; cf red diesel
dyke	see **bund**
dynamic analysis	A method used to control an inherently unstable and dynamic system or process
dynamic elastic constant	see **dynamic stiffness**
dynamic equilibrium	see **equilibrium**
dynamic error	*aka* measurement error. The difference between the measured value of the quantity changing with time and its true value assuming no static error (qv)
dynamic head	In hydraulic calculations, the pressure needed to overcome friction in a flowing system, expressed in meters of fluid; cf static head, static pressure, impact pressure
dynamic loading	In structural engineering, the response of structural components to cyclical loading produced by variable loads
dynamic pressure	*aka* dynamic head (though this use may be deprecated); see **impact pressure**

dynamic range	Range of measurable values; defined formally in BS ISO 2041:2018 Mechanical vibration, shock and condition monitoring. Vocabulary
dynamic response	The behavior of the output of a controlled process or item of equipment as a function of the input with respect to time
dynamic seal	A seal with a moving component. Defined formally in ASME BPE (American Society of Mechanical Engineers: Bioprocessing Equipment)
dynamic sealing pressure rating	The maximum continuous differential pressure at maximum allowable temperature which a seal can withstand with the shaft rotating; defined formally in API RP 682 - Pump Seals; cf static sealing pressure rating
dynamic similarity	One of the types of system similarity used for process scaleup, especially with respect to fluid mechanics; cf geometric similarity, kinetic similarity
dynamic spray device	A moving spray device, producing a moving spray pattern. Defined formally in ASME BPE (American Society of Mechanical Engineers: Bioprocessing Equipment); cf static spray device
dynamic spring constant	see **dynamic stiffness**
dynamic stiffness	*aka* dynamic elastic constant. The ratio of a dynamic force to the resulting dynamic displacement across a range of frequencies. Defined formally in BS ISO 2041:2018 Mechanical vibration, shock and condition monitoring. Vocabulary
dynamic viscosity	*aka* absolute viscosity; qv viscosity
dynamite	A phlegmatized (qv) explosive product, comprised of packaged, absorbed nitroglycerin
Dynasand	Proprietary eponym for a continuous sand filter
dyne	An obsolete metric unit of force

E

E526	see **slaked lime**
E_a	see **activation energy**
EA	see **Environment Agency**
EAA	see **Europe, Africa and Asia**
Eadie-Hofstee plot	An obsolete graphical representation of enzyme kinetics; cf Lineweaver-Burk plot
eagle-eye	The leveling of power line cross-arms (qv) by sight without measuring instruments
EAL	see **environmental assessment level**
earth bonding	*aka* cross bonding (deprecated usage). Electrically connecting (not gluing; cf bond) conductive objects to each other and to earth (ground (qv)), so that they are at the same electrical potential to control static hazards; defined formally in NFPA 30: Flammable and Combustible Liquids Code, 2018 Edition and NFPA 654: Standard for the Prevention of Fire and Dust Explosions from the Manufacturing, Processing, and Handling of Combustible Particulate Solids, 2017 Edition
earthing	Connecting electrically to ground (qv) (called earth in UK English)
earthquake load	The addition to design loading allowing for earthquake conditions
earthwork drawing	A drawing showing requirements for earthworks cut and fill. Defined formally in BS EN ISO 10209:2012 Technical product documentation — Vocabulary — Terms relating to technical drawings, product definition and related documentation
easement	An acquired right allowing an entity limited use of another's property, e.g., the right to run a pipeline through land belonging to another
easily or readily accessible	"A location that can be safely reached by a personnel from the floor, platform, or other permanent work area", according to the EHEDG Glossary
easily or readily removable	"Quickly separated from the equipment with the use of simple hand tools if necessary. The latter are implements normally used by fitters, operating and cleaning personnel such as a screwdriver, a wrench or hammer" according to the EHEDG Glossary
easy cleanability	see **cleanability**
EBC	see **European Brewing Convention**

EBCT	see **empty bed contact time**
Ebonite	Proprietary eponym for a hard, black, highly vulcanized rubber used to line process equipment, especially that containing dilute hydrochloric acid
EC	1. see **eddy current**
	2. see **European Commission**
EC&I	Electrical, control and instrumentation
Ecc	*(drawing notation)* see **eccentric**
eccentric	Not having a common center (especially applied to circular objects); cf concentric
eccentric screw pump	see **progressing cavity pump**
ECHA	see **European Chemicals Agency**
economic pipe diameter	The most economical commercially available pipe diameter, with due consideration of installed cost of pipe and fittings, and the costs of operation, maintenance, and asset depreciation. There are a number of different formulae used to calculate this[35]
economic potential	An academic approach to costing, promoted by Douglas in his 'Conceptual Design of Chemical Processes', in recognition of the lack of knowledge of professional engineering among 'engineering academics'[36]
economical reflux ratio	see **optimum reflux ratio**
economizer	1. Generically, any device used to raise energy efficiency
	2. Most commonly a heat exchanger used to increase boiler efficiency
ECU	see **electrochemical unit**
eddy	Circular movement of fluid or current
eddy current	*aka* EC, Foucault's currents. An eddy (qv) of current induced magnetically in an electrically conductive material
eddy current testing	A number of NDT (qv) methods used to detect cracks and other flaws in metals based on induction of eddy currents (qv)
EDI	see **electrodialysis**
EDMS	see **electronic document management system**
educational occupancy	A term used to describe a building or part of one used for educating at least six people for at least 12 hours a week; defined formally in NFPA 30: Flammable and Combustible Liquids Code, 2018 Edition

---

[35] If indeed it is calculated at all. Pipes are commonly sized based on superficial velocity and lowest capital cost without this analysis being done.

[36] A poor substitute for professional practice even in its most sophisticated form. In the rudimentary form often used in academia it can make selling ice to the Inuit appear economically viable!

eductor	*aka* inductor. A venturi pump; defined formally in NFPA 1142: Standard on Water Supplies for Suburban and Rural Fire Fighting, 2017 Edition and NFPA 11: Standard for Low-, Medium-, and High-Expansion Foam, 2016 Edition
EEA	see **European Environment Agency**
EEBD	see **emergency escape breathing device**
EEL	see **emergency exposure limit**
eels	*aka* line condom (amongst many other things). A term used by linemen for a temporary lightweight insulator for power lines
EER	see **energy efficiency ratio**
EERA	see **escape, evacuation and rescue assessment**
effect	1. *aka* impact. "Any change in the physical, natural or cultural environment brought about by a development project" according to the EC Guidance on EIA EIS Review 2. Also used for an equilibrium stage in distillation
effective coefficient of discharge	A quantity used along with effective discharge area (qv) to estimate pressure relief valve (qv) relief capacity. Defined formally in API RP 520 P1 7th Edition, January 2000 Sizing, Selection, and Installation of Pressure-Relieving Devices in Refineries; Part I - Sizing and Selection
effective discharge area	*aka* effective orifice area. A quantity used along with effective coefficient of discharge (qv) to estimate pressure relief valve (qv) relief capacity. Defined formally in API RP 520 P1 7th Edition, January 2000 Sizing, Selection, and Installation of Pressure-Relieving Devices in Refineries; Part I - Sizing and Selection and API RP 576 - Inspection of Pressure Relieving Devices
effective mass	*aka* apparent mass. The mass of an object, viewed from the perspective of forces acting upon it. If other forces are already acting upon the object, as when for example it is moving through a fluid, its effective mass may differ significantly from its true mass.
effective orifice area	see **effective discharge area**
effective size	A particle size equal to 90% retention size
effective surface	The working outside surface of heat exchanger tubes. Defined formally in API 660 - Shell-and-Tube Heat Exchangers and API RP 661 - Heat Exchangers
effectiveness	1. The degree of success in a quality management context, formally defined in BS EN ISO 9000 Quality management systems Fundamentals and vocabulary. 2. In the context of heat exchangers, the ratio of actual heat transfer rate to the theoretical maximum possible heat transfer rate; cf efficiency

effectiveness factor	The ratio of the real reaction rate of a catalyst particle with an internal diffusion-limited concentration gradient to the theoretical one in the absence of such a gradient
effectivity	A method to control the validity of variations, used with documents and parts. Defined formally in BS EN ISO 10209:2012 Technical product documentation — Vocabulary — Terms relating to technical drawings, product definition and related documentation
efficiency	The ratio of useful results to consumed resources. Defined formally, in a quality management context, in BS EN ISO 9000 Quality management systems Fundamentals and vocabulary
effluent	*aka* efflux. Fluid flow out of a process. Often associated with a waste flow or stream; cf influent, wastewater
efflux	see **effluent**
efflux coefficient	see **coefficient of discharge**
effusion	The pressure driven movement of gas particles through a hole smaller than their mean free path; cf diffusion
EFRD	see **emergency flow restricting device**
EFRT	see **external floating roof tank**
EFW	*(drawing notation)* Electric fusion welded [pipe]
EGSB	see **expanded granular sludge bed**
EHEDG	see **European Hygienic Engineering & Design Group**
EHRC	see **enhanced high rate clarification**
EHS	see **environment health and safety**
EI	see **electronic interface**
EIA	see **environmental impact assessment**
EIA team	see **environmental impact assessment team**
EIC	Energy Industries Council
EIS	see **environmental impact statement**
EIT	1. see **equivalent isothermal temperature** 2. see **engineer in training**
EIV	see **emergency isolation valve**
ejector	*aka* injector. A venturi pump used in refining and petrochemical industry vacuum distillation towers to pull vacuum; on steam turbines to minimize pressure of the exhaust steam, maximizing turbine efficiency; and in the nuclear industry to minimize moving parts needing maintenance in radioactive service. Defined formally, in the context of firepumps, in NFPA 1142: Standard on Water Supplies for Suburban and Rural Fire Fighting, 2017 Edition
El	*(drawing notation)* Elevation

elastic allowable stress	*aka* elastic range allowable stress. Defined formally in API STD 530 - Calculation of Tube Heater Thickness
elastic design pressure	*aka* rupture pressure. The maximum transient pressure that can be contained, commonly related to PRV design. Defined formally in API STD 530 - Calculation of Tube Heater Thickness
elastic modulus	see **Young's modulus**
elastic range allowable stress	see **elastic allowable stress**
elasticity	The ability to deform reversibly in response to an applied force
elastomer	A rubbery, elastic material; defined formally in ASME BPE (American Society of Mechanical Engineers: Bioprocessing Equipment)
elastomeric material	see **elastomer**. (There is an apparently redundant formal definition of this term in ASME BPE (American Society of Mechanical Engineers: Bioprocessing Equipment))
elbow	*aka* ell. A pipe bend which can be of long or short radius; stock versions turn pipe through 90, 45, or 22.5 degrees
Elbolet	A proprietary eponym, an abbreviated form of ELBOw outLET. One of various olet (qv) types
ELD	see **engineering line diagram**
electric motor drive	*aka* EMD. A drive (qv) involving an electric motor
electric vaporizer	An electrically heated LPG vaporizer; defined formally in NFPA 58: Liquefied Petroleum Gas Code, 2017 Edition
electrical and mechanical run out	*aka* slow roll runout. Mechanical runout measures variability of the concentric roundness of the shaft, whilst electrical runout measures variability of electrical conductivity and magnetic permeability of the shaft; defined formally in ISO 3977 Gas turbines - Procurement - Part 3: Design requirements
electrical clearance	The air gap between water sprays and uninsulated live electrical equipment; defined formally in NFPA 15 Standard for Water Spray Fixed Systems for Fire Protection
electrical enclosure	A box in which electrical equipment is installed. Defined formally in API RP 505 2nd Edition, August 2018 Recommended Practice for Classification of Locations for Electrical Installations at Petroleum Facilities Classified as Class I, Zone 0, Zone 1, and Zone 2
electrochemical equivalent	(*symbol* Z *or* Eq) The mass of a substance deposited on an electrode when one Coulomb of electricity is passed through an electrochemical cell; cf equivalent

electrochemical unit	*aka* ECU. A unit of chlor-alkali process (qv) output; being 1 ton of chlorine, 1.13 tons of 100% caustic soda and 0.03 tons of hydrogen
electrode	A conductor which carries electricity to a nonmetallic part of a circuit such as an electrolyte
electrode potential	Electromotive force or reducing potential of an electrode, compared with a standard hydrogen electrode
electrodialysis	*aka* EDI. An electrically driven dialysis process which may be used for water desalination[37]
electrokinetic potential	*aka* zeta potential. A measure of the potential difference in a solution due to unbalanced charges that results in the formation of a double layer; useful in analysis of coagulation of colloids
electrolysis	A process of ionic substances being decomposed by an electric current passed through them in solution or molten form
electrolyte	A substance whose ions conduct electricity either when molten or when dissolved in a polar solvent
electrolytic cell	1. A device for electrolysis (qv) 2. The bimetallic system responsible for electrolytic corrosion
electrolytic corrosion	*aka* bimetallic corrosion, dissimilar metals corrosion. Electrically driven corrosion in two joined metals with different electrochemical potentials in the presence of an electrolyte
electrolytic polishing	see **electropolishing**
electrolytic refining	A method for purifying high value metals in which an impure metal cathode is dissolved and deposited on a thin, very pure metal anode by electrolysis (qv). A type of electrometallurgy (qv)
electromagnetic flowmeter	*aka* magnetic flowmeter. A flowmeter which measures the flow of electrically conducting fluids based on the current induced by their passing through a magnetic field
electromechanical transducer	A mechanically actuated electrical transducer, such as a strain gauge (qv); defined formally in BS ISO 2041:2018 Mechanical vibration, shock and condition monitoring. Vocabulary
electrometallurgy	Producing metals by electrolysis (qv)
electronic document and records management system	see **electronic document management system**
electronic document management system	*aka* EDMS, Electronic document and records management system, digital document management system. A centralized digital electronic system used to manage drawings and other documents

---

[37] Possibly still somewhat experimental/unreliable according to reports

electronic interface	*aka* EI. A method of communication with and between electronic devices, including physical connections such as USB and ethernet connections, and intangible network/internet connections such as websites and portals
electroplating	Producing a thin coherent coating of metal on the surface of a different base metal (qv) by means of electrolysis (qv)
electropolishing	*aka* anodic polishing, electrolytic polishing. Removing a thin surface layer from stainless steel by electrolysis (qv), providing a small improvement in surface roughness. Defined formally in ASME BPE (American Society of Mechanical Engineers: Bioprocessing Equipment)
electrosensitive protective equipment	*aka* ESPE; see **sensitive protective equipment**
electrostatic precipitation	Removing fine particulates (both solid and liquid) from flue gases passed between charged plates. Can be viewed as a form of filtration or dry scrubbing
electrostatic precipitator	*aka* EP, precip. An electrostatic precipitation (qv) device
element	1. In chemistry, a substance which cannot be degraded by chemical reaction 2. In product design, and associated documentation, part of a component; defined formally in BS EN ISO 10209:2012 Technical product documentation – Vocabulary – Terms relating to technical drawings, product definition and related documentation 3. A discrete component, such as a heating element
element bundle	In the context of centrifugal pumps, the rotor and internal stationary parts; defined formally in EN ISO 13709:2003 Centrifugal pumps for petroleum, petrochemical and natural gas industries
elementary flow	1. In fluid dynamics, a basic flow which can be used to construct more complex flows 2. In the context of life cycle assessment, "Material or energy entering the system being studied that has been drawn from the environment without previous human transformation, or material or energy leaving the system being studied that is released into the environment without subsequent human transformation" according to EN ISO 14040:2006 Environmental management -- Life cycle assessment -- Principles and framework

elementary particle	An indivisible particle. Some types are part of an atom (whose name means indivisible, because it used to be thought that the atom was the elementary particle); other types are not part of matter
elementary reaction	A simple chemical reaction with a single reaction step, a single transition state and no intermediates
elephant ear	A high strength strain insulator (qv)
elephant's feet	1. A characteristic buckling mode of tank failure 2. The ends of suction pipes in tankers 3. Horizontal plates on top of towing pins
elevated flare	*aka* elevated flare stack. A flare with a highly elevated burner; defined formally in API Standard 521. Pressure-relieving and Depressuring Systems. Sixth Edition | January 2014 and API RP 537 Flare Details for Petroleum, Petrochemical, and Natural Gas Industries
elevated flare stack	see **elevated flare**
elevation	A side-on view in a drawing. Defined formally in BS EN ISO 10209:2012 Technical product documentation — Vocabulary — Terms relating to technical drawings, product definition and related documentation; cf plan
elevation drawing	A drawing showing elevation; defined formally in BS EN ISO 10209:2012 Technical product documentation — Vocabulary — Terms relating to technical drawings, product definition and related documentation
elimination reaction	In organic chemistry, a type of reaction in which two atoms or groups are removed from a molecule
e-line	see **wireline**
ell	see **elbow**
Ellingham diagram	A plot of Gibbs free energy (qv) vs temperature, illustrating temperature dependence of compound stability; used in practice to evaluate the best reducing agent for metal oxides, amongst other things
ellipsoidal	Ellipse-shaped or less formally, oval. One of the three most common head shapes for pressure vessels; cf torispherical, hemispherical
eluate	A solution produced by elution (qv)
elution	The removal of sorbed species from a sorbent (qv) by washing with solvent
elutriate	The stream of fine particles in fluid produced by elutriation (qv)
elutriation	*aka* classification. Separation of finer (most commonly submicron) particles from coarser particles with an upflow of gas (or less commonly liquid) exceeding their terminal velocity

elutriator	*aka* classifier. A device for elutriation (qv)
eluviation	The removal of dissolved or suspended solids from soil layers by net downward groundwater flow
ELV	see **emission limit value**
embankment	*aka* dike, bank. An elongated raised construction, used to retain water, support a road or rail bed, or for protection
EMC	see **equilibrium moisture content**
EMD	see **electric motor drive**
EMEA	1. see **Europe, Middle East, Africa** 2. see **European Medicines Evaluation Agency**
emergency BA set	*aka* emergency escape set. A device providing breathable air for a short duration (10-30 minutes) for use by a worker in an immediately dangerous to life or health atmosphere
emergency escape breathing device	see **escape BA**
emergency escape set	see **emergency BA set**
emergency exposure limit	*aka* EEL. The maximum allowable exposure to atmospheric contaminant for an unpredictable, short single exposure (usually <1h, never more than 24h); set by the US National Research Council's Committee on Toxicology
emergency flow restricting device	*aka* EFRD. A device such as a check valve or remote control valve which restricts the amount of contents released from a pipeline after loss of containment
emergency isolation valve	*aka* EIV. A valve which prevents process fluid flow from feeding a loss of containment (qv) situation or a fire; cf emergency shutdown valve
emergency motor stop	*aka* EMS, e-stop, manual emergency switch. A latching electrical safety device with an emergency stop function (qv)
emergency operation	1. Features and procedures for preventing or stopping an emergency situation; defined formally in BS EN ISO 12100:2010 Safety of machinery. General principles for design. Risk assessment and risk reduction 2. An operating regime under emergency conditions
emergency relief vent	A physical feature or device intended to automatically release overpressure caused by fire; defined formally in NFPA 30: Flammable and Combustible Liquids Code, 2018 Edition
emergency shutdown	Controlled, (some might argue semi-controlled) rapid shutdown in an emergency situation; defined formally in ISO 3977 Gas turbines - Procurement - Part 3: Design requirements and API Standard 521. Pressure-relieving and Depressuring Systems. Sixth Edition	January 2014

emergency shutdown system	*aka* ESS; see **safety instrumented system**
emergency shutdown valve	A valve used for emergency shutdown (qv), which may or may not be an emergency shutoff valve (qv) and can be remotely operated
emergency shutoff valve	A valve for emergency shutdown (qv) by fuel shutoff. NFPA formal definitions differ in whether this is manual or automatic; see NFPA 58: Liquefied Petroleum Gas Code, 2017 Edition and NFPA 86: Standard for Ovens and Furnaces, 2019 Edition; cf emergency isolation valve
emergency situation	An urgent, hazardous situation; defined formally in BS EN ISO 12100:2010 Safety of machinery. General principles for design. Risk assessment and risk reduction
emergency stop	see **emergency stop function**
emergency stop function	A function initiated by a single human action which terminates an actual or impending emergency situation. Defined formally in BS EN ISO 12100:2010 Safety of machinery. General principles for design. Risk assessment and risk reduction
EMI	Electromagnetic interference
emission limit value	*aka* ELV. "The permissible quantity of a substance contained in the waste gases from a combustion plant which may be discharged into the air during a given period" according to EU Directive 2015/2193
emission value	A quantified emission of noise etc. by a machine; defined formally in BS EN ISO 12100:2010 Safety of machinery. General principles for design. Risk assessment and risk reduction
EMP	Electromagnetic pulse
empirical	Derived[38] from practice and observation of reality rather than theory and speculation
empty bed contact time	*aka* EBCT. A measure of residence time used in GAC system design; hydraulic retention time (qv) based on the space taken up by a bed of media
EMS	1. see **emergency motor stop** 2. see **environmental management system**
emulsification	Making an emulsion (qv)
emulsion	A stable dispersion of fine droplets of one liquid (the dispersed phase (qv)) within another (the continuous phase (qv)) with which it is immiscible
emulsion breaker	A chemical which is dosed to 'break' (destabilize) an emulsion

---

[38] Like engineering!

enabling device	*aka* dead man's handle/switch. A device requiring continuing activation by an operator to allow a machine to function; defined formally in BS EN ISO 12100:2010 Safety of machinery. General principles for design. Risk assessment and risk reduction
enameled steel	see **glass on steel**
ENCL	*(drawing notation)* see **enclosure**
enclosed area	see **confined space**. This term defined formally in API RP 505 2nd Edition, August 2018 Recommended Practice for Classification of Locations for Electrical Installations at Petroleum Facilities Classified as Class I, Zone 0, Zone 1, and Zone 2
enclosed flare	A flare burner enclosure which hides the flame; defined formally in API Standard 521. Pressure-relieving and Depressuring Systems. Sixth Edition | January 2014 and API RP 537 Flare Details for Petroleum, Petrochemical, and Natural Gas Industries
enclosure	*aka* building, container, containment, housing. Defined formally in the context of gas turbines in ISO 3977 Gas turbines - Procurement - Part 3: Design requirements
enclosureless dust collector	A device which removes dust from solids transport air which has unenclosed filter media; defined formally in NFPA 654: Standard for the Prevention of Fire and Dust Explosions from the Manufacturing, Processing, and Handling of Combustible Particulate Solids, 2017 Edition
end of curve	*aka* EOC. 1. In the context of a Q/H curve (qv), the area of the curve where a centrifugal pump should not be running 2. *aka* point of tangency. In drawing notation, the point where a curve (such as one in a road) ends
end of pipe	1. Add-on solutions (as opposed to in-process/resource minimization ones) to environmental releases 2. A drawing notation for the literal end of a pipe
end suction pump	A centrifugal pump with suction nozzle in line with the drive shaft; defined formally in NFPA 20: Standard for the Installation of Stationary Pumps for Fire Protection, 2019 Edition
end use	The normal intended use of a product. Defined formally in EN 12566 - Small wastewater treatment systems for up to 50 PT Part 3: Packaged and/or site assembled domestic wastewater treatment plants

end user	The entity which uses a product or plant; may include those responsible for installation, operation and maintenance. Defined formally in BS 4884-2:1993 Technical manuals. Guide to content
endothermic flare	*aka* fuel assisted flare. A flare which is used when assist or enrichment gas must be added to relief gas, to improve its heating value to sustain combustion. Defined formally in API RP 537 Flare Details for Petroleum, Petrochemical, and Natural Gas Industries
endurance limit	see **fatigue endurance limit**
energy efficiency ratio	*aka* EER. In the context of HVAC (qv), a USA term for the ratio of heating or cooling provided to power consumption in BTU/h per watt. The SI equivalent is coefficient of performance (qv) (COP). Conversion factor is COP of 1.0 = EER of 3.4.
energy flow	Quantified flows of energy across a process boundary. Defined formally in the context of life cycle assessment in EN ISO 14040:2006 Environmental management - Life cycle assessment - Principles and framework
energy recovery wheel	see **rotary heat exchanger**
engagement	Taking an active part in activities furthering collective objectives; defined formally in BS EN ISO 9000 Quality management systems Fundamentals and vocabulary
Engine Oil Licensing and Certification System	A standard for base oils (qv); defined formally in API 1509 Engine Oil Licensing and Certification System
engine speed	In the context of firepumps, engine nameplate speed; defined formally in NFPA 20: Standard for the Installation of Stationary Pumps for Fire Protection, 2019 Edition
engineer	A practitioner of the profession of imagining and bringing into being an artefact which safely, cost effectively and robustly achieves a specified aim[39]. Formally, someone with an accredited Master's degree in engineering (governed by the Washington Accord) and a minimum of 4 years of specified professional engineering experience; cf CEng (UK), Professional Engineer, **Box E1**.
engineer in training	*aka* EIT. The first step in becoming a Professional Engineer (qv) in the USA
engineering	The profession of imagining and bringing into being an artefact which safely, cost effectively and robustly achieves a specified aim

---

[39] Or to put it another way, someone who gets excited about things no-one else cares about!

Box E1. Engineer

What is an engineer?

Within UK engineering institutions, the debate largely centers around reserving the title 'engineer' for a Chartered Engineer, and this is the sense in which I use the term: i.e. a professional engineer, with a UK-accredited degree in engineering, at least four years of experience working as an engineer in the design or technical supervision of full scale-engineering projects, and the letters CEng after their name (a designation which is not quite the same as PE in the USA).

However, anyone can call themselves an engineer in the UK, so the term is frequently used to describe car mechanics, domestic appliance repairers, satellite tv dish installers, etc. Thus, in the UK at present, a Chartered Fellow of a UK engineering institution, recognised as "an engineering professional of distinction", and possibly also as a professional engineer in Europe by FEANI (which covers jurisdictions where the term engineer is protected), has the same professional title as someone with no relevant qualifications or experience or, as a colleague quipped, an ungeneer.

What about the people who have degrees in engineering but never go on to practice as engineers, (i.e. designing or providing technical supervision of full scale-engineering projects)? Are they engineers? Rather like graduates of medical or law degrees, a UK degree in engineering meets "the academic requirement for the formation of a chartered engineer" but that does not mean that an engineering degree graduate is an engineer; rather the graduate is ready to be made into an engineer by other engineers. The institutions consider that the minimum period required to gain an adequate level of understanding of professional practice is four years.

Around the world, the definition of engineer can vary, but a consensus professional hierarchy might look something like this:

engineer = **CEng (UK) / PE / EUR ING** = Accredited Master's degree (governed by the Washington Accord) and minimum 4 years of specified professional experience
engineering technologist = IEng = Apprenticeship/Bachelor's degree (governed by the Sydney Accord) + specified experience
engineering technician = EngTech = Apprenticeship / Bachelor's degree/diploma (governed by the Dublin Accord) + specified experience

Everyone else is an enjuneer!

What about the people who teach and research in university engineering departments? They are members of a different profession, that of academia, though there certainly exists a school of thought that engineering academics can consider themselves engineers. However, academic titles do not map onto the professional hierarchy above. A professor of engineering may well be an enjuneer from a Professional Engineer's point of view!

Related Terms
Ungineer

engineering line diagram	*aka* ELD. One of a number of different types of engineering diagram, such as the electrical SLD (qv), or the process engineer's PFD (qv)
engineering procurement and construction company	*aka* EPC company, contracting company, EPCM company, EPCMM company, EPCMV company. A company which builds plants and usually has detailed design capability. They may offer additional services after construction, such as validation in a hygienic context (EPCMV), or maintenance in a BOOT (qv) contract model (EPCMM)
engineering science	The application of scientific principles to the study of engineering artefacts; cf natural science
engineering technician	Someone with an apprenticeship / Bachelor's degree/diploma (governed by the Dublin Accord) in engineering, plus specified experience; cf engineering technologist, engineer; see **Box E1**.
engineering technologist	Someone with an apprenticeship/Bachelor's degree (governed by the Sydney Accord) in engineering, plus specified experience; cf engineering technician, engineer; see **Box E1**.
engineering tolerance	see **tolerance of dimension**
engineering work package	*aka* EWP. A project management term for a deliverable forming part of a construction work package (qv), itself comprised of a defined set of deliverables, (commonly defined as being those for a single engineering discipline, though there isn't much in engineering which is genuinely unidisciplinary)
engineersplaining	Similar to 'mansplaining', but done by an engineer of any gender to a non-engineer of any gender; explaining something to someone who already understands it (possibly better than the person doing the explaining) in what the explainee considers a condescending, overconfident, oversimplified manner. (Of course, since the explainee isn't an engineer, maybe they don't really understand it at all, and don't want to understand it enough to get past the engineer's lack of social skill.)
enhanced high rate clarification	*aka* EHRC. A generic term for a number of techniques which increase settling velocities in effluent clarifiers
enhanced oil recovery	*aka* EOR. A generic term for a number of techniques which increase the yield of oil from wells, usually by injecting a fluid into the well
enjuneer	*(jocular) aka* inguhneer, ungeneer. A non-engineer[40]

---

[40] "Last yeer I kudn't spel ungeneer. Now I are won"

enlargement scale	*aka* enlargement scale factor. 1. A drawing scale factor less than 1:1 2. The ratio of a dimension on a drawing to its smaller real-world equivalent; cf reduction scale
enlargement scale factor	see **enlargement scale**
enriching section	see **rectifying section**
enrichment	Generically, increasing the fraction of some component of a mixture; one specific example being adding assist gas (qv) to relief gas in an endothermic flare, as defined formally in API Standard 521. Pressure-relieving and Depressuring Systems. Sixth Edition	January 2014 and API RP 537 Flare Details for Petroleum, Petrochemical, and Natural Gas Industries
enthalpy	(*symbol* H). A thermodynamic property: the internal energy of a system, plus its pressure multiplied by its volume. It is not strictly the amount of heat or energy in a system, even though it has the units of joules. In practice the most common usage is ΔH (qv), change in enthalpy
enthalpy balance	A system balance akin to an energy balance, with respect to enthalpy
enthalpy of combustion	see **heat of combustion**
enthalpy of solution	see **heat of solution**
enthalpy wheel	see **rotary heat exchanger**
enthalpy-entropy chart	see **Mollier diagram**
entity	see **object**. However, this term is often used in the context of a legal person, e.g., a business
entrainer	A substance which forms an azeotrope (qv) with more than one component of a distillation feed to enhance recovery of some component
entrainment	Entrapment by a fluid of solid particles, liquid droplets, or gas bubbles, and commonly also the transport of the entrapped material by the fluid
entrance losses	*aka* entry losses. In hydraulics, the head losses associated with fluid entering a pipe or channel, (rather than entering a vessel); cf exit losses
entropy	(*symbol* S) The never decreasing (at a whole system level) degree of disorder of a system or, to put it another way, energy unavailable for work
entry length	*aka* entrance length. The distance down a pipe or channel which a fluid travels in order to develop a stable velocity profile
entry losses	see **entrance losses**
ENVID	see **environmental identification**

Environment Agency	*aka* EA. A public body, sponsored by the UK's Department for Environment, Food & Rural Affairs (qv), which acts as a regulator of environmental matters in England and Wales
environment health and safety	*aka* EHS. A common rolling-together of non-profitmaking obligations by those businesses which take a minimalist approach to such matters[41]. Another form rolls quality in as well; cf QESH
environment load	see **environmental load**
environmental aspect	Aspect of an entity that can affect the environment; defined formally in EN ISO 14040:2006 Environmental management - Life cycle assessment - Principles and framework
environmental assessment level	*aka* EAL. A non-statutory benchmark of concentration for a substance after dispersion into the receiving environment used in the UK as a benchmark to assess whether a proposed development's environmental impact is acceptable
environmental identification	*aka* ENVID. An identification tool like HAZID (qv), but for environmental impacts
environmental impact assessment	*aka* EIA. An assessment of the likely environmental and related socio-economic, cultural and human-health impacts (positive or negative) of a proposed development
environmental impact assessment team	*aka* EIA team. "The team which carries out the environmental studies and prepares the environmental information for submission to the competent authority" according to EC Guidance on EIA EIS Review
environmental impact statement	*aka* EIS. A statement of the likely environmental and related socio-economic, cultural and human-health impacts (positive or negative) of a proposed development
environmental information	Information provided by a developer to the competent authority (qv) on the impacts of a project on air, water, soil, land, flora and fauna, energy, noise, waste and emissions. In the UK there is a statutory right of access to such information held by public authorities
environmental load	*aka* environment load, water depth load, wave load, wind load. The addition to design loading to allow for loadings from wind, waves, current, water depth or ice and snow buildup
environmental management system	*aka* EMS. An organization's set of processes and practices aimed at reducing environmental impacts

---

[41] All too often delegated to a particular subset of sidelined staff who may not be taken seriously in the wider organization.

environmental mechanism	A cause effect chain in pollution abatement; defined formally in EN ISO 14040:2006 Environmental management - Life cycle assessment - Principles and framework
Environmental Permitting (England & Wales) Regulations 2016	*aka* EP regulations. Regulations governing environmental permitting under the Environmental Permitting Programme (qv); the current name for what was formerly known as the IPPC Regulations in the UK
Environmental Permitting Programme	*aka* EPP. An environmental regulatory regime in England & Wales which requires an environmental permit to be issued, granting permission to operate any regulated activity
environmental pollution	According to the UK PPC Act (qv), "pollution which causes harm…includes both harm to land, harm to living organisms and harm to ecological systems of which living organisms form part"
Environmental Protection Agency	*aka* EPA. The USA environmental regulator, equivalent of the UK's Environment Agency (qv)
environmental quality standard	*aka* EQS. A term used in EU/UK pollution control (especially water): "a limit for environmental disturbances, in particular, from ambient concentration of pollutants and wastes, that determines the maximum allowable degradation of environmental media", according to the Glossary of Environment Statistics, Studies in Methods, Series F, No. 67, United Nations, New York, 1997
enzyme	A biological catalyst, almost always a protein
enzyme inhibition	Strictly, reduced enzyme activity, though sometimes used to mean reduced enzyme production
enzyme kinetics	Quantification of the effects of varying reaction conditions on enzyme activity
EO	see **overall plate efficiency**
EOC	*(drawing notation)* see **end of curve**
EOLCS	see **Engine Oil Licensing and Certification System**
EOP	*(drawing notation)* see **end of pipe**
EOR	see **enhanced oil recovery**
EOS	see **equation of state**
EOT	see **extension of time**
Eötvös number	(*symbol* Eo) *aka* Bond number. A dimensionless number measuring the relative importance of gravitational and surface tension forces used to describe the shape of the dispersed phase in a dispersion
EOX	see **extractable organically bound halogens**
EP	1. see **electrostatic precipitator** 2. Environmental protection

EP regulations	see **Environmental Permitting (England & Wales) Regulations 2016**
EPA	1. Environmental Protection Act 1990, a UK law 2. see **Environmental Protection Agency**
EPC company	see **engineering procurement and construction company**
EPCI company	Engineering, procurement, construction and installation company, a variant on engineering procurement and construction company (qv)
EPCM company	Engineering, procurement and construction management company, a variant on engineering procurement and construction company (qv)
EPCMM company	Engineering, procurement and construction management, and maintenance company; a variant on engineering procurement and construction company (qv)
EPCMV company	Engineering, procurement, construction management and validation company, a variant on engineering procurement and construction company (qv)
epm	see **equivalents per million**
EPP	see **Environmental Permitting Programme**
EPR	1. see **Environmental Permitting (England and Wales) Regulations 2016** 2. Environmental Protection (Duty of Care) Regulations (UK)
EPS	Emergency power system
Eq	see **equivalent**
EQS	see **environmental quality standard**
equalization tank	see **flow equalization**
equation of state	*aka* EOS. A thermodynamic equation which describes the properties of a system, usually in terms of pressure, volume and temperature. There are several such equations, none of which is universally applicable across all phases of matter.
equilibrium	*aka* chemical equilibrium. The condition in which there are equal rates of forward and reverse reaction for a reversible chemical reaction
equilibrium constant	*aka* law of chemical equilibrium. Specifically, the molar ratio of reactants and products at chemical equilibrium.
equilibrium distillation	see **flash distillation**
equilibrium moisture	see **equilibrium moisture content**
equilibrium moisture content	*aka* EMC, equilibrium moisture. The moisture content of a hygroscopic material in equilibrium with its surroundings
equilibrium ratio	(*symbol* K) *aka* K-value, vapor liquid equilibrium ratio. The ratio of the mole fraction in the vapor to the mole fraction in the liquid in a vapor liquid system at equilibrium

equilibrium segregation coefficient	see **segregation coefficient**
equilibrium stage	see **theoretical stage**
equilibrium vapor pressure	The pressure exerted by a vapor in thermodynamic equilibrium with its condensed phases in a closed system at a given temperature. Note that this might not be the same as vapor pressure (qv)
equimolar counterdiffusion	The condition where equal numbers of molecules of two substances are diffusing in opposite directions
equipment	The mechanical and electrical assemblies required to achieve an objective. A term which can describe a whole plant, but is more commonly applied to a single process unit, or ancillary machines. Defined formally in BS EN 82079-1:2012 Preparation of instructions for use. Structuring, content and presentation. General principles and detailed requirements and BS EN ISO 10209:2012 Technical product documentation – Vocabulary – Terms relating to technical drawings, product definition and related documentation
equipment isolation valve(s)	The manual shutoff valve(s) for an item of equipment; defined formally in one context in NFPA 86: Standard for Ovens and Furnaces, 2019 Edition
equipment layout	Layout at the level of a single process unit and associated ancillaries: the consideration of other small plant or associated/attendant items around a process unit
equipment list	*aka* equipment schedule. A formal list of all main plant items on a process plant with their most notable characteristics
equipment load	Loading on a structure from the equipment's own weight
equipment schedule	see **equipment list**
equivalent	*aka* Eq. A unit used in water treatment, the product of valency and molarity, often for convenience in the form of milliequivalents (mEq), (one equivalent of a univalent ion is one mole; one equivalent of a divalent ion is half a mole); cf equivalents per million
equivalent dose	see **dose equivalent**
equivalent hydraulic diameter	The diameter of a circular pipe which would give the same headloss as that for a rectangular duct (or sometimes more broadly any conduit with a non-circular cross section)
equivalent isothermal temperature	*aka* EIT. A concept used in hydroprocessing, being an approximation of the isothermal reactor temperature, which would give equivalent performance to a real non-isothermal reactor

equivalent length	An expression of the headloss of fittings as equivalent lengths of straight run pipe
equivalent person	see **population equivalent**
equivalents per million	*aka* epm. Equivalents per million grammes; an oil and gas unit similar, but not always identical to milliequivalents (qv). Identical to mEq for water at 20 °C
equivalent system	A system substituted for another to simplify analysis; defined formally in BS ISO 2041:2018 Mechanical vibration, shock and condition monitoring. Vocabulary
equivalent tube metal temperature	An approximated constant temperature producing the same rate of metal creep damage as a variable temperature; defined formally in API STD 530 - Calculation of Tube Heater Thickness
ER	Electrical resistance
Erbar–Maddox correlation	An empirical correlation of the number of theoretical stages required in a distillation column as a function of reflux ratio
erg	A largely obsolete unit of energy equal to $10-7$ joules (100 nJ); still used in the USA
ergodic process	Any process with statistical properties which can be deduced from a random sampling, because they do not change over time; defined formally in BS ISO 2041:2018 Mechanical vibration, shock and condition monitoring. Vocabulary
Ergun equation	An alternative to the Carman-Kozeny equation (qv) for calculation of headloss through a packed bed
erosion	A reduction in thickness due to abrasion by a fluid. Defined formally in BS EN ISO 13705:2012/ISO 13705:2012(E) Petroleum, petrochemical and natural gas industries. Fired heaters for general refinery service, API STD 530 - Calculation of Tube Heater Thickness and API STD 560 - Fired Heaters for General Refinery Service and API RP 579 - Fitness for Service
erosion/corrosion	The combined effect on a surface by these two mechanisms in concert
ERP	Emergency response plan
error	Inaccuracy; cf accuracy
errors & omissions insurance	see **professional indemnity insurance**
ERW	*(drawing notation)* Electric resistance weld
ES	1. *(drawing notation)* Electronic switch 2. Environmental statement

escape BA	*aka* emergency escape breathing device, EEBD. A device comprising a mask and hood providing 10-15 minutes of oxygen and upper body protection used to immediately leave an area which has become dangerous
escape, evacuation and rescue analysis	see **escape, evacuation and rescue assessment**
escape, evacuation and rescue assessment	*aka* EERA, escape, evacuation and rescue analysis. A systematic investigation intended to ensure risks to personnel are ALARP (qv), in the event of a requirement for them to leave site due to a loss of containment incident
ESD	see **emergency shutdown**
ESDV	see **emergency shutdown valve**
ESPE	Electrosensitive protective equipment; see **sensitive protective equipment**
ESS	Emergency shutdown system; see **safety instrumented system**
ESSA	Emergency systems survivability analysis
estimation	A way of obtaining a 'good enough' answer for intended purposes
e-stop	see **emergency motor stop**
Esw	*(drawing notation)* Eccentric swage
ET	*(drawing notation)* Electric tracing
ETAP	Electrical transient and analysis program
etching	Removing the surface of a material with acid; defined formally in ASME BPE (American Society of Mechanical Engineers: Bioprocessing Equipment)
ethane	A gaseous C2 alkane/paraffin
ethene	A gaseous C2 alkene/olefin
ethical	1. Generically, compatible with professional ethics 2. In the pharmaceutical industry, see **ethical medical products**
ethical medical products	Substances and devices for the diagnosis or treatment of illness available only by prescription from a medical practitioner
ethical pharmaceutical	Substances for the diagnosis or treatment of illness available only by prescription from a medical practitioner; defined formally in ASME BPE (American Society of Mechanical Engineers: Bioprocessing Equipment)
ethyl alcohol of agricultural origin	see **neutral spirit**
ethylene	see **ethene**
ethyne	*aka* acetylene. A gaseous C2 alkyne/acetylene
ETLA	*aka* XTLA. *(jocular)* Extended three letter acronym; i.e., a four letter acronym!
EU ETS	see **European Union Emissions Trading System**

EUE	*(drawing notation)* External upset ends
eukaryotic	A classification of 'higher' forms of life, the cells of which have a membrane-bound nucleus, amongst other things; cf prokaryotic
Euler number	*(symbol* Eu) A dimensionless number used in fluid dynamics which relates the pressure drop across a restriction to the kinetic energy per unit volume of a flowing fluid; N.B. not Euler's number, e
Eulerian fluid dynamics	An approach which defines a control volume and expresses fluid flow properties as fields, ignoring individual particles; cf Lagrangian fluid dynamics
EUR ING	European engineer, a professional engineering designation of FEANI (qv)
Europe, Africa and Asia	*aka* EAA. A commonly used collection of geographical territories
Europe, Middle East, Africa	A commonly used collection of geographical territories
European Brewing Convention	*aka* EBC. A color (qv) measurement scale used in the European brewing industry; cf standard reference measure
European Chemicals Agency	*aka* ECHA. The EU's chemical regulatory authority
European Commission	*aka* EC. The executive branch of the EU
European Environment Agency	*aka* EEA. A European body which provides 'independent' environmental advice; not an environmental regulator like the Environment Agency (qv) in England & Wales
European Hygienic Engineering & Design Group	*aka* EHEDG. A European NGO providing advice on best practice in hygienic engineering and design
European Medicines Evaluation Agency	*aka* EMEA. The European equivalent of the US FDA
European Union Emissions Trading System	*aka* EU ETS. An EU-wide system of trading greenhouse gas emissions
eutectic	see **eutectic mixture**
eutectic mixture	*aka* eutectic, eutectic system. A homogeneous mixture of substances with a melting point lower than that of any constituent
eutectic point	The eutectic temperature (qv) as shown on a phase diagram
eutectic system	see **eutectic mixture**
eutectic temperature	The temperature at which a particular eutectic mixture (qv) freezes or melts

eutrophication	Nutrient enrichment of water, potentially leading to the growth of problematic levels of algae and/or cyanobacteria; qv **harmful algal bloom**
evacuation time	*aka* pumpdown time, time of exhaust. Time required to reduce vacuum system pressure from atmospheric to a given pressure; defined formally in NFPA 86: Standard for Ovens and Furnaces, 2019 Edition
evaluation	In life cycle assessment, an interpretation phase used to establish confidence in analysis results; defined formally in EN ISO 14040:2006 Environmental management - Life cycle assessment - Principles and framework
evaporation	Relatively slow vaporization of a liquid from its surface at a temperature below its boiling point; cf **boiling**
evaporator	A term used for disparate heat exchange devices in which some fluid vaporizes (not usually strictly by evaporation (qv)), for the purpose of heat transfer to the vaporized fluid, or to make the residual fluid more concentrated in non-volatile components. In a heat recovery steam generator (qv), the evaporator is the part where water is boiled (rather than evaporated); defined formally in API RP 534 – Heat Recovery Steam Generators
event tree	A logic diagram used to identify and analyze event sequences leading to an event such as an accident
EVM	Electronic voltmeter
EW	1. *(drawing notation)* Each way 2. *(drawing notation)* Electric weld
EWP	see **engineering work package**
Ex rated equipment	see **explosionproof equipment**
ex works	see **Incoterms**
exa-	(*symbol* E-) SI prefix denoting 10^{18}
excess air	The percentage of air supplied to a burner above that required for complete fuel combustion; defined formally in BS EN ISO 13705:2012/ISO 13705:2012(E) Petroleum, petrochemical and natural gas industries. Fired heaters for general refinery service, API RP 537 Flare Details for Petroleum, Petrochemical, and Natural Gas Industries, API STD 530 - Calculation of Tube Heater Thickness, API STD 560 - Fired Heaters for General Refinery Service and API RP 535 - Burners for Fired Heaters at Refineries
excess flow check valve	see **excess flow valve**

excess flow valve	A valve which prevents excess flow, i.e., which closes when fluid exceeds given flow; defined formally in API RP 2510A - Fire Protection Considerations for the Design and Operation of Liquefied Petroleum Gas (LPG) Storage Facilities and NFPA 58: Liquefied Petroleum Gas Code, 2017 Edition
excess reactants	*aka* excess reagents. Reactants supplied above those required for complete reaction, or sometimes the excess percentage of those reagents
excess temperature limit interlock	A safety device which cuts off a heat source at excess operating temperature; defined formally in NFPA 86: Standard for Ovens and Furnaces, 2019 Edition
excessive penetration	A defect in which an excess of metal protrudes from a weld root; defined formally in ASME BPE (American Society of Mechanical Engineers: Bioprocessing Equipment)
exchange capacity	In the context of ion exchange, the quantity of charge which can be exchanged per mass of resin (or soil in an environmental context)
exchange sites	The reactive groups on an ion exchange medium
exchanger	*(informal)* Equipment which exchanges heat or sometimes ions between process streams
exchanger bed	Ion exchange equipment comprising media, vessel and all local ancillaries
exchanger bundle removal load	Half the weight of a heat exchanger tube bundle, used to consider loads in maintenance activity
excitation	An external force tending to produce forced vibration; cf free vibration
exclusion list	List of criteria for identifying categories of projects not requiring EIA/EIS in the EU; cf FONSI, mandatory list
excursion	1. A process excursion is a deviation of process parameters from acceptable range 2. *aka* total excursion. In the context of vibration, excursion means displacement; defined formally in BS ISO 2041:2018 Mechanical vibration, shock and condition monitoring. Vocabulary
exergy	*aka* stream availability. The maximum useful work available for a system
exergy analysis	Analysis of system exergy (qv) using tools such as exergy balance to detect thermodynamic inefficiencies
exfiltration	Losses of water due to its escaping from underground pipes or sewers into the environment; cf infiltration
exhaust gas mass flow	see **exhaust steam mass flow**

exhaust steam mass flow	Mass flow through a turbine casing into backpressure system or condensing plant; defined formally in ISO14661 Thermal turbines for industrial applications (steam turbines, gas expansion turbines) - General requirements
exhausting section	see **stripping section**
exhaustion	1. The effective phase of ion exchange, in which the resin's capacity for exchange is exhausted, and target ions are removed from solution 2. More generally, the process of loss of sorption (qv) capacity
exit losses	In hydraulics, the head losses associated with fluid exiting a pipe or channel, (rather than entering a vessel); cf entrance losses
exit velocity	1. Generically, the speed at which something leaves something 2. Specifically, flare burner design exit speed; defined formally in API RP 537 Flare Details for Petroleum, Petrochemical, and Natural Gas Industries
expanded bed reactor	A more novel type of AD (qv) reactor based on an expanded bed of carrier material, far smaller (but less robust) than the traditional type
expanded granular sludge bed	*aka* EGSB. A more novel type of AD (qv) reactor based on an expanded bed of sludge granules, far smaller (but less robust) than the traditional type
expansion	1. Generically, increase in volume of a mass of material 2. Specifically, in the case of firefighting foam production, the ratio of foam volume to foam solution volume; defined formally in NFPA 11: Standard for Low-, Medium-, and High-Expansion Foam, 2016 Edition
expansion bend	*aka* expansion loop. A series of bends or a loop installed in a metallic pipe to provide flexibility to accommodate thermal expansion
expansion coefficient	see **coefficient of expansion**
expansion loop	see **expansion bend**
expansion turbine	see **gas expansion turbine**
experimental design	Working backwards from a statistical test to devise a sufficiently rigorous trial, to allow a sufficiently certain answer to a question to be generated. An approach all too infrequently followed when devising plant performance tests, leading to entirely avoidable conflict; qv design of experiments
EXPF	*(drawing notation)* Expander flange

expiration date	*aka* expiry date. The date after which a product is deemed unsafe to use. The relationship between this and shelf life (qv) is somewhat contested, but, according to ASME BPE (American Society of Mechanical Engineers: Bioprocessing Equipment), expiration date is the same as the end of shelf life
expiry date	see **expiration date**
exploded view	A pictorial view of an assembly of parts which shows their physical interrelationship and/or assembly order; defined formally in BS EN ISO 10209:2012 Technical product documentation — Vocabulary — Terms relating to technical drawings, product definition and related documentation
exploratory hole	A dug or bored hole used for visual investigation of ground conditions; defined formally in BS 5930:2015 Code of practice for ground investigations
explosion	A sudden violent expansion which could be caused by the rupture of an enclosure by subsonic internal deflagration (of a low explosive) or supersonic detonation (of a high explosive). The NFPA definition only covers the first of these; see NFPA 654: Standard for the Prevention of Fire and Dust Explosions from the Manufacturing, Processing, and Handling of Combustible Particulate Solids, 2017 Edition
explosion doors	Doors akin to blowout panels, relieving pressure in the event of explosion within a furnace
explosion limit	see **flammable limits**
explosion resistant radiant tube	A radiant tube (qv) system resistant to catastrophic failure under maximum fuel/air deflagration overpressure; defined formally in NFPA 86: Standard for Ovens and Furnaces, 2019 Edition
explosion suppression system	A system to detect and suppress residual fire and explosion hazards using extinguishing agent (qv)
explosionproof	see **explosionproof enclosure**
explosionproof enclosure	*aka* enclosure, explosionproof. An electrical enclosure which prevents any internal fire or explosion from propagating outside the enclosure, and operates at a temperature below the ignition temperature of an expected surrounding fuel/air mixture. (Counterintuitively, according to the API, this differs from a flameproof enclosure (qv) mainly in that it is allowed to be damaged by the explosion); defined formally in API RP 505 2nd Edition, August 2018 Recommended Practice for Classification of Locations for Electrical Installations at Petroleum Facilities Classified as Class I, Zone 0, Zone 1, and Zone 2, and in NEMA 250- Enclosures for Electrical Equipment.

explosionproof equipment	*aka* Ex rated equipment. Equipment with an explosionproof enclosure; cf intrinsically safe apparatus
explosionproof motor	A motor with an explosionproof enclosure; defined formally in NFPA 20: Standard for the Installation of Stationary Pumps for Fire Protection, 2019 Edition
explosive	1. Capable of causing an explosion 2. A material with this property, especially one capable of initiation with a blasting cap; cf blasting agent
explosive gas atmosphere	Any atmosphere whose proportion of gases has explosive properties. Defined formally in API RP 505 2nd Edition, August 2018 Recommended Practice for Classification of Locations for Electrical Installations at Petroleum Facilities Classified as Class I, Zone 0, Zone 1, and Zone 2
explosive limits	see **flammable limits**
explosive range	The range of concentration between LEL (qv) and UEL (qv) under given conditions; cf flammable range
exponential decay	Decreasing at a rate proportional to current value, as a first order reaction does
exponential growth	Increasing at a rate proportional to current value, as the most rapid period of microbiological growth does
exposure	see **dose**
exposure hazard	A habitable structure greater than 9 m^2 in floor area, separated by less than or equal to 15m from another; defined formally in NFPA 1142: Standard on Water Supplies for Suburban and Rural Fire Fighting, 2017 Edition
exposure protection	Protection of structures or equipment from fire damage using water sprays; defined formally in NFPA 15 Standard for Water Spray Fixed Systems for Fire Protection
extended surface	A surface whose area has been increased by the addition of such devices as studs or fins, most usually applied to heat exchange surfaces. Defined formally in BS EN ISO 13705:2012/ISO 13705:2012(E) Petroleum, petrochemical and natural gas industries. Fired heaters for general refinery service, API STD 530 - Calculation of Tube Heater Thickness, and API STD 560 - Fired Heaters for General Refinery Service
extension of time	Increased duration of a construction contract, or a contractual provision allowing this in the event of specified circumstances
extension ratio	1. *aka* stretch ratio. Most commonly, the ratio of a length of stretched material to its unstretched length

	2. Less commonly, the ratio of extended surface (qv) area to unextended 'bare tube' surface area; defined formally in BS EN ISO 13705:2012/ISO 13705:2012(E) Petroleum, petrochemical and natural gas industries. Fired heaters for general refinery service, API STD 530 - Calculation of Tube Heater Thickness and API STD 560 - Fired Heaters for General Refinery Service
extension shaft	A shaft which allows a septic tank or packaged sewage treatment plant to be fitted flush with, or slightly above, ground level[12], defined formally in EN 12566 - Small wastewater treatment systems for up to 50 PT Part 1: Prefabricated septic tanks
extensive variable	A variable whose value is independent of the size of the system; cf intensive variable
extent of reaction	A less common, less useful way to look at equilibrium constant (qv)
external floating roof tank	*aka* EFRT. A floating roof tank without a fixed roof. Defined formally in API MPMS 19.1 Manual of Petroleum Measurement Standards Chapter 19.1 Evaporative Loss from Fixed-roof Tanks; cf covered floating roof tank, internal floating roof tank
external pressure relief valve	A PRV (qv) entirely external to the vessel and piping it serves; defined formally in NFPA 58: Liquefied Petroleum Gas Code, 2017 Edition
external provider	Provider external to an organization; defined formally in BS EN ISO 9000 Quality management systems Fundamentals and vocabulary
external supplier	see **external provider**
externally mounted seal	*aka* outside mounted seal. A pump seal mounted outside a seal chamber
extinction safety time	*aka* flame failure response time. A British term for the time between flame extinction and safety shutoff device (qv) deenergization; defined formally in BS EN 12952-8:2002 Water-tube boilers and auxiliary installations. Requirements for firing systems for liquid and gaseous fuels for the boiler; cf flame failure response time
extinguishing agent	A substance used to put out fires. Different extinguishing agents are better for different fire classifications (qv)
extra extra heavy	*aka* XXH; see **double extra strong**
extra extra strong	*aka* XXS; see **double extra strong**
extra heavy	*aka* XH, see **extra strong**
extra heavy oil	see **asphalt**

---

[12] Something which it is easy to do incorrectly

extra strong	*aka* extra heavy, XH, XS. ANSI 'extra strong' pipe schedule
extract	1. Generically, to remove 2. Commonly, in engineering, to remove something from one phase to another by preferential partition into a solvent. i.e., solvent extraction (qv)
extractable organically bound halogens	A measure of organohalogens in water, soil or sludge used in environmental contexts, similar to adsorbable organically bound halogens (qv)
extractables (polymeric)	Chemicals which can contaminate a product by extraction from wetted polymeric materials. Defined formally in ASME BPE (American Society of Mechanical Engineers: Bioprocessing Equipment)
extraction	1. Generically, removal 2. Commonly in engineering, removal of something from one phase to another by preferential partition into a solvent, or the corresponding unit operation, i.e., solvent extraction (qv)
extraction mass flow	Mass of gas extracted from a turbine at a pressure intermediate between turbine inlet and outlet pressure; defined formally in ISO14661 Thermal turbines for industrial applications (steam turbines, gas expansion turbines) - General requirements
extraction steam conditions	Conditions of steam extracted from a turbine for process purposes, measured at turbine connections. Defined formally in ISO14661 Thermal turbines for industrial applications (steam turbines, gas expansion turbines) - General requirements
extraction turbine	Turbine designed to produce extraction steam for process use; defined formally in ISO14661 Thermal turbines for industrial applications (steam turbines, gas expansion turbines) - General requirements
extractive distillation	Separation of an azeotropic mixture by adding a high-boiling 'separation solvent' which does not form an azeotrope (qv) with the other components
extractor	1. Generically, something which removes something 2. Commonly in engineering, an air moving device, or a device which removes something from one phase to another by preferential partition into a solvent, i.e., solvent extraction (qv)
extrados	The line of the exterior curve of an arch; cf intrados
extraneous material	Something which does not belong where it is found; cf foreign body, foreign object
extrusion	Forcing a fluid (especially a setting fluid) through a die; cf spinning
EXW	Ex works, see **Incoterms**

Eyring equation	*aka* Eyring–Polanyi equation. An equation which relates reaction rate to temperature in a way similar to the Arrhenius equation, but with application to a wider number of phases of matter
Eyring–Polanyi equation	see **Eyring equation**

F

F value	Time in minutes of treatment required to give a set level of reduction in organism numbers (usually 12D); cf 12D process, D value
F&D	*(drawing notation)* Faced (or flanged) and drilled
F&G	Fire and gas
F/F	*(drawing notation)* Face of flange
F:M ratio	see **food to microorganism ratio**
FA	Flue ash or fly ash; see **pulverized flue ash**
face to back configuration	In the context of double mechanical pump seals, having a mating ring mounted between two flexible elements and a flexible element mounted between two mating rings; defined formally in API RP 682 - Pump Seals; cf face to face configuration
face to face configuration	In the context of double mechanical pump seals, having both mating rings mounted between two flexible elements; defined formally in API RP 682 - Pump Seals; cf face to back configuration
facility	1. Generically, a term sometimes used for a process plant 2. Specifically, part of an LPG plant; defined formally in NFPA 59: Utility LP-Gas Plant Code, 2018 Edition
facility hose	Permanently installed hose and couplings used to unload LPG tankers into industrial plant; defined formally in NFPA 58: Liquefied Petroleum Gas Code, 2017 Edition
factor of safety	see **safety factor**
factory acceptance test	*aka* FAT. Testing (usually performed by the customer according to its user requirements) that an assembled item of equipment meets specification before it leaves the factory
facultative anaerobe	In biology, an aerobic organism which has the ability to survive under both aerobic and anaerobic conditions

FAD	1. (Most commonly) see **free air delivered**
	2. (Less commonly) see **failure assessment diagram**
FAFR	Fatal accident frequency rate, see **fatal accident rate**
Fahrenheit scale	A non-SI temperature scale, favored in the USA
fail closed	Equipment designed to automatically shut down/close in the event of failure; cf air-to-open
fail freeze	see **fail in place**
fail in place	aka fail freeze, fail last, fail last position. Equipment designed to automatically maintain its position at point of failure in the event of failure
fail last	see **fail in place**
fail last position	see **fail in place**
fail open	Equipment designed not to automatically shut down/close in the event of failure; cf air-to-close
fail over	Equipment designed to automatically achieve a safe state in the event of failure, by bringing a standby (qv) online
fail safe	*aka* failsafe. Equipment designed to automatically achieve a safe state in the event of failure
fail safe system	see **failsafe system**
fail to danger	see **failure to danger**
fail to safety	A fail safe (qv) design philosophy; cf graceful degradation
failsafe	see **fail safe**
failsafe system	*aka* fail safe system. A system which automatically renders plant or equipment safe in the event of failure; defined formally in EN ISO 10437 Petroleum, petrochemical and natural gas industries - Steam turbines - Special-purpose applications
failure	Inability to meet specified performance on demand; defined formally in BS EN ISO 12100:2010 Safety of machinery. General principles for design. Risk assessment and risk reduction
failure assessment diagram	*aka* FAD. A chart used to evaluate crack-like flaws; defined formally in API RP 579 - Fitness for Service
failure mode and effects analysis	*aka* FMEA. A systems engineering (qv) process that examines failures in products or processes using a standardized method and scoring system
failure to danger	A somewhat contested term; either:
	1. Failure of equipment to a condition that may lead to a hazardous situation or stop effective operation if called upon to prevent or suppress a hazard, or

	2. A situation where plant or equipment failure does not automatically lead to a safe condition; defined formally in BS EN ISO 12100:2010 Safety of machinery. General principles for design. Risk assessment and risk reduction
failures in time	*aka* FIT. Number of failures per 109 hours
Falconbridge process	A process for production of nickel from a nickel-copper mixture by acid-leaching most of the copper, and producing pure nickel from the residue by electrolysis (qv)
fall arrest system	see **fall arrester**
fall arrester	*aka* fall arrest system. A system attached to a harness which automatically arrests any excessive descent speed, in much the same way as an inertia reel seatbelt; cf fall protection
fall protection	Physical measures reducing risk of fall from a height; defined formally in BS EN ISO 14122 Safety of machinery. Permanent means of access to machinery. Working platforms and walkways
falling film evaporator	An evaporator (qv) in which a continuous film of fluid to be evaporated flows downwards under the influence of gravity
falling rate drying	The part of a drying cycle in which the water stops evaporating as from a free water surface, (qv constant rate drying) and the rate of change of moisture content decreases with time
false grain	Undesirably small crystals in sugar refining
fan	The lowest-pressure type of gas compressor
fan coil heater	*aka* fan coil unit. Part of an HVAC system, consisting of a heat exchanger (coil) and a fan to distribute the heated air
fan coil unit	*aka* FCU; see **fan coil heater**
fan curve	Plot of delivered flow vs pressure for a fan
fan laws	see **affinity laws**
fan rated point	Fan flow and head at specified duty points; defined formally in BS EN ISO 13705:2012/ISO 13705:2012(E) Petroleum, petrochemical and natural gas industries. Fired heaters for general refinery service
fan static pressure	Static pressure of the system downstream of a fan; defined formally in BS EN ISO 13705:2012/ISO 13705:2012(E) Petroleum, petrochemical and natural gas industries. Fired heaters for general refinery service
fan total pressure	Differential head across a fan; defined formally in BS EN ISO 13705:2012/ISO 13705:2012(E) Petroleum, petrochemical and natural gas industries. Fired heaters for general refinery service

fan velocity pressure	Pressure at average velocity at fan outlet; defined formally in BS EN ISO 13705:2012/ISO 13705:2012(E) Petroleum, petrochemical and natural gas industries. Fired heaters for general refinery service
fan vendor	Fan manufacturer; defined formally in BS EN ISO 13705:2012/ISO 13705:2012(E) Petroleum, petrochemical and natural gas industries. Fired heaters for general refinery service; cf supplier
Fanning friction factor	The more commonly used friction factor in the Darcy Weisbach equation (qv). A dimensionless number, the ratio of local shear stress to local flow kinetic energy density. 1/4 of the Darcy friction factor (qv)
FAR	see **fatal accident rate**
farad	(*symbol* F) The SI unit of capacitance
farad per meter	(*symbol* F/m) The SI unit of permittivity
Faraday still	A particular type of pot still, named after its inventor[43]
Faraday's law	An equation describing the rate of corrosion (see ASTM G102-89(2004) e1 Standard Practice for Calculation of Corrosion Rates and Related Information from Electrochemical Measurements)
farm cart	see **moveable fuel storage tender**
FAS	Free alongside ship; see **Incoterms** (sometimes incorrectly thought to stand for free along side)
fast neutron reactor	see **fast reactor**
fast reactor	*aka* FNR, fast neutron reactor. A compact nuclear reactor with no moderator, using fast neutrons to sustain the reaction
fast rinse	Part of a backwash/regeneration cycle following the slow rinse for a packed bed such as ion exchange
FAT	see **factory acceptance test**
fat	A hydrocarbon or lipid solid. More solid than a grease (qv), less solid than a wax (qv)
fatal accident rate	*aka* FAR, fatal accident frequency rate. Individual risk in a given industry expressed in estimated number of fatalities per 108 worker exposure hours
fatigue	Damage to a material caused by cyclical stress, which can cause failure at far lower stress values than for a single loading; defined formally in API RP 579 - Fitness for Service

---

[43] Probably not the Faraday you were thinking of!

fatigue endurance limit	*aka* fatigue limit, endurance limit, fatigue strength. Generally, stress level below which fatigue failure does not occur with an infinite number of cycles. Commonly about half tensile strength; defined formally in API RP 579 - Fitness for Service cf fatigue strength
fatigue limit	see **fatigue endurance limit**
fatigue notch factor	see **fatigue strength reduction factor**
fatigue strength	Though used generically as a synonym for fatigue endurance limit (qv), this is properly defined as the stress level below which fatigue failure does not occur with a stated finite number of cycles; defined formally in this context in API RP 579 - Fitness for Service
fatigue strength reduction factor	*aka* fatigue notch factor. Factor accounting for deviation in theoretical fatigue strength associated with things like corrosive environment, surface conditions, mode of loading, and notch effects; defined formally in API RP 579 - Fitness for Service
fats oils and greases	*aka* FOGs. A term used in effluent treatment for floating or entrained matter soluble in hexane or similar solvents. Tends to be applied to biologically derived materials, even though the test also detects hydrocarbon materials and the TPH (qv) test is similar; cf NAPL
fatty acid	Malodorous organic chemical comprising a hydrocarbon chain with a terminal carboxyl group
FAU	see **formazin attenuation unit**
faucet	US English for tap (qv) or spigot
fault	A state of failure; defined formally in BS EN ISO 12100:2010 Safety of machinery. General principles for design. Risk assessment and risk reduction
fault tolerant external control circuit	Fire pump external control circuits which do not prevent the fire pump starting when they are in a fault state; defined formally in NFPA 20: Standard for the Installation of Stationary Pumps for Fire Protection, 2019 Edition
fault tree analysis	A systematic graphical way to estimate probability of an event by placing the root (*aka* top event) of a 'tree of logic' diagram, and each situation-causing effect being added to a branching 'tree' as a series of logic expressions
FB	1. Free base 2. *(drawing notation)* Full body
FBC	Fluidized bed combustion
FBD	see **function block diagram**
FBE	see **fusion bond epoxy**
FBF	*(drawing notation)* Full barrel flange

FC	*(drawing notation)* see **fail closed**
FCA	1. see **future corrosion allowance**
	2. *(drawing notation)* Issued for client approval
	3. Free carrier, see **Incoterms**
FCC	see **fluid catalytic cracker**
FCCU	see **fluid catalytic cracking unit**
FCR	*(drawing notation)* Issued for client review
FCS	*(drawing notation)* Forged carbon steel
FCU	see **fan coil unit**
FDA	Food and Drug Administration, see **US Food and Drug Administration**
FDN	*(drawing notation)* Foundation
FDS	see **functional design specification**
FE	1. *(drawing notation)* Flanged ends
	2. *(drawing notation)* Flow element
FEA	see **finite element analysis**
FEANI	Fédération Internationale d'Associations Nationales d'Ingénieurs (International Federation of National Engineering Institutions)
feasibility study	An early stage in design/project management which may be undertaken by non-engineers. To the extent that there is design content, this will be at conceptual design (qv) level; see **Box S2**.
fed batch culture	see **fed batch process**
fed batch process	*aka* fed batch culture. A regime intermediate between continuous and batch processes; essentially a batch process with continuing feed addition. Reasonably common in bioreactor operation; cf semibatch process
Fédération Internationale des Ingénieurs-Conseils	*aka* FIDIC. International Federation of Consulting Engineers: an international standards organization for consulting engineering and construction best known for its family of contract templates
feed	1. Most commonly, the introduction of a stream into a process
	2. Fairly commonly, short for feedstock (qv).
	3. Rarely, food for animals; defined formally as such in Regulation (EC) No 178/2002 of the European Parliament and of the Council of 28 January 2002
FEED	see **front end engineering design**
feed business	An undertaking producing food for animals; defined formally in Regulation (EC) No 178/2002 of the European Parliament and of the Council of 28 January 2002
feed rate	Rate of feed (qv), usually expressed as mass per unit of time

feed section	*aka* flashing section. The location where feed is introduced into a distillation column; cf stripping section, rectifying section
feed tank level	see **pumping liquid level**
feedback	1. In process control, the signal from a sensor used to control a correcting element (qv) 2. In quality management, user expressions of satisfaction or otherwise with a product or service; defined formally as such in BS EN ISO 9000 Quality management systems Fundamentals and vocabulary
feedback control	A reactive process control method, in which a sensor signal taken downstream of the action point of a correcting element is compared with a setpoint, and the difference used to control the magnitude of a correcting element's control action
feeder	1. An incoming power line; defined formally in the context of firepumps in NFPA 20: Standard for the Installation of Stationary Pumps for Fire Protection, 2019 Edition 2. Also a generic term for the equipment upstream of a key piece of equipment
feedforward control	An anticipatory process control method in which a correcting element's control action is based on measurements of variables that are expected to bear on the future value of the variable to be controlled (rather than its current value, as with feedback control (qv))
feedstock	*aka* raw material. The main input material to a process (usually ignoring water)
FEL	see **front end loading**
female pipe thread	*aka* FPT; see **national pipe thread**
FEMI	Fixed equipment mechanical integrity
femto-	(*symbol* f-) SI unit prefix denoting a factor of 10^{-15}
Fenske equation	A shortcut equation used to estimate the minimum number of theoretical plates required for separation by total reflux distillation of a binary mixture
FERA	see **fire and explosion risk assessment**
FERC	Federal Energy Regulatory Commission (USA)
fermentation	Whilst in biochemistry/academia this term means only anaerobic biological processes, in engineering it is used to describe all biological syntheses of organic compounds; defined formally in ASME BPE (American Society of Mechanical Engineers: Bioprocessing Equipment)
fermenter	see **fermentor**

fermentor	*aka* fermenter, bioreactor. A process vessel used for fermentation; defined formally in ASME BPE (American Society of Mechanical Engineers: Bioprocessing Equipment)
ferralium	*aka* ferralium 255, superduplex stainless steel. An exceptionally strong and hardwearing alloy of steel popular in pulp and paper applications
ferric	1. Generically, a term meaning involving or akin to iron[44]; trivalent iron compounds in particular 2. Most particularly, shorthand for iron (III) chloride and sulfate solutions used as coagulants in water treatment
ferritic	Related to the naturally occurring low-carbon ferrite metallurgical structure of carbon steel; defined formally in API RP 571 - Damage Mechanism Affecting Fixed Refinery Equipment
ferritic stainless steel	Stainless steel with a predominantly ferritic metallurgical structure (and consequently lower strength and corrosion resistance than other stainless steels) such as 405, 409, 430, 442, and 446; defined formally in API RP 571 - Damage Mechanism Affecting Fixed Refinery Equipment
ferrous	1. Generically, a term meaning involving or like iron, divalent iron compounds in particular; cf ferric, copperas 2. Commonly used in water treatment as shorthand for iron (II) sulfate
ferrous metal	Iron and its alloys
FF	1. Film-forming 2. Finish to finish 3. *(drawing notation)* Flat/full face
F-F	*(drawing notation)* Face to face
FFFP	see **film forming fluoroprotein foam concentrate**
FFKM perfluoroelastomer	Chemically resistant, high temperature elastomer (qv) used in pump O-ring(qv)s; defined formally in API RP 682 - Pump Seals
FFMEA	see **functional failure mode and effects analysis**
FFRT	see **flame failure response time**
FFS	1. see **fit for service** 2. (profane) An expression of exasperation, perhaps at something or someone considered unfit for service
FGD	see **flue gas desulfurization**
FGR	see **flue gas recirculation**
FIA	see **flow injection analysis**

---

[44] I'd have written iron-y, but as everyone knows, Americans don't understand irony!

FIBC	see **flexible intermediate bulk container**
fiber	*aka* fibre. A strand or length of material (or indigestible food component)
fiber reinforced plastic	*aka* FRP; see **glass reinforced plastic**
fiberglass	see **glass reinforced plastic**
fibre	see **fiber**
fibre reinforced plastic	*aka* FRP; see **glass reinforced plastic**
fibreglass	see **glass reinforced plastic**
Fick's laws of diffusion	The simplest equations (similar in form to Fourier's law (qv)) describing the commonest kind of diffusion, which state that molar flux due to diffusion is proportional to the concentration gradient, and rate of change of concentration at a point in space due to diffusion is proportional to the second derivative of concentration with space
fictitious force	see **inertial force**
FID	see **final investment decision**
fiddle factor	*(informal) aka* fudge factor. A factor which corrects theoretical predictions for real world use
FIDIC	see **Fédération Internationale des Ingénieurs-Conseils**
field changeable	Capable by design of being modified after installation; defined formally in EN ISO 10437 Petroleum, petrochemical and natural gas industries - Steam turbines - Special-purpose applications
field erected container	In an LPG context, a container (qv) at least partially fabricated close to its installed position; defined formally in NFPA 59: Utility LP-Gas Plant Code, 2018 Edition
field metallographic replication	*aka* FMR. A nondestructive testing (qv) technique in which the surface of a component is polished and etched, and its grain structure recorded on adhesive tape
field mounted instrument	*aka* locally mounted instrument. An instrument mounted separately out on the plant, usually in an individual enclosure local to monitored equipment; cf **panel mounted instrument**
field operator	A process plant operator based in the field. Often the most junior grade of staff, particularly in very hot climates where field work can be arduous
field oxide	*aka* FOX. An oxide layer of importance in semiconductor manufacture
fieldbus	Generic term for a group of networking protocols used for control and data acquisition in processing facilities, standardized under IEC 61784/61158
FIFO	1. see **first in first out**; 2. see **fly in fly out**

Fikentscher K-value	An empirically-derived K-value (qv) used for viscosity-based estimation of the molecular weight of polymers
file drawing	A reference as-built drawing (qv); defined formally in BS EN ISO 10209:2012 Technical product documentation — Vocabulary — Terms relating to technical drawings, product definition and related documentation
fill	Anthropogenic ground (qv) comprising material previously selected, placed and compacted to a formal specification; defined formally in BS 5930:2015 Code of practice for ground investigations
filler	see **diluent**
filler metal	Additive welding metal; defined formally in API STD 620 - Low Pressure Storage Tanks
filler valve	A one-way valve used solely to fill or pressurize a container; defined formally in NFPA 58: Liquefied Petroleum Gas Code, 2017 Edition
fillet weld	A triangular weld used to join two perpendicular items; defined formally in API RP 579 - Fitness for Service
filling density	In the context of LPG, the percentage of a container's water capacity (as mass at 16°C) of a given fill of LPG (as mass, no temperature specified); defined formally in NFPA 59: Utility LP-Gas Plant Code, 2018 Edition
film	Usually refers to a coherent layer of fluid, often coating a surface, though may also refer to a thin plastic sheet, or surface fouling
film boiling	Boiling which takes place at a surface much hotter than a liquid's boiling point producing an insulating film of vapor which reduces heat transfer efficiency. This is known as the Leidenfrost effect (qv); cf nucleate boiling
film concept	see **film model**
film condensation	*aka* filmwise condensation. Condensation on a wettable surface, resulting in a film of condensate which rapidly falls under the influence of gravity; cf dropwise condensation
film diffusion	Movement of ions through a fluid film, of importance in ion exchange
film formation	The formation of a film (qv), often a consequence of a positive spreading coefficient (qv); defined formally in the context of firefighting foams in NFPA 11: Standard for Low-, Medium-, and High-Expansion Foam, 2016 Edition

film forming fluoroprotein foam concentrate	A foam concentrate with fluorinated protein surfactants used for fighting hydrocarbon fires; defined formally in NFPA 11: Standard for Low-, Medium-, and High-Expansion Foam, 2016 Edition
film forming foam	A foam type used for fighting hydrocarbon fires; defined formally in NFPA 11: Standard for Low-, Medium-, and High-Expansion Foam, 2016 Edition
film model	*aka* film concept, film theory, two film theory. A mathematical model of mass transfer in which all resistance is deemed to exist in a thin film layer at the boundary
film theory	see **film model**
filmwise condensation	see **film condensation**
filter	A device which removes solid particles from a fluid on the basis of their size; defined formally in the context of fired heaters in API RP 535 - Burners for Fired Heaters at Refineries
filter cake	A layer of filtered particles built up on a filtration medium, which may itself become the filtration medium in cake filtration (qv)
filter mud	*aka* cuchasa. Insoluble matter extracted from sugar cane juice during processing
filter press	A compressed stack of recessed plates covered in a filtration medium; a robust Victorian technology still in widespread use
filter stage	A term specific to gas turbines. A filter designed to remove specified contaminants at specified efficiency; defined formally in ISO 3977 Gas turbines - Procurement - Part 3: Design requirements
filtrate	A liquid which has passed through a filter, and consequently has a reduced level of solids
filtration	The unit operation of passing through a filter; cf **Box L1**.
FIM	Full indicator movement, see **total indicator reading**
final certificate	A document recording definitive acceptance of a plant by a client as meeting contractual requirements, which usually triggers a final stage payment. Defined formally in BS EN ISO 10209:2012 Technical product documentation — Vocabulary — Terms relating to technical drawings, product definition and related documentation; cf completion certificate, interim certificate
final consumer	The person who eats or drinks a food or drink, or more broadly, uses a product; defined formally in the context of food and drink in Regulation (EC) No 178/2002 of the European Parliament and of the Council of 28 January 2002

final investment decision	*aka* FID. The last stage gate: commitment to invest in a project, having obtained finance etc.
final moisture content	The moisture content at the end of a drying process as a percentage w/w
final order	Findings of fact and orders for appropriate relief issued by the US Pipeline and Hazardous Materials Safety Administration, as described in 49 CFR 190.213.
final settlement	see **secondary settlement**
final settlement tank	see **final settling tank**
final settling tank	Most commonly, a gravity settling tank or clarifier used to remove biological solids from the effluent from an aeration basin in the activated sludge process. The equivalent for biofilters is a humus tank (qv)
finding of no significant impact	*aka* FONSI. A USA official document based on an initial environmental assessment, demonstrating that a proposed action would have no significant environmental impact, therefore no EIA (qv) is required; cf exclusion list, mandatory list
finding Waldo	*(informal)* Nitpicking, especially spotting small errors (or usually more accurately, differences from how the nitpicker would do things) in someone else's design. From 'Where's Waldo', the US equivalent of the UK's 'Where's Wally' children's book series
fine chemicals	The sector producing relatively low volume/high value chemical products, and the products themselves; cf bulk materials
fines	Usually, undesirable small particulate materials
Fin-fan	A genericized trademark for a type of heat exchanger
finish marks	Marks or codes on a drawing which show the required surface finish of a component
finishing marks	Material surface texture caused by finishing processes; defined formally in ASME BPE (American Society of Mechanical Engineers: Bioprocessing Equipment)
finite element analysis	A form of computer simulation, used to model the stresses and thermal gradients in a solid (usually a structural element)
fire	A surprisingly contested term: see **Box F1**.
fire and explosion hazard analysis	see **fire and explosion risk assessment**
fire and explosion hazard assessment	see **fire and explosion risk assessment**

F1. Fire/flame/combustion/explosion proof/resistant/retardant

There is a cluster of contested terms around fires and explosions (and the degrees of resistance to these), beginning most obviously with the term "inflammable", the use of which as a synonym of "flammable" has become almost universally officially deprecated.

Starting with flammable, I define this term as "combustible and/or easily ignited. Not strictly an absolute term: there are degrees of flammability". I also note the degrees of flammability described in the NFPA standards.

Combustible means burnable, although some differentiate it from flammable. (The term combustible is rather more controversial than I expected: I discovered that the NFPA held a ballot on about twenty different definitions of it and the API also adopted the somewhat wordy 'winner').

Burnable is not a term I have chosen to define. I define combustion as "burning", plus the sub-types flaming combustion, glowing combustion and smoldering combustion.

Meanwhile, a combustible dust is "liable to catch fire or explode on ignition when dispersed in air or other oxidizing media", whilst a combustible liquid is defined by its flashpoint (and further classified by NFPA based upon ranges of flashpoint).

So, we have catch fire, explode, combust, burn, flame, glow, smolder, plus we have ignition (of which I further define three kinds: autoignition, selfignition and spontaneous ignition, some of which have their own controversies)

The difference between fire and explosion is not very sharply defined, but the general consensus is that a fire is slower than a flash fire, which in turn is about as fast as (subsonic) deflagration, which is slower than (supersonic) detonation. Explosions might start somewhere around deflagration.

Next, there is the difficulty of defining degrees of resistance to fire, flame, explosion etc. Explosionproof enclosures and equipment are proof against both fire and explosion. Explosive limits are usually identical to flammable limits. A flameproof enclosure on the other hand prevents any internal fire or explosion from propagating outside the enclosure, without sustaining damage, but unlike an explosionproof enclosure it is not required to operate at a temperature below the ignition temperature of the expected surrounding fuel air mixture. The meaning of flameproof in flameproof enclosure is not the same as flame resistant.

Something which is flame resistant (or fire resistant) is designed to prevent ignition and/or burning, usually with the qualification of a time for which the effect is guaranteed. Fireproof is a somewhat deprecated term for fire resistant (or worse still, fire retardant). Fire retardant materials delay (rather than prevent) ignition and/or combustion.

I could go on working outwards into further levels of potential confusion, but suffice it to say that just because engineers use terms every day, don't assume they are using them to mean the same thing.

Related terms
conflagration, deflagration, flash fire

fire and explosion risk assessment	*aka* FERA, fire explosion and risk assessment, fire and explosion hazard assessment, fire and explosion hazard analysis. An assessment of fire and explosion risks, involving identification of possible loss of containment incidents, potential grades of release (qv), characterization of flammability properties of potentially escaping material, and proposal of mitigation measures to ALARP (qv) standard
fire area	A type of firecell (qv); part of a building entirely separated from the remainder by barriers with an hour or more of fire resistance. Defined formally in NFPA 15 Standard for Water Spray Fixed Systems for Fire Protection and NFPA 30: Flammable and Combustible Liquids Code, 2018 Edition
fire barrier wall	*aka* fire partition. A wall with a fire resistance rating, but which is not a firewall (qv), largely by virtue of not being structurally self-sufficient; defined formally in NFPA 654: Standard for the Prevention of Fire and Dust Explosions from the Manufacturing, Processing, and Handling of Combustible Particulate Solids, 2017 Edition
fire cell drawing	A drawing showing how a building is divided into firecells (qv); defined formally in BS EN ISO 10209:2012 Technical product documentation — Vocabulary — Terms relating to technical drawings, product definition and related documentation
fire classification	A number of competing fire classifications exist; see **Box F2**.
fire code	There are many fire codes in jurisdictions which tend towards codification. Most commonly when used as a generic, the NFPA® 1 Fire Code, 2018 Edition
fire damper	A damper (qv) used in HVAC to prevent/mitigate the spread of fire inside ductwork. Often located where ductwork passes through fire compartments, e.g., fire resistant walls and floors
fire department	A private or public organization responsible for providing a firefighting and rescue service; defined formally in NFPA 1142: Standard on Water Supplies for Suburban and Rural Fire Fighting, 2017 Edition
fire department connection	A pipe connection for a fire department (qv) to use to add supplementary water into a site's fixed fire protection system; defined formally in NFPA 24: Standard for the Installation of Private Fire Service Mains and Their Appurtenances, 2019 Edition

Box F2. Fire Classification

This mismatch may have even greater potential to cause dangerous confusion than those covered in **Box F1**. on fire and flammability, because the lettered classification of fires is often marked on fire extinguishers. Using the wrong fire extinguisher can have life-threatening results.

In the USA, UK, Europe and Australia, a fire involving solid materials such as wood, paper or textiles is a called a class A fire, while a class D fire is a fire involving metals.

However, the definition of a class B fire varies. In the USA, it is a flammable fluid (gas or liquid) fire, but it can only be a flammable liquid fire in the UK, Europe and Australia.

In the USA, a class C fire involves electrical equipment (a type of fire which Australians and others informally call class E); but in the UK, Europe and Australia, a class C fire involves flammable gases (which is covered by class B in the USA).

In Australia, a class E fire involves live electrical apparatus. The USA and Europe have no Class E. Europe doesn't even have a letter for electrical fires, which is probably why in the UK, we informally co-opt "E for electrical" for convenience, in the context of fire extinguishers.

A class F fire in the UK, Europe and Australia involve cooking oils such as in deep-fat fryers, (F for fat?) but the USA calls this a class K fire. I'm not sure what happened to all the letters between E and K in the USA. Perhaps they are following the example set by vitamins?

Related Terms
extinguishing agent

fire diamond	*aka* safety square. A tetrahedral graphic illustrating, for emergency personnel, the hazardous properties of a material; defined formally in NFPA 704: Standard System for the Identification of the Hazards of Materials for Emergency Response; cf fire triangle, fire tetrahedron
fire explosion and risk assessment	see **fire and explosion risk assessment**
Fireye	*aka* (wrongly) fire eye. A genericized trademark for flame sensors/detectors
fire eye	see **Fireye**
fire hazard	Conditions which might plausibly initiate or sustain a fire which might harm life or property; defined formally in NFPA 654: Standard for the Prevention of Fire and Dust Explosions from the Manufacturing, Processing, and Handling of Combustible Particulate Solids, 2017 Edition
fire ice	see **methane clathrate**

fire partition	see **fire barrier wall**
fire point	The lowest temperature at which a flammable liquid will ignite and burn in the ASTM D92 test; defined formally in NFPA 30: Flammable and Combustible Liquids Code, 2018 Edition and API RP 2001 - Fire Protection at Refineries
fire proof	see **fireproof**
fire protection	Systems for prevention, detection, and suppression of fire; defined formally in NFPA 58: Liquefied Petroleum Gas Code, 2017 Edition
fire pump	A (most commonly, water) pump used for fire protection purposes; defined formally in NFPA 20: Standard for the Installation of Stationary Pumps for Fire Protection, 2019 Edition and NFPA 24: Standard for the Installation of Private Fire Service Mains and Their Appurtenances, 2019 Edition
fire pump unit	Complete fire pump assembly including pump, drive, and controller; defined formally in NFPA 20: Standard for the Installation of Stationary Pumps for Fire Protection, 2019 Edition
fire resistant	*aka* flame resistant. Preventing ignition and/or burning, usually with the qualification of the time for which such an effect is guaranteed. See also **Box F1**.
fire retardant	Delaying ignition and/or combustion; cf fire resistant, fireproof. See also **Box F1**.
fire suppression system	A system to detect and suppress residual fire hazards using extinguishing agent (qv); cf explosion suppression system
fire tetrahedron	An illustration of the four elements a fire needs to be sustained: fuel, heat, an oxidizing agent and a chain reaction; cf fire triangle, fire diamond
fire triangle	An illustration of the three elements a fire needs to occur: fuel, heat and an oxidizing agent; cf fire diamond, fire tetrahedron
fire tube boiler	see **firetube boiler**
fire wall	see **firewall**
fireball	An approximately spherical cloud of burning fuel-air mixture from a flash fire or BLEVE (qv), with energy release mostly as radiant heat
firecell	Part of a building entirely separated from the remainder by fire barriers of some type

fired heater	Generically, any direct-fired heat exchanger, though this is a contested term. The term 'fired heater' is generally used to describe those used for simple heating of a process stream, whilst the term furnace (qv) usually refers to a more complex process involving additional chemical and physical changes in the process streams. There is a specific formal definition in API 560 Fired Heaters for General Refinery Service
firedamp	A methane based gas mixture found only in mines; cf group I gases
fireproof	*aka* fire proof. A somewhat deprecated term for fire resistant (qv) (or worse still, fire retardant). See **Box F1**.
fireproofing	Fire resistant insulation; defined formally in API RP 2510A - Fire Protection Considerations for the Design and Operation of Liquefied Petroleum Gas (LPG) Storage Facilities
firetube boiler	*aka* fire tube boiler. A boiler with hot combustion gases passing through the tubes, usually smaller and with lower operating pressures than a watertube boiler (qv)
firetube HRSG	A heat recovery steam generator with shell side steam generation, and hot gases in the tubes; defined formally in API RP 534 – Heat Recovery Steam Generators
firewall	1. *aka* fire wall. A structurally stable fire resistant wall between buildings or parts of a building intended to prevent spread of fire; defined formally in NFPA 654: Standard for the Prevention of Fire and Dust Explosions from the Manufacturing, Processing, and Handling of Combustible Particulate Solids, 2017 Edition 2. Also, by analogy, a system which prevents computer systems from unauthorized interference from outside
firing ports	Fuel tip orifices; defined formally in API RP 535 - Burners for Fired Heaters at Refineries
firing rate	Rate of fuel supply to a burner or heater expressed in units of heat per hour; defined formally in API RP 535 - Burners for Fired Heaters at Refineries
firing system	All plant involved in storage, preparation, and combustion of fluid fuels; defined formally in BS EN 12952-8:2002 Water-tube boilers and auxiliary installations. Requirements for firing systems for liquid and gaseous fuels for the boiler
firing system heat input	Estimated combustion chamber heat input based on fuel mass flow rate; defined formally in BS EN 12952-8:2002 Water-tube boilers and auxiliary installations. Requirements for firing systems for liquid and gaseous fuels for the boiler
firm fixed price	see **firm price**

firm price	*aka* firm fixed price, fixed firm price. A price which will not vary (usually for a specified period)
firmly bound CO_2	see **combined CO_2**
first in first out	*aka* FIFO. 1. A pattern of equipment/material storage in warehouses or storage areas
	2. A method of inventory valuation; cf last in first out; last in last out
first law of thermodynamics	The first law of thermodynamics is that energy cannot be created or destroyed in an isolated system[45]; cf second law of thermodynamics, third law of thermodynamics
first order	1. Generically, something increasing linearly with respect to the driver
	2. Specifically for reactions, having a rate of reaction increasing proportionally to the concentration of the reactant; cf order of reaction
first stage regulator	In the context of LPG, a device designed to regulate delivery pressure from a container to a maximum of 69 kPag; defined formally in NFPA 58: Liquefied Petroleum Gas Code, 2017 Edition
Fischer–Tropsch process	Catalytic conversion of carbon monoxide and hydrogen into liquid hydrocarbons
fish	*(informal)* In oil and gas drilling, a term for anything broken off or lost in a drill hole which must be retrieved through fishing (qv)
fish box hook	A length of heavy wire with a hook in the end; used for moving items around the deck of oil and gas drilling support vessels
fish plate	A metal bar that is bolted to the ends of two rails to join them together in a track, often a small copper or nickel silver plate that slips onto both rails to provide the functions of maintaining alignment and electrical continuity
fishbone diagram	*aka* cause effect chart/diagram, Ishikawa diagram. A systematic cause and effect analysis tool with roots in quality assurance, that produces a diagram shaped rather like a fish skeleton
fishing	*(informal)* Retrieving objects from a borehole, such as a broken drill string (qv), or tools
fishing tool	*(informal)* Any tool used to catch a fish (qv), such as a bulldog
fissile material	Material capable of undergoing nuclear fission, sometimes qualified by a requirement that it can sustain a chain reaction

---

[45] Or, to put it another way, "You can't get anything without working for it."

fission	Breaking into two or more subunits, as with atoms in nuclear reactions and cells in microbiological growth. The opposite of fusion (qv)
FIT	see **failures in time**
fit for purpose	An ambiguous term used by lawyers, the meaning of which can range from "making the client happy, irrespective of what was asked for in contract documentation" to "just about works, such that total failure and/or prosecution is averted", depending on who is being asked[46]; cf **Box R1**.
fit for service	*aka* FFS. Not the same as fit for purpose (qv): fitness for service is to do with a component, such as a pressure vessel, having quantitatively proven structural integrity
FITA	Finger in the air: the basis of a very approximate estimate, somewhere between a WAG (qv) and a SWAG (qv)
fitness for service assessment	A method for generating quantitative proof of structural integrity; defined formally in the context of pressure vessels in API RP 579 - Fitness for Service
fittings	In the context of pipework, all fluid-carrying components other than straight pipe lengths, such as bends
fixed bed	A layer of solid particles which is stationary when in service (though it may be mobilized as part of a cleaning or regeneration cycle)
fixed cannon	see **fixed monitor**
fixed costs	Business costs which do not vary with production volume; cf variable costs
fixed firm price	*aka* firm fixed price; see **firm price**
fixed foam discharge outlet	*aka* foam chamber. A foam discharge device fixed to a containment structure; defined formally in NFPA 11: Standard for Low-, Medium-, and High-Expansion Foam, 2016 Edition
fixed foam maker	see **foam nozzle**
fixed guard	A machine safety guard which is not readily openable or removable; defined formally in BS EN ISO 12100:2010 Safety of machinery. General principles for design. Risk assessment and risk reduction
fixed head heat exchanger	A type of shell and tube heat exchanger in which straight tubes are secured at both ends to tubesheets welded to the shell; cf floating head heat exchanger

---

[46] Unless the contractor was unwise enough to sign a contract with a 'fitness for purpose' clause, in which case they may have effectively agreed to make the client happy, whatever it takes). Generally speaking, applicable to the provision of equipment, rather than professional services, but everything about this term is really in 'ask a lawyer' territory. (It may be a bad sign that you are looking it up!)

fixed ladder	A ladder which is fixed, including inclined 'ships ladders'; defined formally in BS EN ISO 14122 Safety of machinery. Permanent means of access to machinery. Working platforms and walkways and OSHA 1910.29 - Fall protection systems and falling object protection-criteria and practices
fixed ladder system	*aka* ladder system. A fixed ladder or ladders with associated fall protection, landings etc.; defined formally in BS EN ISO 14122 Safety of machinery. Permanent means of access to machinery. Working platforms and walkways
fixed liquid level gauge	In the context of LPG containers, a level indicator with vent valve which indicates when liquid level has reached the indicator; defined formally in NFPA 58: Liquefied Petroleum Gas Code, 2017 Edition
fixed maximum liquid level gauge	In the context of LPG containers, a level indicator with vent valve which indicates when liquid level has reached the maximum permitted filling limit; defined formally in NFPA 58: Liquefied Petroleum Gas Code, 2017 Edition
fixed monitor	*aka* fixed cannon. A stationary mounted firefighting foam delivery device; defined formally in NFPA 11: Standard for Low-, Medium-, and High-Expansion Foam, 2016 Edition
fixed price	see **firm price**
fixed roof tank	A tank with a self-supporting external fixed roof, with or without internal support columns
fixed system	A complete firefighting foam system with central foam production, permanent pumps and fixed monitors; defined formally in NFPA 11: Standard for Low-, Medium-, and High-Expansion Foam, 2016 Edition
fixture marks	Imperfections at the point where the electrical connection was made to an electropolished component; defined formally in ASME BPE (American Society of Mechanical Engineers: Bioprocessing Equipment)
FKM	*aka* Viton, FPM. A common fluoroelastomer mechanical seal O-ring material; defined formally in API RP 682 - Pump Seals. Neither the US- favored FKM nor the European equivalent FPM are acronyms or initialisms. FKM is the ASTM D1418 abbreviation for 'fluoroelastomer', whilst FPM is the ISO 1629 abbreviation for 'fluoroelastomer'[47]

---

[47] No, I don't get it either, in either case!

flagged off	In the context of manufacturing equipment, a synonym of 'dispatched'. Also applied more generally to project initialization[48]
flaking	Producing solid flakes of product, most commonly from a melt by cooling, or in the case of food products by crushing between rollers
flame	The visible combustion of gases, (including gases arising from the vaporization of liquid or solid fuels)
flame arrester	*aka* flame arrestor, flashback arrester, backfire arrester. A device fitted in a duct containing a flammable gas/air mixture to prevent flame propagation back through supply duct by passive quenching; defined formally in API RP 2210 - Flame Arrestors and NFPA 86: Standard for Ovens and Furnaces, 2019 Edition
flame arrestor	see **flame arrester**
flame curtain	A line burner which ignites flammable gases exiting a furnace and reduces air ingress; defined formally in NFPA 86: Standard for Ovens and Furnaces, 2019 Edition
flame curtain pilot	The pilot light of a flame curtain; defined formally in NFPA 86: Standard for Ovens and Furnaces, 2019 Edition
flame detection system	*aka* flame detector. Safety system based on a flame sensor to verify flame is present in a flare or furnace; defined formally in API RP 537 Flare Details for Petroleum, Petrochemical, and Natural Gas Industries and NFPA 86: Standard for Ovens and Furnaces, 2019 Edition
flame detector	see **flame detection system**
flame establishing period	see **trial for ignition period**
flame failure response time	*aka* FFRT, extinction safety time. USA term for the time from detection of loss of flame to the deenergizing of safety shutoff valves; defined formally in NFPA 86: Standard for Ovens and Furnaces, 2019 Edition; cf extinction safety time
flame front generator	A flare pilot ignition device. Defined formally in API RP 537 Flare Details for Petroleum, Petrochemical, and Natural Gas Industries
flame monitor	A flame detection system which monitors for break-away of the flame as well as its presence; defined formally in BS EN 12952-8:2002 Water-tube boilers and auxiliary installations. Requirements for firing systems for liquid and gaseous fuels for the boiler
flame propagation	The spread of flame through a combustible fuel–air mixture

---

[48] A term used more commonly by non-native speakers of English, especially from the Indian sub-continent

flame propagation rate	*aka* flame speed, flame velocity. The speed of flame propagation (qv); defined formally in NFPA 86: Standard for Ovens and Furnaces, 2019 Edition; cf burning velocity
flame resistant	see **fire resistant**
flame retention device	A device intended to prevent flare flame blowoff; defined formally in API RP 537 Flare Details for Petroleum, Petrochemical, and Natural Gas Industries and API Standard 521. Pressure-relieving and Depressuring Systems. Sixth Edition | January 2014
flame rod	A type of flame sensor using a rod which extends into the flame; defined formally in NFPA 86: Standard for Ovens and Furnaces, 2019 Edition
flame speed	see **flame propagation rate**
flame stabilization point	A burner's continuous ignition zone; defined formally in API RP 535 - Burners for Fired Heaters at Refineries
flame stabilizer	A flame-stabilizing combustion air stream restriction; defined formally in API RP 535 - Burners for Fired Heaters at Refineries
flame temperature	Approximately, the steady state temperature of a burner flame; defined formally in API RP 535 - Burners for Fired Heaters at Refineries in a way which is somewhat more complex, calling for consideration of multiple factors
flame trap	see **flame arrester**
flame velocity	see **flame propagation rate**
flameless thermal oxidizer	A combustion system which heats heat storage media to VOC destruction temperature. VOCs are subsequently introduced into the preheated media to be destroyed flamelessly; defined formally in NFPA 86: Standard for Ovens and Furnaces, 2019 Edition; cf afterburner, direct thermal oxidizer
flameproof	Generally, synonymous with flame resistant (qv). However, flameproof enclosure (qv) uses a different meaning; see **Box F1**.
flameproof enclosure	*aka* enclosure, flameproof. An electrical enclosure which prevents any internal fire or explosion from propagating outside the enclosure, without sustaining damage. Unlike an explosionproof enclosure (qv), it is not required to operate at a temperature below the ignition temperature of the expected surrounding fuel-air mixture (according to the API); defined formally in API RP 505 2nd Edition, August 2018 Recommended Practice for Classification of Locations for Electrical Installations at Petroleum Facilities Classified as Class I, Zone 0, Zone 1, and Zone 2, and in NEMA 250- Enclosures for Electrical Equipment

flaming combustion	Combustion with visible flames; cf glowing combustion, smoldering combustion
flammability limits	see **flammable limits**
flammable	A contested term, but generally similar in meaning to combustible and/or easily ignited. Not an absolute term: there are degrees of flammability. The term is defined formally in API RP 2001 - Fire Protection at Refineries and API RP 505 2nd Edition, August 2018 Recommended Practice for Classification of Locations for Electrical Installations at Petroleum Facilities Classified as Class I, Zone 0, Zone 1, and Zone 2. There are also definitions of degrees of flammability, and between flammable and combustible liquids on the basis of flashpoint in NFPA 30: Flammable and Combustible Liquids Code, 2018 Edition. OSHA however adopt the UN Globally Harmonized System for Classification and Labelling of Chemicals Rev 8 which classes all liquids with a flashpoint below 93°C as flammable liquid (qv), and does not recognize the term combustible liquid (qv) which NFPA use for liquids with a flashpoint over 37.8°C; see **Box F1**
flammable (class I) liquid	A liquid with a flashpoint lower than 37.8°C; according to the formal definition in NFPA 30: Flammable and Combustible Liquids Code, 2018 Edition, NFPA 11: Standard for Low-, Medium-, and High-Expansion Foam, 2016 Edition and NFPA 30: Flammable and Combustible Liquids Code, 2018 Edition
flammable gas detection equipment	Equipment which detects flammable gas concentration (usually as a percentage of flammable limits); defined formally in NFPA 15 Standard for Water Spray Fixed Systems for Fire Protection
flammable highly volatile liquid	see **highly volatile liquid**
flammable limits	The maximum (UFL - qv) and minimum (LFL - qv) concentrations of a gas in air in which a flame can be propagated at given temperature and pressure; defined formally in API RP 505 2nd Edition, August 2018 Recommended Practice for Classification of Locations for Electrical Installations at Petroleum Facilities Classified as Class I, Zone 0, Zone 1, and Zone 2, and NFPA 86: Standard for Ovens and Furnaces, 2019 Edition. Used interchangeably with explosive limits, in which case LEL and UEL replace LFL and UFL

flammable liquid	According to the NFPA, any liquid with a flash point below 37.8°C; defined formally by classes in NFPA 30: Flammable and Combustible Liquids Code, 2018 Edition and NFPA 15 Standard for Water Spray Fixed Systems for Fire Protection. OSHA however adopt the UN Globally Harmonized System for Classification and Labelling of Chemicals Rev 8 which classes all liquids with a flashpoint below 93C as flammable liquids, and does not recognize the term combustible liquid (qv) which NFPA use for liquids with a flashpoint over 37.8C. See **Box F1**.
flammable liquid (class I liquid)	This term appears in API RP 505 2nd Edition, August 2018 Recommended Practice for Classification of Locations for Electrical Installations at Petroleum Facilities Classified as Class I, Zone 0, Zone 1, and Zone 2; cf flammable liquid, flammable (class I) liquid
flammable mist	Fine droplets of flammable liquid suspended in air; defined formally in API RP 505 2nd Edition, August 2018 Recommended Practice for Classification of Locations for Electrical Installations at Petroleum Facilities Classified as Class I, Zone 0, Zone 1, and Zone 2
flammable mixture	see **ignitable mixture**
flammable range	The range of concentration between LFL and UFL under given conditions. Defined formally in API RP 2001 - Fire Protection at Refineries; cf explosive range
flammable special atmosphere	A special atmosphere for use in a furnace comprising gases which are flammable when mixed with air; defined formally in NFPA 86: Standard for Ovens and Furnaces, 2019 Edition; cf nonflammable special atmosphere, indeterminate special atmosphere, inert special atmosphere
flange	Generically, a ridge on a circular component which adds strength. Most commonly used to describe a circular plate on the end of a pipe used to bolt pipes together; qv ANSI flanges
flare	System for environmentally compliant discharge of relief gases (and to some extent liquids) by combustion; defined formally in API Standard 521. Pressure-relieving and Depressuring Systems. Sixth Edition	January 2014 and API RP 537 Flare Details for Petroleum, Petrochemical, and Natural Gas Industries
flare boom	see **flare stack**

flare burner	*aka* flare tip. The burner (qv) of a flare; defined formally in API Standard 521. Pressure-relieving and Depressuring Systems. Sixth Edition	January 2014 and API RP 537 Flare Details for Petroleum, Petrochemical, and Natural Gas Industries
flare header	A piping system which feeds a flare; defined formally in API Standard 521. Pressure-relieving and Depressuring Systems. Sixth Edition	January 2014 and API RP 537 Flare Details for Petroleum, Petrochemical, and Natural Gas Industries
flare stack	*aka* flare boom, flare tower. The support structure of an elevated flare burner
flare tip	see **flare burner**
flare tower	see **flare stack**
flared gas	see **relief gas**
flaring	1. Using a flare burner (qv) 2. Widening a pipe
flash	1. A term used for many rapid processes (qv flash distillation, flash drying, flash electropolish, flash evaporation, flash freezing, flash vaporization) 2. A theoretical approach to vapor-liquid equilibrium
flash distillation	Flash evaporation (qv) followed by vapor condensation, the most common application being multistage flash distillation of seawater for potable water production
flash drum	*aka* breakpot, KO drum, knockout drum, knockout pot, compressor suction drum or compressor inlet drum. Generically, a vapor/liquid separator (qv). Specifically, may be associated with flash evaporation (qv)
flash drying	Rapid drying of powder, granules or slurry using brief exposure to a high-temperature gas stream. This method reduces product clumping and allows heat-labile materials to be dried
flash electropolish	Short duration, low current density electropolishing (qv) which does not significantly alter the material surface; defined formally in ASME BPE (American Society of Mechanical Engineers: Bioprocessing Equipment)
flash evaporation	*aka* flash vaporization, flash vaporization. Rapid evaporation caused by a sudden pressure drop through a throttling device to below saturation pressure
flash fire	A rapidly spreading fire in a diffuse fuel-air mixture which does not produce a damaging pressure increase; defined formally in NFPA 654: Standard for the Prevention of Fire and Dust Explosions from the Manufacturing, Processing, and Handling of Combustible Particulate Solids, 2017 Edition; cf deflagration
flash freezing	see **blast freezing**

flash point	The minimum temperature at which an ignitable vapor is produced by a liquid. Defined formally in API RP 2001 - Fire Protection at Refineries and NFPA 30: Flammable and Combustible Liquids Code, 2018 Edition and API RP 505 2nd Edition, August 2018 Recommended Practice for Classification of Locations for Electrical Installations at Petroleum Facilities Classified as Class I, Zone 0, Zone 1, and Zone 2
flash vaporisation	*aka* flash vaporization, see **flash evaporation**
flash vaporization	*aka* flash vaporisation, see **flash evaporation**
flashback	A flame front near-instantaneously propagating through a fuel-air mixture, against the direction of feed, due to burner nozzle feed velocity being less than flame velocity; defined formally in API Standard 521. Pressure-relieving and Depressuring Systems. Sixth Edition	January 2014, API RP 537 Flare Details for Petroleum, Petrochemical, and Natural Gas Industries and API RP 535 - Burners for Fired Heaters at Refineries
flashback arrester	see **flame arrester**
flashback arrestor	see **flame arrester**
flashing	1. Generally, see **flash evaporation** 2. In the context of pump seals, defined formally in API RP 682 - Pump Seals
flashing hydrocarbon	Liquid hydrocarbon with vapor pressure above ambient pressure under working conditions; defined formally in API RP 682 - Pump Seals
flashing section	see **feed section**
flaw	An identified[49] defect, as defined formally in API RP 579 - Fitness for Service
flex	*(drawing notation)* Flexitallic, a brand of gasket
flexible connecting shaft	A pump drive shaft with flexible joints and telescopic element. Defined formally in NFPA 20: Standard for the Installation of Stationary Pumps for Fire Protection, 2019 Edition
flexible connector	A short length of flexible piping with connections at either end; defined formally in NFPA 58: Liquefied Petroleum Gas Code, 2017 Edition
flexible coupling	A torque transmitting connection between prime mover and pump which allows for minor misalignments; defined formally in NFPA 20: Standard for the Installation of Stationary Pumps for Fire Protection, 2019 Edition

---

[49] Though presumably it was in some sense a flaw *before* it was identified, (in the manner of a tree which falls in the forest...)

flexible element	An assembly which allows for axial movement in a pump seal; defined formally in API RP 682 - Pump Seals
flexible graphite	Material used in mechanical pump seal static gaskets; defined formally in API RP 682 - Pump Seals
flexible hose connector	An LPG hose; defined formally in NFPA 58: Liquefied Petroleum Gas Code, 2017 Edition
flexible intermediate bulk container	*aka* FIBC, big bag, builder's bag. A cuboidal bag of similar dimensions to an RIBC, used to transport solid material; defined formally in NFPA 654: Standard for the Prevention of Fire and Dust Explosions from the Manufacturing, Processing, and Handling of Combustible Particulate Solids, 2017 Edition; cf rigid intermediate bulk container
flexible metallic connector	A metallic flexible connector (qv) for LPG; defined formally in NFPA 58: Liquefied Petroleum Gas Code, 2017 Edition
Flexipac	Proprietary eponym. Structured packing used in packed towers
flexural vibration	Vibration which causes deformation; defined formally in BS ISO 2041:2018 Mechanical vibration, shock and condition monitoring. Vocabulary
flight	1. A set of steps between landings according to the formal definition in BS EN ISO 14122 Safety of machinery. Permanent means of access to machinery. Working platforms and walkways 2. see **flighting**
flighting	*aka* flight. The helical element of an auger, Archimedean screw pump, or ribbon mixer
Flixborough disaster	An explosion at a chemical plant on 1 June 1974 which killed 28 people and seriously injured 36 as a result of lack of management of change (qv); cf unconfined vapor cloud explosion
FLNG	see **floating liquified natural gas**
float	1. *aka* slack. Most commonly, a term used in project programming - how late a task can be without delaying project completion 2. Less commonly, the moving part in sight glasses, some level gauges or rotameters 3. The floating material in a flotation process 4. Rarely, a shorthand for froth flotation (qv)
float cell	*aka* cell. A single flotation unit in mineral processing
float gauge	*aka* magnetic gauge. A gauge with a float inside a vessel which transmits its position to an external liquid level gauge; defined formally in the context of LPG in NFPA 58: Liquefied Petroleum Gas Code, 2017 Edition

floater	see **floating production, storage and offloading**
floating bushing	A loose-fitting pump seal shaft bushing; defined formally in API RP 682 - Pump Seals
floating control	see **on/off control**
floating head heat exchanger	A type of shell and tube heat exchanger in which tubes are secured at only one end to a tubesheet welded to the shell; cf fixed head heat exchanger
floating liquefied natural gas	*aka* FLNG. See **floating production, storage and offloading**
floating production system	A deprecated term for floating production, storage and offloading (qv)
floating production, storage and offloading	*aka* FPSO. A floating hydrocarbon processing plant. Variants include FSO (floating storage and offloading), FDPSO (floating drilling and production, storage and offloading) FLNG (floating liquefied natural gas) and FSRU (floating storage regasification unit)
floating roof	A type of tank roof which floats on the liquid surface; defined formally in API MPMS 19.1 Manual of Petroleum Measurement Standards Chapter 19.1 Evaporative Loss from Fixed-roof Tanks
floating roof tank	A type of bulk tank for volatile hydrocarbons; defined formally in API MPMS 19.1 Manual of Petroleum Measurement Standards Chapter 19.1 Evaporative Loss from Fixed-roof Tanks
floating storage and offloading unit	*aka* FSO, FSU. See **floating production, storage and offloading**
floating storage and regasification unit	*aka* FSRU. See **floating production, storage and offloading**
floc	A composite particulate formed by flocculation (qv) from smaller particles, large enough to separate from the bulk liquid phase
flocculant	A chemical additive which encourages the production of suitably sized, sufficiently dense and strong particles for an associated separation process via flocculation (qv)
flocculation	Using a flocculant (qv) to produce a floc; cf coagulation (commonly confused)
flogging	see **slogging**
flogging spanner	see **slogging spanner**

flooded suction	When the level of feed liquid in an open tank is always higher than the centerline of a pump fed from it, the pump is said to have a flooded suction (cf suction lift). There is a more formal definition in NFPA 20: Standard for the Installation of Stationary Pumps for Fire Protection, 2019 Edition, which is more akin to net positive suction head (qv)
flooding	see **column flooding**
floorhand	see **roughneck**
flotation	The removal of suspended particles or oil droplets from a liquid by assisting their flotation with air bubbles, froth flotation (qv) and dissolved air flotation (qv) being the most common variants
flow	Streaming movement (noun or verb)
flow assurance	The relatively new discipline in the oil and gas industry of ensuring, through engineering analysis, the flow of hydrocarbons all the way from source to end user, often focusing on reservoir engineering
flow balancing	see **flow equalization**
flow boiling	*aka* forced convection boiling. A form of boiling found in a system in which fluid is driven past a heat exchange surface by a pump or natural buoyancy
flow chart	see **flowsheet**
flow coefficient	1. (*symbol* ϕ) A dimensionless coefficient derived from the affinity laws (qv); 2. *aka* Av, Kv, valve flow coefficient, Cv factor, Cv. The relationship between flow and pressure drop for a valve. Kv is a British standard/metric version of this, Cv is a USA/ US customary units (qv) version of the same thing. Av is a slightly different related parameter, also a British Standard/metric version
flow conditioner	A device to reduce swirl in a flow, and redistribute velocity profile (cf flow straightener) reducing the required number of pipe diameters upstream or downstream of a disturbance; defined formally in BS EN ISO 5167-1:2003 Measurement of fluid flow by means of pressure differential devices inserted in circular cross-section conduits running full
flow distributor	see **distributor**
flow equalization	*aka* flow balancing. Averaging a variable incoming flow by allowing accumulation in a tank of flows in excess of average, and their release from that tank at lower than average flows

flow hydrant	Fire hydrant used in a flow test; defined formally in NFPA 24: Standard for the Installation of Private Fire Service Mains and Their Appurtenances, 2019 Edition
flow induced vibration	Vibration caused by fluctuating fluid flow; defined formally in BS ISO 2041:2018 Mechanical vibration, shock and condition monitoring. Vocabulary
flow injection analysis	*aka* FIA. An approach to chemical analysis in which a sample is injected into a carrier stream feeding an analytical device
flow rate	Mass or volume of flow per unit of time
flow regulating valve	*aka* FRV. A valve which regulates the flow passing through it by degree of closure, and hence pressure drop, usually of a type with a suitable flow coefficient (qv) for that duty
flow straightener	A device to reduce swirl in a flow, reducing required number of pipe diameters upstream or downstream of a disturbance; defined formally in BS EN ISO 5167-1:2003 Measurement of fluid flow by means of pressure differential devices inserted in circular cross-section conduits running full; cf flow conditioner
flow switch	A switch activated by fluid flow; defined formally in NFPA 86: Standard for Ovens and Furnaces, 2019 Edition
flowline	A pipe that connects a single wellhead to process equipment or manifold. Some sources state that it connects a bell nipple (qv) to a possum belly (qv)
flowline trap	see **possum belly**
flowsheet	*aka* flow chart. A term covering a wide range of different diagrams of shapes, connected by lines to represent a sequential series of operations or processes; used by doctors as well as software, electrical and process engineers. A deprecated term for a PFD (qv) for this reason[50]
flue ash	see **pulverized flue ash**
flue gas	A mixture of combustion gases and excess air sent to stack; defined formally in API STD 530 - Calculation of Tube Heater Thickness BS EN ISO 13705:2012/ISO 13705:2012(E) Petroleum, petrochemical and natural gas industries. Fired heaters for general refinery service and API STD 560 - Fired Heaters for General Refinery Service
flue gas desulfurization	*aka* FGD. The removal of SOx (many only reference SO_2) from flue gases; a subset of desulfurization

---

[50] The term is popular in academia, often used to describe non-professional textbook illustrations or modeling program outputs

flue gas passes	Either the number of times flue gas passes through a boiler heat exchanger before exiting, or the flue gas pipework in which this happens
flue gas recirculation	*aka* FGR. Flue gas return from flame zone exit to inlet; defined formally in BS EN 12952-8:2002 Water-tube boilers and auxiliary installations. Requirements for firing systems for liquid and gaseous fuels for the boiler
fluid	1. Flowing easily 2. A substance which flows easily (liquid or gas)
fluid catalytic cracker	*aka* FCC, fluid catalytic cracking unit. Unit in which fluid catalytic cracking takes place
fluid catalytic cracking	Breaking down long chain hydrocarbons into shorter ones with the aid of temperature and a catalyst, making heavy oils into more valuable medium distillates; cf hydrocracking
fluid catalytic cracking unit	*aka* FCCU; see **fluid catalytic cracker**
fluid hammer	*aka* hammer, hydraulic shock, water hammer, surge or transient. Potentially destructive pressure waves (which can be positive and/or negative), caused by momentum change in fluid. Water hammer (qv) is the special case where water is the fluid.
fluid logic	see **fluidics**
fluid mechanics	The academic study of fluid behavior; cf hydraulics
fluidics	*aka* fluid logic. Using flows of fluid to perform logical operations[51]
fluidization	Generically, passing a fluid through a bed of granular material to produce a fluid-like state in the granular material. There are a number of classified subtypes of fluidization intensity: as fluid velocity increases, the bed goes from packed, to expanded, to minimum fluidization, to smooth fluidization, to bubbling fluidization, to turbulent fluidization, to fast fluidization, to solids conveying
fluidized bed	A bed of granular solids which has been fluidized, as in e.g., a fluidized bed reactor
fluke	1. Proprietary eponym for digital multimeter 2. A blade on a pump impeller
flume	1. A type of open channel used for conveying water by gravity, used to carry solid materials (as in a log flume) 2. A hydraulic structure used for flow measurement purposes (such as a Parshall flume)
fluoridation	see **fluoridization**

---

[51] Not used in practice, but popular in academia

fluoridization	*aka* fluoridation. The addition of fluoride compounds to drinking water to promote dental health[52]
fluorination	A chemical reaction which adds fluorine to a compound
fluoropolymer	A fluorocarbon based polymer; defined formally in ASME BPE (American Society of Mechanical Engineers: Bioprocessing Equipment)
fluoroprotein foam concentrate	A firefighting foam concentrate with a fluorinated protein surfactant; defined formally in NFPA 11: Standard for Low-, Medium-, and High-Expansion Foam, 2016 Edition
flush	1. In the context of quality of finish, being mounted level with a surface 2. Cleaning with a flow of fluid (especially water). A specific subtype of this, in the context of pump seals, is the flow of fluid though the process side of the seal chamber intended to prevent cross contamination; this sense is defined formally in API RP 682 - Pump Seals
flush plan	Complete electromechanical system used to flush pump seals; defined formally in API RP 682 - Pump Seals
flush type full internal pressure relief valve	A flush fitted internal PRV (qv); defined formally in NFPA 58: Liquefied Petroleum Gas Code, 2017 Edition
flushing	see **flush** (in sense of cleaning); defined formally in ASME BPE (American Society of Mechanical Engineers: Bioprocessing Equipment)
flushing test	Cleaning of a firemain system prior to bringing into service of construction stage debris using high fluid flowrates; defined formally in NFPA 24: Standard for the Installation of Private Fire Service Mains and Their Appurtenances, 2019 Edition
flutter	A type of self-excited structural vibration; defined formally in BS ISO 2041:2018 Mechanical vibration, shock and condition monitoring. Vocabulary
flux	1. Most commonly, the flow of something across an area 2. Something that reduces the melting point of an alloy 3. A material used to assist welding
fly ash	Fine solid particles of ash carried into the air during combustion, especially in power stations; cf pulverized flue ash

---

[52] Or for mind control purposes, according to some sections of the internet!

fly in fly out	*aka* FIFO. A non-residential working pattern common in remote locations, such as mines, having few facilities on site. Workers are flown in and out on a rotating schedule, such as '8:4', for 8 periods of continuous work on site, followed by 4 equal periods at home[53]
flyings	Solid particles, typically formed from fibers, nominally greater than 500μm in size. May present an explosion hazard if combustible when suspended in air
FMA	see **free mineral acidity**
FMCG	Fast-moving consumer goods
FMEA	see **failure mode and effects analysis**
FMR	see **field metallographic replication**
FNPT	Female national pipe thread; see **national pipe thread**
FNR	Fast neutron reactor, see **fast reactor**
FNU	see **formazin nephelometric unit**
FO	1. *(drawing notation)* see **fail open** 2. see **forward osmosis**
foam	A stable aggregation of bubbles; defined formally in NFPA 11: Standard for Low-, Medium-, and High-Expansion Foam, 2016 Edition
foam chamber	see **fixed foam discharge outlet**
foam concentrate	1. A concentrated liquid foaming agent as supplied by the manufacturer; defined formally in NFPA 11: Standard for Low-, Medium-, and High-Expansion Foam, 2016 Edition 2. Overheads from a flotation vessel
foam concentrate type	A classification of firefighting foam concentrate by chemical composition, and by the type of fires it can be used with; defined formally in NFPA 11: Standard for Low-, Medium-, and High-Expansion Foam, 2016 Edition
foam generating methods	Methods of firefighting foam generation; defined formally with a list of methods in NFPA 11: Standard for Low-, Medium-, and High-Expansion Foam, 2016 Edition
foam generators	Aspirator or blower type firefighting foam generators; defined formally in NFPA 11: Standard for Low-, Medium-, and High-Expansion Foam, 2016 Edition
foam hose stream	Low capacity foam supply from a handline; defined formally in NFPA 11: Standard for Low-, Medium-, and High-Expansion Foam, 2016 Edition

---

[53] These periods might refer to days or weeks so it is important to be clear which it is before signing up!

foam monitor stream	High capacity foam supply from a foam monitor; defined formally in NFPA 11: Standard for Low-, Medium-, and High-Expansion Foam, 2016 Edition
foam nozzle	*aka* fixed foam maker. A nozzle which aspirates air into foam solution to make firefighting foam; defined formally in NFPA 11: Standard for Low-, Medium-, and High-Expansion Foam, 2016 Edition
foam solution	A water/foam concentrate mixture which will form firefighting foam when mixed with air; defined formally in NFPA 11: Standard for Low-, Medium-, and High-Expansion Foam, 2016 Edition
FOB	1. *(drawing notation)* Flat on bottom 2. *(drawing notation)* Flow on bottom 3. Free on board, see **Incoterms**
FOE	*(drawing notation)* Flange one end
FOGs	see **fats oils and greases**
folding frequency	see **Nyquist frequency**
foldout	An oversize leaf in a bound publication, arranged so that it folds up inside the publication when not being viewed; cf throwclear
FONSI	see **finding of no significant impact**
food	"Any substance or product, whether processed, partially processed or unprocessed, intended to be, or reasonably expected to be ingested by humans" according to Regulation (EC) No 178/2002 of the European Parliament and of the Council of 28 January 2002. Sometimes extended to cover animal food, usually known as feed (qv) as in the case of BS EN 1672-2:2005+A1:2009 Food processing machinery. Basic concepts. Hygiene requirements
food area	Machinery surfaces in contact with food (or animal feed); defined formally in BS EN 1672-2:2005+A1:2009 Food processing machinery. Basic concepts. Hygiene requirement
food hygiene	"All conditions and measures necessary to ensure the safety and suitability of food at all stages of the food chain" according to the EHEDG Glossary. Other formal definitions in BS EN 1672-2:2005+A1:2009 Food processing machinery. Basic concepts. Hygiene requirements and BS EN ISO 14159:2008 Safety of machinery. Hygiene requirements for the design of machinery
food safe area	Part of a plant clearly separated from other production areas that is intentionally kept clean and uncontaminated. For example, incoming potatoes may be covered in mud, and must be cleaned and peeled before entering a food safe area

food safety	"Assurance that food will not cause harm to the consumer when it is prepared and/or eaten according to its intended use" according to the EHEDG Glossary
Food Safety Modernization Act	*aka* FSMA. USA food law passed in 2011 intended to prevent rather than react to food contamination
food steam	see **culinary steam**
food suitability	"Assurance that food is acceptable for human consumption according to its intended use. Note: suitability is now preferred to wholesomeness" (qv) according to the EHEDG Glossary
food to microorganism ratio	*aka* F:M ratio. A key design parameter in biological sewage treatment: the ratio of mass of cBOD/d to total mass of mixed liquor suspended solids (qv) in an aeration tank
foot	A US Customary Units (qv) measure of length (a former imperial/British units measure)
foot pound second	The basis of several systems of units, most notably the almost obsolete imperial units (qv), and the still current US Customary Units (qv)
foot valve	An NRV (qv) installed on the inlet of a water pipe, upstream of a pump, especially in a borehole or suction lift application
footing	International English equivalent of plinth (qv)
for construction	Most commonly, the final stage of process plant design prior to construction. See **Box S2**.
for construction drawing	Drawings produced at the for construction (qv) stage of process plant design; cf construction drawing, working drawing, IFC
force	(*symbol* F) An influence that tends to change the motion of an object; defined formally in BS ISO 2041:2018 Mechanical vibration, shock and condition monitoring. Vocabulary
force transducer	A device which produces an electrical or electronic output proportional to an input force; defined formally in BS ISO 2041:2018 Mechanical vibration, shock and condition monitoring. Vocabulary
forced convection	Heat transfer in a fluid which is being moved mechanically (e.g., by a fan); cf natural convection
forced convection boiling	see **flow boiling**
forced draft	A mechanically induced pressure difference; defined formally in API RP 535 - Burners for Fired Heaters at Refineries

forced draught heater	A heater for which combustion air is mechanically supplied; defined formally in BS EN ISO 13705:2012/ISO 13705:2012(E) Petroleum, petrochemical and natural gas industries. Fired heaters for general refinery service, API STD 530 - Calculation of Tube Heater Thickness and API STD 560 - Fired Heaters for General Refinery Service
forced vortex	A vortex caused by mechanical stirring; cf free vortex
forcing type pressure foam maker	see **pressure foam maker**
fore pump	see **holding pump**
forecourt separator	see **interceptor**
foreign body	*aka* foreign object. Something which does not belong where it is found. In the context of food, something that the consumer perceives as alien to food. Note the emphasis on consumer perception in this context; cf extraneous material
foreign object	see **foreign body**
foreign object damage	In the context of gas turbines, damage caused to a turbine component as a result of a foreign object/body passing through the turbine; defined formally in ISO 3977 Gas turbines - Procurement - Part 3: Design requirements
foreman	*aka* foreperson/ forewoman. Holder of the second-lowest level front-line supervisory position on a process plant; cf chargehand
forensic engineering	Engineering failure investigation, often in the service of legal or insurance claims
foreperson	see **foreman**
foreshots	see **heads**
forewoman	see **foreman**
Form 1 panel	One of the four 'forms' of electrical panel defined formally in BS EN 61439-1:2011 Low-voltage switchgear and controlgear assemblies. General rules, a Form 1 (*aka* 'wardrobe style') panel has no separation of busbars (qv), terminals and functional units
Form 2 panel	One of the four 'forms' of electrical panel defined formally in BS EN 61439-1:2011 Low-voltage switchgear and controlgear assemblies. General rules, a Form 2 panel has functional units separated from busbars (qv)
Form 3 panel	One of the four 'forms' of electrical panel defined formally in BS EN 61439-1:2011 Low-voltage switchgear and controlgear assemblies. General rules, a Form 3 panel has busbars (qv) separated from functional units; and functional units separated from each other

Form 4 panel	One of the four 'forms' of electrical panel defined formally in BS EN 61439-1:2011 Low-voltage switchgear and controlgear assemblies. General rules, a Form 4 panel has busbars (qv) separated from both functional units and terminals; and functional units separated from each other
formation fluids	The fluids within the reservoir being drilled into by an oil or gas well; a mixture of gases, crude oil, water, and solids
formazin attenuation unit	*aka* FAU; see **formazin nephelometric unit**
formazin nephelometric unit	*aka* FNU, formazin attenuation unit (FAU), formazin turbidity unit (FTU). A unit of turbidity; this term most often used when referencing the ISO 7027 (European) turbidity method, identical in practice to NTU (qv), FTU and FAU
formazin turbidity unit	*aka* FTU; see **formazin nephelometric unit**
formula	1. A mathematical expression 2. A notation to describe a chemical reaction or structure 3. A recipe for creating something
formwork	*aka* shuttering. A temporary mold, usually made of plywood, into which concrete is poured
Fortran	A venerable, but still quite widely used computer programming language[54]
forward mixing	The mixing of reactants with reaction products; cf backmixing
forward osmosis	Not the same as 'osmosis'. A term most commonly used to describe a largely theoretical alternative to reverse osmosis for water desalination using membranes, with a highly concentrated 'draw' solution; cf osmosis, reverse osmosis
fossil fuel	A form of fuel based on carbon sequestered millennia ago, the lavish use of which has given us the world we live in today
FOT	*(drawing notation)* Flat on top
Foucault's currents	see **eddy current**
foul sewer	*aka* sanitary sewer. A sewer containing wastewater (especially sewage) rather than uncontaminated surface water; cf drainage, process sewer
fouling	Deposition of a coating of relatively insoluble material on a surface, interfering with process efficiency; cf scaling

---

[54] Originally written as FORTRAN, as it could not code for lower case letters, despite being short for something like FORmula TRANslation (opinions differ as to precisely what)

fouling allowance	*aka* fouling factor. Usually, a factor to allow for increased pressure drop due to fouling in a heat exchanger; defined formally in BS EN ISO 13705:2012/ISO 13705:2012(E) Petroleum, petrochemical and natural gas industries. Fired heaters for general refinery service, API STD 530 - Calculation of Tube Heater Thickness and API STD 560 - Fired Heaters for General Refinery Service
fouling factor	see **fouling allowance**
fouling resistance	Usually, a factor to allow for reduced heat transfer coefficient due to fouling in a heat exchanger; defined formally in BS EN ISO 13705:2012/ISO 13705:2012(E) Petroleum, petrochemical and natural gas industries. Fired heaters for general refinery service, API STD 530 - Calculation of Tube Heater Thickness and API STD 560 - Fired Heaters for General Refinery Service
foundation	The ultimate support for a building or structure; defined formally in BS ISO 2041:2018 Mechanical vibration, shock and condition monitoring. Vocabulary
Fourier number	(*symbol* Fo) A dimensionless number useful for analysis of transient heat conduction
Fourier transform	A mathematical transform useful in analysis of waveforms. Commonly used in vibration analysis; defined formally in BS ISO 2041:2018 Mechanical vibration, shock and condition monitoring. Vocabulary
Fourier's law	Law (similar in form to Fick's law of diffusion (qv)) which states that the rate of heat conduction is proportional to the temperature gradient and the area through which heat flows
FOX	see **field oxide**
fox bolt	An anchor bolt (qv) with a split end to receive a fox wedge (qv) for use in blind holes
fox stick	A long stick used to unlock the "hook" that holds a drill string (qv) so that it can be rotated
fox wedge	Steel wedge used to expand the end of a fox bolt, splitting mating surfaces or leveling machinery and structures before fixing in place
FPC	see **frac pump control**
FPM	see **FKM**
FPS	1. see **floating production systems** 2. see **foot pound second**
FPSO	see **floating production, storage and offloading**
FPT	*(drawing notation)* Female pipe thread; see **national pipe thread**
frac fleet	see **frac spread**

frac fluid	*aka* fracking fluids, fracturing fluids. Water based composite fluids used for fracking (qv)
frac pump control	*aka* FPC, well fracturing pump control, fracturing pump control. A shorthand term for the control system of a hydraulic fracturing pump
frac spread	*aka* frac fleet. A collection of above-ground equipment such as high-pressure pumps used in fracking (qv)
fracking	The hydraulic fracturing of rock, usually for oil and gas production
fracking fluids	see **frac fluid**
fractional crystallization	Producing pure crystals from a mixed melt or solution of substances, to separate components based on solubility product
fractional distillation	*aka* fractionation. The separation of a mixture of components into more than two fractions with different boiling points by distillation; cf binary distillation
fractionating column	see **distillation column**
fractionation	see **fractional distillation**
fracturing fluids	see **frac fluid**
fracturing pump control	see **frac pump control**
Frasch process	A near-obsolete process for the recovery of elemental sulfur by in situ melting with superheated water. Elemental sulfur is now almost exclusively produced as a byproduct of oil sweetening (qv)
FRBR	A questionable abbreviation for radial flow fixed bed reactor (qv)
free air delivered	The rating of a compressor as the volume of air delivered at atmospheric pressure
free along side	Erroneous version of free alongside ship (qv), see **Incoterms**
free alongside ship	*aka* FAS, free alongside vessel; see **Incoterms**
free alongside vessel	see **free alongside ship**
free carrier	*aka* FCA; see **Incoterms**
free chlorine residual	In the context of water disinfection, the residue of chlorine available for disinfection in the form of dissolved hypochlorite ions, hypochlorous acid or chlorine gas
free CO_2	In the context of carbonate buffering, a term for undissociated carbonic acid
free convection	see **natural convection**
free energy	see **thermodynamic free energy**
free mineral acidity	*aka* FMA. Acidity from mineral (hydrochloric, nitric and/or sulfuric) acids
free moisture	Unbound moisture, very similar to adventitious moisture (qv); cf bound moisture

free on board	*aka* FOB; see **Incoterms**
free radical	An atom or molecule with an unpaired valence electron. Free radicals are often intermediates in reactions such as polymerization or oxidation
free surface	Generically, a stationary liquid/gas or, more rarely, solid/gas interface. In practice, almost always the part of a liquid or solid exposed to the air
free vibration	Vibration which persists in the absence of either excitation (qv) or restraint; defined formally in BS ISO 2041:2018 Mechanical vibration, shock and condition monitoring. Vocabulary
free vortex	A fluid 'circling the drain' naturally as it leaves a vessel; cf forced vortex, vortex breaker
free water knockout	*aka* FWKO. In the oil and gas industry, a pressurized gravity separator, separating well fluids into two streams of free (unemulsified) water and an oil/gas mixture (two phase free water knockout), or three streams of free water, oil and gas (three phase free water knockout). cf KO drum
freeboard	The distance between the maximum fill level and upper edge of a vessel shell
freeze drying	*aka* lyophilization, cryodesiccation. Drying by sublimation of water from a frozen wet solid; cf desiccation
freezing point	The temperature at which a liquid solidifies
freezing point depression	The reduction in freezing point (qv) produced by mixing or dissolution of another material in a liquid
French key	see **Stillson**
Freon	Proprietary eponym for CFC (qv) refrigerants
frequency	The reciprocal of the period of a wave; defined formally in BS ISO 2041:2018 Mechanical vibration, shock and condition monitoring. Vocabulary
frequency inverter	see **inverter**
frequency response analysis	A control system analysis technique based on characterizing systems by the gain and phase shift they exhibit at different frequencies[55]
frequency response function	*aka* FRF, $H(\omega)$ or $H(f)$. In the context of control engineering, a quantitative measure of the response to excitation of a system
Freundlich adsorption isotherm	One of the more popular types of empirical adsorption isotherm; cf Langmuir adsorption isotherm
FRF	see **frequency response function**

---

[55] Much favored by control system academics, but of limited practical use in chemical engineering, where processes and system components exhibit inconvenient non-linearities and discontinuities

friction	A force between two surfaces in contact with each other and resisting movement[56]
friction factor	One of a number of dimensionless fiddle factors (qv) based on empirical experiment used at the fluid mechanics/hydraulics interface
friction head	see **frictional head**
frictional head	*aka* friction head. The dynamic head required to overcome friction in a piping system
fridge	Proprietary eponym (from FrigidaireTM), often used for domestic refrigerators, and commonly used professionally as a shorthand for a refrigeration (qv) utility
Friedel-Crafts reaction(s)	A series of reactions used commercially for alkylation or acylation of aromatics
frig	*(informal)* Both verb and noun for the temporary bypass of safety features or forcing of control signals or actions; used colloquially on UK sites (N.B. alternate offensive non-engineering meaning)
frigging out	*(informal)* The temporary bypass of safety features or forcing of control features during commissioning, used colloquially on UK sites (N.B. alternate offensive non-engineering meaning)
frittage	see **sintering**
frog	A device permitting the wheels on one rail of a track to cross an intersecting rail
frog's eye perspective	A one-point perspective from below; defined formally in BS EN ISO 10209:2012 Technical product documentation — Vocabulary — Terms relating to technical drawings, product definition and related documentation; cf bird's eye perspective
front end	The first stages of a process, the opposite of tail end (qv)
front end engineering design	*aka* FEED. The second stage of process plant design, also known as a preliminary or basic engineering study; an early-stage plant design exercise. Commonly, this follows an initial feasibility study (qv) and/or conceptual design (qv) study, giving a progressively closer definition of the final intent with a greater clarity of cost and program; see **Box S2**.
front end loading	*aka* FEL. Adding resources early in a project, especially addressing detailed design issues early in a project
frostproof hydrant	see **dry barrel hydrant**

---

[56] But defined in *API RP2001 - Fire Protection at Refineries* as *the heat produced by* two surfaces in contact with each other and resisting movement, which is just plain wrong!

froth	Generically, almost a synonym for foam (qv), but the term froth is often used to imply unstable bubbles, and possibly a larger/more variable bubble size
froth flotation	*aka* ore flotation. A mineral processing operation in which a froth of air bubbles is used to separate particles by density and size; cf dissolved air flotation, induced air flotation
froth flow	see **churn flow**
frothover	*aka* froth-over. When a vessel of hot (but not burning) oil overflows as a result of continuous low intensity frothing caused by boiling of water; defined formally in API RP 2001 - Fire Protection at Refineries; cf slopover, boilover
froth-over	see **frothover**
Froude number	(*symbol* Fr) A dimensionless number which compares flow inertia with 'local field' (usually gravity) with many useful applications
FRP	Fiber/fibre reinforced plastic, see **glass reinforced plastic**
FRS	see **functional requirements specification**
FRV	see **flow regulating valve**
FS	1. Finish to start 2. Forged steel
FSD	1. *(drawing notation)* Flat side down 2. see **functional specifications document**
FSMA	see **Food Safety Modernization Act**
FSO	see **floating, storage, and offloading unit**
FSRU	see **floating storage and regasification unit**
FST	1. Forged steel 2. see **final settling tank**
FSTNR	*(drawing notation)* Fastener
FSU	1. *(drawing notation)* Flat side up 2. see **floating storage and offloading unit**
FTF	see **fundamental train frequency**
FTG	*(drawing notation)* Fitting
FTU	see **formazin turbidity unit**
FUBAR	*(informal, offensive)* Fucked/fouled up beyond all recognition. Part of a spectrum (SNAFU, FUMTU, TARFU, FUBAR - qv), conveying increasing degrees of disorder/deviation from correct practice or operation and the absence of a means to resolve it
fudge factor	*(informal)* see **fiddle factor**
fuel	Combustible or flammable material; defined formally (and slightly differently) in API RP 2001 - Fire Protection at Refineries and API RP 535 - Burners for Fired Heaters at Refineries

fuel assisted flare	see **endothermic flare**
fuel cell	A device that converts fuels (especially hydrogen) directly into electricity
fuel efficiency	The ratio of total heat absorbed to heat produced by fuel combustion (based on LHV (qv)); defined formally in BS EN ISO 13705:2012/ISO 13705:2012(E) Petroleum, petrochemical and natural gas industries. Fired heaters for general refinery service, API STD 530 - Calculation of Tube Heater Thickness and API STD 560 - Fired Heaters for General Refinery Service
fuel gas	*aka* refinery fuel gas (RFG). Gas used as fuel, including natural gas (qv), manufactured gas (qv), biogas (qv), LPG (qv), LP gas-air mixtures, petroleum gas itself, and any mixture of these; defined formally in NFPA 86: Standard for Ovens and Furnaces, 2019 Edition
fuel injector	The nozzle and valve arrangement through which fuel is sprayed into a combustion chamber
fuel object	*aka* fuel package. An (often finely divided) solid which might fuel a fire or deflagration; defined formally in NFPA 654: Standard for the Prevention of Fire and Dust Explosions from the Manufacturing, Processing, and Handling of Combustible Particulate Solids, 2017 Edition
fuel oil	*aka* bunker oil, bunker fuel oil, BFO, furnace oil, heavy oil, heavy fuel oil, HFO, marine diesel oil, MDO, marine fuel etc. 1. Generically, a broad class of petroleum distillates, heavier than gasoline and naphtha, and used for a range of purposes 2. Specifically, a liquid fuel; defined formally in ASTM D396, Standard Specifications For Fuel Oils and NFPA 86: Standard for Ovens and Furnaces, 2019 Edition
fuel package	see **fuel object**
fuel rod	A fuel assembly used in some types of nuclear reactors
fuel tank inerting	Filling the headspace of a fuel tank with an inert gas to prevent formation of a flammable atmosphere.
fugacity	A 'pseudo-pressure' fiddle factor (qv) that allows one to modify the ideal gas law (qv) to match the real world
fugacity coefficient	*aka* activity coefficient. A fiddle factor (qv) used to fit the ideal gas law to real gases

fugitive emission	An unintended gas (especially flammable gas) release to environment during normal process operation; defined formally in API RP 505 2nd Edition, August 2018 Recommended Practice for Classification of Locations for Electrical Installations at Petroleum Facilities Classified as Class I, Zone 0, Zone 1, and Zone 2 and NFPA 30: Flammable and Combustible Liquids Code, 2018 Edition
Fukushima Daiichi nuclear disaster	The March 2011 nuclear disaster at the Fukushima Daiichi nuclear power plant, Japan
full bore	1. Generally, operating at maximum speed or capacity 2. In the context of valves, having no obstruction to flow or reduction in bore in the flow path
full bore safety valve	A safety valve having no obstruction to flow or reduction in bore in the flow path
full indicator movement	see **total indicator reading**
full internal pressure relief valve	A recessed PRV (qv); defined formally in NFPA 58: Liquefied Petroleum Gas Code, 2017 Edition
full lift safety valve	*aka* vollhub safety valve. A safety valve with a discharge area which is not set by disk position
full penetration weld	A weld with weld metal through an entire component's depth; defined formally in API 660 - Shell-and-Tube Heat Exchangers, API RP 661 - Heat Exchangers and ASME BPE (American Society of Mechanical Engineers: Bioprocessing Equipment)
full port	see **full bore**
full retention interceptor	see **interceptor**
full size	*aka* full-scale. Scale factor 1:1.; defined formally in BS EN ISO 10209:2012 Technical product documentation — Vocabulary — Terms relating to technical drawings, product definition and related documentation
full twat	A degree of opening for valves and pumps; cf half twat, quarter twat
Fuller's earth	Absorbent fine-grained minerals, largely synonymous with bentonite (qv); cf diatomaceous earth
full-scale	see **full size**
fully developed flow	In fluid mechanics and hydraulics, the point along a flow at which the velocity profile does not change
fulvic acids	The class of brown carboxylic acids formed by decomposition of biological matter often present in surface water which remain soluble when a strong base extraction is performed; cf humic acids

fume incinerator	A process exhaust thermal fume destruction device; defined formally in NFPA 86: Standard for Ovens and Furnaces, 2019 Edition
fume(s)	A gas, or a suspension (of fine solid particles, liquid droplets or both in air), produced by a process, especially if obnoxious or toxic; cf smoke
FUMTU	*(informal, offensive)* Fucked/fouled up more than usual; part of a spectrum (SNAFU, FUMTU, TARFU, FUBAR - qv), conveying increasing degrees of disorder/deviation from correct practice or operation and the absence of a means to resolve it
function block diagram	*aka* FBD. A PLC (qv)/DCS (qv) graphical programming language, an alternative to ladder logic (qv)
functional design specification	*aka* FDS. A contested term, either: 1. A specification (not confined to control systems), that details the required functionality, but not the method of implementation, akin to a user requirement specification (qv)/functional requirements specification (qv); or 2. A control philosophy (qv), in the sense of a description in words of what the process engineer wants the control system to do
functional failure mode and effects analysis	*aka* FFMEA, system failure modes and effects analysis (SFMEA). A subtype of failure mode and effects analysis (qv), as applied to design
functional group	1. Most commonly, a group of atoms which give a chemical compound its characteristic activity 2. A combination of technical documentation elements as per BS EN ISO 10209:2012 Technical product documentation – Vocabulary – Terms relating to technical drawings, product definition and related documentation
functional requirements specification	*aka* FRS, functional spec, functional specifications document (FSD), spec, user requirements specification (URS), functional design specification (FDS). This term is not contested, though it has many other names. It is a document containing an early stage, high level description of what a proposed project needs to be able to do, rather than how it does it
functional safety	An approach that uses active intervention techniques to achieve a safe process, as distinct from passive safety systems (e.g., bunding); usually used in reference to electrical/electronic systems (trips and alarms). Promulgated by the IEC 61508 and related standards
functional spec	see **functional requirements specification**

functional specifications document	see **functional requirements specification**
fundamental constants	Parameters that do not change throughout the known universe (e.g., charge on an electron; speed of light in a vacuum; gravitational constant)
fundamental dimensions	*aka* fundamental units. Strictly, the units of amount of light, amount of matter, electric current, length, mass, temperature and time used as the basis of a system of derived units. Sometimes mass, length and time are referred to as fundamental units/dimensions
fundamental frequency	The lowest natural frequency of an undamped system; defined formally in **BS ISO 2041:2018 Mechanical vibration, shock and condition monitoring. Vocabulary**
fundamental period	The smallest time in which a periodic function repeats; defined formally in **BS ISO 2041:2018 Mechanical vibration, shock and condition monitoring. Vocabulary**
fundamental train frequency	*aka* FTF, cage frequency, cage failing frequency. The number of turns which a rolling element bearing cage makes per shaft rotation; cf ball pass frequency
fundamental units	see **fundamental dimensions**
fungus	A kingdom of eukaryotic organisms, including yeasts and molds, widely used in both traditional (e.g., beer brewing) and advanced (e.g., pharmaceutical) biotechnology
furnace	Generically, any direct-fired heat exchanger, though this is a contested term. The term fired heater (qv) is generally used to describe those used for simple heating of a process stream, whilst furnace usually refers to a more complex heating process involving additional chemical and physical changes in the process streams. Furnace is also sometimes held to be any heater operating above 1000°F (538°C), whilst oven (cf) is used to refer to any heater operating below this temperature
furnace oil	see **fuel oil**
fusel alcohol	see **fusel oil**
fusel oil	*aka* fusel alcohol, fuselol. An oily mixture of undesirable alcohols, (mainly amyl alcohol) produced during fermentation
fuselol	see **fusel oil**
fusion	1. The opposite of fission (qv) 2. The melting of a solid, especially in welding; as defined formally in ASME BPE (American Society of Mechanical Engineers: Bioprocessing Equipment) and API STD 620 - Low Pressure Storage Tanks

fusion bond epoxy	aka FBE. A thermosetting powder coating applied to steel panels using an electrostatic process, then heat cured
fusion welding	Joining by fusion (qv); defined formally in ASME BPE (American Society of Mechanical Engineers: Bioprocessing Equipment)
future corrosion allowance	Design corrosion allowance which considers future operating conditions and service life; defined formally in API RP 579 - Fitness for Service
fuzzy logic	Computer logic based in analogue 'degrees of truth', rather than binary true/false[57]
FW	1. *(drawing notation)* Field weld 2. *(drawing notation)* Fillet weld 3. Fire water
FWKO	see **free water knockout**

---

[57] Some say truly one of the least useful forms of academic nonsense ever devised!

G

g	see **acceleration due to gravity**
GA	see **general arrangement drawing**
GAC	see **granular activated carbon**
GAD	see **general arrangement drawing**
gaffer tape	see **Duck tape**
gage	Deprecated archaic USA spelling of gauge
gain	A term used in process control: 'process gain' describes how a process variable responds to a change in manipulated variable; 'controller gain' is a tuning constant in a PID controller; in engineering, gain is not the opposite of loss (qv)
Galilei number	(*symbol* Ga), *aka* Galileo number. A dimensionless number, relating gravitational and viscous forces; used in analysis of condensers etc.
galling	*aka* cold welding. Wearing of sliding metallic surfaces as a result of transitory adhesion; defined formally in API RP 576 - Inspection of Pressure Relieving Devices
gallon	Either the obsolete imperial units (qv) measure of volume, or the US Customary units (qv) measure with the same name, but a different magnitude: 1 imperial gallon being around 1.2 US gallons
gallon (imperial)	Obsolete imperial units (qv) measure of volume
gallon (US)	A US customary units (qv) measure of volume, equal to about 230 In3 or 3.8L
GALV	*(drawing notation)* Galvanized
galvanic corrosion	see **dissimilar metals corrosion**
galvanization	The coating of steel with zinc for corrosion protection, most commonly achieved by dipping a steel component in molten zinc, a process known as 'hot-dip galvanizing'
galvanized mild steel	*aka* GMS. A common and economical material of construction for items such as access ladders, decking, bridges, etc.
gangue	Commercially valueless components of metallic ore, the proportion of which is reduced by beneficiation (qv); cf dross, slag, tailings, overburden
gantry	A synonym for pipebridge (qv) or more generally, bridge-like overhead support
Gantt chart	*aka* GANTT chart (sic). A graphical representation of a project program/schedule

GANTT chart	Incorrect[58] version of Gantt chart (qv)
garbage	see **municipal solid waste**
Gardner color	A color (qv) scale used to measure the yellowness of liquids; defined formally in ASTM D1544-04(2018) Standard Test Method for Color of Transparent Liquids (Gardner Color Scale)
gas air mixer	In the context of LPG, a system which mixes LPG vapor with air; defined formally in NFPA 58: Liquefied Petroleum Gas Code, 2017 Edition
gas analyzer	Generically, a device with various industrial uses such as chemical analysis of reaction products and fire detection. Specifically defined formally as a device associated with a fired heater which measures concentration of key components in feed or flue gas, see NFPA 86: Standard for Ovens and Furnaces, 2019 Edition
gas ballast pump	*aka* vented exhaust mechanical pump. A ballast gas (qv) pump; defined formally in NFPA 86: Standard for Ovens and Furnaces, 2019 Edition
gas ballast valve	A valve on an oil sealed vacuum pump which allows a little compressed gas to be expelled, taking with it condensation which might otherwise contaminate its sealing oil and compromise its performance
gas centrifuge	A centrifuge used to separate mixtures of gases based on density, most notably in purifying radioisotopes
gas condensate	see **natural gas liquids**
gas conditioning tower	see **conditioning tower**
gas constant	(*symbol* R) *aka* ideal gas constant, molar gas constant, universal gas constant. A constant which relates the average relative kinetic energy of particles in a gas to the temperature of the gas, in energy per increment of temperature per mole
gas cutting	see **oxygen cutting**
gas cyclone	*aka* cyclone separator, cyclonic separator. A cyclone separator (qv) removing solid particles from a gas (usually air) by vortex separation; cf hydrocyclone

---

[58] Because there was a Mr Gantt

gas dispersion and smoke ingress analysis	*aka* GDSI, smoke and toxic gas dispersion analysis, smoke and toxic gas dispersion analysis study, smoke and gas ingress analysis (SGIA). A safety study identifying and mitigating hazards associated directly or indirectly with a loss of containment of hazardous gas, especially any hazards which might interfere with non-process buildings, escape/evacuation routes and temporary refuges
gas expansion turbine	*aka* expansion turbine or turboexpander. A device used to turn the expansion of a gas into useful work, such as driving a compressor; defined formally in ISO14661 Thermal turbines for industrial applications (steam turbines, gas expansion turbines) - General requirements
gas formation volume factor	(*symbol* B_g) *aka* gas FVF. An estimate of gas volume in an oil deposit/reservoir/formation, based on gas volume produced at the surface corrected for the difference between surface and reservoir conditions (some say between STP (qv) and reservoir conditions)
gas FVF	see **gas formation volume factor**
gas gathering station	see **gathering station**
gas grouping	*aka* gas grouping. The hazardousness of gases from a flammability point of view. ATEX (qv) gas groups are as follows (in increasing ease of flammability): Group 1 e.g., methane; Group 2A e.g., propane; Group 2B e.g., ethylene and Group 2C e.g., hydrogen. There is also a group 3 for dusts. Defined formally in API RP 505 2nd Edition, August 2018 Recommended Practice for Classification of Locations for Electrical Installations at Petroleum Facilities Classified as Class I, Zone 0, Zone 1, and Zone 2. There are however competing definitions. In the e.g., in the USA, hydrogen is classified as gas group B whereas in the EU it is group IIC. In the USA there are gas groups A to G, whilst in the EU there are only IIA, IIB and IIC.
gas groups	see **gas grouping**
gas hydrate	see **methane clathrate**
gas jet mixer	An eductor (qv) driven by fuel gas used to entrain and mix combustion air to feed a burner; defined formally in NFPA 86: Standard for Ovens and Furnaces, 2019 Edition
gas laws	Boyle's law, Charles's law and Guy-Lussac's law (qv) which relate the pressure, volume and temperature of ideal gases, and the ideal gas law which combines them. Avogadro's law (cf), which relates this to moles of substance, is also sometimes included under this heading

gas lift	The operating principle of a simple type of low head pump, using compressed gas to reduce the bulk density of a fluid to cause it to rise up a tube; an air lift (qv) is an example, with air as the gas
gas lift well	A type of oil well where flow from the reservoir is enhanced by injecting gas into production tubing, usually via the space between the casing and piping, making the well into a gas lift (qv) pump
gas oil	1. Diesel (qv) fuel or similar material, usually the components (which might need further processing) which go into blending finished diesel 2. In the UK, a term for red diesel (qv)
gas oil separation plant	*aka* GOSP. A separator (qv) in the oil and gas sense, i.e.; a number of types of pressurized gravity separator, separating well fluids (qv) into various streams of oil, water and gas, including but not limited to the test separator (qv)
gas quenching	The cooling of hot gases by direct contact with a colder fluid, or cooling work inside a furnace with gas; defined formally in NFPA 86: Standard for Ovens and Furnaces, 2019 Edition
gas seal	see **air seal**
gas sweetening	The process of removing hydrogen sulfide from sour gas (qv)
gas to liquids	*aka* GTL. A process which converts methane rich gas into useful liquid products such as LPG, Diesel, gasoline, etc.
gas tungsten arc welding	*aka* GTAW. A process which produces a high-quality weld in thin stainless steel workpieces; consequently, favored in hygienic industries; defined formally in ASME BPE (American Society of Mechanical Engineers: Bioprocessing Equipment)
gas turbine	*aka* combustion turbine. A rotary thermal power unit driven by high pressure gas, as opposed to a steam turbine (qv), or to put it another way, essentially a jet engine whose shaft is coupled to a compressor or generator
gas void fraction	The fraction of the system volume occupied by gas in a multiphase system
gas welding	see **oxy-acetylene welding**
gas wetness	*aka* wetness. The mass fraction of liquid in a wet gas (qv-commonest sense); defined formally in the case of steam in ISO14661 Thermal turbines for industrial applications (steam turbines, gas expansion turbines) - General requirements

gas/oil ratio	*aka* GOR. The volumetric ratio of gas to oil in well fluids at standard conditions, measured after gas has come out of solution. Commonly expressed in cubic feet of gas per barrel of oil, though it can be expressed as a dimensionless ratio if consistent units are used
gasification	The production of fuel gas from liquid or solid feedstocks
gasket	A static seal between mating surfaces made from deformable material; defined formally in a hygienic context in ASME BPE (American Society of Mechanical Engineers: Bioprocessing Equipment)
gasohol	A liquid fuel made of GASoline and alcOHOL (petrol and ethanol)
gasoline	A proprietary eponym used in the USA for what is known as petrol in the UK
gastight	*aka* hermetically sealed. Impermeable to gas; cf bubbletight, liquidtight
gate valve	*aka* sluice valve. A valve closed with a movable barrier or 'gate', useful for isolating duty in materials such as slurries which might cause problems with valve seating. The knife gate valve is particularly suited to slurries or plastic solids
gathering line	Pipelines which pass oil, gas or mixed well fluids (qv) from wellheads to a transmission line; defined formally for gas gathering lines in Guidelines for the Definition of Onshore Gas Gathering Lines API Recommended Practice 80 First Edition, April 2000
gathering station	*aka* gas gathering station, gathering test station, GGS, group gathering station. Plant which separates the mixture of well fluids (qv) arising from oil wells to recover natural gas byproducts, and compress the gas for further transport via a transmission line; a formal definition for gases can be found in 49CFR 192.3 and for well fluids (referred to as petroleum) in 49CFR 195.2
gathering test station	see **gathering station**
gauge pressure	(*symbols* bar g, psig *etc.*) Absolute pressure minus atmospheric pressure; cf absolute pressure
Gay-Lussac's law	One of the gas laws (qv) relating temperature and pressure
GDSI	see **gas dispersion and smoke ingress analysis**
gear pump	A type of positive displacement pump with interlocking gears as the pump head; defined formally in NFPA 20: Standard for the Installation of Stationary Pumps for Fire Protection, 2019 Edition

gel	A jelly (Jell-O) like mixture containing a polymeric substance which 'sets' to a more solid form. IUPAC offer a formal definition of the term in Definitions of terms relating to the structure and processing of sols, gels, networks, and inorganic-organic hybrid materials (IUPAC Recommendations 2007)
Geldart group	The classification of powders by decreasing ease of fluidization, where Group A is easy, and D is hard
GEMBA	Common false acronym, see **gemba**
gemba	*aka* GEMBA, genba. 'Crime scene' in Japanese. 'Gemba walks' involve looking at and understanding the production process by direct observation and asking questions. A fundamental tool of lean management philosophy and an example of management by walking around (qv)
genba	see **gemba**
general arrangement drawing	*aka* GA, GAD, plot plan. A surprisingly contentious term, with a range of relationships with the term 'plot plan' (qv). General arrangement drawing is defined formally in BS EN ISO 10209:2012 Technical product documentation – Vocabulary – Terms relating to technical drawings, product definition and related documentation, as a drawing of construction works, though this does not address the range of meanings the term is given: 1. The term almost always refers to a scale drawing which shows the layout in space of a plant 2. Usually, it is a drawing which shows the layout of equipment and pipework of a plant, usually to scale, which may in addition be dimensioned, commonly also known as a plot plan 3. Some however consider that the term general arrangement should be used in reference to a piping layout, whereas a plot plan is a type of equipment-only GA 4. A minority view is that plot plan is a plan view of the plant area whereas a GA includes plan, elevation or sectional views
general assembly drawing	A whole-product assembly drawing (qv) for a product or item of equipment; defined formally in BS EN ISO 10209:2012 Technical product documentation – Vocabulary – Terms relating to technical drawings, product definition and related documentation
general purpose turbines	A term used for certain smaller turbines used as prime movers in the oil and gas industry; defined formally in EN ISO 10437 Petroleum, petrochemical and natural gas industries - Steam turbines - Special-purpose applications

general purpose warehouse	From a fire safety point of view, a warehouse which is classified as 'low' or 'ordinary' hazard occupancy by building codes and NFPA 101; defined formally in NFPA 30: Flammable and Combustible Liquids Code, 2018 Edition
generated special atmosphere	A special atmosphere (qv) for use in a fired heater generated from reaction air and/or gas; defined formally in NFPA 86: Standard for Ovens and Furnaces, 2019 Edition
generator	1. Most generically used to describe a device which converts mechanical into electrical energy 2. Less commonly used to describe a device which produces a gas 3. Least commonly, the water and steam heating section of a heat recovery steam generator (qv); defined formally in API RP 534 – Heat Recovery Steam Generators
generic trademark	see **proprietary eponym**
genericized trademark	see **proprietary eponym**
geographic information system	*aka* GIS. A type of electronic mapping system
geometric similarity	One of the types of system similarity used for process scaleup. This one involves systems of the same shape, which differ only in size i.e., based upon a scale model; cf kinematic similarity, dynamic similarity
geometrically safe	Incapable of supporting a self-sustaining nuclear chain reaction by virtue of component geometry
geometry pig	*aka* caliper pig. A device to measure internal pipe ovality
GEP	see **good engineering practice(s)**
GFS	Glass fused to steel, see **glass on steel**
GG	Gauge glass
GGS	Group gathering station, see **gathering station**
GHP	1. Gas horse power 2. Gross horse power 3. see **good hygiene practice(s)**
Gibbs free energy	(*symbol* G) Thermodynamic energy free to do work at prevailing conditions; cf Helmholtz free energy
Gibbs phase rule	*aka* Gibbs' phase rule, phase rule. A rule of thermodynamics which relates the number of independent components and phases to the number of independent variables
Gibbs triangle	see **triangular diagram**
giga-	(*symbol* G-) The SI unit prefix denoting a factor of 10^9
Gilliland correlation	An empirically generated correlation used to approximately estimate number of distillation stages

Gilliland–Sherwood correlation	An empirically generated dimensionless correlation used to approximately estimate gas-liquid mass transfer
girth weld	A circumferential butt weld (qv) around a cylinder or cone; defined formally in API RP 579 - Fitness for Service
GIS	see **geographic information system**
giveaway	The amount by which a product property exceeds its specification
GJ	*(drawing notation)* Ground joint
gland package	see **stuffing box**
gland plate	1. Most commonly, a plate associated with a sealed cable entry (gland) 2. Less commonly, the endplate between a stationary mechanical seal assembly and seal chamber, as defined in API RP 682 - Pump Seals
Glasgow screwdriver	*(jocular)* see **Manchester screwdriver**
glass coated steel	see **glass on steel**
glass on steel	*aka* glass coated steel, glass fused to steel, enameled steel. A term used for panels of glass fused to steel sheet; used for economical sectional bolted tank manufacture
glass reinforced plastic	*aka* GRP; fiber reinforced plastic, fibre reinforced plastic, FRP, fiberglass, fibreglass. A composite material of glass fibers and polymeric resin
global positioning system	*aka* GPS. A satellite-based system for location mapping
global warming potential	*aka* GWP. The heat absorbed by given mass of a greenhouse gas (qv) in the environment, expressed as a multiple of that of CO_2; cf carbon dioxide equivalent
globe valve	A valve with a globular external appearance (cf ball valve) which is closed with a moveable disc, like a domestic faucet. It is used for both isolating and throttling duty in the oil and gas industry specifically, despite its high headloss in control applications and difficulty in sealing
glove box	*aka* glovebox. A sealed chamber with gloves passing through the wall, allowing manipulation of objects inside the box; used with highly hazardous substances
glovebox	see **glove box**
glowing combustion	Incandescent combustion without flames; cf smoldering combustion, flaming combustion
GLP	see **good laboratory practice(s)**
GLV	*(drawing notation)* see **globe valve**

GMP	1. see **good manufacturing practice(s)** 2. Guaranteed maximum price: a contract term used in conjunction with cost plus/cost reimbursable contracts; cf **cost reimbursable**
GMP facility	A facility (usually hygienic) which is designed, built and operated in accordance with GMP (qv); defined formally in ASME BPE (American Society of Mechanical Engineers: Bioprocessing Equipment)
GMS	see **galvanized mild steel**
goatskin	Tarpaulin
go-byes	see **go-bys**
go-bys	*(informal) aka* go-byes. A term for templates/cheat sheets used in the oil and gas industry
going	The horizontal distance between access step nosings; defined formally in BS EN ISO 14122 Safety of machinery. Permanent means of access to machinery. Working platforms and walkways
gold hydrogen	Hydrogen produced by adding bacteria to spent oil wells (qv **hydrogen colors**)
golden rain	see **roman candle effect**
Goldox	A proprietary oxygen-enhanced cyanide gold leaching process
GOME	see **gross order of magnitude estimate**
good engineering practice(s)	*aka* GEP. A concept used in the hygienic industries, a subset of good manufacturing practice(s) (qv), similar to RAGAGEP (qv)
good hygiene practice(s)	*aka* GHP. "Measures applicable throughout the food chain (including primary production through to the final consumer), to achieve the goal of ensuring that food is safe and suitable for human consumption. Note 1: GHP are prerequisite programs as defined in ISO 22000... Note 2: Application of GHP is a prerequisite before any HACCP (qv) study" according to the EHEDG Glossary
good laboratory practice(s)	*aka* GLP. "The means by which laboratory work is planned, performed, monitored and recorded to ensure accuracy and reliability of results, safety and efficiency in the laboratory" according to the EHEDG Glossary
good manufacturing practice(s)	*aka* GMP, *(jocular)* great mountains of paper. "All procedures, processes, practices and activities aimed at ensuring that the suitability and safety objectives are met consistently" according to the EHEDG Glossary
gooseneck	A term applied to a number of types of thin, curved terminal connections, such as the connection between a rotary hose and a swivel

GOR	see **gas/oil ratio**
GOSP	see **gas oil separation plant**
gouge	An undesired discontinuity far longer than it is wide in a rigid material, caused by mechanical removal of surface material. Defined formally in API RP 579 - Fitness for Service; cf **cracks**
governor controlled valve	A device controlling turbine steam flow in line with speed governor output; defined formally in EN ISO 10437 Petroleum, petrochemical and natural gas industries - Steam turbines - Special-purpose applications
GPM	Gallons per minute (most commonly US gallons nowadays)
GPS	1. see **global positioning system** 2. see **ISO Geometrical Product Specification**
GPSA	Gas Processors Suppliers Association
GR	*(drawing notation)* see **grade**
graceful degradation	A philosophy of design robustness in which, in the event of component failure, the system automatically defaults to the best available mode, allowing continued (perhaps impaired) operation; cf **fail to safety**
grade	1. Most commonly refers to local ground level and/or slope 2. Less commonly, a specification or ranking, as defined in the context of quality in BS EN ISO 9000 Quality management systems Fundamentals and vocabulary, or more broadly to product specifications, e.g.; 'winter grade gasoline', 'on-grade or off-grade product'
grades of release	An approach in which potential releases of flammable/explosive materials are assigned one of three grades: continuous, primary, and secondary. In general, these sources lead directly to zonal classifications, namely: continuous: Zone 0; primary: Zone 1; secondary: Zone 2. Formal definitions may be found in EN 60079-14:2014 Explosive atmospheres. Electrical installations design, selection and erection and API RP 505 2nd Edition, August 2018 Recommended Practice for Classification of Locations for Electrical Installations at Petroleum Facilities Classified as Class I, Zone 0, Zone 1, and Zone 2
Graetz number	*(symbol* Gz*)* A dimensionless number which can be used to analyze the degree of development of heat transfer in laminar flow in conduits
Graham's law of diffusion	see **Graham's law of effusion**
Graham's law of effusion	*aka* Graham's law of effusion. The rate of effusion (and by extension diffusion) is inversely proportional to the molecular weight of the effusing substance

grain boundary	In metallurgy, the interface between two adjacent grains in a metal; defined formally in ASME BPE (American Society of Mechanical Engineers: Bioprocessing Equipment)
grains per gallon	An imperial units (qv) measure of concentration, not entirely obsolete but restricted to use as a measure of water hardness (qv) as a calcium carbonate equivalent
gram	(*symbol* g) *aka* gramme. One thousandth of the SI fundamental unit of mass (the kilogram, Kg)
gramme	see **gram**
granular activated carbon	*aka* GAC. Activated carbon (qv) particles large enough to be retained by a 50-mesh sieve; cf powdered activated carbon
granular bed filter	see **depth filter**
granulation	Forming granules, usually by agglomeration (qv) of a fine powder (as is common in hygienic industries), or less commonly by particle size reduction (the commonest meaning in environmental applications)
granule	The product of granulation (qv)
Grashof number	(*symbol* Gr) A dimensionless number relating buoyant and viscous forces; used to study natural convective heat transfer
grasshopper pump	see **nodding donkey**
grassroots design	A completely new design on a new site, as opposed to a modification of an existing design on an existing site; cf brownfield
gravity	An oil and gas term for a measure of mass per unit volume, a multiple of specific gravity as in API gravity (qv) etc.
gravity flow	1. Generically, fluid flow driven solely by gravity, rather than being pumped 2. Less commonly, in the context of layout design, lines may be labelled 'gravity flow' on a P&ID (qv) to indicate a need to avoid pockets or deadlegs (qv) in the pipe
gravity retaining wall	A wall holding back soil solely by virtue of its mass
gravity separation	see **sedimentation**
gravity tank	A storage tank in an elevated position which drives gravity flow (qv); defined formally in NFPA 22: Standard for Water Tanks for Private Fire Protection, 2018 Edition
gravity wall	1. A bridgewall (qv) 2. A gravity retaining wall (qv)
gray	(*symbol* Gy) An SI derived unit for absorbed dose of ionizing radiation
gray hydrogen	*aka* grey hydrogen. Hydrogen produced using fossil fuels; qv greenwash, hydrogen colors

gray iron	*aka* grey iron. Cast iron (qv) with a graphitic microstructure; cf ductile iron
gray list	*aka* grey list: 1. Generically, a list slightly less crucial than a black list (or sometimes a red list) 2. Specifically, most commonly refers to the gray list of marine pollutants in the Convention on the Prevention of Marine Pollution by Dumping of Wastes and Other Matter 1972, covering wastes containing arsenic, copper, cyanides, lead, organosilicon compounds, pesticides, non-blacklist radioactive matter and zinc, as well as bulky debris which might obstruct fishing or navigation
gray per second	(*symbol* Gy/s) The absorbed ionizing radiation dose rate
gray water	Domestic wastewater other than that arising from toilets; defined formally in EN 12566 - Small wastewater treatment systems for up to 50 PT Part 1: Prefabricated septic tanks; cf black water
GRE	Glass reinforced epoxy, a type of GRP (qv)
grease	A hydrocarbon or lipid semisolid. More solid than an oil (qv), less solid than a fat (qv) or wax (qv)
green chemistry	An academic initiative aimed at minimizing the HSE (qv) impacts of process chemistry[59]
green claim	see **comparative assertion**
green hydrogen	Hydrogen produced from water using renewable power; qv hydrogen colors
green tea	Tea which has not been deliberately oxidized in the production process; defined formally in ISO 3103 Tea - Preparation of liquor for use in sensory tests
green utilities	Mostly synonymous with clean utilities (qv), but sometimes used in the same way as it is outside the profession to describe 'green' electricity, etc.
green vitriol	see **copperas**
greenfield	The design of a complete new plant. Also known as grassroots or generic plant design. Commonly used (confusingly) to refer to development on previously undeveloped land; cf brownfield
greenhouse gas	A gas whose release to environment has the potential effect of increasing the Earth's temperature; e.g., carbon dioxide, methane, etc.
greenhushing	Not publicizing an organisation's environmental targets, or degree of compliance with such targets

---

[59] Not appreciating that chemical engineering already does this!

greensand	Naturally occurring granular material with ion exchange and catalytic properties, used to remove iron and manganese ions in water treatment
greenwash	A contested term, but generically, a term for initiatives intended to appear 'green' for PR purposes; see **Box G1**.
grey hydrogen	see **gray hydrogen**
grey iron	see **gray iron**

Box G1. Greenwash

Are plastic or wooden forks better in your work canteen? Many engineers know that there is very little difference in environmental impact, (unless you are under the impression that wood biodegrades in a landfill?) - but switching to wood can make a nice PR story. What about those little signs in hotels urging us to reuse towels and switch the lights off? The main beneficiary of these practices will be the hotel's profit margin, but they provide a nice bit of green PR at less than no cost. This is the essence of greenwash: initiatives intended to appear 'green' for PR purposes; or false or misleading claims of "environmental friendliness" (whatever that means), "sustainability" (ditto) etc. These terms are essentially meaningless, or to put it another way, they can mean anything you like.

In the UK, at least, green claims for a product or service must be relevant to anyone buying or using it, clearly and accurately stated, and justifiable. Failure to follow the guidance can lead to being reported to the Advertising Standards Authority, or the Trading Standards office.

Engineers will sometimes need to investigate potential greenwash initiatives. Ideally, we will use a rational scheme for this, such as for example the IChemE sustainability metrics. We will always have to balance apples and oranges, but at least metrics provide a transparent, quantified and auditable basis for doing so, rather than simply spouting vague terms (clean coal, anyone?)

It should be noted however that the outcomes of techniques such as environmental accounting and life cycle assessment etc. can be very sensitive to how we balance those apples and oranges. Even reasonably well-defined procedures, such as those used in IPPC applications, are vulnerable to bias.

Of course, with such contested terms, any rational approaches may themselves be considered to be greenwash by people with a different political agenda. That said, if the politicians ever do decide the planet needs saving, it will be engineers who get it done.

Related Terms
carbon abatement technology, zero liquid discharge, comparative assertion

grey list	see **gray list**

grid region	see **grid zone**
grid zone	*aka* distributor zone, grid region. The bottom part of a fluidized bed into which fluid is introduced
grinder	see **pulverizer**
grinding	In common with the term milling (qv), this has non-identical meanings in mechanical and chemical engineering. The mechanical engineering senses are associated with removal of metal from a workpiece by two different mechanisms. The differences between milling and grinding in chemical engineering processes (both always meaning a size reduction process) are to some extent dependent on industry sector.
grist	1. Generically, feed for a mill of foodstuffs, such as cleaned grain 2. In sugar refining, sugar grist is a type of particle size analysis
grit channel	A preliminary sewage treatment process which removes larger, denser particles by gravity settlement
grit trap	*aka* Detritor. A preliminary sewage treatment process which removes larger, denser particles, by enhanced gravity settlement
grizley	see **grizzly**
grizzly	*aka* grizley. A type of very coarse screen comprising bars placed in chutes and bins, to prevent oversize material passing through and to allow for collection
groove	Undesired thin spot far longer than it is wide in a rigid material, caused by erosion or corrosion; defined formally in API RP 579 - Fitness for Service; cf gouge
groovelike flaw	A groove or gouge; defined formally in API RP 579 - Fitness for Service
grooving corrosion	see **preferential weld corrosion**
gross economic potential	see **economic potential**
gross heating value	see **higher heating value**
gross operated basis	see **8/8ths**
gross order of magnitude estimate	*aka* GOME, rough order of estimate. An experience-based plug number (qv) used in the absence of real cost data or a quote when estimating costs; cf RE estimate
gross structural discontinuity	see **major structural discontinuity**

ground	1. Somewhat tautologically, everything underground, (perhaps more helpfully, below grade) as defined formally in BS 5930:2015 Code of practice for ground investigations 2. The state of the product of grinding 3. US English for electrical earth
ground flare	A flare at grade, usually enclosed; defined formally in API Standard 521. Pressure-relieving and Depressuring Systems. Sixth Edition | January 2014 and API RP 537 Flare Details for Petroleum, Petrochemical, and Natural Gas Industries; cf elevated flare
ground investigation	Various forms of site investigation focusing on ground conditions; defined formally in BS 5930:2015 Code of practice for ground investigations
ground model	A model of the subsurface composition of a site, in terms of the distribution of various ground components; defined formally in BS 5930:2015 Code of practice for ground investigations
ground snow load	*aka* 50-year ground snow load. The environmental loading due to snow to be used in structural design of buildings as per ASCE 7, Minimum Design Loads for Buildings and Other Structures and defined formally in NFPA 58: Liquefied Petroleum Gas Code, 2017 Edition
grounding	see **earthing and earth bonding**; this term defined formally in NFPA 30: Flammable and Combustible Liquids Code, 2018 Edition and NFPA 654: Standard for the Prevention of Fire and Dust Explosions from the Manufacturing, Processing, and Handling of Combustible Particulate Solids, 2017 Edition
groundwater	Underground water; defined formally in BS 5930:2015 Code of practice for ground investigations and NFPA 20: Standard for the Installation of Stationary Pumps for Fire Protection, 2019 Edition; cf surface water
groundwater control	*aka* dewatering. Defined formally in BS 5930:2015 Code of practice for ground investigations
group 1 gases	Atmospheres containing firedamp (qv); defined formally in API RP 505 2nd Edition, August 2018 Recommended Practice for Classification of Locations for Electrical Installations at Petroleum Facilities Classified as Class I, Zone 0, Zone 1, and Zone 2; cf gas grouping

group 2 gases	Atmospheres containing aboveground flammable gases, subdivided into 2A, 2B and 2C; defined formally in API RP 505 2nd Edition, August 2018 Recommended Practice for Classification of Locations for Electrical Installations at Petroleum Facilities Classified as Class I, Zone 0, Zone 1, and Zone 2; cf gas grouping
group 2A gases	Atmospheres containing gases such as acetone, ammonia, ethanol, methane, or propane; defined formally in API RP 505 2nd Edition, August 2018 Recommended Practice for Classification of Locations for Electrical Installations at Petroleum Facilities Classified as Class I, Zone 0, Zone 1, and Zone 2; cf gas grouping
group 2B gases	Atmospheres containing gases such as acetaldehyde or ethylene; defined formally in API RP 505 2nd Edition, August 2018 Recommended Practice for Classification of Locations for Electrical Installations at Petroleum Facilities Classified as Class I, Zone 0, Zone 1, and Zone 2; cf gas grouping
group 2C gases	Atmospheres containing gases such as acetylene or hydrogen; defined formally in API RP 505 2nd Edition, August 2018 Recommended Practice for Classification of Locations for Electrical Installations at Petroleum Facilities Classified as Class I, Zone 0, Zone 1, and Zone 2; cf gas grouping
group gathering station	see **gathering station**
growth curve	A plot of organism numbers or mass over time
growth media	see **growth medium**
growth medium	*aka* media, growth media. A semisolid or liquid containing nutrients, used as a substrate for biological culture
GRP	see **glass reinforced plastic**
GRPG	see **grains per gallon**
GS hydro	A proprietary cold flaring flange formation system
GTAW	see **gas tungsten arc welding**
GTL	see **gas to liquids**
guarantee point	An operating condition specified in process guarantees; defined formally in ISO14661 Thermal turbines for industrial applications (steam turbines, gas expansion turbines) - General requirements
guard	A physical protective barrier designed into a machine; defined formally in BS EN ISO 12100:2010 Safety of machinery. General principles for design. Risk assessment and risk reduction
guard locking device	A safety device which holds a guard in the locked position until hazardous circumstances no longer exist

guarded	Having a guard (qv). A lengthier formal definition may be found in NFPA 86: Standard for Ovens and Furnaces, 2019 Edition
guarded motor	A guarded (qv) open motor (qv). A lengthier formal definition may be found in NFPA 20: Standard for the Installation of Stationary Pumps for Fire Protection, 2019 Edition
guardrail	A thigh-high guard against accidental sideways fall, found on working platforms and walkways; defined formally in BS EN ISO 14122 Safety of machinery. Permanent means of access to machinery. Working platforms and walkways
guardrail system	The OSHA equivalent of guardrail (qv); defined formally in OSHA 1910.29 - Fall protection systems and falling object protection-criteria and practices
gubbins	*(informal)* Somewhat dismissive UK term for miscellaneous electrical and mechanical components
guess and check	A mathematical problem-solving methodology which is far more commonly used in engineering than rigorous calculation[60]; cf trial and error
guesstimate	*aka* guestimate. A rough estimate where professional judgment fills in for missing data; cf SWAG
guestimate	see **guesstimate**
guide rings	see **rider bands**
guided type fall arrester on rigid anchorage line	A permanent fall arrester (qv) installation; defined formally in BS EN ISO 14122 Safety of machinery. Permanent means of access to machinery. Working platforms and walkways
guidevane	Adjustable vanes installed on the suction side of a turbocompressor to allow manipulation of the flow
guideword	Terms such as 'no/not' 'other' 'more' 'less' 'as well as' 'part of' and 'reverse' which are permutated with parameters like 'flow', 'composition' etc. in a HAZOP (qv) study to try to identify non-obvious failure modes
guillotine	*aka* guillotine damper, isolation blind. A linear damper (qv) used for isolation duties; defined formally in BS EN ISO 13705:2012/ISO 13705:2012(E) Petroleum, petrochemical and natural gas industries. Fired heaters for general refinery service, API STD 530 - Calculation of Tube Heater Thickness and API STD 560 - Fired Heaters for General Refinery Service
guillotine damper	see **guillotine**

---

[60] And more than engineers may like to admit!

Guldberg rule	*aka* Guldberg's law. Heuristic which states that the normal boiling point of most liquids measured in kelvin is approximately two-thirds of their critical temperature measured on the same scale
Guldberg's law	see **Gulberg rule**
gunite	*aka* shotcrete, sprayed concrete, spraycrete. Proprietary eponym for concrete applied by a spray of dry components (with water sometimes admixed at point of discharge)
guyed flare	An elevated flare stabilized by guy lines; defined formally in API RP 537 Flare Details for Petroleum, Petrochemical, and Natural Gas Industries
GV	*(drawing notation)* see **gate valve**
GWP	see **global warming potential**
gyratory crusher	A type of primary crusher used in mineral processing; cf jaw crusher

H

H	1. see **enthalpy**
	2. The chemical symbol for hydrogen
H&MB	see **heat and material balance**
H(f)	see **frequency response function**
H(ω)	see **frequency response function**
H/A ratio	see **hardness to alkalinity ratio**
H/C ratio	see **hydrogen to carbon ratio**
HAB	see **harmful algal bloom**
Haber process	*aka* Haber-Bosch process. The main commercial method for the production of ammonia, by means of catalytic synthesis from nitrogen and hydrogen
Haber-Bosch process	see **Haber process**
HAC	see **hazardous area classification**
HAC study	see **hazardous area classification study**
HACCP	see **hazard analysis critical control point**
HACR	Heating, air conditioning and refrigeration
Hagen–Poiseuille equation	*aka* Hagen–Poiseuille law, Poiseuille equation, Poiseuille law. An equation which estimates pressure drop for an incompressible Newtonian fluid in laminar flow in a straight pipe of uniform circular cross section[61]
Hagen–Poiseuille law	see **Hagen–Poiseuille equation**
half bridge scraper	A rotating bridge scraper (qv) which spans only the radius of a circular tank, as opposed to a fixed full bridge scraper
half cell reaction	Half of a redox reaction (qv) - either reduction or oxidation - to describe the electrochemical potential between two ions/an ion and an element
half cut	A drawing representing a symmetrical object, showing half in view (qv) and half in cut sectional view (qv); defined formally in BS EN ISO 10209:2012 Technical product documentation – Vocabulary – Terms relating to technical drawings, product definition and related documentation[62]

---

[61] A rare set of circumstances in professional engineering
[62] N.B. alternate offensive non-engineering meaning

half section	A drawing representing a symmetrical object, showing half in view (qv) and half in sectional view (qv); defined formally in BS EN ISO 10209:2012 Technical product documentation – Vocabulary – Terms relating to technical drawings, product definition and related documentation
half twat	A degree of opening for valves and pumps; cf full twat, quarter twat
half-bound CO_2	see **semicombined CO_2**
halflife	The time taken for the value of something to halve. The most common application is in nuclear physics, but useful for characterizing any exponentially decaying process
Hall–Héroult smelting process	The most common process used to produce aluminum commercially, involving electrolysis of aluminum oxide dissolved in molten cryolite
halogenation	Chemically adding a halogen atom or atoms to a compound
hammer	1. see **fluid hammer** 2. *aka* Manchester screwdriver (qv)
hammer mill	A type of mill (qv) used for size reduction of feeds as diverse as grain and scrap cars, using the impact of mechanical hammers; particularly favored for fibrous feeds
hammer rash	*(informal, jocular)* Markings on a silo (see **Figure H1**) caused by hammering to clear blockages or percussive maintenance[63] (qv)
handholes	A small sealed hole in a vessel or boiler allowing access for a hand
handline	A handheld firehose or safety line; defined formally in the case of a fire hose in NFPA 11: Standard for Low-, Medium-, and High-Expansion Foam, 2016 Edition
handrail	The top of a walkway guardrail, used for support; defined formally in BS EN ISO 14122 Safety of machinery. Permanent means of access to machinery. Working platforms and walkways and OSHA 1910.29 - Fall protection systems and falling object protection-criteria and practices
handshake	An interlocked sequence of signals between connected components, in which each component waits for the acknowledgement of its previous signal before proceeding with its action, such as data transfer
hangers	see **supports**
Hansen parameter	*aka* Hansen solubility parameter, HSP. A measure of polymer solubility

---

[63] Also to be found on the hands of inexpert hammer operators!

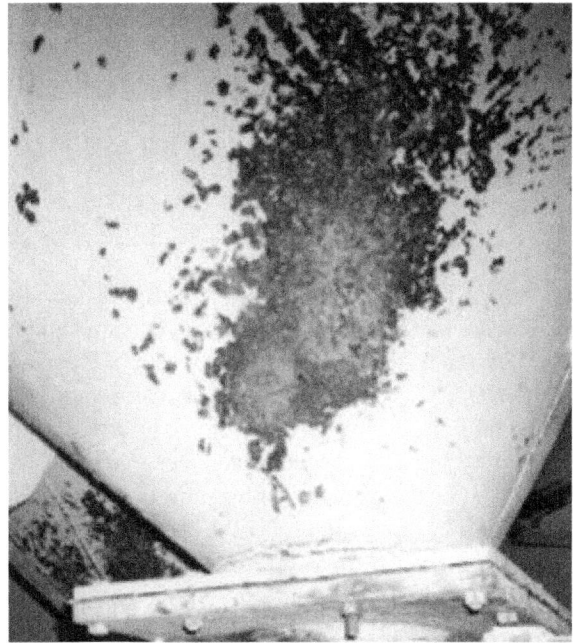

Figure H1 'Hammer rash' markings (Courtesy R. Farnish)

Hansen solubility parameter	see **Hansen parameter**
HAP	see **hazardous air pollutant**
hard copy	*aka* hardcopy. A physical version of a document; defined formally in BS EN 82079-1:2012 and BS EN ISO 10209:2012 Technical product documentation – Vocabulary – Terms relating to technical drawings, product definition and related documentation; cf soft copy
hardcopy	see **hard copy**
hardness	1. Generically, the resistance of a material to permanent deformation by scratching and indentation 2. The rebound height of a hammer following impact 3. In the context of water chemistry, the amount of dissolved calcium and magnesium, often expressed as equivalent mg/l $CaCO_3$; see **Table H1**

Table H1 Conversion Table for Hardness and Alkalinity Parameters

Water Hardness Classification	American Degrees (mg/L CaCO$_3$)	German Degrees (dGH/°dH)	English/Clark Degrees (°Clark,°e or e)	French Degrees (°fH or °f)	American gpg	Milliequivalents/L (meq/L mval/L)
Soft	0-60	0-3.4	0-4.2	0-6	0-3.5	0-3
Moderately hard	61-120	3.4-6.7	4.2-8.4	6.1-12	3.6-7.0	3-6
Hard	121-180	6.7-10.1	8.4-12.6	12.1-18	7.0-10.5	6-9
Very hard	≥ 181	≥ 10.1	≥ 12.6	≥ 18	≥ 10.5	>9

hardness to alkalinity ratio — *aka* H/A ratio. The ratio of hardness (qv) to alkalinity (qv); used as a metric in water engineering (especially ion exchange) and some water-based industries such as brewing

Hardoll fitting — *aka* Avery Hardoll fitting. A proprietary eponym for a class of tanker couplings/connections; cf Bauer coupling

hardstanding — Parking area for heavy vehicles

hardwired — Physically or electrically connected/controlled using wires; defined formally in NFPA 86: Standard for Ovens and Furnaces, 2019 Edition

harm — An induced loss of function or quality. The consensus may be that harm is something that happens to people, whilst damage (qv) covers other induced losses of function or quality in line with the formal definition in BS EN ISO 12100:2010 Safety of machinery. General principles for design. Risk assessment and risk reduction. This is not however a universal view. Harm is physical damage to human health, property or environment; according to BS EN 82079-1:2012 Preparation of instructions for use. Structuring, content and presentation. General principles and detailed requirements, and BS EN 1050:1997 Safety of machinery. Principles for risk assessment

harmful algal bloom — see **algal bloom**

harmonic — Despite its meaning in other contexts, use of this term as a synonym of overtone is deprecated according to BS ISO 2041:2018 Mechanical vibration, shock and condition monitoring. Vocabulary

harmonic excitation — A sinusoidally variable force applied to a system, and/or the excitation produced in the system by this; defined formally in BS ISO 2041:2018 Mechanical vibration, shock and condition monitoring. Vocabulary

harp tank — A geometrically safe (qv) radioactive liquid storage tank whose shape is reminiscent of a harp

HART — see **highway addressable remote transducer protocol**

harvesting	Biomass collection/separation processes, especially in biochemical engineering; defined formally in ASME BPE (American Society of Mechanical Engineers: Bioprocessing Equipment)
Hastelloy	One of a range of grades of a slightly exotic and expensive nickel-molybdenum-chromium-tungsten superalloy (qv), all of which have better corrosion resistance than the highest grades of stainless steel under most circumstances
haunches	The outside areas between the springline (qv) and the bottom of a pipe
HAW	see **higher activity waste**
Hayden-O'Connell	A thermodynamic model for high pressure gases
HAZ	see **heat affected zone**
HAZAN	see **hazard analysis**
hazard	A somewhat contested term; see **Box H1**.
hazard analysis	*aka* HAZAN, process hazard analysis (qv) (PHA), qualitative risk analysis (qv) (QRA). Defined formally in NFPA 15 Standard for Water Spray Fixed Systems for Fire Protection; see **Box H1** and **Box R2** for discussion of hazard and risk
hazard analysis critical control point	*aka* HACCP. "A system which identifies evaluates and controls hazards that are significant for food safety. Note: A HACCP study must be performed during the development of new products and processes, covering thus new equipment, and when changes are made on existing lines or to products" according to the EHEDG Glossary; see **Box H1**.
hazard and operability study	*aka* HAZOP. A 'what-if' exercise or risk study applied to a fairly advanced process design, no earlier than FEED (qv) stage, in order to disclose unforeseen but reasonably likely interactions between systems which have adverse effects on safety or operability. Carried out correctly, it is considered to be the most rigorous of the risk evaluation studies applied to a plant design. Individual unit operations and/or equipment/equipment strategies may be evaluated using FMEA (qv), HACCP (qv), or similar risk evaluation processes. The use of a proven risk assessment process like HAZOP is a common expectation of regulators. See **Box H1** and **Box R2** for discussion of hazard and risk
hazard assessment	The identification of the hazards of a given design, with an estimation of the probability and severity of occurrence. See **Box H1** and **Box R2** for discussion of hazard and risk

Box H1. Hazard

Legislation may provide slightly different definitions depending on the jurisdiction but, generally speaking, a hazard is a source of potential damage to people, property or environment, and is thus closely associated with risk.

The meaning of hazard may however be restricted in the engineering profession to potential sources of harm. Most (but not all) people agree that harm is something that happens to people, whilst damage is not, as discussed in the entry for harm.

The hygienic industries tend to define hazard entirely in the context of food safety (see the EHEDG glossary). Process safety (and therefore consideration of damage and harm to people other than consumers) can therefore tend to be a lesser priority than food safety, to the extent that HACCP studies may be believed to render HAZOP studies redundant.

As far as the oil and gas industry is concerned, API's definition of hazard in RP2001 seems to imply that a hazard is either a property inherent to a material which can cause harm; or the process conditions which can lead to harm. API differentiates this from risk in the standard; risk being the potential for exposure to hazard. However, as discussed in **Box R2**, risk is an even more contested term than hazard.

Note that plants may be located in one regulatory area but need to conform to another (for example in the pharmaceutical industry where plants worldwide must manufacture to US Food and Drug Administration (FDA) and/or European Medicines Agency (EMA) standards). Likewise, in the oil and gas industry, both local and end-user regulation may need to be followed.

In summary, it is good to be clear what people mean when they use the term hazard. The worst-case scenario is that they mean 'risk'.

Related Terms
HACCP, HAZOP, HAZAN, HAZID, hazard analysis, process hazard analysis

hazard identification and risk analysis	*aka* HIRA. A term mostly used in the USA for the all aspects of formal techniques for the identification, assessment and mitigation of risks to employees, public and the environment of a plant
hazard identification study	*aka* HAZID. An exercise undertaken early in design to identify the main hazards to be considered as the design progresses; cf ENVID; see **Box H1** and **Box R2** for discussion of hazard and risk
hazard zone	The space around machinery where hazards are present; defined formally in BS EN ISO 12100:2010 Safety of machinery. General principles for design. Risk assessment and risk reduction; see **Box H1** and **Box R2** for discussion of hazard and risk

hazardous (classified) location	A place where there may be fire or explosion hazards due to the presence of flammable, combustible, or ignitable materials; defined formally in API RP 505 2nd Edition, August 2018 Recommended Practice for Classification of Locations for Electrical Installations at Petroleum Facilities Classified as Class I, Zone 0, Zone 1, and Zone 2; qv zone 0, zone 1, zone 2, zone 20, zone 21, zone 22
hazardous air pollutant	Formally, one of 188 substances specified in the USA version of the Clean Air Act (qv)
hazardous area classification	*aka* HAC, zoning study. The classification of areas of a plant based on how much of the time they are expected to contain a flammable atmosphere under foreseeable conditions. This ensures that an appropriate amount of care is taken, proportional to the fraction of the time that a flammable or explosive atmosphere is present, to ensure potential ignition sources are controlled
hazardous area classification study	*aka* HAC study. A study identifying the class of fire and explosion hazard of areas of a facility under ATEX 137 (1999/92/EC), IEC 60079-10-1 Area Classification – Gases and Vapors or IEC 60079-10-2 Area Classification – Hazardous Dusts; qv zone 0, zone 1, zone 2, zone 20, zone 21, zone 22
hazardous chemical	see **hazardous material**
hazardous chemical reaction	see **hazardous reaction**
hazardous event	A potentially harmful occurrence; defined formally in BS EN 1050:1997 Safety of machinery. Principles for risk assessment and BS EN ISO 12100:2010 Safety of machinery. General principles for design. Risk assessment and risk reduction
hazardous material	*aka* hazmat, HAZMAT. Generally, a material which presents a hazard. However, the NFPA official definition restricts the term to materials presenting non-fire hazards; see NFPA 30: Flammable and Combustible Liquids Code, 2018 Edition
hazardous materials storage locker	A prefabricated structure brought to site for the purpose of hazardous materials storage (more or less), according to the NFPA; a lengthier formal definition may be found in NFPA 30: Flammable and Combustible Liquids Code, 2018 Edition
hazardous reaction	Generally, a reaction which presents a hazard. However, the NFPA official definition restricts the term to materials presenting non-fire hazards; see NFPA 30: Flammable and Combustible Liquids Code, 2018 Edition

hazardous situation	Generically, the obvious broad meaning (qv hazard). However, the formal definition, in BS EN ISO 12100:2010 Safety of machinery. General principles for design. Risk assessment and risk reduction references the narrow definition of hazard in the same document
hazardous waste	Waste (qv) which can present a hazard (qv). As both of these terms are contested, a somewhat vague categorization. In the UK, the EA and HSE both offer formal definitions of the term applicable to their respective fields. In the USA, the EPA offers a similar formal definition. What all of these definitions have in common is that they consider hazard synonymous with harm (qv) (in the broader sense: physical damage to human health, property or environment)
HAZASS	Hazard assessment: a hazard evaluation/identification technique favored by Mecklenbergh which never really took off
HAZCON	A specific risk assessment identifying the hazards associated with a construction activity
haze	1. In the case of air or water clarity, turbidity caused by fine particles 2. Reduced metal surface finish brightness, as defined in ASME BPE (American Society of Mechanical Engineers: Bioprocessing Equipment)
Hazen	*aka* HU, Hazen unit. A unit of color (qv)
Hazen unit	see **Hazen**
HAZID	see **hazard identification study**
hazmat	see **hazardous material**
hazmat suit	*aka* HAZMAT suit. A type of personal protective equipment: a whole-body covering designed to protect the wearer from hazardous materials
HAZOP	see **hazard and operability study**
HB	see **Brinnell hardness number**
HBF	*(drawing notation)* Heavy barrel flange
HC	*(drawing notation)* Hose coupling
HCA	see **high consequence area**
HCFC	see **hydrochlorofluorocarbon**
HCGO	Heavy coker gas oil
HCO	Heavy cycle oil
HCPs	see **host cell proteins**
HDPE	High density polyethylene, a material of construction; cf polyethylene
HDxx	In the context of plastics, high density xx, as in HDPE for high density polyethylene

HE	1. *(drawing notation)* see **head** (meaning 2) 2. see **heat exchanger**
head	1. A pressure expressed as the height of the fluid which would exert that pressure at its base, e.g., mWG (qv). Defined formally in NFPA 20: Standard for the Installation of Stationary Pumps for Fire Protection, 2019 Edition 2. The end cap of a pressure vessel, commonly ellipsoidal (qv), hemispherical (qv) or torispherical (qv) in shape
head coefficient	(*symbol* ψ) A dimensionless characteristic coefficient derived from affinity laws (qv) for centrifugal pump scaleup
head end	*aka* headend. Preliminary reception and processing of spent nuclear fuel
head height	1. Commonly used in professional discussions to mean the height above local grade at which a person's head would be 2. Defined formally in BS EN ISO 14122 Safety of machinery. Permanent means of access to machinery. Working platforms and walkways as more akin to 'headroom' in everyday language, i.e., the clear space between floor and an overhead obstruction/roof
headend	see **head end**
header	1. see **gravity tank** 2. A connecting pipe linking multiple parallel pieces of equipment and/or a common electrical/fluid line from which equipment draws
header box	Enclosures for header return bends (qv), which prevent leakage at the inspection plugs at the point of the U-bend turn. Defined formally in API STD 530 - Calculation of Tube Heater Thickness, API STD 560 - Fired Heaters for General Refinery Service and BS EN ISO 13705:2012/ISO 13705:2012(E) Petroleum, petrochemical and natural gas industries. Fired heaters for general refinery service
header return bend	U-turns at the end of heated tubes in fired heaters. Defined formally in BS EN ISO 13705:2012/ISO 13705:2012(E) Petroleum, petrochemical and natural gas industries. Fired heaters for general refinery service, API STD 530 - Calculation of Tube Heater Thickness and API STD 560 - Fired Heaters for General Refinery Service
headloss	Pressure drop expressed as head
headroom	see **head height**
heads	*aka* foreshots. The equivalent of top product (qv) in potable alcohol distillation: the lower boiling fractions, the opposite of tails (qv)

headspace	The upper part of an enclosed vessel which is not filled with liquid
health and safety audit	see **safety audit**
Health and Safety Executive	*aka* HSE. The UK health and safety regulator
health care occupancy	A building, or part of one, used for inpatient treatment of at least four patients who will require assistance from others to escape in an emergency. Defined in NFPA 30: Flammable and Combustible Liquids Code, 2018 Edition
heat absorption	As well as the commonplace meaning, there is a specific meaning in the context of fired heaters, which applies only to coils, and which corrects for combustion air preheat. Defined formally in BS EN ISO 13705:2012/ISO 13705:2012(E) Petroleum, petrochemical and natural gas industries. Fired heaters for general refinery service, API STD 530 - Calculation of Tube Heater Thickness and API STD 560 - Fired Heaters for General Refinery Service
heat affected zone	*aka* HAZ. The unmelted but nevertheless heat affected base material adjacent to a weld; defined formally in API RP 571 - Damage Mechanism Affecting Fixed Refinery Equipment, API RP 579 - Fitness for Service and ASME BPE (American Society of Mechanical Engineers: Bioprocessing Equipment)
heat and mass balance	see **heat and material balance**
heat and material balance	*aka* H&MB, heat and mass balance. The practical application of conservation laws to understanding mass and energy flows for a process plant[64]
heat capacity	(*symbol* C) *aka* thermal capacity. The amount of heat required to produce a specified temperature change in a specified mass of substance; cf specific heat capacity
heat exchanger network	*aka* HEN. A system of interacting heat exchangers within a process design intended to reduce the overall energy requirements of a facility, though often uneconomic/impractical to pursue in practice
heat exchanger unit	The heat exchanger or exchangers which meet a given duty. Defined formally in API 660 - Shell-and-Tube Heat Exchangers and API RP 661 - Heat Exchangers

---

[64] But if you didn't already know that, you are reading the wrong dictionary!

heat integration	The attempt to reduce the overall energy requirements of a process by identifying opportunities to exchange heat between process streams, as opposed to exclusively using utilities for heating and cooling. Often used to refer to formalized approaches such as pinch analysis (qv) in academic environments
heat labile	Susceptible to alteration or destruction at elevated temperatures
heat number	*aka* HN. An ID coupon number stamped on a rolled metal plate, referencing its material test report (qv); defined formally in ASME BPE (American Society of Mechanical Engineers: Bioprocessing Equipment)
heat of combustion	*aka* heating value, enthalpy of combustion. Energy released by complete combustion to stable products of a given quantity of a substance; defined formally in API RP 2001 - Fire Protection at Refineries; cf higher heating value, lower heating value
heat of dilution	see **heat of solution**
heat of solution	*aka* enthalpy of solution, heat of dilution. Total heat liberated or absorbed during formation of an infinitely dilute solution, often expressed in kJ/mol. Defined formally in API RP 2001 - Fire Protection at Refineries
heat of x	Many 'heat of x' type terms are used in the physical sciences, but this dictionary includes only the two with sufficient relevance to professional practice to appear in commonly used codes and standards; qv heat of combustion, heat of solution
heat pump	A device which transfers energy from a lower temperature location to a higher temperature location, usually through the application of mechanical work. Common examples include refrigeration and mechanical recompression evaporators
heat rate	A measure of efficiency in power generation, being the amount of heat used by an electrical generator to produce a kWh of electricity. Defined formally in ISO14661 Thermal turbines for industrial applications (steam turbines, gas expansion turbines) - General requirements
heat recovery steam generator	*aka* HRSG. A four-part heat exchange system which makes steam from a hot gas stream which can be used for process purposes or power generation via a turbine. HRSGs comprise an economizer (qv), evaporator (qv), superheater (qv) and water preheater (qv). Defined formally in API RP 534 – Heat Recovery Steam Generators
heat recovery wheel	see **rotary heat exchanger**

heat release	An estimate (based on LHV -qv) of heat liberated by combustion of relief gases, commonly expressed in BTU/h in the oil and gas industry. Defined formally in API Standard 521. Pressure-relieving and Depressuring Systems. Sixth Edition	January 2014, API RP 537 Flare Details for Petroleum, Petrochemical, and Natural Gas Industries and API RP 535 - Burners for Fired Heaters at Refineries
heat tint	Undesirable discoloration of stainless steel as a result of oxygen exposure during hot work, especially welding. Defined formally in ASME BPE (American Society of Mechanical Engineers: Bioprocessing Equipment); cf heat affected zone
heat transfer coefficient	(*units* W. (m^2 K)$^{-1}$) The proportionality constant between the heat flux (heat flow per unit area) and the thermodynamic driving force for the flow of heat (i.e., the temperature difference, ΔT). Value is based on the physical characteristics of the heat exchange system and its associated materials and the physical properties of the fluids involved
heat transfer fluid	*aka* heating medium, heat transfer medium. A fluid used to transfer heat e.g., cooling water, steam etc.; defined formally in NFPA 30: Flammable and Combustible Liquids Code, 2018 Edition; N.B. only includes liquids in the definition
heat transfer medium	see **heat transfer fluid**
heater	Usually refers to equipment which heats a process stream with condensing steam or sometimes electrical heating. There are also fired heaters (both direct and indirect)
heating medium	see **heat transfer fluid**
heating value	see **heat of combustion**
heating ventilation and cooling	*aka* HVAC, heating, ventilation and air conditioning. Technology for the control of air quality inside buildings, mostly associated with maintaining worker safety and comfort
heavier than air gases	Gases with a density greater than air; defined formally in API RP 505 2nd Edition, August 2018 Recommended Practice for Classification of Locations for Electrical Installations at Petroleum Facilities Classified as Class I, Zone 0, Zone 1, and Zone 2; cf heavy gases
heavy chemicals	see **bulk chemicals**
heavy fuel oil	*aka* HFO, heavy oil, fuel oil. A heavy grade of fuel oil (qv) formerly burned in power stations but since largely replaced with cleaner fuels
heavy gas oil	*aka* HGO, heavy gasoil. A petroleum distillate used to produce gasoline by catalytic cracking (qv). Perhaps confusingly, neither a grade of gas oil (qv), nor the same as heavy fuel oil (qv)

heavy gases	Gases with densities more than 130% of air, such as LPG (qv); defined formally in BS EN 12952-8:2002 Water-tube boilers and auxiliary installations. Requirements for firing systems for liquid and gaseous fuels for the boiler; cf heavier than air gases
heavy gasoil	see **heavy gas oil**
heavy key	see **key component**
heavy medium separation	see **dense medium separation**
heavy metal	Loosely, a fairly toxic metallic element with a fairly high density. Formal definitions differ so widely that only three elements (bismuth, lead and mercury) are covered by all of them
heavy oil	see **heavy gas oil**
heavy oil upgrading	*aka* HOU, upgrading. Processing heavy oil into lighter, more valuable products
heavy phase	In liquid–liquid separation, the denser of the two liquids
heavy stiffener	A substantial support in a welded pressure vessel; defined formally in PD 5500:2018+A1:2018 Specification for unfired fusion welded pressure vessels
heavy water	Water with deuterium (qv) in place of the more common isotope of hydrogen (protium)
heavyweight drill pipe	*aka* HWDP. An oil and gas industry term for the grade of pipe commonly used for transition pipe (qv)
hecto-	(*symbol* h-) The SI unit prefix denoting a factor of 100
heel	1. Unspent feed remaining in a reactor below the agitator (or sometimes bottom port), or the liquid remaining in a tank after 'emptying' 2. In fermentation, the broth (qv) that is left in a fermenting vessel to inoculate the next batch 3. Also used for the bottom layer of solid on the filter cloth (not normally discharged with the batch) of a pressure filter dryer
heel tool	1. A blunt blade on a pole used for manual removal of the heel (qv) from a pressure filter dryer 2. (*informal*) In drilling/completions, a term used by production engineers in meetings when speaking out of turn
height equivalent to a theoretical plate	*aka* HETP. The real height of internals/packing in a separation column equivalent to a theoretical stage (qv)
height of a transfer unit	*aka* HTU. A measure of column efficiency being the (dimensionless) ratio of the height equivalent to a theoretical plate (qv) to the number of transfer units (qv)
Heisler charts	Graphs used in analysis of heat transfer[65]

---

[65] Back when we used graphs for such things, rather than computers

helical mixer	1. A static mixer (qv) with a helical mixing element 2. A vessel agitator with a helical shape
helical strake	see **strake**
Helmholtz free energy	A measure of a system's ability to do work, based on the internal energy; cf Gibbs free energy
hemispherical	*aka* spherical (deprecated). In the context of pressure vessel end caps, shaped like a half-sphere; cf torispherical, ellipsoidal
hemp and paste	see **hemp and white**
hemp and white	A pipe thread sealing method, using hemp thread and paste
HEN	see **heat exchanger network**
Henderson-Hasselbalch equation	An equation to relate pKa, pH and the acid/salt concentrations in a buffer solution
henry	(*symbol* H; *units* $m^2.kg.s^{-2}.A^{-2}$) The SI derived unit of electrical inductance. The inductance of a circuit is 1H if the rate of change of current in a circuit is one ampere per second ($A.s^{-1}$) and this results in an electromotive force of one volt (V)
henry per meter	(*symbol* H/m; *units* $kg.m.s^{-2}.A^{-2}$) The SI derived unit of magnetic permeability
Henry's law	The gas law relating the amount of dissolved gas in a liquid to its partial pressure above the liquid
Henry's law constant	(*symbol* KH) Temperature dependent ratio of a gas's partial pressure above a liquid of a gas to concentration dissolved in the liquid; cf Henry's law
HEPA filter	see **high efficiency particulate air filter**
Heras	A proprietary eponym for temporary mesh site fencing panels
hermetically sealed	see **gastight**
Herschel–Bulkley fluid	A theoretical model of non-Newtonian fluids
hertz	(*symbol* Hz) The SI unit of frequency, equal to one cycle per second
HESIA	Health, safety & environmental impact analysis
Hess's law	A law which states that total enthalpy (qv) change for any chemical reaction is the same, irrespective of intermediate stages
heterogeneous	Having non-uniform composition. The opposite of homogeneous (qv)
heterogeneous mixture	A non-uniform mixture of different substances and/or phases
heterotroph	see **heterotrophic organism**
heterotrophic organism	*aka* heterotroph. In biology, an organism that requires organic compounds as a carbon source to make complex organic molecules; cf autotrophic organism
HETP	see **height equivalent to a theoretical plate**

heuristics	Practical, imperfect problem-solving techniques[66]; cf problem, task; see **Box H2**
hex	1. *(drawing notation)* Hexagonal bolt, head or plug 2. see **uranium hexafluoride**
HF	Hydrogen fluoride (hydrofluoric acid)
HFC	see **hydrofluorocarbon**
HFO	see **heavy fuel oil**
HGO	see **heavy gas oil**
HH	1. see **higher heating value** 2. *(drawing notation)* see **high high alarm**
HHPS	see **hot high-pressure separator**
HHV	see **higher heating value**
HHx	see **high high alarm**
HIC	1. see **hydrogen induced cracking** 2. see **hydrophobic interaction chromatography**
hierarchical control	Process control achieved by means of controllers/computers controlling other controllers. In its more advanced forms, popular in oil and gas and academia, but less so in other sectors
Higbie's penetration theory	An entirely theoretical approach to predicting unsteady state mass transfer
high backpressure type pressure foam maker	see **pressure foam maker**
high care areas	see **zoning**
high Conradson feed	see **Conradson**
high consequence area	*aka* HCA. Defined in pipeline safety regulations as a place where loss of containment could have significant health, safety or environmental consequences
high efficiency particulate air filter	*aka* HEPA filter. A filter with a minimum efficiency of 99.97% for 0.3 μm particle size, as determined by test; used in unidirectional airflow benches, air handlers, and as terminal air supply filters in cleanrooms
high fuel pressure switch	A fuel pressure activated burner safety shutdown switch. Defined formally in NFPA 86: Standard for Ovens and Furnaces, 2019 Edition
high hazard level 1 contents	Materials which present detonation hazards. Defined formally with illustrative examples in NFPA 30: Flammable and Combustible Liquids Code, 2018 Edition
high hazard level 2 contents	Materials which present deflagration hazards. Defined formally with illustrative examples in NFPA 30: Flammable and Combustible Liquids Code, 2018 Edition

---

[66] Therefore, all real-world problem-solving techniques - despite assertions to the contrary!

Box H2. Heuristics and rigor

Professional engineers use heuristics every day: practical methods, not guaranteed to be optimal, perfect, logical, or rational, but sufficient for reaching an immediate goal.

Heuristics are at the heart of engineering, giving the lie to the idea that engineering is a simple application of pure science and mathematics. However, heuristics are more or less absent from academic curricula, and indeed from most textbooks (mine and a small number of others excluded). Where they are included in academic curricula, they are commonly misrepresented as being derived from first principles.

We usually learn heuristics in practice, from practitioners. Even when using a 'rigorous' method to estimate something, we check it against heuristics. Thus, heuristics are a systematization of engineering experience and common sense, of know-how (and more importantly know-how-not), key steps on the path to mastery. An engineer's professional judgment allows them to semi-intuitively discern approaches to problems, in perhaps a similar way to the experienced chess player, by ignoring the blind alleys which a beginner would waste time exploring; and including options which beginners would be unlikely to think of. They will know which simple calculations will allow them to choose quickly between classes of solution. Consequently, experts can quickly achieve outcomes which less experienced practitioners might never arrive at.

However, heuristics have their limitations: they are approximations with a limited range of application - not universally applicable, precise truths. If their underlying assumptions are not valid, they are at the very least highly questionable. Parametric statistics, for example, are only valid with data which is (amongst other things) 'continuous', i.e., 'not restricted to defined separate values, but can occupy any value over a continuous range'. However, people regularly present the arithmetic mean of integers in professional practice. The old '2.4 children' trope is a statistics joke, and here is another: 'Three statisticians are out hunting when they see a deer. The first statistician takes aim and shoots, but the bullet goes past the deer's nose by 9 inches. The second statistician takes aim and fires, but the bullet goes past the deer's rump by 9 inches. The third statistician exclaims: "We got him!"'

Indeed, the ultimate cause of many engineering disasters has been found to be the stretching of a heuristic beyond its proper application. It should always be borne in mind that the safety margin/safety factor is itself a heuristic. Aircraft are apparently only overdesigned for strength by a factor of 1.2, which is considered sufficient because of the extraordinary degree of quality control in building and operation, but this is not an appropriate factor for bridge building.

Familiarity plus predictability mean individual professional judgment is likely to provide a 'right-enough' answer. If the stakes are low, we can self-validate; if high, then professional engineers should look to sanity-check with other professional engineers.

Related terms

fiddle factor, safety margin, safety factor, guess and check, trial and error, Trouton's rule, inappropriate rigor, spurious precision, olfactorithmetic

high hazard level 3 contents	Materials which present physical hazards or readily support combustion. Defined formally with illustrative examples in NFPA 30: Flammable and Combustible Liquids Code, 2018 Edition
high high alarm	*aka* HHx. An instrument alarm condition which normally signifies the shut-down threshold, where x is the parameter, e.g., HHL would be a high high level alarm in a vessel; cf LLx
high integrity pipeline protection system	see **high integrity pressure protection system**
high integrity pressure protection system	*aka* HIPPS, high integrity pipeline protection system, high integrity protection system (HIPS). A type of safety instrumented system (qv), the main purpose of which is prevention of system overpressurization
high integrity protection system	*aka* HIPS. An alternate USA term for high integrity pressure protection system (qv)
high intensity burner	A burner with combustion intensity greater than 1M BTU/h/ft3 (around 10M W/m³). Defined formally in API RP 535 - Burners for Fired Heaters at Refineries
high level waste	*aka* HLW. Solid and liquid wastes produced during nuclear fuel reprocessing; cf highly active waste
high lift safety valve	see **standard safety valve**
high performance liquid chromatography	see **high pressure liquid chromatography**
high point of pavement	see **high point paving**
high point of paving	see **high point paving**
high point paving	*aka* HPP, high point of pavement/paving. A drawing notation used on plot plans and grade elevations
high pressure fuel gas system	A system which uses an eductor driven by a fuel gas gauge pressure of 1 psi (around 7 kPa) or greater to entrain combustion air. Defined formally in NFPA 86: Standard for Ovens and Furnaces, 2019 Edition; cf low pressure fuel gas system
high pressure gas quenching	Cooling with gases at a gauge pressure of 15 psi (around 1 bar) or greater. Defined formally in NFPA 86: Standard for Ovens and Furnaces, 2019 Edition
high pressure liquid chromatography	*aka* HPLC, high performance liquid chromatography. A separation technique based on a solid stationary phase and a liquid mobile phase
high pressure processing	*aka* HPP, ultra high pressure processing, ultra high hydrostatic pressure processing (UHHPP), pascalization, cold pasteurization. The 'cold pasteurization' of food products by subjecting them to high pressures (300–600MPa)

high pressure regulator	In the context of LPG, a device designed to regulate delivery pressure from a container to a maximum of 6.9 Kpag; defined formally in NFPA 58: Liquefied Petroleum Gas Code, 2017 Edition
high pressure separator	*aka* HPS. 1. A hot high pressure separator (qv) 2. A cold high pressure separator (qv)
high pressure spool	In the context of gas turbines, a high-pressure rotor assembly driven independently from low-pressure stages; defined formally in ISO 3977 Gas turbines - Procurement - Part 3: Design requirements
high purity water	Water which has been purified to reduce chemical and biological entities, conforming to USA or EU pharmacopeia monographs or equivalent; cf water for injection, purified water
high rate filter	In water and wastewater treatment, a filter - most commonly a depth filter (qv) or biofilter (qv) - operating at a 'high' loading. Loading in this case usually means surface loading (qv), though it may refer to organic loading (qv) in the case of biofilters. The threshold for 'high' loading however differs greatly from application to application
high rise building	A building where the highest occupiable floor level is more than 23m above the lowest fire vehicle access level. Defined formally in NFPA 20: Standard for the Installation of Stationary Pumps for Fire Protection, 2019 Edition; cf very tall building
high shear vacuum mixer	see **Scanima**
high temperature device	A device operating at more than 80% of the ignition temperature of the atmosphere it operates in. Defined formally in API RP 505 2nd Edition, August 2018 Recommended Practice for Classification of Locations for Electrical Installations at Petroleum Facilities Classified as Class I, Zone 0, Zone 1, and Zone 2
high temperature, short time	*aka* HTST. A bioburden (qv) reduction process in which raw materials/products are heated at a high temperature for a short period of time, resulting in a higher retention of quality characteristics; cf pasteurization, sterilization
high test	Meeting a high standard
high test gasoline	High octane gasoline
high test hypochlorite	Pure, solid calcium hypochlorite
high vacuum	In the context of fired heaters, refers to a system pressure $1 \times 10-3$ to $1 \times 10-5$ mmHg (around 1 x 10-6 Kpa); defined formally in NFPA 86: Standard for Ovens and Furnaces, 2019 Edition; cf low vacuum

high volume low concentration odorous gas	Especially in the context of pulp mill processes, collected air with low concentrations of odorous gases, consistently below their flammable limit. Defined formally in a pulp mill context in BS EN 12952-8:2002 Water-tube boilers and auxiliary installations. Requirements for firing systems for liquid and gaseous fuels for the boiler
higher activity waste	see **highly active waste**
higher alloy	Metals (especially steels) with a high concentration of alloying elements; defined formally in ASME BPE (American Society of Mechanical Engineers: Bioprocessing Equipment)
higher heating value	Total heat yielded by complete combustion of fuel at a starting temperature of 16°C, often expressed in US customary units (qv). Defined formally and almost identically in API RP 535 - Burners for Fired Heaters at Refineries, API STD 560 - Fired Heaters for General Refinery Service, API RP 537 Flare Details for Petroleum, Petrochemical and Natural Gas Industries, API STD 530 - Calculation of Tube Heater Thickness and BS EN ISO 13705:2012/ISO 13705:2012(E) Petroleum, petrochemical and natural gas industries. Fired heaters for general refinery service; cf lower heating value
highly active waste	*aka* higher activity waste. Radioactive waste products of the nuclear industry which are unsuitable for disposal in a low level waste repository. These unsurprisingly include high level waste, but also include intermediate level waste (qv) and some types of low level waste (qv)
highly purified water	*aka* HPW. A high-quality water used in pharmaceutical manufacturing
highly volatile liquid	*aka* HVL. A liquid with vapor pressure exceeding 276 kPa at 37.8°C, according to the formal definition in API RP 505 2nd Edition, August 2018 Recommended Practice for Classification of Locations for Electrical Installations at Petroleum Facilities Classified as Class I, Zone 0, Zone 1, and Zone 2
highway addressable remote transducer protocol	*aka* HART. A proprietary eponym for electronic communications using the named protocol
hindered drying	Drying of a wet solid after the free liquid has been removed and mass transfer limitations start to dominate
hindered settling	The settling of higher concentrations of solids in water at less than terminal velocity as interparticle forces predominate cf discrete sedimentation
HIPPS	see **high integrity pressure protection system**
HIPS	see **high integrity protection system**

HIRA	see hazard identification and risk analysis
HL	see lower heating value
HLPS	see hot low pressure separator
HLR	see hydraulic loading rate
HLW	see high level waste
HMI	see human machine interface
HN	see heat number
HO	see null hypothesis
HO	Hot oil
Hock process	see cumene process
Hoechst-Wacker process	see Wacker process
Hofstee plot	see Eadie–Hofstee plot
hogger	see hogger jet
hogger ejector	see hogger jet
hogger jet	*aka* hogger, hogger ejector, rapid evacuation ejector. A large air ejector discharging to atmosphere; used for initial vacuum system pressure reduction
hold down bolts	*aka* holding down bolts. Bolts securing equipment to substrate; defined formally in EN ISO 10437 Petroleum, petrochemical and natural gas industries - Steam turbines - Special-purpose applications
hold to run control device	*aka* dead man's handle, dead man's switch etc. A device which starts and maintains machine operation only for as long as it is positively actuated by the operator. Defined formally in BS EN ISO 12100:2010 Safety of machinery. General principles for design. Risk assessment and risk reduction
hold up time	A slightly contested term: sometimes considered identical to hydraulic residence time. More commonly, and probably better understood to mean the time it would take for a vessel to empty from 'working' to 'low' level if feed were stopped, which may well not be the same thing; cf surge time
holding down bolts	see hold down bolts
holding pump	*aka* backing pump, fore pump, vacuum fore pump, mechanical backing pump. A pump used to maintain oil diffusion pump efficiency, as opposed to a roughing pump (qv) which reduces pressure to the point where the diffusion pump can be used. Defined formally in NFPA 86: Standard for Ovens and Furnaces, 2019 Edition
holding tank	A storage vessel, mostly open topped; sometimes considered synonymous with break tank (qv)

holdup volume	The residue of liquid in a system after it has drained; defined formally in ASME BPE (American Society of Mechanical Engineers: Bioprocessing Equipment); cf heel
holiday	*aka* jeep. A discontinuity in a coating system such as glass on steel; defined formally in NFPA 22: Standard for Water Tanks for Private Fire Protection, 2018 Edition
holiday detector	A device to detect a holiday (qv); cf jeeping
hollow bar	A thick-walled tube
hollow bodies	see **hollow body**
hollow body	*aka* hollow bodies. "Void spaces, inaccessible to cleaning, which may become sources of contamination", according to the EHEDG Glossary
hollow fiber bioreactor	*aka* hollow fibre bioreactor. A membrane bioreactor using hollow fiber membranes
hollow fiber membrane	*aka* hollow fibre membrane. A small (<5mm) bore self-supporting cylindrical tubular semi-permeable polymeric membrane, used primarily in water treatment. The ratio of surface area to internal volume is very large; cf tubular membrane
hollow fibre bioreactor	see **hollow fiber bioreactor**
hollow fibre membrane	see **hollow fiber membrane**
homogeneous	*aka* homogenous. Having uniform composition; the opposite of heterogeneous (qv)
homogenization	1. A process for making a fluid (e.g., milk) more homogeneous (qv) at a fine scale 2. Processes used to make the particle size distribution of solids more uniform, e.g., milling (qv)
homogenizer	Usually, a device which makes a fluid homogeneous (qv), especially one used to produce consistent particle or droplet size in an emulsion; cf mixer
homogenous	see **homogeneous**
honeycomb coke	An intermediate grade of petcoke (qv)
hook sleeve	*aka* hook-type sleeve. A hooked sleeve used (more commonly in the past) to protect a pump shaft; defined formally in API RP 682 - Pump Seals
Hooke's law	A law governing elastic materials whose deformation is related to the force applied, or a term for such materials
hook-type sleeve	see **hook sleeve**
HOP	Human and organizational performance
hopper	A structure which holds bulk materials prior to release to a chute or conveyor, a prime site for hammer rash (qv)
horizontal flow tank	A type of settlement tank used in effluent treatment

Horizontal Guidance 1 (H1) (UK)	Former UK guidance for environmental permitting assessment, strictly superseded, but still partially in use
horizontal pump	A pump (usually centrifugal) with a horizontal shaft; defined formally in NFPA 20: Standard for the Installation of Stationary Pumps for Fire Protection, 2019 Edition
horizontal pumps total head	(*symbol* H) The total head (qv) of a horizontal pump (qv). A very complex but ultimately no more informative formal definition is in NFPA 20: Standard for the Installation of Stationary Pumps for Fire Protection, 2019 Edition
horizontal split case pump	A pump (usually centrifugal) with a horizontal (parallel to shaft) split casing; defined formally in NFPA 20: Standard for the Installation of Stationary Pumps for Fire Protection, 2019 Edition
horizontal thermosyphon reboiler	*aka* HTS. A horizontally oriented thermosyphon reboiler (qv), far more popular in oil and gas than a vertical thermosyphon reboiler (qv)
horse-collar	A self-inflatable life vest
horsehead	The part on the end of the arm of a nodding donkey (qv) which looks rather like a horse's head[67]
horsehead pump	see **nodding donkey**
horsepower	(*symbol* hp) A unit of power, which has a number of varieties, including a metric supplementary unit equal to around 740W. Most commonly however associated with superseded imperial measurements, or their still current US cousins, equal to around 746W. It should be noted that whilst the various types of horsepower are not exactly equivalent, all but one are equal to approximately 740W. Boiler horsepower is the exception being around 13 times as large as this figure.
Horton sphere	see **Hortonsphere**
Hortonsphere	*aka* Horton sphere. Proprietary eponym for a large spherical pressure vessel used to hold compressed gases
hose	A flexible tube or pipe used to convey liquids; see **Box P1**.
hose house	An enclosure near a fire water supply, containing all necessary hoses, tools and equipment to allow the supply to be used for firefighting by site staff and the fire department. Defined formally in NFPA 24: Standard for the Installation of Private Fire Service Mains and Their Appurtenances, 2019 Edition
host cell proteins	*aka* HCPs. Trace protein contaminants in biotherapeutic products, derived from cultured host cells (qv)

---

[67] Also that of a dinosaur, bird or donkey, judging by the various *akas* for nodding donkey

host cells	Cells into which recombinant DNA is inserted so that the protein it codes for can be produced in large quantities
hot commissioning	*aka* live commissioning. Startup with process fluids, under load, and at normal operating temperatures, especially, but not exclusively in the nuclear industry (whose version requires the introduction of fissile material); cf cold commissioning
hot face layer	The fired heater refractory surface which is in contact with the hottest gases. Defined formally in BS EN ISO 13705:2012/ISO 13705:2012(E) Petroleum, petrochemical and natural gas industries. Fired heaters for general refinery service, API STD 530 - Calculation of Tube Heater Thickness and API STD 560 - Fired Heaters for General Refinery Service
hot face temperature	Temperature at the hot face layer (qv). Defined formally in BS EN ISO 13705:2012/ISO 13705:2012(E) Petroleum, petrochemical and natural gas industries. Fired heaters for general refinery service, API STD 530 - Calculation of Tube Heater Thickness and API STD 560 - Fired Heaters for General Refinery Service
hot gas ignition temperature	*aka* minimum hot gas ignition temperature. The minimum temperature of a defined stream of hot gas which will ignite a substance
hot gas path temperatures	*aka* hot-gas-path temperatures. Gas turbine combustion gas temperatures, sometimes used to mean the maximum of these. Defined formally in ISO 3977 Gas turbines - Procurement - Part 3: Design requirements
hot high pressure separator	*aka* HHPS. An additional separator found on newer hydrocracker designs which feeds a cold high pressure separator (qv) and a hot low pressure separator (qv)
hot isostatic processing	see **hot pressing**
hot leaching process	One of a range of beneficiation (qv) processes involving leaching (qv) at elevated temperatures
hot low pressure separator	*aka* HLPS. An additional separator found on newer hydrocracker designs; fed by a hot high pressure separator (qv), the bottoms of which are sent to the fractionator
hot pressing	1. *aka* hot isostatic processing. A process for forming shapes from powdered metals or ceramics 2. The use of a heated hydraulic press for vegetable oil extraction, especially in cocoa processing
hot standby	see **idle**
hot surface ignition temperature	*aka* minimum surface ignition temperature. The minimum temperature of a defined surface which will ignite a substance

hot tapping	Making a connection to a pipe while it is filled with pressurized fluid
hot work	Work involving use of an ignition source, such as welding. Defined formally in NFPA 654: Standard for the Prevention of Fire and Dust Explosions from the Manufacturing, Processing, and Handling of Combustible Particulate Solids, 2017 Edition
hotel	A building with sleeping facilities for more than 16 transient people, according to NFPA 30: Flammable and Combustible Liquids Code, 2018 Edition
hot-gas-path temperatures	see **hot gas path temperatures**
HOU	see **heavy oil upgrading**
houdriforming	A proprietary catalytic reforming process with a platinum catalyst
housing	see **enclosure**
HP	1. High pressure
	2. see **horsepower**
HPLC	see **high pressure liquid chromatography**
HPP	1. see **high pressure processing**
	2. *(drawing notation)* see **high point paving**
HPS	see **high pressure separator**
HPW	see **highly purified water**
HRSG	see **heat recovery steam generator**
HRT	see **hydraulic retention time**
H-S chart	see **Mollier diagram**
HSAS	Heat stable amine salts
HSE	1. Health, safety and environment
	2. see **Health and Safety Executive**
HSFO	High sulfur fuel oil
HSLA	High strength low alloy
HSP	Hansen solubility parameter, see **Hansen parameter**
HTF	see **heat transfer fluid**
HTH	see **high test hypochlorite**
HTS	see **horizontal thermosyphon reboiler**
HTST	see **high temperature, short time**
HTU	see **height of a transfer unit**
HU	Hazen unit; see **Hazen**

huddling chamber	An annular chamber in a PRV (qv) which assists the valve in lifting. Near-identical formal definitions in API RP 520 P1 7th Edition, January 2000 Sizing, Selection, and Installation of Pressure-Relieving Devices in Refineries; Part I - Sizing and Selection, API Standard 521 Pressure-relieving and Depressuring Systems Sixth Edition	January 2014 and API RP 576 - Inspection of Pressure Relieving Devices
human factor	A human characteristic which might affect an object, especially in a health and safety context. Defined formally in BS EN ISO 9000 Quality management systems Fundamentals and vocabulary
human machine interface	*aka* HMI, man-machine interface. A user interface, usually a screen which allows human interaction with and control of the onboard electronics of an item of equipment or instrumentation
humectants	Substances added to reduce loss of moisture
humic acids	The class of brown carboxylic acids formed by decomposition of biological matter often present in surface water which coagulate when a strong base extraction is performed; cf fulvic acids
humid heat	(*symbol* Cs) Other than the commonplace climatic comfort meaning - the heat capacity per given mass of dry air for humid air, at constant pressure and percentage humidity
humid volume	The volume of humid air (per given mass of dry air), at constant pressure and percentage humidity
humidification	Increasing the humidity of something (most commonly air)
humidity	The mass of water in humid air (per given mass of dry air), at constant pressure; often expressed as a percentage
humus tank	A final settling tank downstream of a biological filter in sewage treatment
Hv	see **velocity head**
HVAC	see **heating ventilation and cooling**
HVAC drawing	Heating, ventilation and cooling (qv) drawing
HVACR	Heating, ventilation, air conditioning and refrigeration
HVCA	Heating and Ventilating Contractors Association (United Kingdom)
HVGO	Heavy vacuum gas oil
HVL	see **highly volatile liquid**
HVO	*aka* HVO100. Hydrotreated vegetable oil
HVO100	see **HVO**
HW	*(drawing notation)* Hot water
HWDP	see **heavyweight drill pipe**
HX	*(drawing notation)* Heat exchanger

Hx alarm	High instrument alarm condition, where x is the parameter; this is normally the alarm threshold e.g., HL would be a high level alarm in a vessel; cf Lx
hybrid	The offspring derived from hybridization (qv)
hybrid mixture	A potentially explosive heterogeneous mixture with a particular ratio of gas and suspended solid or liquid particulates; defined formally in NFPA 654: Standard for the Prevention of Fire and Dust Explosions from the Manufacturing, Processing, and Handling of Combustible Particulate Solids, 2017 Edition
hybrid refrigeration	*aka* hybrid cooling. A cooling system either having two different cooling sources, such as dry and evaporative cooling; or one storing heat transfer fluid cooled using off-peak power for later use
hybridization	1. The combining of genetic traits from different organisms to confer some desired characteristic on the hybrid (qv) offspring; used in agriculture and biotechnology 2. Also used to mean conceptually similar techniques applied in machine learning algorithms
HyCO	*aka* synthesis gas. A hydrogen-carbon monoxide mixture
hydrant	An exterior valved fire water hose connection point; defined formally in NFPA 24: Standard for the Installation of Private Fire Service Mains and Their Appurtenances, 2019 Edition
hydrant butt	A hydrant hose connection point; defined formally in NFPA 24: Standard for the Installation of Private Fire Service Mains and Their Appurtenances, 2019 Edition
hydrate	1. Most commonly, a crystalline compound of a low-boiling-point hydrocarbon and water. Defined formally in API Standard 521. Pressure-relieving and Depressuring Systems. Sixth Edition| January 2014 2. Less commonly, a term used by chemists to describe hydrated salts
hydrated lime	see **slaked lime**
hydration	*aka* degree of hydration. The amount of water incorporated into a particular crystal structure
hydraulic classification	Generically, classification (qv), specifically applied to the tendency of smaller particles to rise to a packed bed surface during backwashing
hydraulic depth	see **mean hydraulic depth**
hydraulic diameter	see **mean hydraulic diameter**
hydraulic fracturing	see **fracking**

hydraulic jump	A turbulent region where shallow supercritical flows decelerate rapidly to deeper subcritical flows (as, e.g., when flow passes over a weir or approaches a flume)
hydraulic loading rate	A process design heuristic for settlement tanks and similar processes in water treatment, defined as the nominal average upflow velocity, i.e., the incoming volumetric flow divided by the surface area
hydraulic mean depth	see **mean hydraulic depth**
hydraulic mean diameter	see **mean hydraulic diameter**
hydraulic power recovery turbine	A turbine which recovers power from a stream of fluid; defined formally in EN ISO 13709:2003 Centrifugal pumps for petroleum, petrochemical and natural gas industries
hydraulic radius	The ratio of cross-sectional area to wetted perimeter for something which contains a flow, most commonly an open channel; cf equivalent hydraulic diameter
hydraulic retention time	*aka* HRT, residence time. A process design heuristic in general usage, being the nominal average retention time of fluid, i.e., the working volume divided by the incoming volumetric flow; cf solids retention time, space time, space velocity
hydraulic roughness	A measure of resistance to flow in channels, affected by many factors other the surface roughness of the channel wall; cf roughness
hydraulic shock	see **fluid hammer**
hydraulically calculated water demand flow rate	Estimated flow rate for a firewater system or hose; defined formally in NFPA 24: Standard for the Installation of Private Fire Service Mains and Their Appurtenances, 2019 Edition
hydraulics	1. Systems driven by pressurized liquids 2. The engineering science of systems driven by pressurized liquids, especially the practical applications of fluid mechanics (qv) to plant and equipment sizing
hydrocarbon	An organic chemical consisting solely of hydrogen and carbon
hydrochlorofluorocarbon	*aka* HCFC. A refrigerant gas
hydrocracking	A two-stage catalytic cracking (qv) and hydrogenation (qv) process used in the oil and gas industry to break down long chain hydrocarbons into shorter ones, making heavy oils into more valuable medium distillates; cf fluid catalytic cracking
hydrocyclone	A cyclone separator (qv) which removes solids from water (or sometimes other liquids, despite the meaning of 'hydro-')
hydrodealkylation	see **dealkylation**
hydrodesulfurization	A catalytic sweetening process used for hydrocarbon gas and liquid fuels

hydrodynamic bearing	A hydrodynamically lubricated bearing in a pump or turbine. Defined formally in EN ISO 13709:2003 Centrifugal pumps for petroleum, petrochemical and natural gas industries and EN ISO 10437 Petroleum, petrochemical and natural gas industries - Steam turbines - Special-purpose applications
hydrofining	*aka* hydrorefining. A catalytic hydrogenation (qv) and desulfurization (qv) process used in hydrocarbon processing, especially to produce gas and liquid fuels
hydrofluorocarbon	*aka* HFC. A refrigerant gas
hydroformylation	*aka* oxo process, oxo synthesis. A catalytic process for production of aldehydes from alkenes
hydrogasification	The production of synthetic natural gas (qv) from coal or biomass using hydrogen, with or without steam
hydrogen assisted cracking	see **hydrogen induced cracking**
hydrogen blistering	Subsurface cavities caused by diffusing molecular hydrogen, (mostly occurring with low strength metals) which reduce mechanical strength. Defined formally in API RP 579 - Fitness for Service
hydrogen bond	An electrostatic attraction between molecules
hydrogen cation exchanger	An acid regenerated cation exchange resin (qv) which exchanges hydrogen ions
hydrogen colors	*aka* colors of hydrogen. Hydrogen production methods are primarily assigned colors as follows: grey hydrogen – produced by reforming hydrocarbons with unabated CO_2 emissions; brown (*aka* black) hydrogen – produced from coal gasification with unabated CO_2 emissions; blue hydrogen – pairs hydrocarbon reforming with carbon capture and storage; green hydrogen – produced by water electrolysis using renewable energy; less well known categories are gold hydrogen (qv), orange hydrogen, pink hydrogen, purple hydrogen, red hydrogen, turquoise hydrogen, white hydrogen, yellow hydrogen (qv), plus rose tinted hydrogen (qv)
hydrogen cycle	A cation exchange resin (qv) acid regeneration cycle
hydrogen embrittlement	see **hydrogen induced cracking**
hydrogen induced cracking	*aka* HIC, hydrogen assisted cracking, hydrogen embrittlement, hydrogen stress cracking. A development of hydrogen blistering (qv), in which cracks connect the blisters. Defined formally in API RP 571 - Damage Mechanism Affecting Fixed Refinery Equipment, API RP 579 - Fitness for Service and API RP 932B Corrosion Air Coolers

hydrogen service	Loosely, any equipment used for hydrogen-containing fluids. Formal definitions require a minimum partial pressure of hydrogen of 700 Kpa; see for example API 660 - Shell-and-Tube Heat Exchangers and API RP 661 - Heat Exchangers
hydrogen stress cracking	see **hydrogen induced cracking**
hydrogen to carbon ratio	*aka* H/C ratio. The molar H/C ratio of a hydrocarbon fuel. Defined formally in API RP 535 - Burners for Fired Heaters at Refineries
hydrogenation	Reacting hydrogen with an unsaturated hydrocarbon or other material (e.g., hydrogenating edible oils to make partially hydrogenated fats, some of which will be in the trans isomer, which has adverse health effects)
hydroisomerization	Catalytic isomerization of alkanes via intermediate alkenes
hydrolysis	A chemical process in which a molecule is cleaved into two parts by the addition of a molecule of water. The reverse of condensation reaction (qv)
hydrometallurgy	Beneficiation (qv) using aqueous solutions
hydrometer	An instrument for measuring the specific gravity of liquids
hydromethane	see **methane clathrate**
hydrophilic media	Especially in the context of effluent treatment, media with a high surface affinity for water and a low affinity for oils; cf hydrophobic media
hydrophobic interaction chromatography	*aka* HIC. A chromatographic purification technique used in pharmaceuticals that exploits hydrophobic regions within a molecule
hydrophobic media	Especially in the context of effluent treatment, media with a low surface affinity for water and a high affinity for oils; cf hydrophilic media
hydroprocessing	Catalytic processing of oil distillates with hydrogen, including hydrotreating (qv) and hydrocracking (qv)
hydroprocessing unit	A unit operation involving hydroprocessing (qv); defined formally in API RP 932B Corrosion Air Coolers
hydropyrolysis	Pyrolysis (qv) in a hydrogen-rich atmosphere
hydrorefining	see **hydrofining**
hydroskimming	A simple oil refining process comprised mainly of primary processing units (atmospheric distillation (qv) and naphtha reforming), found mainly in developing countries
hydrostatic head	see **head**
hydrostatic pressure	see **pressure**

hydrostatic test	*aka* hydrotest. Static pressure testing of a system of piping and vessels; a late stage of mechanical commissioning. Defined formally in NFPA 24: Standard for the Installation of Private Fire Service Mains and Their Appurtenances, 2019 Edition and ASME BPE (American Society of Mechanical Engineers: Bioprocessing Equipment
hydrotest	see **hydrostatic test**
hydrotreating	A catalytic process reacting hydrogen with hydrocarbons to remove chlorine, nitrogen, oxygen, and sulfur
hydroxide cycle	An anion exchange resin (qv) hydroxide regeneration cycle
hygiene	see **food hygiene**
hygiene areas	see **zoning**
hygienic	In the context of process equipment and piping; design, materials, and operation which promote hygiene (qv). Defined formally in ASME BPE (American Society of Mechanical Engineers: Bioprocessing Equipment) see **Box C2**.
hygienic clamp joint	A clamped union fitting designed with a hygienic wetted contact surface. Defined formally in ASME BPE (American Society of Mechanical Engineers: Bioprocessing Equipment); other standards are applicable (e.g., BS 4825, DIN32676, ISO2852 & ISO1127) but are not all mutually compatible
hygienic design and engineering	1. Generically, a system of design that meets standards, specification, codes, regulatory and industrial guidelines, and acceptable engineering design methods to reach a degree of sanitation required by food, pharmaceutical, and cosmetics processing 2. Specifically, the "design and engineering of equipment and premises assuring the food is safe and suitable for human consumption", according to the EHEDG Glossary
hygienic equipment class I	"Equipment that can be cleaned in place (CIP) and can be freed from relevant microorganisms without dismantling" according to the EHEDG Glossary; qv cleaning in place
hygienic equipment class II	"Equipment that is cleanable after dismantling and can be freed from relevant microorganisms after reassembly" according to the EHEDG Glossary; qv cleaning out of place
hygienic integration	"The process of combining or arranging two or more pieces of equipment or components to work together in a hygienic manner" according to the EHEDG Glossary
hygienic joint	A union fitting designed with a hygienic wetted contact surface; defined formally in ASME BPE (American Society of Mechanical Engineers: Bioprocessing Equipment)

hygienic piping systems	Systems that provide for the maintenance of cleanliness so that products transferred and/or conveyed through them will not have their identity, strength, quality, purity, or potency compromised
hygienic weld	see **sanitary weld**
hygroscopic	Tending to absorb moisture from the air
Hypalon	Proprietary eponym for a chemically resistant synthetic elastomer
hyperalloy	A non-existent alloy even more super than a superalloy (qv) used to build the combat chassis of the T-800 series of Terminators
hyperfiltration	see **ultrafiltration**; cf **Box L1**.
hypergolic	Igniting spontaneously on mixing, like 'T-stoff and Z-stoff'[68]
hypertonic	Having a higher osmotic pressure than another solution; cf hypotonic
hypochlorite	A generic term for aqueous solutions of sodium hypochlorite, potassium hypochlorite or calcium hypochlorite: oxidizing agents used for disinfecting and bleaching
hypothesis	An informal but still correct definition would be 'a proposed explanation', akin to the misuse of the word 'theory' in everyday English. The related formal meaning of hypothesis requires testability, amongst other things
hypothetical component	see **pseudocomponent**
hypotonic	Having a lower osmotic pressure than another solution; cf hypertonic
hysteresis	1. A lagging of effect behind cause, especially in control engineering 2. Differences in behavior based on direction of travel - e.g., a valve may have a different position for the same actuator position depending on whether it was opening or closing, or a reacting system may operate at different rates at the same temperature depending on whether it has warmed up or cooled down to arrive at that temperature
hysteresis damping	*aka* structural damping, hysteretic damping. Damping by intra-structure frictional energy losses; defined formally in BS ISO 2041:2018 Mechanical vibration, shock and condition monitoring. Vocabulary
hysteretic damping	see **hysteresis damping**
Hysys	Proprietary process modelling software developed for the oil and gas industry but widely used in academia for other purposes

---

[68] Look it up, it's worth it!

I

I/O	Inputs/outputs, in a control context
I/P transducer	*aka* current-to-pressure transducer. A control system component that produces a pressure output proportional to a 4-20mA input
IBBM	Iron body bronze mounted (valve)
IBC	see **intermediate bulk container**
IBC tank	see **intermediate bulk container**
IBC tote	see **intermediate bulk container**
IBV	see **inlet butterfly valve**
IC4	Isobutane
ICC	(USA) Interstate Commerce Commission
ice load	see **environmental load**
ice point	The freezing point of water at one atmosphere, i.e., 0°C
IChemE	Institution of Chemical Engineers. A UK body similar in some ways to the more widely recognized US AIChE (qv)
icicles	Weld defects from localized excessive penetration (qv); defined formally in ASME BPE (American Society of Mechanical Engineers: Bioprocessing Equipment)
ICP	see **industrial control panel**
ID	1. see **inside diameter** (especially of pipes) 2. see **identification**
ideal gas	A simplified theoretical model of a gas based on assumptions, most notably assuming no interparticle interactions; cf real gas
ideal gas constant	see **gas constant**
ideal gas law	A simplified theoretical equation of state for an ideal gas (qv)
ideal mixture	A theoretical mixture which forms an ideal solution (qv)
ideal polytropic gas	see **polytropic gas**
ideal solution	A simplified theoretical solution which follows Raoult's law (qv) under all conditions
ideal stage	see **theoretical stage**
identification	*aka* ID. Most commonly, a synonym of identifier (qv). However, ISO 10209:2012 Technical product documentation — Vocabulary — Terms relating to technical drawings, product definition and related documentation gives a formal definition which is closer to, or at least includes elements of 'specification'

identifier	Usually, a label. Slightly different formal definition in ISO 10209:2012 Technical product documentation — Vocabulary — Terms relating to technical drawings, product definition and related documentation, although the standard uses the term in the usual meaning throughout
identifying block	The place on a drawing where reference designations (qv) should feature; defined formally in ISO 10209:2012 Technical product documentation — Vocabulary — Terms relating to technical drawings, product definition and related documentation
IDF	see **induced draft fan**
idle	*aka* hot standby, warm standby. A term applied to equipment which is energized, but not doing anything useful; cf standby
idler	A non-driven roller, usually used underneath a weightometer (qv)
IDLH	Immediately dangerous to life or health
idling	The state of idle (qv) equipment
IET	Institute of Engineering & Technology (UK)
IFA	Issued for approval
IFC	Issued for construction
IFD	Issued for design
IFH	Issued for HAZOP (qv)
IFI	Issued for information
IFR	Issued for review
IFRT	see **internal floating roof tank**
IGC Code	International Gas Carrier Code, *aka* International Code of the Construction and Equipment of Ships Carrying Liquefied Gases in Bulk
IGCC	see **integrated gasification combined cycle**
IGCI	Industrial Gas Cleaning Institute
IGF	see **induced gas flotation**
ignitable mixture	*aka* flammable mixture. A mixture of gas and air within the flammable limits (qv) of the gas. Defined formally in API RP 505 2nd Edition, August 2018 Recommended Practice for Classification of Locations for Electrical Installations at Petroleum Facilities Classified as Class I, Zone 0, Zone 1, and Zone 2
ignition	The initiation of combustion, or 'lighting' in everyday English
ignition ports	Fired heater burner tip orifices, which divert part of the fuel to maintain flame stability; defined formally in API RP 535 - Burners for Fired Heaters at Refineries

ignition safety time	In the context of fired heater startup, the delay between fuel entering a combustion chamber and the deenergization of a quick acting shutoff device (qv); defined formally in BS EN 12952-8:2002 Water-tube boilers and auxiliary installations. Requirements for firing systems for liquid and gaseous fuels for the boiler; cf ignition time, extinction safety time
ignition source	see **sources of ignition**
ignition temperature	The lowest surface temperature required to ignite a gas-air mixture; defined formally in NFPA 86: Standard for Ovens and Furnaces, 2019 Edition and API RP 505 2nd Edition, August 2018 Recommended Practice for Classification of Locations for Electrical Installations at Petroleum Facilities Classified as Class I, Zone 0, Zone 1, and Zone 2; though other standards define the term differently; cf autoignition temperature
ignition time	In the context of fired heater startup, the delay between fuel entering a combustion chamber and flame detection; defined formally in BS EN 12952-8:2002 Water-tube boilers and auxiliary installations. Requirements for firing systems for liquid and gaseous fuels for the boiler; cf ignition safety time
ignitor	A device which ignites a burner; defined formally in API RP 535 - Burners for Fired Heaters at Refineries
IGPM	Imperial gallons per minute
IGU	International Gas Union
IGV	see **inlet guide vane**
ILW	see **intermediate level waste**
imbibition	Adding water in a sugar refinery to increase extraction; cf maceration
imbibition water	Water added for imbibition (qv)
immediate repair condition	The presence of a pipeline defect requiring immediate repair action, according to relevant safety guidance
immersible pump	A pump mounted vertically, with the motor above the liquid level and the impeller below the liquid level; cf submersible pump
immersion type vaporizer	see **waterbath vaporizer**
immiscible	(Usually of two liquids) incapable of being mixed to stable homogeneity; cf miscible
immobilized cell bioreactor	A bioreactor in which cells are fixed onto or inside a solid phase
impact	see **effect**

impact category	A classification of environmental impact used in life cycle assessment; defined formally in EN ISO 14040:2006 Environmental management - Life cycle assessment - Principles and framework
impact category indicator	*aka* category indicator, indicator. Quantity associated with an impact category (qv) for the purposes of life cycle assessment; defined formally in EN ISO 14040:2006 Environmental management - Life cycle assessment - Principles and framework
impact pressure	*aka* dynamic pressure. The pressure exerted by a fluid on a surface upon which it impinges; cf static pressure
impeding device	A physical obstacle retarding personnel access to a hazard zone (qv); defined formally in BS EN ISO 12100:2010 Safety of machinery. General principles for design. Risk assessment and risk reduction cf barrier
impeller	*aka* impellor. A bladed device which converts rotation into radial flow, used for agitation in chemical engineering; cf propeller
impellor	see **impeller**
imperial ton	see **long ton**
imperial units	An almost obsolete British non-SI measurement system similar to but not identical to US Customary Units (qv). May be used incorrectly to describe US Customary Units; see **Box B1**.
impingement	In the context of process engineering, almost always means the impaction of droplets or particles carried in a gas with a surface
impingement separator	A device which separates by impingement (qv)
implied BAT	Implied best available techniques: the implied duty on an operator under the PPC Act (qv) to use best available techniques (qv) for pollution control. This does not form part of new licenses under the Environmental Permitting Programme (qv), but is retained on pre-existing ones
implied precision	see **significant figures**; cf spurious precision, **Box A1**.
implied resolution	see **significant figures**, **Box A1**.
implosion	The rapid inward collapse of the walls of a vessel caused by excessive internal vacuum; defined formally in BS EN 12952-8:2002 Water-tube boilers and auxiliary installations. Requirements for firing systems for liquid and gaseous fuels for the boiler and NFPA 86: Standard for Ovens and Furnaces, 2019 Edition
important building	A building which must not be lost to fire; defined formally in NFPA 30: Flammable and Combustible Liquids Code, 2018 Edition

improvement	A process/performance enhancement, or action towards that end; defined formally in BS EN ISO 9000 Quality management systems Fundamentals and vocabulary
impulse	The integral of a force over its time of application
impulse line	Small bore piping connecting a pressure sensor to a monitored system; defined formally in NFPA 86: Standard for Ovens and Furnaces, 2019 Edition
in parallel	see **parallel**
in place cleanability	"The suitability to be easily cleaned without dismantling", according to the EHEDG; qv cleaning in place
in series	see **series**
in sight from	*aka* within sight, within sight from. In the context of fire safety, visible from, and no more than 15m away from; defined formally in NFPA 20: Standard for the Installation of Stationary Pumps for Fire Protection, 2019 Edition
inadequate ventilation	Insufficient ventilation to prevent the accumulation of significant quantities of fuel/air mixtures at >20% of their LEL (qv) /LFL (qv), defined formally in API RP 505 2nd Edition, August 2018 Recommended Practice for Classification of Locations for Electrical Installations at Petroleum Facilities Classified as Class I, Zone 0, Zone 1, and Zone 2; cf adequate ventilation
inappropriate rigor	A failing of a certain type of ungeneer (qv); see **Box A1**.
inboard	A term used to describe components close to the drive coupling of a pump or motor; cf outboard
incandescence	The emission of electromagnetic radiation (especially visible light) as a result of heating
incident	An unexpected occurrence that does not lead to death or injury, or does not have a high probability of causing these; cf accident
incidental liquid use	*aka* incidental storage. A term for the condition in which use of flammable liquids in an area is a subordinate activity to that upon which occupancy (qv) is classified under NFPA 30, where this term is formally defined
incidental storage	see **incidental liquid use**
incineration	*aka* thermal treatment, thermal recycling (where there is energy recovery). The 'destruction' of waste by burning
incinerator	see **thermal oxidizer**
inclined tube manometer	A simple device to measure small pressure differences
inclusion	Foreign material particles, often non-metallic particles in a metallic matrix, especially a weld. Defined formally in API RP 579 - Fitness for Service and ASME BPE (American Society of Mechanical Engineers: Bioprocessing Equipment)

Incolloy	Misspelling of Incoloy (qv)
Incoloy	*aka* Incolloy (sic). A trademarked name for a series of austenitic chromium/nickel based oxidation corrosion resistant superalloys (qv)
incombustible	Not burnable; cf combustible, flammable, inflammable
incomplete fusion	*aka* lack of fusion. A weld defect characterized by incomplete melting and coalescence; defined formally in ASME BPE (American Society of Mechanical Engineers: Bioprocessing Equipment) and API RP 579 - Fitness for Service
incomplete penetration	*aka* lack of penetration. A weld defect characterized by incomplete penetration of the weld through a joint; defined formally in ASME BPE (American Society of Mechanical Engineers: Bioprocessing Equipment) and API RP 579 - Fitness for Service
incompressible fluid	A fluid where any density changes as a result of compression are not practically significant (to put it another way, a liquid); cf compressible fluid
Inconel	A trademarked name for a series of superaustenitic chromium/nickel based stainless steel, oxidation corrosion resistant superalloys (qv)
Incoterms	Standard international delivery terms, which are summarized in **Table I1**.
indemnity cover	see **professional indemnity insurance**
indeterminate special atmosphere	A special atmosphere for use in a furnace whose flammability is not reliably predictable. Defined formally in NFPA 86: Standard for Ovens and Furnaces, 2019 Edition; cf flammable special atmosphere, nonflammable special atmosphere, inert special atmosphere
indicating gate valve	see **indicating valve**

Table I1 Summary of Incoterms

Group 1 Incoterms (which apply to any mode of transport)	Group 2 Incoterms (which apply to sea and inland waterway transport)
EXW Ex works	FAS Free alongside ship
FCA Free carrier	FOB Free on board
CPT Carriage paid to	CFR Cost and freight
CIP Carriage and insurance paid to	CIF Cost, insurance, and freight (qv)
DAT Delivered at terminal	
DAP Delivered at place	
DDP Delivered duty paid (qv)	

indicating valve	A valve with a visual indication of whether it is open or closed; defined formally in NFPA 24: Standard for the Installation of Private Fire Service Mains and Their Appurtenances, 2019 Edition
indication	An anomalous area of a weld, the acceptability of which has not yet been determined; defined formally in ASME BPE (American Society of Mechanical Engineers: Bioprocessing Equipment)
indicative BAT	Indicative best available techniques; those based in specified benchmarks, standards, or techniques; cf implied BAT
indicator	Something which indicates a state, such as an impact category indicator (qv), indicator microorganisms (qv), or a chemical indicator (qv)
indicator diagram	see **pressure/volume diagram**
indicator microorganisms	"Microorganisms whose presence indicates a failure of a GHP", according to EHEDG Glossary. More generally, organisms which are relatively easy to test for, but which cause one to suspect the likely presence of undesirable organisms which are harder to detect
indirect air preheater	*aka* indirect APH. A shell and tube heat exchanger (qv) used to preheat boiler combustion air (qv), often using flue gas as heat source. Defined formally in API STD 530 - Calculation of Tube Heater Thickness, API STD 560 - Fired Heaters for General Refinery Service and BS EN ISO 13705:2012/ISO 13705:2012(E) Petroleum, petrochemical and natural gas industries. Fired heaters for general refinery service
indirect APH	see **indirect air preheater**
indirect discharger	A producer of industrial effluent who discharges effluent into a public sewer for a municipal undertaker to treat prior to discharge to environment; cf direct discharger
indirect electric vaporizer	A subtype of waterbath vaporizer (qv) for LPG; defined formally in NFPA 58: Liquefied Petroleum Gas Code, 2017 Edition
indirect fired heating system	In the context of fired heaters, a system where flue gas does not enter the heating chamber; defined formally in NFPA 86: Standard for Ovens and Furnaces, 2019 Edition
indirect fired internal heating system	In the context of fired heaters, a system of burners separated from the oven atmosphere by enclosure inside gastight radiators; defined formally in NFPA 86: Standard for Ovens and Furnaces, 2019 Edition
indirect fired vaporizer	see **indirect vaporizer**

indirect fluid to air heat transfer device	see **indirect air preheater**
indirect tank heater	A device used to heat an LPG tank by recirculating its contents through a heat exchanger and back to the tank; defined formally in NFPA 58: Liquefied Petroleum Gas Code, 2017 Edition; cf direct tank heater
indirect vaporizer	An LPG vaporizer heated by a fluid, rather than directly; defined formally in NFPA 58: Liquefied Petroleum Gas Code, 2017 Edition
individual risk of fatality	The expected frequency at which individuals will suffer a fatal injury as a result of a business's activities; defined formally in UK HSE Reducing Risk, Protecting People
induced air flotation	Flotation of solids and liquids (especially those lighter than water) on water, using relatively coarse air bubbles entrained in a water flow; cf dissolved air flotation, froth flotation
induced draft	In the context of fired heaters, mechanically produced pressure difference across the system; defined formally in API RP 535 - Burners for Fired Heaters at Refineries; cf natural draft
induced draft fan	*aka* IDF. A fan producing an induced draft
induced draught heater	A fired heater with its combustion air induced (and flue gases removed) by a fan; defined formally in API STD 530 - Calculation of Tube Heater Thickness, API STD 560 - Fired Heaters for General Refinery Service and BS EN ISO 13705:2012/ISO 13705:2012(E) Petroleum, petrochemical and natural gas industries. Fired heaters for general refinery service; cf natural draft heater
induced environment	An environment local to a system caused by its operation; defined formally in BS ISO 2041:2018 Mechanical vibration, shock and condition monitoring. Vocabulary
induced gas flotation	*aka* dispersed gas flotation, IGF. The flotation of solids and liquids (especially those lighter than water) on water, using natural gas or nitrogen bubbles entrained in a water flow; cf induced air flotation
induction	Instructing a new employee, contractor or visitor to an industrial site on safety matters, such as rules of behavior on the site, possible dangers, routes of evacuation and assembly points
induction heating system	A system based on heat produced by electrical eddy currents in a conductor; defined formally in NFPA 86: Standard for Ovens and Furnaces, 2019 Edition

induction mass flow	Fluid flow induced to a turbine expressed in mass units; defined formally in ISO14661 Thermal turbines for industrial applications (steam turbines, gas expansion turbines) - General requirements
induction steam conditions	Conditions of the steam entering a turbine; defined formally in ISO14661 Thermal turbines for industrial applications (steam turbines, gas expansion turbines) - General requirements
inductor	see **eductor**
industrial complex	Multiple process plants on one site; defined formally in ISO 10209:2012 Technical product documentation — Vocabulary — Terms relating to technical drawings, product definition and related documentation
industrial control panel	*aka* ICP. An American term for controlgear (qv), very similar in meaning to motor control center (qv), the preferred term in the UK. A formal definition of the term can be found in the US National Electrical Code (NEC) Section 409.2
industrial effluent	Effluent (qv), especially liquid effluent, arising from industrial activity
Industrial Emissions Directive	European legislation which forms the basis of issue of environmental emissions permits for larger process plants
industrial occupancy	A term used to describe a building or part of one used to manufacture or repair products; in other words, a factory. Defined formally in NFPA 30: Flammable and Combustible Liquids Code, 2018 Edition
industrial plant	As well as the commonplace meaning, there is a specific definition: an LPG storage facility with containers larger than 4000 Gal (15.2 m³). Defined formally in NFPA 58: Liquefied Petroleum Gas Code, 2017 Edition
industrial type steam turbine	A steam turbine in an industrial application; defined formally in ISO14661 Thermal turbines for industrial applications (steam turbines, gas expansion turbines) - General requirements
inert gas	1. see **inert special atmosphere** 2. see **noble gases**
inert special atmosphere	In the context of fired heaters, a special atmosphere of non-flammable gases with <1% oxygen. Defined formally in NFPA 86: Standard for Ovens and Furnaces, 2019 Edition; cf indeterminate special atmosphere, flammable special atmosphere, nonflammable special atmosphere, purge gas
inert substance	Almost always, a chemically unreactive substance (under conditions of interest), though it also has a meaning in thermodynamics

inertial force	*aka* fictitious force. An apparent force exerted by a mass opposing acceleration; defined formally in BS ISO 2041:2018 Mechanical vibration, shock and condition monitoring. Vocabulary
inertial mist eliminator	Usually, a vane-type mist eliminator, as in the formal definition in the context of gas turbines in ISO 3977 Gas turbines - Procurement - Part 3: Design requirements; cf wire mesh type mist eliminator
inerting	see **blanketing**
inerting agent	A gas such as nitrogen, helium or carbon dioxide used for inerting (qv)
infiltration	Water leaking into underground pipes; cf exfiltration
inflammable	A strongly deprecated term, formerly synonymous with flammable (qv). It is however Spanish for 'flammable'[69]; see **Box F1**.
influent	see **influx**; cf effluent
influx	*aka* influent. Fluid flow into a process; cf efflux
information	Meaningful data
information for use	*aka* instructions for use. Detailed instructions on how to use a product safely. Defined formally in BS EN ISO 12100:2010 Safety of machinery. General principles for design. Risk assessment and risk reduction, BS EN 82079-1:2012 Preparation of instructions for use. Structuring, content and presentation. General principles and detailed requirements and IEC/IEEE 82079-1:2019 Preparation of information for use (instructions for use) of products – Part 1: Principles and general requirements
infranatant	A liquid lying below a precipitate; cf supernatant, subnatant
infrasonic	In the context of sound, having a frequency of less than 20Hz
infrastructure	A somewhat vague term, which usually means general civil engineering site structures such as buildings and roads, plus the central services (qv) and utilities (qv) which support the main process. However, offsites (qv) are sometimes considered a subset of infrastructure, and sometimes there may be a differentiation between process engineering offsites and civil engineering infrastructure. As well as the general civil engineering site structures which are always in this category, onsite effluent treatment plant may be included rather than being considered process plant, utilities, central services or offsites

---

[69] As The Simpson's Dr Nick says: "Inflammable means flammable? What a country!"

infused leaf	Used tea leaves; defined formally in ISO 3103 Tea - Preparation of liquor for use in sensory tests
ingot	A cast bar or block, usually of relatively pure metal
ingress protection rating	*aka* IP rating, IP*xx* (where *xx* is a number). A coded evaluation of equipment's ability to operate safely in the presence of solids and moisture; defined formally in BS EN 60529:1992+A2:2013 Degrees of protection provided by enclosures (IP Code), amongst other places
inguhneer	see **enjuneer**
inherent moisture	Collective term for the four kinds of bound moisture (qv) in coal; cf adventitious moisture
inherent safety	Eliminating highly hazardous features of a design, rather than trying to control their consequences; cf intrinsic safety
inherently safe design measure	A measure which achieves inherent safety (qv); defined formally in BS EN ISO 12100:2010 Safety of machinery. General principles for design. Risk assessment and risk reduction
inhibition	A reduction of reaction rate
initial charge	1. The first charge given to a new battery prior to use 2. An up-front fee for the purchase of shares or investments
initial operations	A term used in some industries to describe the entire process of precommissioning, commissioning, initial startup, steady production and performance testing of a plant; see **Box C3**.
initial startup	A term used in commissioning for when feedstocks are introduced to a plant for the express purpose of producing a product for the first time; see **Box C3**.
initial steam or gas conditions	Steam or gas conditions (qv) at a turbine stop valve's inlet; defined formally in ISO14661 Thermal turbines for industrial applications (steam turbines, gas expansion turbines) - General requirements
initiating event	The first event in a sequence that leads to, for example, a pipeline accident; cf top event, root cause, ultimate cause
initiator	Something which starts a chain reaction
injection grid	In the context of flue gas NOx control, a system of reactant injection nozzles in the flue gas stream; defined formally in API RP 536 - NOx control on Fired Heaters at Refineries
injection regime	A term used in distillation for a condition in which the liquid above the plate is in the form of individual drops dispersed in the vapor, so that there is virtually no mixing in the main bulk of the liquid
injector	see **ejector**
inlet	An opening for an intake (e.g., of air); cf outlet
inlet air filter	Compressor ancillaries which remove particles from inlet air

inlet butterfly valve	1. In the context of centrifugal compressors, a valve used to throttle suction (less efficiently than an inlet guide vane (qv)) 2. May also be used to mean any butterfly valve on the inlet side of something, whether used for control or isolation
inlet connections	The incoming connection point on a turbine stop valve or casing; defined formally in ISO14661 Thermal turbines for industrial applications (steam turbines, gas expansion turbines) - General requirements
inlet guide vane	An efficient way of throttling suction to a centrifugal compressor; cf inlet butterfly valve
inlet plenum	The plenum on a compressor inlet; defined formally in ISO 3977 Gas turbines - Procurement - Part 3: Design requirements
inlet size	Usually, the size of a PRV (qv) inlet connection expressed as NPS (qv); defined formally in API RP 520 P1 7th Edition, January 2000 Sizing, Selection, and Installation of Pressure-Relieving Devices in Refineries; Part I - Sizing and Selection and API RP 576 - Inspection of Pressure Relieving Devices
inlet velocity pressure	The fan pressure required to accelerate incoming flow to required velocity; the difference between total pressure and static pressure; defined formally in BS EN ISO 13705:2012/ISO 13705:2012(E) Petroleum, petrochemical and natural gas industries. Fired heaters for general refinery service
inline	Strictly, an analysis using an instrument mounted in the line being sampled, but may be used as a synonym of online; cf online, atline, offline
inline balanced pressure proportioning	A firefighting foam dosing system, using a concentrate dosing pump or bladder tank and integrated pressure balancing valve, to balance water and concentrate pressures. Defined formally in NFPA 11: Standard for Low-, Medium-, and High-Expansion Foam, 2016 Edition
inline eductor	A venturi, especially when used as a metering device, e.g., when used to make up firefighting foam, defined formally in NFPA 11: Standard for Low-, Medium-, and High-Expansion Foam, 2016 Edition
inline mixer	*aka* motionless mixer, static mixer. A device without moving parts in a pipe or channel which mixes fluids passing through it using energy derived from its headloss
inline pump	A pump with inlet and outlet flanges on a common centerline. A more specific definition is given in NFPA 20: Standard for the Installation of Stationary Pumps for Fire Protection, 2019 Edition

inner seal	The inner seal of an arrangement 2 seal (qv) or arrangement 3 seal (qv) i.e., the one closest to the pump impeller; defined formally in API RP 682 - Pump Seals
innovation	Something real engineers avoid as much as possible (but no more than that)
inoculation	The addition of a small quantity of a live culture of organisms or seed crystals (qv), depending on context
input	The flow into a unit process, especially in the context of LCA (qv); defined formally in EN ISO 14040:2006 Environmental management - Life cycle assessment - Principles and framework
insert ring	see **consumable insert**
in-service margin	*aka* buckling factor, safety margin. The degree to which an operating condition lies within the failure assessment diagram (qv) failure envelope; defined formally in API RP 579 - Fitness for Service
inside diameter	Heat exchanger tube 'inside diameter' used in design calculations, without corrosion allowance. May well differ from ID (qv), and actual inside diameter (qv). Defined formally in API STD 530 - Calculation of Tube Heater Thickness
inside liquid storage area	An NFPA occupancy (qv), a separate enclosure for storage of containers of flammable liquids. Defined formally in NFPA 30: Flammable and Combustible Liquids Code, 2018 Edition
inside mounted seal	see **internally mounted seal**
inspecting authority	An entity responsible for inspection and verification. Defined formally in PD 5500:2018+A1:2018 Specification for unfired fusion welded pressure vessels
inspection	Establishing compliance with specification. Defined formally in BS EN ISO 9000 Quality management systems Fundamentals and vocabulary
inspection door	see **peephole**
inspector	A suitably qualified purchaser/owner representative, responsible for inspection (qv). Defined formally in API STD 620 - Low Pressure Storage Tanks
inspector's delegate	A person delegated by an owner's inspector (qv). Defined formally in ASME BPE (American Society of Mechanical Engineers: Bioprocessing Equipment)
inspirator	An eductor (qv) driven by fuel gas pressure used to premix fuel and air in premix burners. Defined formally in API RP 535 - Burners for Fired Heaters at Refineries

installation	1. Most commonly, the activities associated with or process of fixing in place and making ready for use of equipment 2. Less commonly, installed equipment. This is the general sense in which the EU IPPC Directive defines "a stationary technical unit which carries out one or more activities listed in Annex I of the European Industrial Emissions Directive"
installation diagram	A diagram of the locations and interconnections between electrical installation components. Defined formally in ISO 10209:2012 Technical product documentation – Vocabulary – Terms relating to technical drawings, product definition and related documentation; cf installation drawing
installation drawing	A drawing of an item's general configuration, showing required information to allow installation. Defined formally in ISO 10209:2012 Technical product documentation – Vocabulary – Terms relating to technical drawings, product definition and related documentation; cf installation diagram
instantaneous selectivity	Selectivity (qv) expressed as the ratio of production rates of desired and undesired products
instantaneous value	The value at a given instant of a variable
instantaneous release	see **puff**
instruction manual	An O&M manual (qv) for a product; defined formally in BS EN 82079-1:2012 Preparation of instructions for use. Structuring, content and presentation
instructions for use	see **information for use**
instrument	*aka* instrumentation (qv). Device or devices which measure a variable (and may also display, or less commonly control, the variable). Defined formally in BS1646-1 Symbolic Representation for Process Measurement Control Functions and BS 1646-3:1984 Symbolic representation for process measurement control functions and instrumentation. Specification for detailed symbols for instrument interconnection diagrams Instrumentation Part 1: Basic Requirements
instrument air	Compressed air of a quality suitable for use to drive pneumatic valves
instrumentation	The sensors and actuators of a process control system
INSUL	*(drawing notation)* Insulate or insulation; N.B. insulated has a very different meaning in English from isolated (qv)

insulated	1. Generally, provided with electrical or thermal insulation (qv) 2. A specific definition, purely with respect to thermal insulation of steel to limit its maximum temperature in an expected maximum duration of fire exposure can be found in NFPA 15 Standard for Water Spray Fixed Systems for Fire Protection
insulating RIBC	A RIBC (qv) made of insulating materials such as plastic, incapable of electrical grounding. Defined formally in NFPA 654: Standard for the Prevention of Fire and Dust Explosions from the Manufacturing, Processing, and Handling of Combustible Particulate Solids, 2017 Edition cf type A FIBC
insulation	1. The opposite of conduction (qv), sometimes confused by non-native speakers with isolation (qv) 2. A material with low conductivity with respect to heat, electricity or sound, or the application of such material
Intalox saddle	A proprietary type of random packing (qv)
intangible asset	An identifiable non-monetary asset without physical substance, e.g., licenses, patents, trademarks, computer software, know-how and goodwill (company name and reputation)
integral 2 psi service regulator	An LPG pressure regulator unit integrating a high-pressure regulator (qv) and a 2 psi service regulator (qv); defined formally in NFPA 58: Liquefied Petroleum Gas Code, 2017 Edition
integral action control	In the context of process control, using controller output proportional to the integral of the difference between set point and measured value. The 'I' in PID control (qv), with a zero reset action
integral liquid or salt media quench-type tank	An integrated tank of quenchant (qv) used to rapidly cool furnace workpieces, thus maintaining work under protective atmosphere until it enters the tank. Defined formally in NFPA 86: Standard for Ovens and Furnaces, 2019 Edition; cf open liquid or salt media quench-type tank
integral two stage regulator	An LPG pressure regulator unit, integrating a high-pressure regulator and second-stage regulator. Defined formally in NFPA 58: Liquefied Petroleum Gas Code, 2017 Edition
integrated gasification combined cycle	*aka* IGCC. A coal gas production process
integrated pollution prevention and control	*aka* IPPC. The European Union IPPC Directive (2008/1/EC) formed the basis of issue of environmental emissions permits for larger process plants until it was replaced in 2010 by the Industrial Emissions Directive
integrated remote operating center	*aka* IROC. A central control room (qv) which is not colocated on site

integrity operating windows	*aka* IOW. Limits within which equipment can be safely operated without the risk of harming its physical integrity
intelligent pig	*aka* smart pig. A pig (qv) with onboard instrumentation and data logging facilities used to inspect pipelines
intelligent transmitter	A transmitter comprising two major components: (a) a sensor module which comprises the process connections and sensor assembly, and (b) a two-compartment electronics housing with a terminal block and an electronics module that contains signal conditioning circuits and a microprocessor
intended conditions of use	see **conditions for intended use**
intended use	Use of equipment as per instructions, for the purpose it was designed to serve. Defined formally in BS EN 82079-1:2012 Preparation of instructions for use. Structuring, content and presentation. General principles and detailed requirements and BS EN ISO 12100:2010 Safety of machinery. General principles for design. Risk assessment and risk reduction
intensive variable	A variable whose value is additive for subsystems; cf extensive variable
interceptor	*aka* petrol water separator, forecourt separator. A UK term for a packaged oil water separator, used to exclude light nonaqueous phase liquids from collected surface water before discharge. There are various subtypes: a class 1 interceptor achieves < 5ppm LNAPL in its discharge, suitable for direct discharge and a class 2 up to 100 ppm, suitable for indirect discharge. Full retention interceptors are designed for all foreseeable levels of rainwater. Bypass interceptors are designed for only 10% of this flow, though that (perhaps surprisingly) covers 99% of rainfall events
interconnection diagram	A diagram showing connections between an installation's units. Defined formally in ISO 10209:2012 Technical product documentation — Vocabulary — Terms relating to technical drawings, product definition and related documentation and BS 1646-3:1984 Symbolic representation for process measurement control functions and instrumentation. Specification for detailed symbols for instrument interconnection diagrams
interconnector	A conductive, gas-tight component connecting single cells in a fuel cell stack
intercooler	A heat exchanger between compression stages

interested party	*aka* stakeholder. This particular term used in EN ISO 14040:2006 Environmental management - Life cycle assessment - Principles and framework and BS EN ISO 9000 Quality management systems Fundamentals and vocabulary
interface	A boundary, most commonly between physical phases, or between humans and machines; cf HMI
interface drawing	A drawing showing how two parts fit together for assembly. Defined formally in ISO 10209:2012 Technical product documentation — Vocabulary — Terms relating to technical drawings, product definition and related documentation
interface temperature	As well as the generic meaning, a term for a calculated temperature at the interface between two layers of refractory, as defined formally in BS EN ISO 13705:2012/ISO 13705:2012(E) Petroleum, petrochemical and natural gas industries. Fired heaters for general refinery service, API STD 530 - Calculation of Tube Heater Thickness and API STD 560 - Fired Heaters for General Refinery Service
interference model	A model illustrating space requirements and possible clashes (qv). Defined formally in ISO 10209:2012 Technical product documentation — Vocabulary — Terms relating to technical drawings, product definition and related documentation
interim certificate	A document authorizing payment for work completed as of a specified date. Defined formally in ISO 10209:2012 Technical product documentation — Vocabulary — Terms relating to technical drawings, product definition and related documentation; cf completion certificate, final certificate
interlock	see **interlocking device**
interlocking device	*aka* interlock. A safety device which makes the operation of hazardous machinery dependent upon certain circumstances. Defined formally in BS EN ISO 12100:2010 Safety of machinery. General principles for design. Risk assessment and risk reduction
interlocking guard	A guard (qv) with integrated interlocking device (qv); defined formally in BS EN ISO 12100:2010 Safety of machinery. General principles for design. Risk assessment and risk reduction
interlocking guard with a start function	see **control guard**
interlocking guard with guard locking	A guard with integrated interlocking device (qv) and guard locking device (qv). Defined formally in BS EN ISO 12100:2010 Safety of machinery. General principles for design. Risk assessment and risk reduction

intermediate bulk container	*aka* IBC, IBC tank, IBC tote, pallet tank. Most commonly, a cuboidal rigid plastic container with integral pallet of 1000-1500L capacity; defined more widely in NFPA 30: Flammable and Combustible Liquids Code, 2018 Edition - most notably allowing up to 3000L capacity
intermediate flow	Flows of mass or energy between unit processes; defined formally in EN ISO 14040:2006 Environmental management - Life cycle assessment - Principles and framework
intermediate landing	A horizontal platform for resting between consecutive staggered access ladder flights rated for one person. Defined formally in BS EN ISO 14122-4 2016 Safety of machinery. Permanent means of access to machinery. Working platforms and walkways
intermediate level waste	*aka* ILW, intermediate nuclear waste. Moderately radioactive nuclear waste, such as used reactor components other than fuel assemblies; cf high level waste, low level waste, highly active waste
intermediate nuclear waste	see **intermediate level waste**
intermediate platform	A horizontal platform for resting between consecutive access ladder flights rated for more than one person; defined formally in BS EN ISO 14122-4 2016 Safety of machinery. Permanent means of access to machinery. Working platforms and walkways
intermediate product	Material flowing between unit processes requiring further transformation to create final product. Defined formally in EN ISO 14040:2006 Environmental management - Life cycle assessment - Principles and framework
intermittent pilot	A fired heater burner pilot (qv), lit only during main burner light-off and firing. Defined formally in NFPA 86: Standard for Ovens and Furnaces, 2019 Edition; cf interrupted pilot
internal circulating device	*aka* pumping ring. A device within a pump seal chamber which circulates seal chamber fluid through a cooler or fluid reservoir. Defined formally in API RP 682 - Pump Seals
internal diffusion	Most commonly, diffusion within a catalyst particle or biofilm which can limit a reaction rate (intraparticle diffusion)
internal energy	(*symbol* U) The total energy of a thermodynamic system
internal excess flow valve	An excess flow valve (qv) with all parts critical to closure contained within the LPG vessel. Defined formally in NFPA 58: Liquefied Petroleum Gas Code, 2017 Edition

internal floating roof tank	*aka* IFRT. A storage tank with a lightweight floating roof inside a heavier external fixed roof. Defined formally in API MPMS 19.1 Manual of Petroleum Measurement Standards Chapter 19.1 Evaporative Loss from Fixed-roof Tanks; cf external floating roof tank, covered floating roof tank
internal reflux	see **reflux**
internal resistance	Ohmic resistance inside a fuel cell, measured between the current collectors
internal spring type pressure relief valve	A PRV (qv) differing from a full internal pressure relief valve (qv) in being only partly recessed, with its wrenching pads and seating section outside the vessel wall. Defined formally in NFPA 58: Liquefied Petroleum Gas Code, 2017 Edition
internal valve	A remotely closeable LPG container shutoff valve, incorporating an internal excess flow valve (qv). Defined formally in NFPA 58: Liquefied Petroleum Gas Code, 2017 Edition
internally mounted seal	*aka* inside mounted seal. A pump seal mounted within a seal chamber; defined formally in API RP 682 - Pump Seals; cf externally mounted seal
international pound	see **pound**
interpolation	A type of estimation by creating new data points between known data points
interrupted pilot	A fired heater burner pilot (qv), lit only during main burner light-off and shut off automatically at end of a trial for ignition period (qv); defined formally in NFPA 86: Standard for Ovens and Furnaces, 2019 Edition; cf intermittent pilot
interstiffener collapse	Collapse of a pressure vessel or pipe between stiffeners. Defined formally in the case of pressure vessels in PD 5500:2018+A1:2018 Specification for unfired fusion welded pressure vessels
interstitial	Of the interstices: the space between particles
interstitial space	*aka* service floor, utility floor, technical floor, plant floor. An (often reduced height) floor between floors in a building used to house services
interstitial volume	*aka* void volume. The space between particles
intrados	The line of the interior curve of an arch; cf extrados
intraparticle diffusion	The diffusion of a reagent inside catalyst particles. In some cases, it could be the rate-limiting step influencing product yield and reaction selectivity
intrinsic safety	Making an electrical system suitable for use in a flammable atmosphere by limiting electrical and thermal energy below that required for ignition; cf inherent safety

intrinsically safe	The condition of an electrical system suitable for use in a flammable atmosphere as its electrical and thermal energy has been limited below that required for ignition
intrinsically safe apparatus	Apparatus which is intrinsically safe (qv); defined formally in API RP 505 2nd Edition, August 2018 Recommended Practice for Classification of Locations for Electrical Installations at Petroleum Facilities Classified as Class I, Zone 0, Zone 1, and Zone 2; cf explosionproof equipment
intrinsically safe circuit	A circuit which is intrinsically safe (qv). Defined formally in API RP 505 2nd Edition, August 2018 Recommended Practice for Classification of Locations for Electrical Installations at Petroleum Facilities Classified as Class I, Zone 0, Zone 1, and Zone 2
intrinsically safe equipment	see **intrinsically safe apparatus**
intrinsically safe system	An electrical system which is intrinsically safe (qv). Defined formally in API RP 505 2nd Edition, August 2018 Recommended Practice for Classification of Locations for Electrical Installations at Petroleum Facilities Classified as Class I, Zone 0, Zone 1, and Zone 2; cf inherent safety
intumescent paint	Paint which swells to produce a controlled thickness of insulating char in a fire
inventory	*aka* stock, stocktake, stocktaking (UK English). A quantified listing of items in hand, the items themselves, or a process for quantifying them
inventory difference	see **material unaccounted for**
inverse thixotropy	see **rheopecty**
inversion	1. The hydrolysis of sucrose into a glucose/fructose mixture called invert sugar 2. Entrapment of cold air under a layer of warm air in the atmosphere, which prevents the dispersal of airborne pollutants
invert	The bottom of the internal bore of a pipe, directly below the crown (qv)
invert level	The level above datum of the bottom of the internal bore of a pipe
invert sugar	A glucose/fructose mixture produced from sucrose by inversion (qv)
inverter	*aka* frequency inverter, variable frequency drive (VFD), variable speed drive (VSD). A device which produces an AC output of selected frequency from a fixed input frequency. The output can be used to control the speed of rotation of electrical motors, as this is dependent on supply frequency

inviscid fluid	*aka* superfluid. A fluid with zero viscosity
invitation to tender	*aka* ITT. A formal invitation to submit a tender (qv) to undertake works. Defined formally in ISO 10209:2012 Technical product documentation — Vocabulary — Terms relating to technical drawings, product definition and related documentation
involuted feed	A spiral smooth feed (as opposed to turbulent tangential feed)
IOC	International oil company
ion exchange	*aka* IX. A process in which ions in aqueous solution are exchanged for similarly charged ions on a solid carrier under certain conditions, and the process subsequently reversed under regeneration (qv) conditions
ion leakage	Residual ions in water demineralized by ion exchange (qv)
ionic strength	An estimate of concentration of ions in a solution, taken in an ion exchange context as half of the sum of ion concentrations multiplied by the square of their charge
ionisation	see **ionization**
ionization	*aka* ionisation. A process which produces ions, for example under the influence of ionizing radiation. A term which may be used in error to describe dissociation (qv) of a solute
IOW	see **integrity operating windows**
IP	1. Ingress protection 2. Intermediate pressure 3. Intellectual property
IP rating	see **ingress protection rating**
IPP	Independent power plant
IPPC	see **integrated pollution prevention and control, IPPC directive**
IPPC Directive	**Integrated Pollution Prevention and Control Directive**; the European Union IPPC Directive (2008/1/EC)
IPS	Iron pipe size
IPxx	(where xx is a number) see **ingress protection rating**
IR	Infrared (radiation)
IRIS	Internal rotating inspection system; see API RP 571 - Damage Mechanism Affecting Fixed Refinery Equipment
IROC	see **integrated remote operating center**
irradiance	(*symbol* E, *units* w/m^2) The SI unit of radiant flux per unit area
irradiation	Exposure to ionizing radiation, in engineering contexts most commonly as a food preservation technique
irreversible	Generally, incapable of being reversed, especially in the context of reaction; cf reversible
irritant	A substance, contact with which causes inflammation in living tissue, but is not corrosive (qv)

IS	see **isolated point**
IS&Y	*(drawing notation)* Inside screw and yoke
ISBL	Inside battery limits (qv)
isentropic	*aka* isoentropic. Having the same entropy; in thermodynamics, a process which takes place with constant entropy (qv), and by implication, is therefore both reversible and adiabatic (qv)
Ishikawa diagram	see **fishbone diagram**
ISO	International Standards Organisation
iso	see **isometric drawing**
ISO 14001	A series of international standards to do with environmental performance, inspired by ISO 9000 (qv)
ISO 9000	The most popular international quality management systems standard, derived from BS 5750 Quality systems - Specification for design/development, production, installation and servicing
ISO Geometrical Product Specification	*aka* ISO GPS. A specification for workpieces (as opposed to global positioning system)
ISO GPS	see **ISO Geometrical Product Specification**
isobar	A line on a graph (or map) of constant pressure
isobaric	Having the same pressure. In thermodynamics, a process which takes place under constant pressure
isochoric	1. Having equal volume 2. In thermodynamics, a constant volume process
isoentropic	see **isentropic**
isogram	see **isopleth**
isohypse	see **level contour line**
isolated	The condition of being disconnected from something, most commonly applied to power supply. A term commonly confused with insulated (qv) by some non-native speakers
isolated point	*aka* IS. An unmeasurably small indication (qv) which cannot be characterized by ultrasonic data alone. Defined formally in PD 5500:2018+A1:2018 Specification for unfired fusion welded pressure vessels
isolated vapor pocket	*aka* IVP, vapor pocket; see **vapor pocket**
isolating switch	*aka* isolator. A switch used for isolation (qv) of an electric circuit from its power supply. Defined formally in NFPA 20: Standard for the Installation of Stationary Pumps for Fire Protection, 2019 Edition
isolation	Being disconnected from something, most commonly applied to power supply; cf insulation
isolation blind	see **guillotine**
isolation valve	A valve (usually manual) located at equipment, used to disconnect it from piping; cf root valve

isolator	1. Most commonly, an informal term for an isolating switch 2. A shock- and vibration-attenuating support structure, as defined formally in BS ISO 2041:2018 Mechanical vibration, shock and condition monitoring. Vocabulary
isoline	see **isopleth**
isomerization	Generically, the process of making isomers. Most commonly refers to the conversion of straight chain hydrocarbons (paraffins) to branched hydrocarbons (isoparaffins)
isomers	Chemical compounds which have the same chemical composition but different structures and/or characteristics
isometric axonometry	An orthogonal axonometry (qv) where projection lines form equal angles with respect to coordinate axes. Defined formally in ISO 10209:2012 Technical product documentation – Vocabulary – Terms relating to technical drawings, product definition and related documentation
isometric drawing	*aka* iso. A dimensioned drawing used to define arrangements of pipework and fittings for fabrication and pricing purposes. They are more like diagrams than drawings: pipes are shown as single lines and symbols are used to represent pipe fittings, valves, pipe gradients and welds
isometric representation	A drawing projection method with coordinate axes all inclined at the same angle to the projection plane. Defined formally in ISO 10209:2012 Technical product documentation – Vocabulary – Terms relating to technical drawings, product definition and related documentation
isopieste	A line on a graph of constant moisture content
isopleth	*aka* isoline, isogram. A contour line joining locations of equal values on a meteorological map or diagram, such as an isotherm (qv)
isotach	see **isovel**
isotherm	A line on a graph of constant temperature
isothermal	Having equal temperature. In thermodynamics, a process which takes place at constant temperature
isothermal efficiency	The ratio of theoretical work to compress a gas isothermally to the work actually done; a measure of compressor efficiency
isotonic	Exerting the same osmotic pressure
isotropic	Uniform in all directions, the opposite of anisotropic (qv)
isovel	*aka* isotach. A line representing points of equal velocity, commonly used to indicate the distribution of flow velocities in channels
ISPE	International Society for Pharmaceutical Engineering
item	see **object**

item number	1. Generically, an identifying number assigned to an item 2. In the context of heat exchangers, the purchaser's identifying number; defined formally in API 660 - Shell-and-Tube Heat Exchangers and API RP 661 - Heat Exchangers
ITEQ	International toxicity equivalents
iteration	A process of refinement of approximation by repeated calculation, the output of each calculation being the input to the next
ITT	see **invitation to tender**
IUE	*(drawing notation)* Internal upset ends
IVP	Isolated vapor pocket; see **vapor pocket**
IX	see **ion exchange**

J

J	see **joule**
J factor	see **Chilton–Colburn analogy**
jack up rig	*aka* self-elevating unit. An offshore platform used for windfarms and exploratory drilling for oil and gas, which can be floated into place and then temporarily lifted clear of the sea by means of moveable legs
jacket	*aka* steam jacket. A sealed secondary external vessel surrounding a process vessel (called a jacketed vessel), the cavity of which is used for circulating heating or cooling fluid
jacketed vessel	A vessel with a jacket (qv)
Jafarey, Douglas and McAvoy correlation	see **McAvoy correlation**
jarosite process	A process for removal of iron from zinc sulfate leach liquors by precipitation as a basic iron sulfate
jaw crusher	A type of primary crusher used in mineral processing; cf gyratory crusher
JB	*(drawing notation)* Junction box
JDA	see **joint development agreement**
jeep	see **holiday**
jeeping	*(informal)* Using a holiday detector (qv) to verify coating integrity on a pipeline, prior to backfill
jerk	A term for rate of change of acceleration, according to BS ISO 2041:2018 Mechanical vibration, shock and condition monitoring. Vocabulary (N.B. alternate offensive non-engineering meaning)
jerry-built	Built as cheaply as possible; cf jerry-rigged
jerry-rig	see **kludge**
jerry-rigged	*aka* jury-rigged. Repaired provisionally with the tools and materials at hand, possibly involving Duck tape (qv)
Jersey	*aka* Jersey barrier, Jersey bump, Jesey curb, Jersey wall, K-rail. A concrete or waterfilled-plastic modular barrier used as a median barrier in a road, amongst other things
Jersey barrier	see **Jersey**
Jersey bump	see **Jersey**
Jersey curb	see **Jersey**
Jersey wall	see **Jersey**

jet fire	A burning jet caused by the ignition of a leak of a pressurized flammable fluid. Defined formally in API Standard 521. Pressure-relieving and Depressuring Systems. Sixth Edition|January 2014
jet fuel	*aka* aviation turbine fuel (ATF). Fuel for aviation turbines (jet engines), a hydrocarbon distillate
jet milling	The use of high velocity fluid, usually air, to cause solid particles to hit each other and break up into smaller pieces
jet reactor	1. Most commonly, refers to the Joint European Torus (JET) fission reactor 2. Less commonly, refers to a jet loop reactor, a type of loop reactor (qv) used in biochemical engineering
JHA	see **job hazard analysis**
jig	1. A bespoke tool used in machining 2. An abbreviation of jig separator or jig concentrator, a device used for ore beneficiation (qv)
job costing	In product manufacture, dividing total direct costs (plus a fraction of indirect production costs) by number of items; cf process costing
job hazard analysis	*aka* JHA. The identification of safety and environmental hazards, prior to job commencement, with a view to their mitigation and communication to workers. Defined formally in Australian Standard – Petroleum Pipelines - AS2885.0:2018
jobshop	Production facilities used in the manufacture of small batches of specialized products, as in contract manufacture of fine chemicals, or in a tool and die shop
jockey pump	see **pressure maintenance pump**
joint	A junction between materials. Defined formally in BS EN 1672-2:2005+A1:2009 Food processing machinery. Basic concepts. Hygiene requirements and BS EN ISO 14159:2008 Safety of machinery. Hygiene requirements for the design of machinery
joint development agreement	A formal contract between two or more companies to work together, usually to develop a new technology or process
joint penetration	Weld depth; defined formally in API STD 620 - Low Pressure Storage Tanks and ASME BPE (American Society of Mechanical Engineers: Bioprocessing Equipment)
joint venture	*aka* JV. A formal contract between two or more companies to work together, usually for the purposes of bidding for large engineering design and build contracts
joule	(*symbol* J) The SI base unit of energy
joule per cubic meter	The SI derived unit of energy density
joule per kelvin	The SI derived unit of heat capacity

joule per kilogram	The SI unit of energy
joule per kilogram kelvin	The SI derived unit of specific heat capacity
joule per mole	The SI derived unit of molar energy
joule per mole kelvin	The SI derived unit of molar heat capacity
Joule-Kelvin effect	see **joule-Thomson effect**
Joule-Thomson effect	Cooling caused by expansion of a gas without work or heat transfer
JPL chlorinolysis process	A coal desulfurization process involving oxidation with chlorine
jubilee clip	Proprietary eponym for a worm-drive hose clip, see **Figure J1**
jump	A sudden change such as a hydraulic jump (qv); or in vibration monitoring, a sudden change in response to a small frequency change, as defined formally in BS ISO 2041:2018 Mechanical vibration, shock and condition monitoring. Vocabulary
jump over	Pipework interconnecting a heater coil section; defined formally in BS EN ISO 13705:2012/ISO 13705:2012(E) Petroleum, petrochemical and natural gas industries. Fired heaters for general refinery service, API STD 530 - Calculation of Tube Heater Thickness and API STD 560 - Fired Heaters for General Refinery Service

Figure J1 Jubilee clip

jurisdiction 1. The power to make and enforce judgements
2. An area over which given courts or government have authority to make binding judgments; defined formally in this sense in API RP 579 - Fitness for Service
jury-rig see **kludge**
JV see **joint venture**
JW *(drawing notation)* Jacket water

K

kaizen	A Japanese word meaning 'change for the better', used for a business philosophy of continuous improvement, and a formalized set of principles to achieve this
Kalrez	Proprietary eponym for a FFKM perfluoroelastomer (qv), used to make O-rings etc.
KaMOS gasket	A patented means to pressure-test a joint between flanges without breaking other flanges
Kappa number	Chemists use kappa numbers for things which they call denticity and compressibility, but in engineering contexts, kappa number is usually a measure of paper/pulp lignin content using a standard potassium permanganate solution
Karl Fischer titration	A laboratory technique that measures the amount of water in a (usually fairly dry) sample
Karlovitz numbers	The inverse of Damköhler numbers (qv)
Kármán vortex street	*aka* von Kármán vortex street. An often undesirable repeating pattern of pairs of vortices, formed when a fluid passes a certain type of obstruction at a certain range of velocities
karor	see **crore**
kat	see **katal**
kat/m^3	Katal per cubic meter
katal	(*symbol* kat; *units* mol/s) The SI derived unit of catalytic activity
Kek	Proprietary eponym for a rotary sifter
kelvin	(*symbol* K) The SI base unit of thermodynamic temperature
kerogen	The insoluble organic solids in oil shale and coal; cf vitrinite
kerosene	A US English term for a hydrocarbon distillate fuel; known as paraffin (qv) in the UK
ketone	Molecule containing the -CO- group
kettle reboiler	A type of evaporator used to vaporize fluid at the bottom of a distillation column (qv)
key component	A concept used in shortcut methods for the design of multicomponent distillation columns. 'Light' (more volatile) and 'heavy' (less volatile) key components of the mixture are selected as the basis for the design

K-factor	A term with a range of applications, all distinct from K-value (qv): 1. Most commonly used in the context of hydraulic calculations as a local loss coefficient for a fitting 2. *aka* clearing factor. In centrifugation, used to determine the relative pelleting efficiency of a given rotor at maximum speed 3. *aka* characterization factor, UOP K-factor, UOPK. In oil and gas, used to classify crude oil according to its aromatic, intermediate, naphthenic or paraffinic nature
kg	see **kilogram**
kg/m^2	see **kilogram per square meter**
kg/m^3	see **kilogram per cubic meter**
kgr	Kilograins: an almost obsolete measure of mass, equal to 1000 grains
kieselguhr	see **diatomaceous earth**
kieselgur	see **diatomaceous earth**
kill	1. In the context of steelmaking, to deoxidize 2. In the context of biological or hygienic processes, kill has the common meaning, or something close to it
kill tank	1. Another term for a quench tank (qv) 2. A high HRT (qv) tank used on an anaerobic digester to ensure pathogen removal
kill wing valve	*aka* KWV. A valve typically located on the left-hand side of a Christmas tree (qv) used to inject fluids, and act as a barrier during well interventions
kiln	One of various types of refractory lined fired vessels, such as rotary kilns used for minerals calcining
kilo-	(*symbol* k) The SI unit prefix denoting a factor of 10^3
kilogram	(*symbol* kg) The SI unit of mass
kilogram per cubic meter	(*symbol* kg/m^3) The SI derived unit of density
kilogram per square meter	(*symbol* kg/m^2) A non-SI unit of pressure; cf pascal
kilomole	(*symbol* kmol) One thousand moles
kilovolt ampere	(*symbol* kVA) A measure of the amount of electrical power used by a system
kilowatt hour	(*symbol* kWh) A derived non-SI unit of energy, equal to 3600 kJ; commonly used as the basis of energy supply bills
kindling point	see **autoignition temperature**

kinematic similarity	A type of system similarity, where velocity at any point in a smaller scale model is consistently proportional to the velocity at the corresponding point in a larger one; used for process scaleup, especially with respect to fluid mechanics; cf dynamic similarity, geometric similarity
kinematic viscosity	The ratio of dynamic viscosity (qv) to density
kinetic energy	The work needed to accelerate a body from rest to its current velocity
kinetic modelling	A mathematical representation of a process, taking into account variations as a function of time, particularly in the context of reaction engineering
kinetic parameters	Parameters associated with reaction kinetics such as activation energy, pre-exponent and reaction order
kinetics	The physical chemistry of rates of reaction, i.e., chemical reaction kinetics
kiosk	A weatherproof secure enclosure often used to house equipment in the field, most commonly prefabricated in GRP (qv)
Kirkbride equation	A way to determine the best column feed introduction point in shortcut distillation methods
Kirpichev number	(*symbol* Ki) A dimensionless number used in the analysis of drying processes
KISS principle	'Keep it simple, Stupid', or less commonly, 'Keep it stupid, Simple': a mnemonic for designers to remember that simpler is usually better when it comes to design; often attributed to Kelly Johnson
kit	1. (*informal*) A UK English synonym for equipment in general 2. A complete set of components intended to be assembled into an item of equipment in situ; defined formally in EN 12566 - Small wastewater treatment systems for up to 50 PT Part 4: Septic tanks assembled in situ from prefabricated kits
kLa	see **volumetric oxygen transfer coefficient**
kludge	*aka* jerry-rig, jury-rig. An inelegant, expedient, 'quick and dirty' solution to an urgent problem, related to a bodge (qv) or lashup (qv)
kN	Kilonewton
KN	Kjeldahl nitrogen; cf TKN
knee rail	A guardrail (qv) placed lower than and parallel with a handrail; defined formally in BS EN ISO 14122 Safety of machinery. Permanent means of access to machinery. Working platforms and walkways

knife gate	A type of valve particularly good for slurries containing fibrous or plastic solids
knife-line attack	see **preferential weld corrosion**
knocking	Mistimed combustion in a fuel/air engine due to its not being ignited by the spark plug; controlled in the case of gasoline with a minimum octane number (qv) specification
knockout drum	*aka* knockout pot, KO, KO drum, KO pot, flash drum, breakpot, knock-out drum, knock-out pot, compressor suction drum, separator drum, suction scrubber, compressor inlet drum, vent scrubber. A vessel used to remove and store liquid droplets present in a gas stream; defined formally in API RP 535 - Burners for Fired Heaters at Refineries, API Standard 521. Pressure-relieving and Depressuring Systems. Sixth Edition	January 2014 and API RP 537 Flare Details for Petroleum, Petrochemical, and Natural Gas Industries
knockout pot	see **knockout drum**
knuckle radius	Radius of the tangent torus in a torispherical (qv) head
Knudsen diffusion	*aka* Knudsen flux. Diffusion controlled by interactions with system boundaries, rather than other particles, due to the scale of a system being similar to or smaller than the mean particle free path
Knudsen flux	see **Knudsen diffusion**
KO	Knock out; qv knockout drum
KO drum	see **knockout drum**
KO pot	Knock out pot, see **knockout drum**
koji	A Japanese word for cooked rice/soy beans; traditionally used as a solid fermentation medium for Aspergillus oryzae, used in the production of miso, mirin, soy sauce, sake and citric acid
Kolmogorov eddies	see **Kolmogorov microscale of turbulence**
Kolmogorov microscale of turbulence	*aka* Kolmogorov eddies. The smallest scale of turbulent flow, at which viscosity dominates and kinetic energy is dissipated as heat
Kopp's rule	*aka* Kopp-Neumann law. A heuristic which states that the molecular heat capacity of a solid at around room temperature is approximately the sum of the atomic heat capacities of the elements composing it
Kopp-Neumann law	see **Kopp's rule**
koti	see **crore**
KOW	see **octanol/water partition coefficient**
Kozeny-Carman equation	see **Carman-Kozeny equation**
kPa	Pressure in kilopascals

kPag	Gauge pressure in kilopascals
KPI	Key performance indicator
Kraft process	A process used for the conversion of wood into wood pulp
K-rail	see **Jersey**
Kremser equation	A commonly (mis)used, often unreliable way to calculate the number of theoretical stages in absorption and extraction systems
Kroll process	A process for producing titanium metal by magnesium reduction of titanium ore
kVA	Kilovolt-amperes
K-value	Used in a range of applications, all quite distinct from k-factor (qv), most notably: 1. In the context of thermal conductivity 2. In vapor–liquid equilibria, the ratio of vapor to liquid concentration at equilibrium 3. There are several empirically-derived K-values used for viscosity-based estimation of the molecular weight of polymers, e.g., the Fikentscher K-value (qv); defined formally in DIN EN ISO 1628-1 Plastics — Determination of the viscosity of polymers in dilute solution using capillary viscometers
Kyoto wheel	see **rotary heat exchanger**

L

LA	1. *(drawing notation)* see **level alarm**
	2. see **Local Authority**
lab scale	1. A weighing scale used in a lab
	2. Processes at a scale too small for Chemical Engineers to concern themselves with
label	An identifying mark attached to a product displaying characteristic information, sometimes indicating compliance with standards; defined formally in BS EN 82079-1:2012 Preparation of instructions for use. Structuring, content and presentation. General principles and detailed requirements; cf **CE Marking**
labeled	*aka* labelled. Generally, bearing a label. Defined in the sense of demonstration of standard compliance in NFPA 11: Standard for Low-, Medium-, and High-Expansion Foam, 2016 Edition
labelled	see **labeled**
labile	Easily changeable; commonly applied to things which are degraded under certain conditions, e.g.; 'heat labile' things are degraded by heat
labor capacity ratio	see **capacity ratio**
lac(s)	see **lakh**
lack of fusion	see **incomplete fusion**
lack of fusion after reflow	A type of weld discontinuity, defined formally in ASME BPE (American Society of Mechanical Engineers: Bioprocessing Equipment)
lack of penetration	see **incomplete penetration**
LACT	see **lease automatic custody transfer**
ladder flight	The continuous part of a fixed ladder; defined formally in BS EN ISO 14122 Safety of machinery. Permanent means of access to machinery. Working platforms and walkways
ladder logic	A PLC (qv)/DCS (qv) graphical programming language, an alternative to a function block diagram (qv)
ladder system	see **fixed ladder system**
ladder with one stile	A stationary ladder with rungs attached to both sides of a central stile; defined formally in BS EN ISO 14122 Safety of machinery. Permanent means of access to machinery. Working platforms and walkways

ladder with two stiles	A stationary ladder with rungs attached between two stiles; defined formally in BS EN ISO 14122 Safety of machinery. Permanent means of access to machinery. Working platforms and walkways
LAER	see **lowest achievable emission rate**
lag phase	The initial period of low growth in a bioreactor as organisms adapt to conditions; cf log phase
lag time	1. In biochemical engineering, the duration of the lag phase (qv) 2. In process control, the time after the dead time (qv) which a process variable takes to move 63.3% of its final value after a step change; *aka* capacity element 3. In project scheduling, the gap between activities
lagging	UK term for thermal insulation wrapped around pipes and vessels
lagoon	A relatively large, shallow, natural or semi-natural body of water for water treatment which may function as a bioreactor, a settlement tank, or both
Lagrangian fluid dynamics	An approach to fluid dynamics which considers individual particles; cf Eulerian fluid dynamics
lakh	*aka* lac, lacs. A traditional Indian number multiplier equal to hundreds of thousands (10.5). Also used in the metals market as a multiplier of a troy ounce (qv) of precious metals
lamella clarifier	*aka* lamella settler, plate settler. A clarifier which uses inclined plates to create a larger hindered settling zone, effectively increasing settling area for the clarifier; cf tube settler
lamella plate	An inclined plate used to create an increased settling area in a lamella clarifier (qv), reflux classifier (qv) or API oil separator (qv)
lamella settler	see **lamella clarifier**
lamellar tearing	see **lamellar tears**
lamellar tears	*aka* lamellar tearing. Fibrous fractures in weld base metal at points of high stress concentration; defined formally in ASME BPE (American Society of Mechanical Engineers: Bioprocessing Equipment)
laminar flow	*aka* streamline flow. Fluid flow in parallel layers of constant velocity, without mixing or disruption of layers; cf turbulent flow

lamination(s)	1. Generally, layering: in food, the creation of layers (e.g.; butter between pastry in a croissant or puff pastry); in material science, the bonding of different layers to achieve a desired outcome (e.g.; printable moisture-resistant packaging) 2. In the context of welding, elongated defects, as defined formally in ASME BPE (American Society of Mechanical Engineers: Bioprocessing Equipment)
landfill	1. A lined and covered pit filled with solid waste, nowadays generally only biodegradable waste such as MSW (qv). Produces landfill gas (qv) with a significant methane content, and landfill leachate (qv), a high strength liquid waste 2. Material destined for disposal at a landfill site
landfill gas	A gas with a significant methane content produced by anaerobic digestion in a landfill
landfill gas operation	Operating a gas turbine with landfill gas as a fuel, as defined formally in ISO 3977 Gas turbines - Procurement - Part 3: Design requirements
landfill leachate	*aka* leachate. A high strength liquid waste produced by water passing through landfill contents
landing	A horizontal platform at the end of a flight of stairs, defined formally in BS EN ISO 14122 Safety of machinery. Permanent means of access to machinery. Working platforms and walkways
landscape drawing	A drawing showing required groundworks, roads, plantings etc. Defined formally in BS EN ISO 10209:2012 Technical product documentation – Vocabulary – Terms relating to technical drawings, product definition and related documentation
Lang index	A venerable process plant preliminary cost estimation factor, relating the bought-in cost of equipment to total project cost
Langelier saturation index	*aka* LSI. An estimate of calcium carbonate saturation in water, calculated from pH, temperature and the concentrations of alkalinity, calcium ion, and total dissolved solids. Used by some to provide guidance on whether a water is likely to be scaling or aggressive. As with the Ryznar stability index (qv) and Oddo-Tomson index (qv) it is not an exact science, and experts may differ on its validity
Langmuir adsorption isotherm	One of the more popular types of empirical adsorption isotherm; cf Freundlich adsorption isotherm
lap joint	A (usually welded) joint between two overlapping items; defined formally in API STD 620 - Low Pressure Storage Tanks; cf butt joint

Laplace transform	A way to transform ordinary differential equations into algebraic equations, popular in academic exercises, but rarely of practical utility
large diameter hose	90mm NB or larger hose, in the context of firefighting; defined formally in NFPA 1142: Standard on Water Supplies for Suburban and Rural Fire Fighting, 2017 Edition
Larson ratio	*aka* LR, Larson Skold index. An index of limited validity, commonly used to predict aggressiveness of once-through cooling waters which is based on observed corrosion of mild steel pipework carrying a single natural water type. It is calculated from the ratio of equivalents per million (epm) of sulfate and chloride to the epm of bicarbonate and carbonate
Larson Skold index	see **Larson ratio**
lashup	A jerry-rigged (qv) fix
last in, first out	*aka* LIFO. 1. A pattern of equipment/material storage in warehouses or storage areas 2. A method of inventory valuation 3. A principle traditionally used to determine staff redundancies, which may be illegal in some jurisdictions
last in, last out	*aka* LILO. A pattern of equipment/material storage in warehouses or storage areas
latent heat	(*symbol* L) The heat required to achieve a change of phase; cf sensible heat
lateral(s)	Subdivisions of a pipe, pipeline or sewer. There is a closely related, but highly specifically defined term in a pressure relief context, given in API Standard 521. Pressure-relieving and Depressuring Systems. Sixth Edition | January 2014
Latrolet	*aka* LOL. A proprietary eponym, an abbreviated form of butt welded 45 degree LATeral pipe branch OutLET connection. One of various olets (qv)
law	1. Generally, a system of rules to regulate behavior, or one of those rules 2. In a scientific context, a universal principle which describes the fundamental nature of something 3. Rather commonly in engineering, an approximate heuristic
law of chemical equilibrium	see **equilibrium constant**
law of partial or additive volumes	see **Amagat's law**

laws of conservation	*aka* conservation laws. A set of physical laws stating that a given parameter, such as mass or energy, does not change, i.e.; it is conserved
laws of motion	Three laws which describe the relationship between an object's motion and forces acting upon it
layer	A self-contained set of data, especially in the context of an electronic drawing file; defined formally in BS EN ISO 10209:2012 Technical product documentation — Vocabulary — Terms relating to technical drawings, product definition and related documentation
layered bed	*aka* stratified bed. A packed bed in which particles have been selected by size and density, such that they reform into discrete layers after backwashing
layers of protection analysis	*aka* LOPA. A structured risk assessment technique which considers whether the results of a risk matrix are acceptable, and if not, how to mitigate the risk; most commonly used to determine the required safety integrity level (qv)
layflat hose	*aka* bagging (qv). A low-pressure thin-walled hose used to deliver water on site. Called 'layflat' as it does not have sufficient mechanical strength to maintain a round cross section without internal pressure
layoff	see **negative recruitment period**
layout	The arrangement in space of plant and equipment
layout drawing	*aka* location drawing. A drawing showing plant and equipment layout; defined formally in BS EN ISO 10209:2012 Technical product documentation — Vocabulary — Terms relating to technical drawings, product definition and related documentation
LC	*(drawing notation)* Locked closed
LC 50	The concentration of an air- or water- borne toxin which kills 50% of exposed population with a single exposure; cf LD50
LCA	see **life cycle analysis**, life cycle assessment
LCGO	see **light coker gas oil**
LCI	Life cycle inventory, see **life cycle inventory analysis**
LCI result	see **life cycle inventory analysis result**
LCIA	see **life cycle impact assessment**
LCO	see **light cycle oil**
LD 50	The dose of radiation or ingested toxin which kills 50% of exposed population with a single acute exposure; cf LC 50
LDAR	Leak detection and repair
LDPE	Low density polyethylene, see **polyethylene**
LDs	see **liquidated damages**

LDT	Line designation table, see **line schedule**
Le Chatelier's principle	The principle that any changes in the temperature, pressure, volume, or concentration of a system in a state of chemical equilibrium result in a new state of equilibrium opposing those changes
leachables (polymeric)	Substances which can leach out of polymeric materials of construction, contaminating process fluids, defined formally in ASME BPE (American Society of Mechanical Engineers: Bioprocessing Equipment)
leachate	Generally, liquid which has passed through media and has picked up another component; often used in the context of landfill (qv) sites where rainwater enters into a landfill and acquires contaminants, e.g.; metals, organics, oxygen depleting compounds, color, solids/particulates, etc.
leaching	*aka* lixivation. Extracting soluble components from a solid with a solvent
leader line	A line joining text and the item it refers to on a drawing; defined formally in BS EN ISO 10209:2012 Technical product documentation — Vocabulary — Terms relating to technical drawings, product definition and related documentation
leadership in energy and environmental design	*aka* LEED. A worldwide environmental standard for buildings
leaf filter	A type of batch pressure filter in which stacked coarse filters, contained in a pressure vessel, are coated with a cake of fine solids which act as filter media
leak check	Checking that an LPG piping system is leak free; defined formally in NFPA 58: Liquefied Petroleum Gas Code, 2017 Edition
leak test pressure	Static pressure at inlet specified for a PRV seat leak test; defined formally in API RP 520 P1 7th Edition, January 2000 Sizing, Selection, and Installation of Pressure-Relieving Devices in Refineries; Part I - Sizing and Selection and API RP 576 - Inspection of Pressure Relieving Devices
leakage	*aka* loss of containment. Generally used to mean an undesirable escape of something; this might be an escape past a seal of process fluid, escape of ions from incompletely regenerated ion exchange resin, or a fugitive emission (qv) from a process
leakage concentration	Concentration of a fugitive emission (qv) to environment; defined formally in the context of pump seals in API RP 682 - Pump Seals

leakage rate	The quantity of fugitive emission (qv) per unit time; defined formally in the context of pump seals in API RP 682 - Pump Seals
lean	Generically, a term applied to a process stream with a lower concentration of a component; cf rich
lean manufacturing	A manufacturing waste minimization strategy derived from practices developed at Toyota
lean phase conveying	A method of pneumatic solids conveying with a consistent low concentration of conveyed solids; cf dense phase conveying
lease automatic custody transfer	*aka* LACT, automatic custody transfer, ACT. A device for accurate measurement, sampling and transfer of petroleum products, described in API MPMS 6.1 Manual of Petroleum Measurement Standards Chapter 6.1 Lease Automatic Custody Transfer (LACT) Systems
Lee's disk	A method of measuring thermal conductivity
LEED	see **leadership in energy and environmental design**
Lee-Kesler equation of state	An equation which allows saturated vapor pressure of a mixture of gases at a given temperature to be estimated, if critical pressure and temperature are known. Commonly used in the context of condensate well heads
Legionnaire's disease	*aka* Legionellosis. A human illness caused by an organism which thrives in warm water such as that found in some cooling systems
Leibig condenser	see **Leibig spool**
Leibig spool	*aka* Leibig condenser, Leibig tube. A double pipe heat exchanger
Leibig tube	see **Leibig spool**
Leidenfrost effect	When a liquid droplet separated from a surface far hotter than liquid boiling point by an insulating layer of vapor does not boil rapidly
LEL	see **lower explosive limit**
Lessing ring	A type of random packing (qv)
let down	see **letdown**
letdown	*aka* let down, let-down. The thinning of a viscous fluid, especially paint, or the mixture of solvents and other materials added for this purpose; cf makedown
letdown valve	A pressure reduction valve
lethal dose	The amount of toxin or radiation required to kill a given fraction of a population of organisms, or an individual organism, as in LD 50 (qv)

lettering	Either drawing characters or graphic symbols, or the process of adding these to a drawing; defined formally in BS EN ISO 10209:2012 Technical product documentation – Vocabulary – Terms relating to technical drawings, product definition and related documentation
level	1. Most commonly, the height of fluid in a process vessel 2. In the context of vibration monitoring, the logarithm of the ratio of a quantity to a similar reference, defined formally in BS ISO 2041:2018 Mechanical vibration, shock and condition monitoring. Vocabulary
level alarm	*aka* LA. An alarm state warning operators of an undesirably high or low level in a process vessel
level contour line	*aka* contour line, contour, isohypse. A line of constant elevation above datum (qv) level in a topographical projection; defined formally in BS EN ISO 10209:2012 Technical product documentation – Vocabulary – Terms relating to technical drawings, product definition and related documentation
level control	Automatic control of the level of a process fluid
level gauge	A visually readable field level indicator
levelling drawing	A drawing recording levelled points, formally defined in BS EN ISO 10209:2012 Technical product documentation – Vocabulary – Terms relating to technical drawings, product definition and related documentation
lever rule	A way to use a binary equilibrium phase diagram to determine the mole or mass fraction of each phase of mixture
Lewis fugacity rule	see **Lewis-Randall rule**
Lewis number	A dimensionless group, the ratio of thermal diffusivity to mass diffusivity; useful in the context of simultaneous heat and mass transfer
Lewis–Matheson	A relatively straightforward trial-and-error method for calculating the required number of stages in multicomponent distillation by hand, an application of the more complex Lewis-Sorel method
Lewis–Randall rule	*aka* Lewis fugacity rule. A rule which states that the fugacity (qv) of a component in an ideal solution or mixture of gases is directly proportional to its mole fraction
Lewis-Sorel method	*aka* Sorel-Lewis method. A method to calculate the number of theoretical plates in a distillation column
Lewis-Whitman two film theory	see **Whitman two film theory**
LFL	see **lower flammable limit**
LGO	Light gas oil

LHV	see **lower heating value**
LID	see **lowest ineffective dilution**
LIDAR	Light detection and ranging
life cycle	All stages of a product's life as they affect the environment, from raw material extraction to eventual disposal; defined formally in EN ISO 14040:2006 Environmental management - Life cycle assessment - Principles and framework
life cycle analysis	A systematic approach to quantifying environmental impact of a product. A less favored term for life cycle assessment (qv)
life cycle assessment	*aka* LCA. A systematic approach to quantifying environmental impact of a product, the stages of which have unhelpfully similar-sounding names and initialisms (qv life cycle impact assessment, life cycle interpretation, life cycle inventory). Defined formally in EN ISO 14040:2006 Environmental management - Life cycle assessment - Principles and framework
life cycle impact assessment	Part of life cycle assessment (qv) considering the impacts on human health and environment of the LCA; defined formally in EN ISO 14040:2006 Environmental management - Life cycle assessment - Principles and framework
life cycle interpretation	*aka* LCIA. Part of life cycle assessment (qv) following and interpreting the result of life cycle inventory analysis (qv); defined formally in EN ISO 14040:2006 Environmental management - Life cycle assessment - Principles and framework
life cycle inventory	see **life cycle inventory analysis**
life cycle inventory analysis	*aka* life cycle inventory (LCI). Essentially a mass and energy balance over the lifetime of a product; defined formally in EN ISO 14040:2006 Environmental management - Life cycle assessment - Principles and framework
life cycle inventory analysis result	The product of a life cycle inventory analysis (qv); defined formally in EN ISO 14040:2006 Environmental management - Life cycle assessment - Principles and framework
life of asset	*aka* LOA. 1. The design life of a piece of equipment or plant 2. The remaining financially viable ore body in a mine
LIFO	see **last in first out**

lift	1. Most commonly (ignoring everyday meanings), the vertical height through which water is raised by a pump, (especially when applied to the suction side of the pump, in the term suction lift); as defined in NFPA 1142: Standard on Water Supplies for Suburban and Rural Fire Fighting, 2017 Edition 2. The distance traveled from closed to relieving positions by a NRV's disc; as defined in API RP 520 P1 7th Edition, January 2000 Sizing, Selection, and Installation of Pressure-Relieving Devices in Refineries; Part I - Sizing and Selection, API Standard 521. Pressure-relieving and Depressuring Systems. Sixth Edition	January 2014 and API RP 576 - Inspection of Pressure Relieving Devices
lift station	An underground structure which lifts effluent to a higher elevation; usually a sump fitted with submersible pumps
light and power board	An electrical cabinet (usually separate from the switchboard which contains the equipment for the control of process equipment) containing switchgear (qv) for lighting and power sockets
light coker gas oil	*aka* LCGO. Gas oil (qv) from the distillation section of a delayed coker unit
light cycle oil	*aka* LCO. The gas oil (qv) product of a fluid catalytic cracking unit
light ends	A term used in oil refining for low boiling point hydrocarbon fractions
light gases	Gases with a relative density (qv) less than 1.3, such as natural gas; defined formally in BS EN 12952-8:2002 Water-tube boilers and auxiliary installations. Requirements for firing systems for liquid and gaseous fuels for the boiler
light hydrocarbon	Any liquid hydrocarbon which boils at ambient conditions; defined formally in API RP 682 - Pump Seals
light key	see **key component**
light non-aqueous phase liquid	*aka* LNAPL. A lighter than water nonaqueous phase liquid; an environmental engineering term for a liquid immiscible with and insoluble in water which floats on water
light off	Initial fuel ignition; defined formally in API RP 535 - Burners for Fired Heaters at Refineries
light phase	A lower density fluid, especially in the context of liquid–liquid extraction
light stiffener	A thickened section stiffening a pressure vessel to resist external pressure; defined formally in PD 5500:2018+A1:2018 Specification for unfired fusion welded pressure vessels

light vacuum gas oil	*aka* vacuum gas oil. A gas oil (qv) that is produced from a vacuum distillation unit (qv) fed by a crude distillation unit (qv)
light virgin gas oil	*aka* LVGO. A gas oil (qv) that is produced from the crude oil distillation unit
lighter than air gases	Gases with a relative density (qv) less than 1; defined formally in API RP 505 2nd Edition, August 2018 Recommended Practice for Classification of Locations for Electrical Installations at Petroleum Facilities Classified as Class I, Zone 0, Zone 1, and Zone 2; cf light gases
lighting drawing	A drawing which specifies lighting requirements; defined formally in BS EN ISO 10209:2012 Technical product documentation — Vocabulary — Terms relating to technical drawings, product definition and related documentation
lightning	Very large scale static electricity. API RP 2001 - Fire Protection at Refineries offers a formal definition and a short discussion on the difficulties of protecting against direct strikes
lignin	The substance which makes woody material strong and resistant to degradation. Its partial degradation yields humic acid (qv) and fulvic acid (qv)
LILO	see **last in last out**
lime	Leaving aside the everyday meanings of a citrus fruit and its associated color, this is a vague term, used to describe a range of calcium oxides and hydroxides of various degrees of purity; most commonly used for a number of products derived from limestone, differentiated from each other with terms such as quicklime (qv), slaked lime (qv)
lime softening	*aka* lime-soda softening, Clark's process. Water softening by precipitation using lime (qv) (calcium hydroxide in this case) and soda ash (sodium carbonate)
lime-soda softening	see **lime softening**
limit analysis	A method to estimate load at failure; defined formally in API RP 579 - Fitness for Service
limit analysis collapse load	The maximum load, as estimated by limit analysis (qv); defined formally in API RP 579 - Fitness for Service
limit of detection	*aka* LOD, lower limit of detection. The lowest quantity of a substance that can be distinguished from an absence of that substance at a given confidence level by an analytical technique
limit of quantification	The lowest quantity or concentration of a substance which a given test method can reliably quantify

limit switch	1. Generally, an electrical switch actuated by the position of a device component. 2. In the context of fired heaters, defined specifically in NFPA 86: Standard for Ovens and Furnaces, 2019 Edition
limited movement control device	A safety device which limits movement of a machine when actuated; defined formally in BS EN ISO 12100:2010 Safety of machinery. General principles for design. Risk assessment and risk reduction
limiter(s)	*aka* limiting device(s). 1. Generally, safety devices which prevent operation of a machine above a set value of a measured parameter 2. In the context of fired heaters, defined specifically in BS EN 12952-8:2002 Water-tube boilers and auxiliary installations. Requirements for firing systems for liquid and gaseous fuels for the boiler
limiting device(s)	Generally, a safety device which prevents operation of a machine above a set value of a measured parameter; defined formally in BS EN ISO 12100:2010 Safety of machinery. General principles for design. Risk assessment and risk reduction
limiting factor	A parameter limiting a rate of reaction, population growth etc., or in management accounting, limiting production expansion (e.g., size of factory)
limiting oxidant concentration	*aka* LOC. The oxidant concentration above which deflagration can occur; defined formally in NFPA 86: Standard for Ovens and Furnaces, 2019 Edition
limiting oxygen index	*aka* LOI. The lowest percentage oxygen in an oxygen/nitrogen mixture which will sustain combustion of a polymer in safety testing. Fire retardant (qv) polymers have an LOI greater than atmospheric oxygen concentration. Defined formally in ASTM D 2863, Standard Test Method for Measuring the Minimum Oxygen Concentration to Support Candle-Like Combustion of Plastics (Oxygen Index)
limiting reactant	*aka* limiting reagent. A reactant which is completely consumed in the course of a reaction, which therefore determines the extent of reaction; cf excess reactants, excess air
limiting reagent	see **limiting reactant**
limiting substrate	A nutrient used in biological culture which determines the extent of growth, analogous to a limiting reagent (qv); often expressed as limiting substrate concentration
Linde process	An obsolete process for the production of oxygen from air

line	1. An alternate term for a pipe 2. A one-dimensional geometrical object 3. A graphic convention to show a connection between devices on a flowsheet. Meanings 2 and 3 are defined formally in BS EN ISO 10209:2012 Technical product documentation – Vocabulary – Terms relating to technical drawings, product definition and related documentation and BS EN 61082-1:1993, IEC 61082-1:1991 respectively
line burner	A burner with a linear flame; defined formally in NFPA 86: Standard for Ovens and Furnaces, 2019 Edition
line condom	see **eels**
line designation table	*aka* LDT; see **line schedule**
line diagram	see **single line diagram**
line list	see **line schedule**
line pressure regulator	A gas line pressure regulator used to reduce inlet pressure in LPG vapor service without integral overpressure protection device. Different formal definitions in NFPA 58: Liquefied Petroleum Gas Code, 2017 Edition and NFPA 86: Standard for Ovens and Furnaces, 2019 Edition
line schedule	*aka* line list, line designation table (LDT). A tabulated list of all pipes on a plant giving size, specification, temperature and pressure conditions; the content is variable and may include information on type of fluid, operational temperature and pressure and test conditions
line segment	In technical drawing (rather than geometry), the dots, dashes, etc. of a non-continuous line such as a centerline; defined formally in BS EN ISO 10209:2012 Technical product documentation – Vocabulary – Terms relating to technical drawings, product definition and related documentation
linear damper	*aka* damper. This specific term used for: 1. A duct isolating plate with purely linear travel such as a guillotine (qv) 2. A shock absorber providing linear damping, commonly unidirectional (i.e., one direction of motion is undamped); cf rotary damper (both senses)
linear damping	*aka* viscous damping. Damping by an opposing force proportional to the velocity; defined formally in BS ISO 2041:2018 Mechanical vibration, shock and condition monitoring. Vocabulary

linear damping coefficient	*aka* viscous damping coefficient. Damping force to velocity ratio; defined formally in BS ISO 2041:2018 Mechanical vibration, shock and condition monitoring. Vocabulary
linear dimension	The distance between two points in a straight line; defined formally in BS EN ISO 10209:2012 Technical product documentation – Vocabulary – Terms relating to technical drawings, product definition and related documentation
linear motion sensor	see **rectilinear transducer**
linear optimization	see **linear programming**
linear porosity	A linear arrangement of weld porosity; defined formally in ASME BPE (American Society of Mechanical Engineers: Bioprocessing Equipment)
linear programming	*aka* linear optimization. The special case of mathematical programming (qv) which applies to linear mathematical models; an optimization methodology
linear system	A system where response magnitude is proportional to excitation magnitude; defined formally in BS ISO 2041:2018 Mechanical vibration, shock and condition monitoring. Vocabulary
linear transducer	A transducer whose output and input quantities are linearly related; defined formally in BS ISO 2041:2018 Mechanical vibration, shock and condition monitoring. Vocabulary
linear vibration	see **rectilinear vibration**
Lineweaver–Burk plot	An obsolete graphical representation of enzyme kinetics; cf Eadie-Hofstee plot
lining	A protective layer applied to an inner surface, especially of pipes and vessels
Linton-Sherwood correlation	The basis of an equation used in mass transfer calculations
lipid	Any water insoluble biological molecule, not just triglycerides
liquation	An obsolete ore processing technique
liquefaction	1. Making into a liquid, such as liquefied natural gas (qv) by means of cooling and/or compression 2. Solids fluidization
liquefied gas	A gas in a liquefied form under ambient conditions; defined formally in NFPA 30: Flammable and Combustible Liquids Code, 2018 Edition
liquefied natural gas	*aka* LNG. Natural gas which has been liquefied, most commonly to facilitate transport
liquefied petroleum gas	*aka* LPG. Any liquefied mixture of propane, propylene, butane, isobutane and butylenes. More formal definition in NFPA 58: Liquefied Petroleum Gas Code, 2017 Edition

liquefied refinery gas	Various liquefied hydrocarbon gases produced in refineries during the processing of oil. Components may include (inter alia) ethane, propane, butane, isobutane, and ethylene, propylene, butylene, and isobutylene
liquid	As well as the everyday meaning, there is a range of mutually exclusive technical meanings in different contexts. NFPA 20: Standard for the Installation of Stationary Pumps for Fire Protection, 2019 Edition defines the term solely as liquids for fire protection purposes. NFPA 30: Flammable and Combustible Liquids Code, 2018 Edition on the other hand defines liquids as non-solids by means of standard tests
liquid distributor	A system to spread a liquid over the upper surface of a packed bed (qv)
liquid filtration	see **Box L1**.
liquid fuels	Fuels which are liquid as used, including in some definitions those which must be preheated for use; defined formally in BS EN 12952-8:2002 Water-tube boilers and auxiliary installations. Requirements for firing systems for liquid and gaseous fuels for the boiler
liquid paraffin	see **mineral oil**
liquid penetrant indication	*aka* LPI, dye penetrant inspection, liquid penetrant testing, penetrant testing, PT. An inspection method for surface-breaking defects in non-porous materials; defined formally in ASME Boiler & Pressure Vessel Code
liquid penetrant testing	see **liquid penetrant indication**
liquid pocket	An undrained low point in pipework which collects draining liquid cf deadleg, dead space
liquid seal	*aka* water seal. A water trap (equivalent to a toilet U-bend) on a flare stack, which prevents air getting into the flare stack or feed pipework. Defined more formally in API Standard 521. Pressure-relieving and Depressuring Systems. Sixth Edition	January 2014 and API RP 537 Flare Details for Petroleum, Petrochemical, and Natural Gas Industries
liquid storage room	A room for storing flammable liquids, defined formally in NFPA 30: Flammable and Combustible Liquids Code, 2018 Edition
liquid warehouse	A warehouse for storing flammable liquids, defined formally in NFPA 30: Flammable and Combustible Liquids Code, 2018 Edition
liquid/liquid extraction	*aka* LLE; see **solvent extraction**

Box L1. Liquid Filtration Spectrum

There are several membrane technologies used in water treatment, named according to the size of particles which they remove. Relatively coarse microfiltration and ultrafiltration membranes remove fine solid particles, whilst the finest reverse osmosis membranes allow water and dissolved ions to be separated.

As **Figure L1** shows, the names of the various separation processes have a degree of overlap, so the illustration is only one of many possible (though I think **Figure L1** reflects most closely the international consensus). American engineers might make the microfiltration bar overlap well into ultrafiltration on the left and particle filtration on the right, because the term microfiltration is used far more loosely in the USA (and scarcely at all in the UK). Some add "tight ultrafiltration" between ultrafiltration and nanofiltration. Others have ultrafiltration covering the whole range I have shown for microfiltration.

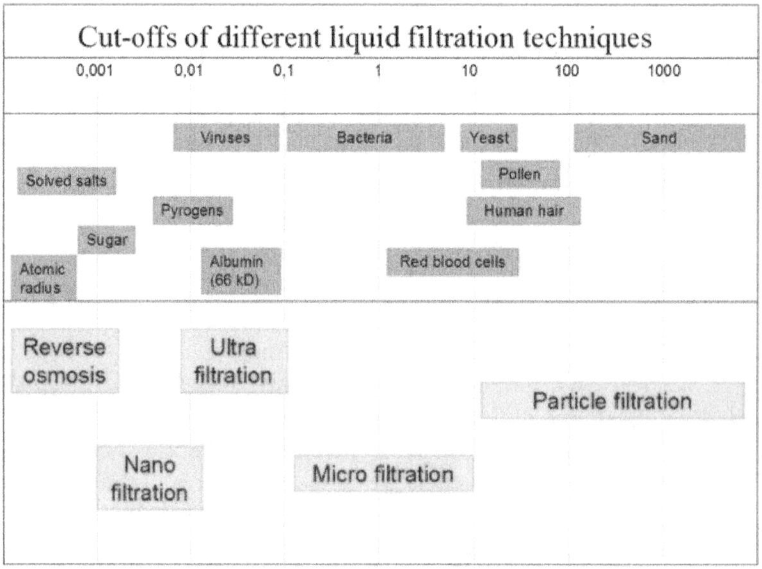

Figure L1 Liquid Filtration Techniques

To differentiate the names by duties, ultrafiltration (UF) or microfiltration (MF) give an absolute barrier against the passage of particles smaller than a given size, for color removal, removal of pathogens (especially Cryptosporidium) and physical disinfection prior to discharge, amongst other things. Nanofiltration (NF) and reverse osmosis (RO) (*aka* hyperfiltration) on the other hand allow us to selectively remove dissolved ions from a solution, used to produce high quality water from seawater or industrial effluent.

Related Terms
membrane filtration, depth filtration, absolute filter, nominal filter

liquidated damages	*aka* LDs. Pre agreed penalties for failing to meet the program or performance stipulations of a contract
liquid-liquid separator	*aka* decanter, solvent-water separator. Usually refers to the unit operation of gravity separation of immiscible liquids with different densities
liquidtight	Generically, impervious to the passage of liquids; cf gastight, bubbletight, watertight. There are various closely related and, in some cases, quantified formal definitions, such as that in NFPA 30: Flammable and Combustible Liquids Code, 2018 Edition
liquidus	The temperature boundary on a phase diagram between the liquid phase and another. Practically, the temperature at which all material should be in a liquid form
liquor	Generally, an aqueous extraction or washing liquid used in a process. (Or sometimes, tea, as defined in ISO 3103 Tea -- Preparation of liquor for use in sensory tests)
listed	Present on an official list, such as one of acceptable or suitable equipment, materials, or services. A long formal definition can be found at NFPA 11: Standard for Low-, Medium-, and High-Expansion Foam, 2016 Edition
liter	(*symbol* l) *aka* litre, cubic decimeter. A non-SI metric unit of volume
litre	see **liter**
live commissioning	see **hot commissioning**
live load	1. In structural engineering, the loading on platforms and floors as a result of operation and maintenance activity, ignoring weight of plant, piping and materials; a minimum figure of 250 kgf m^2 is recommended in this context 2. More generally used in a way synonymous with variable load (qv)
live steam	Steam supplied direct from the boiler at full pressure, especially when added direct to a process
lixiviation	see **leaching**
LJ(F)	*(drawing notation)* Lap joint (flange)
LLC	Liquid level controller
LLE	see **liquid-liquid extraction**
LLW	see **low level waste**
LLWR	see **low level waste repository**
LLx	Low low instrument alarm level condition, where x is the parameter; which normally signifies the shut-down threshold, e.g., LLL would be a low low level alarm in a vessel; cf HHx
LMTD	see **logarithmic mean temperature difference**
LNAPL	see **light non-aqueous phase liquid**

LNG	see **liquefied natural gas**
LNOL	see **lower normal operating limit**
LO	1. *(drawing notation)* Locked open
	2. Lube oil
LOA	see **life of asset**
lobe pump	see **rotary lobe pump**
LOC	1. see **limiting oxidant concentration**
	2. see **loss of containment**
local	In the context of instrument position, field mounted, i.e.; near the point of measurement; defined formally in EN ISO 10437 Petroleum, petrochemical and natural gas industries - Steam turbines - Special-purpose applications; cf remote
Local Authority	*aka* LA. A UK local government body
local primary membrane stress	A category of stress in a pressure vessel; defined formally in API RP 579 - Fitness for Service
local section	A partial sectional representation; defined formally in BS EN ISO 10209:2012 Technical product documentation — Vocabulary — Terms relating to technical drawings, product definition and related documentation
local structural discontinuity	A structural flaw which does not have a significant effect on the structure as a whole; defined formally in API RP 579 - Fitness for Service
local thin area	*aka* LTO, locally thin area. In the context of welding, a discrete area of surface metal loss approximately as long as it is wide; defined formally in API RP 579 - Fitness for Service, which also uses the term 'locally thin area' as if synonymous, but does not define it
localized corrosion	Corrosion where most loss of metal occurs in a discrete area; defined formally in NFPA 59: Utility LP-Gas Plant Code, 2018 Edition
locally mounted instrument	*aka* field mounted instrument. An instrument mounted in the field, local to its point of measurement, rather than in a control panel; defined formally in BS1646-1 Symbolic Representation for Process Measurement Control Functions and Instrumentation Part 1: Basic Requirements and BS 1646-3:1984 Symbolic representation for process measurement control functions and instrumentation. Specification for detailed symbols for instrument interconnection diagrams; cf panel mounted instrument
locally thin area	see **local thin area**
location drawing	see **layout drawing**

lockout	A procedure for locked physical isolation requiring both manual intervention and a key or keys to reverse, especially when applied to energy supply; almost always lockout tagout (qv). Defined formally in BS EN 12952-8:2002 Water-tube boilers and auxiliary installations. Requirements for firing systems for liquid and gaseous fuels for the boiler.
lockout tagout	*aka* LOTO. A procedure for locked physical equipment isolation requiring both manual intervention and a key or keys to reverse, especially when applied to energy supply. The tagout part of the term refers to a visible tag applied to the locked out item of equipment, usually incorporating multiple locking points to facilitate multiple required interventions by various disciplines/ authorities
LOD	see **limit of detection**
log phase	The period of maximum growth in a bioreactor after organisms have adapted to conditions; cf lag phase
logarithmic mean temperature difference	*aka* LMTD. A measure of the driving force for heat transfer
LOI	see **limiting oxygen index**
Lokring	Proprietary eponym for a mechanical piping joint process
LOL	*(drawing notation)* see **Latrolet**
London Brent	see **Brent crude**
long imperfections	Longer weld imperfections, defined in various ways by various standards, usually relative to overall weld length. One formal definition may be found in PD 5500:2018+A1:2018 Specification for unfired fusion welded pressure vessels
long radius	A term applied to bends, elbows, and ells whose radius of curvature is 1.5 times pipe outside diameter; cf short radius
long residue	see **atmospheric residue**
long section	A drawing showing a profile of a ground surface along a route, commonly the route of a pipeline. The elevation of both surface and pipeline are usually shown, facilitating hydraulic calculations
long term exposure limit	*aka* LTEL. According to UK health and safety regulation, the acceptable 8-hour time weighted average concentrations of certain hazardous substances in workplace air; cf maximum exposure limit, short term exposure limit
long ton	*aka* British ton, imperial ton, displacement ton, weight ton, W/T. The obsolete imperial ton: 2240 lb.; cf short ton
longitudinal	In the direction of travel, the 'long axis'
longitudinal weld	A weld on the longitudinal axis of a pipe or pressure vessel; defined formally in API RP 579 - Fitness for Service

loop reactor	One of various types of reactor which allows continuous circulation of outlet flow to inlet
LOPA	see **layers of protection analysis**
LOQ	see **limit of quantification**
loss	Dogen said that "Gain is delusion; loss is enlightenment", but the enlightened chemical engineer's view is that loss is delusion: head is not lost, it is transformed into heat (qv headloss); materials are not lost, they escape to environment (qv loss of containment); money is not lost, someone else gains it. Furthermore, in engineering, gain (qv) is not usually the opposite of loss
loss factor	A measurement of 'losses' from tankage in the oil and gas industry; defined formally in API MPMS 19.1 Manual of Petroleum Measurement Standards Chapter 19.1 Evaporative Loss from Fixed-roof Tanks
loss of containment	*aka* LOC. The most common type of process safety (qv) incident, a euphemism for 'leakage' of hazardous materials
loss of phase	The loss of one phase of a three phase electricity supply; defined formally in NFPA 20: Standard for the Installation of Stationary Pumps for Fire Protection, 2019 Edition
loss prevention	see **process safety**
lost time accident	*aka* LTA; see **lost time incident**
lost time case	see **lost time incident**
lost time incident	*aka* LTI, lost time accident (LTA), lost time injury, lost time case. An incident which results in a worker being unable to complete their normal duties on their next scheduled shift; cf days away, restricted, or transferred
lost time injury	An Australian term for lost time incident (qv)
lot of rupture disks	A production run of rupture disks of a common specification; defined formally in API RP 520 P1 7th Edition, January 2000 Sizing, Selection, and Installation of Pressure-Relieving Devices in Refineries; Part I - Sizing and Selection
LOTO	see **lockout tagout**
louvre damper	A damper (qv) comprising multiple linked pivoting blades; defined formally in BS EN ISO 13705:2012/ISO 13705:2012(E) Petroleum, petrochemical and natural gas industries. Fired heaters for general refinery service, API STD 530 - Calculation of Tube Heater Thickness and API STD 560 - Fired Heaters for General Refinery Service
Lovibond scale	Proprietary eponym for a color (qv) measurement scale used in the brewing industry

low alloy steel	Steels with a low percentage of alloying additions; defined formally in API RP 571 - Damage Mechanism Affecting Fixed Refinery Equipment
low and vacuum pressure safety valve	*aka* LVPSV; see **vent/vac valve**
low care areas	see **zoning**
low emission transfer	A method of LPG transfer which limits loss to environment; defined formally in NFPA 58: Liquefied Petroleum Gas Code, 2017 Edition
low fuel pressure switch	A switch which shuts down a fired heater burner system safely if fuel pressure is low; defined formally in NFPA 86: Standard for Ovens and Furnaces, 2019 Edition
low grade heat	see **waste heat**
low level waste	*aka* LLW. Nuclear waste which does not fall into a formally defined category, though this does not necessarily imply that it has a low level of radioactivity; cf intermediate level waste, high level waste
low level waste repository	*aka* LLWR. The UK's only low level waste repository, where all its low level waste (qv) is stored, is near Drigg in Cumbria
low lift safety valve	A safety valve with a discharge area determined by disk position
low NOx burner	A fired heater burner designed to reduce NOx formation; defined formally in API RP 535 - Burners for Fired Heaters at Refineries
low oxygen oven	A solvent evaporation oven with a low-oxygen atmosphere; defined formally in NFPA 86: Standard for Ovens and Furnaces, 2019 Edition
low pressure distillation	see **vacuum distillation**
low pressure fuel gas system	A system which uses an eductor (qv) driven by fuel gas at a gauge pressure of less than 1 psi (around 7 kPa) to entrain combustion air; defined formally in NFPA 86: Standard for Ovens and Furnaces, 2019 Edition; cf high pressure fuel gas system
low pressure safety valve	*aka* LPSV. A PSV (qv) operating at system pressures close to ambient
low pressure separator	*aka* LPS. Either a hot low pressure separator (qv) or a cold low pressure separator (qv)
low pressure spool	In the context of gas turbines, a low pressure rotor assembly; defined formally in ISO 3977 Gas turbines - Procurement - Part 3: Design requirements

low pressure tank	A storage tank at a low internal pressure, between 1.0 psig and 15 psig measured at top of tank, according to the formal definition in NFPA 30: Flammable and Combustible Liquids Code, 2018 Edition cf atmospheric tank
low temperature distillation	see **cryogenic distillation**
low temperature rack	In the context of a hybrid refrigeration (qv) system, the compressor set which handles the refrigerant operating at the lower temperature; i.e.; the refrigerant which removes heat from the process and transfers it to the medium temperature refrigerant
low vacuum	In the context of fired heaters, system pressure between 760 and 1×10^{-3} mmHg, (around 1×10^{-4} kPa) according to the formal definition in NFPA 86: Standard for Ovens and Furnaces, 2019 Edition; cf high vacuum
lower explosion level	Identical in value to lower flammable limit (qv) or lower explosive limit (qv), this term is used and defined in ISO 3977 Gas turbines - Procurement - Part 3: Design requirements
lower explosive limit	*aka* LEL; see **lower flammable limit**
lower flammability limit	see **lower flammable limit**
lower flammable limit	*aka* lower flammability limit. The concentration of vapor (sometimes including dust) in air (sometimes broadened to oxidizing atmosphere) below which a mixture cannot be ignited. Identical in value to the lower explosive limit (qv). Various closely related formal definitions can be found in ASTM E681 - 09(2015) Standard Test Method For Concentration Limits Of Flammability Of Chemicals (Vapors And Gases), NFPA 30: Flammable and Combustible Liquids Code, 2018 Edition, API RP 2510A - Fire Protection Considerations for the Design and Operation of Liquefied Petroleum Gas (LPG) Storage Facilities and NFPA 654: Standard for the Prevention of Fire and Dust Explosions from the Manufacturing, Processing, and Handling of Combustible Particulate Solids, 2017 Edition

lower heating value	*aka* net heating value. The higher heating value (qv) minus the latent heat of vaporization of water from combustion of hydrogen contained in a fuel. Defined formally and almost identically in BS EN ISO 13705:2012/ISO 13705:2012(E) Petroleum, petrochemical and natural gas industries. Fired heaters for general refinery service, API STD 530 - Calculation of Tube Heater Thickness, API RP 535 - Burners for Fired Heaters at Refineries, API RP 537 Flare Details for Petroleum, Petrochemical, and Natural Gas Industries and API STD 560 - Fired Heaters for General Refinery Service
lower limit of detection	see **limit of detection**
lower master valve	A manually operated gate valve that provides a backup function for the upper master valve (qv) used to isolate and control the flow of fluids from the wellbore
lower normal operating limit	*aka* LNOL. Often used as a set point for alarms to avoid reaching the lower safe operating limit (qv); frequently applied to pressure, temperature and flowrate; cf upper normal operating limit, Lx alarm
lower safe design limit	*aka* LSDL, safe design limit. A term often applied to lower design limits of pressure, temperature and flowrate; cf upper safe design limit
lower safe operating limit	*aka* LSOL. The lower acceptable operating limit, often used as a set point for trips to avoid reaching lower safe design limit. Often applied to pressure, temperature and flowrate; cf upper safe operating limit
lowest achievable emission rate	*aka* LAER. The basis of determination of emission acceptability by the US Environmental Protection Agency (qv)
lowest ineffective dilution	A water quality parameter; defined formally in ISO 15088:2007(en) Water quality — Determination of the acute toxicity of waste water to zebrafish eggs (Danio rerio)
lowest permissible suction pressure	A term used in NFPA 20: Standard for the Installation of Stationary Pumps for Fire Protection, 2019 Edition to refer to the lowest fire pump suction pressure which that standard permits
LP	1. Low pressure 2. Line pipe 3. see **liquefied petroleum gas**
LP gas	see **liquefied petroleum gas**
LP gas system	A system which stores and dispenses LPG in a controlled manner; defined formally in NFPA 58: Liquefied Petroleum Gas Code, 2017 Edition

L-path	A drag conveyor which has a horizontal entry run leading into a vertical run to the discharge, making it L-shaped
LPG	see **liquefied petroleum gas**
LPI	see **liquid penetrant indication**
LPS	see **low pressure separator**
LPSV	see **low pressure safety valve**
LR	1. see **Larson ratio**
	2. *(drawing notation)* Long radius
LRG	see **liquefied refinery gas**
LSDL	see **lower safe design limit**
LSFO	Low sulfur fuel oil
LSI	see **Langelier saturation index**
LSOL	see **lower safe operating limit**
LSTK	Lump sum turnkey; a type of contract
LTA	1. see **local thin area**
	2. Lost time accident; see **lost time incident**
LTEL	see **long term exposure limit**
LTI	see **lost time incident**
LTO	see **local thin area**
lube oil	Lubricating oil
lube oil consoles	Compressor ancillaries which provide a supply of clean cool oil to bearings and driver
Ludwig–Soret effect	see **Soret effect**
luff	To move up and down; cf slew, traverse
Luikov number	see **Lykov number**
lumen	*(symbol* Lm*)* The SI derived unit of luminous flux
lumped mass	see **deadweight**
Lurgi process	The most commonly used coal gasification (qv) process
luster	*aka* lustre. Shining by reflected light; defined formally in ASME BPE (American Society of Mechanical Engineers: Bioprocessing Equipment)
lustre	see **luster**
lute	A 'U' shape in process pipework filled with liquid which prevents gas flow; similar in principle to the 'trap' on a domestic sink
Lutz pump	Proprietary eponym for a barrel pump
lux	*(symbol* Lx*)* The SI derived unit of illumination
LVGO	1. see **light virgin gas oil**
	2. see **light vacuum gas oil**
LVOC	Large volume organic chemicals
LVPSV	see **low and vacuum pressure safety valve**
LWNF	*(drawing notation)* Long weld neck flange

Lx alarm	Low instrument alarm level condition, where x is the parameter; this is normally the alarm threshold, e.g.; LLA would be a low level alarm in a vessel; cf Hx
lye	see **caustic soda**
Lykov number	(*symbol* Lu) *aka* Luikov number. A dimensionless group, the ratio of mass diffusivity to thermal diffusivity; useful for thermal drying calculations
lyophilization	see **freeze drying**
lyophobic	In the context of colloids, having no affinity for the dispersion liquid
lysis	The breaking open of microorganisms

M

M&R station	see **metering and regulating station**
mAb	see **monoclonal antibody**
MAB	see **monoclonal antibody**
MABP	see **molal average boiling point**
MAC	see **maximum allowable concentration**
MacArthur-Forrest process	see **cyanide process**
maceration	1. Most commonly, steeping, especially steeping cane in juice during sugar making; cf **imbibition**
	2. Rarely, see **comminution**
macerator	see **comminutor**
Mach number	(*symbol* M or Ma) The ratio of a fluid's velocity to sonic velocity in that fluid; defined formally in API Standard 521. Pressure-relieving and Depressuring Systems. Sixth Edition	January 2014 and API RP 537 Flare Details for Petroleum, Petrochemical, and Natural Gas Industries
machine	see **machinery**
machine welding	Welding with equipment requiring constant operator attention; defined formally in API STD 620 - Low Pressure Storage Tanks and ASME BPE (American Society of Mechanical Engineers: Bioprocessing Equipment); cf **automatic welding**, **manual welding**
machinery	An assembly of components driven in a controlled way; defined formally in BS EN ISO 14159:2008 Safety of machinery. Hygiene requirements for the design of machinery and BS EN ISO 12100:2010 Safety of machinery. General principles for design. Risk assessment and risk reduction
macroporous resin	*aka* macroreticular resin. A highly crosslinked ion exchange resin in the form of small porous particles
macroreticular resin	see **macroporous resin**
MACT	see **maximum achievable control technology**
made ground	Anthropogenic ground (qv) placed in a random / uncontrolled fashion; defined formally in BS 5930:2015 Code of practice for ground investigations
mag-drive pump	see **magnetic drive pump**
Magflow	Proprietary eponym for an electromagnetic flowmeter
magma	In sugar refining, a mixture of sugar crystals and liquid produced by mingling (qv)

Magnaforming	A proprietary catalytic reforming process with a platinum catalyst
Magnehelic	Proprietary eponym for a low range differential pressure gauge
magnetic drive pump	*aka* mag-drive pump. A pump driven via a magnetic linkage, with no possible path from drive to process fluid
magnetic flowmeter	see **electromagnetic flowmeter**
magnetic flux leakage	*aka* MFL, MFL-ILI. An inline (qv) inspection technique used to identify pipeline defects
magnetic gauge	see **float gauge**
magnex process	A magnetic coal beneficiation (qv) technique
Maillard reaction	A chemical reaction between amino acids and reducing sugars that makes food taste grilled. One of the types of non-enzymatic browning (qv), the other being caramelization (qv)
main burner safety time	The time for a main burner to ignite before a flame detector (qv) triggers a quick acting shut off device (qv). Defined formally in BS EN 12952-8:2002 Water-tube boilers and auxiliary installations. Requirements for firing systems for liquid and gaseous fuels for the boiler
main sewer	see **sewer main**
mainline valve	*aka* sectionalizing block valve (in natural gas lines). A pipeline emergency line section shutoff valve; defined in the USA in the Code of Federal Regulations 49CFR 192.179 and 49CFR 195.260
mains chart	A drawing which shows utility mains; defined formally in BS EN ISO 10209:2012 Technical product documentation – Vocabulary – Terms relating to technical drawings, product definition and related documentation
maintainability	A property of machinery, allowing it to be kept in a safely operable state using routine, specified procedures. Defined formally in BS EN ISO 12100:2010 Safety of machinery. General principles for design. Risk assessment and risk reduction
maintenance	Keeping machinery in a safely operable state. When it uses routine, specified procedures, this is known as routine maintenance (qv). When it does not, it is called reactive maintenance, unplanned maintenance or unscheduled maintenance (qv) amongst other things (the jocular term for the extreme version, involving hitting with a hammer, is percussive maintenance (qv)) Defined formally in BS EN 82079-1:2012 Preparation of instructions for use. Structuring, content and presentation. General principles and detailed requirements

maintenance access	The space required to service and calibrate equipment safely in situ, as well as to remove parts or whole equipment for offsite repair
maintenance hole	see **manhole**
maintenance management system	*aka* MMS. A system which tracks and directs activities to maintain equipment and tools to ensure their availability for manufacturing whilst scheduling for periodic or preventive maintenance. Also provides the response (alarms) to immediate problems and maintains a history of past events or problems to aid in diagnosing problems
maintenance manual	*aka* operating and maintenance manual, O&M manual. A document containing routine specified procedures for maintenance, amongst other things. Defined formally in BS EN ISO 10209:2012 Technical product documentation – Vocabulary – Terms relating to technical drawings, product definition and related documentation
major structural discontinuity	*aka* gross structural discontinuity. In the context of mechanical design of structures such as vessels, a relatively large and structurally significant area of stress or strain intensification, such as a junction; defined formally in API RP 579 - Fitness for Service
make up pump	see **pressure maintenance pump**
makedown	Initial dispersion/dilution/thinning of solid or liquid concentrates, especially of polymers and other chemical additives, fertilizers etc. Not usually applied to paint production, which prefers the near equivalent letdown (qv)
makeup	Adding something to a process (most commonly water) to make up for its loss from the system
male pipe thread	*aka* MPT; see **national pipe thread**
malfunction	Failure to function as intended; defined formally in BS EN ISO 12100:2010 Safety of machinery. General principles for design. Risk assessment and risk reduction
M-Alk	see **methyl orange alkalinity**
malleable	Capable of being shaped (especially into a thin sheet) without fracturing by hammering, rolling or pressing; cf ductile
malsynchronization	A phase mismatch between an alternator and the electrical system it is connected to; defined formally in ISO 3977 Gas turbines - Procurement - Part 3: Design requirements
man and boy	A term for a type of redundancy (qv) involving a full size duty unit and a reduced size assist unit
man made vitreous fiber	*aka* MMVF. A type of synthetic insulation fiber used in fired heaters. Defined formally in BS EN ISO 13705:2012/ISO

	13705:2012(E) Petroleum, petrochemical and natural gas industries. Fired heaters for general refinery service, API STD 530 - Calculation of Tube Heater Thickness and API STD 560 - Fired Heaters for General Refinery Service
man riding hoist	see **man riding winch**
man riding winch	*aka* manriding hoist. A winch used for personnel access and egress at height or to below grade confined spaces; cf rescue winch
management by walking around	*aka* MBWA, management by wandering around. Enhancing management of a production process based upon wandering around a facility having chats with random workers you encounter; cf gemba
management by wandering around	see **management by walking around**
management of change	*aka* MOC; see **change control** (qv). N.B. not the same as change management (qv)
management speak	Corporate jargon frequently intended to make something seem more complex, impressive or ethical than it is, to move a discussion from the rational into the emotional sphere, as thought-terminating cliches, or to demonstrate that the manager is au fait with current management fashion. Examples[70] include:

- ballpark figure
- be proactive, not reactive
- blue-sky thinking
- bring it to the table
- client focus
- core competencies
- critical mass
- deep dive
- deliverables
- going forward
- I hear what you are saying
- ideation
- incentivize
- knowledge base
- mission-critical
- move the goalposts
- organic
- pushing the envelope
- reach out
- run it up the flagpole
- take it to the next level
- there is no 'I' in team
- think outside the box
- touch base
- win-win situation

---

[70] Such phrases can be used to construct bingo cards, allowing engineers bored by a pointless management briefing to play 'bullshit bingo', though this is not recommended. Managers tend to like to be taken seriously.

management system	The policies, procedures and processes of an organization which are intended to systematically ensure fulfilment of objectives, especially quality, health, safety and environmental objectives. Defined formally in BS EN ISO 9000 Quality management systems Fundamentals and vocabulary
Manchester screwdriver	*(jocular) aka* American screwdriver, Birmingham screwdriver, Glasgow screwdriver, Paisley screwdriver. Strictly, a hammer used for driving screws in; cf adjuster
mandatory list	*aka* positive list. A list of criteria for identifying projects always requiring EIA/EIS in the EU; cf exclusion list
manesty	Proprietary eponym for a tableting machine
maneuverable platform	A permanent, configurable surface used for operation and maintenance. Defined formally in BS EN ISO 14122 Safety of machinery. Permanent means of access to machinery. Working platforms and walkways
maneuverable walkway	A permanent, configurable surface used for walking. Defined formally in BS EN ISO 14122 Safety of machinery. Permanent means of access to machinery. Working platforms and walkways
manhole	1. see **manway** 2. *aka* maintenance hole, sewer box, sewer hole, utility hole. An access chamber inserted at intervals and at changes of direction in a sewer
manifold	A branched chamber or collection of pipes and fittings which blend fluids from many (commonly parallel) paths into one, or split fluids from one path into many. Various near-identical definitions in BS EN ISO 13705:2012/ISO 13705:2012(E) Petroleum, petrochemical and natural gas industries. Fired heaters for general refinery service, API Standard 521. Pressure-relieving and Depressuring Systems. Sixth Edition | January 2014, API RP 537 Flare Details for Petroleum, Petrochemical, and Natural Gas Industries, API STD 530 - Calculation of Tube Heater Thickness, API STD 530 - Calculation of Tube Heater Thickness, and API STD 560 - Fired Heaters for General Refinery Service
manifold ASME container	An API ASME container (qv) for LPG consisting of multiple pressure vessels, associated manifold and bracing, certified as a single pressure vessel. Defined formally in NFPA 58: Liquefied Petroleum Gas Code, 2017 Edition
manipulated variable	*aka* MV. A controller output
man-machine interface	see **human machine interface**

manual	1. Achieved by hand (as opposed to automatically) 2. An operating and maintenance instruction document for a product; defined formally in BS EN 82079-1:2012 Preparation of instructions for use. Structuring, content and presentation. General principles and detailed requirements
manual cleaning	In a hygienic context, implies non-automatic soil removal from machinery in an open, or at least partially disassembled state. Defined formally in BS EN ISO 14159:2008 Safety of machinery. Hygiene requirements for the design of machinery; cf cleaning in place, cleaning out of place
manual control	Process control directly by means of operator observation and intervention; cf open loop control
manual emergency switch	Another name for an emergency motor stop (qv), used in the USA. This version is defined formally in NFPA 86: Standard for Ovens and Furnaces, 2019 Edition
manual shutdown	A manually initiated or controlled machine shutdown. Defined formally in the context of gas turbines in ISO 3977 Gas turbines - Procurement - Part 3: Design requirements
manual transfer switch	A manual switch which controls the transfer of something, such a fluid transfer operation or transfer of load conductor connections between power sources; defined in NFPA 20: Standard for the Installation of Stationary Pumps for Fire Protection, 2019 Edition
manual valve	*aka* MV. A valve controlled by hand
manual welding	Welding achieved entirely by means of a hand-held device, requiring a high degree of operator skill, and therefore requiring a welder. Defined formally in API STD 620 - Low Pressure Storage Tanks and ASME BPE (American Society of Mechanical Engineers: Bioprocessing Equipment)
manufactured gas	A generic term for all gases manufactured industrially, including coal gas (qv), producer gas (qv), synthesis gas (qv), biogas (qv), and the various hydrogen colors (qv)
manufacturer	An entity which designs, constructs and tests a product such as a vessel. Slightly different definitions in PD 5500:2018+A1:2018 Specification for unfired fusion welded pressure vessels and API STD 620 - Low Pressure Storage Tanks
manufacturing cell	*aka* cell. A grouping of machines and resources which facilitates efficient manufacturing

manufacturing design range	The rated pressure range marked on a rupture disk (qv). Defined formally in API RP 520 P1 7th Edition, January 2000 Sizing, Selection, and Installation of Pressure-Relieving Devices in Refineries; Part I - Sizing and Selection and API RP 576 - Inspection of Pressure Relieving Devices; cf marked burst pressure
manufacturing drawing	A drawing with sufficient information for the production of a product. Defined formally in BS EN ISO 10209:2012 Technical product documentation — Vocabulary — Terms relating to technical drawings, product definition and related documentation
manufacturing execution system	*aka* MES. A collective term used to describe the functional activities essential for the management and control of production and manufacturing operations in a given organization
manway	*aka* manhole. An access hole in a vessel large enough for person entry. Sizing is subjected to a number of codes and regulations to assure access and egress
MAOP	see **maximum allowable operating pressure**
MAQ	see **maximum allowable quantity**
margin of safety	*aka* safety margin. According to DOE/NV/25946-1891, the margin required in order to ensure safety; in engineering, the margin of safety is the factor of safety (strength of the material divided by the anticipated stress) minus one
Margules activity	see **Margules' activity model**
Margules activity coefficient model	see **Margules' activity model**
Margules' activity model	*aka* Margules activity coefficient model, Margules activity, activity coefficient model. A simple and rather old-fashioned way to estimate excess Gibbs free energy (qv) of a mixture of liquids
marine brass	*aka* naval brass. A strong, hard, corrosion-resistant copper-tin-zinc alloy used in wastewater treatment pumps, heat exchanger tubes, etc.
marine bronze	*aka* aluminum bronze. A strong, corrosion-resistant biostatic copper-aluminum alloy cf duralumin
marine diesel oil	*aka* MDO; see **fuel oil**
marine fuel	see **fuel oil**
Mariotte's law	see **Boyle's law**

marked burst pressure	*aka* rated burst pressure. The tested burst pressure value of a rupture disk (qv) at the given temperature with which it is marked. Defined formally in API RP 520 P1 7th Edition, January 2000 Sizing, Selection, and Installation of Pressure-Relieving Devices in Refineries; Part I - Sizing and Selection API Standard 521. Pressure-relieving and Depressuring Systems. Sixth Edition	January 2014, and API RP 576 - Inspection of Pressure Relieving Devices; cf manufacturing design range, specified burst pressure
martensite	A very hard crystalline form of steel; cf austenite
martensitic	Having the form of martensite (qv). Defined formally in API RP 571 - Damage Mechanism Affecting Fixed Refinery Equipment; cf austenitic
martensitic stainless steel	Hard, corrosion resistant steels such as the AISI 400 series. Defined formally in API RP 571 - Damage Mechanism Affecting Fixed Refinery Equipment
masking	The coating of a catalyst surface with something which renders it inactive; defined formally in API RP 536 - NOx control on Fired Heaters at Refineries
mass balance	*aka* material balance. The practical application of the laws of conservation (qv) of mass, which imply that the total mass of input to a process is equal to the total mass of output of the process. For chemical reactions, a mole balance (qv) is normally carried out and converted to mass units
mass diffusivity	see **diffusion coefficient**
mass fire	A massive fire resulting from the merging of fires started by many simultaneous ignitions
mass flow	A mass of material transferred in a given time period (e.g., kg3/hr.); cf volumetric flow
mass flow diagram	*aka* MFD, block flow diagram. A simplified and highly informal PFD (qv), more popular in academia and management than engineering
mass flowmeter	Instruments which can directly measure the rate of mass flow through a pipe. Typically, these are based on thermal flow principles for gases and Coriolis effect (qv) principles for liquids
mass spectrometry	*aka* MS. An instrument used to identify chemicals in a substance by the mass and charge of their ions

mass transfer	The transport of mass from one point to another; one of the main pillars of transport phenomena (qv). Mass transfer may take place in a single phase or over phase boundaries in multiphase systems. In the vast majority of engineering problems, mass transfer involves at least one fluid phase (gas or liquid), although it may also be described in solid-phase materials
massecuite	French for 'cooked mass': in sugar refining, a mixture of crystals and mother liquor (qv) in vacuum pan crystallization.
massic heat capacity	see **specific heat capacity**
master batch record	*aka* MBR, master production record (MPR), master manufacturing formula (MFF). Complete written instructions for producing a pharmaceutical product under GMP (qv)
master cell bank	A portion of a single pool of cells prepared from a selected cell clone, dispensed into multiple containers and stored under defined conditions; used as the source of all working cell banks (qv)
master document file	A file containing original signed documents or procedures
master fuel trip	A rapid automatic boiler fuel supply- and igniter- deactivation switch. Defined formally in BS EN 12952-8:2002 Water-tube boilers and auxiliary installations. Requirements for firing systems for liquid and gaseous fuels for the boiler
master manufacturing formula	*aka* MMF; see **master batch record**
master production record	*aka* MPR; see **master batch record**
MAT	see **minimum allowable temperature**
material balance	see **mass balance**
material impact energy	*aka* Charpy vee notch energy. Energy absorbed at material fracture as determined by the Charpy impact test (qv). Defined formally in PD 5500:2018+A1:2018 Specification for unfired fusion welded pressure vessels
material impact test temperature	The temperature at which a material's Charpy impact test (qv) was conducted. Defined formally in PD 5500:2018+A1:2018 Specification for unfired fusion welded pressure vessels
material manufacturer	Just what it sounds like; a convoluted formal definition may be found in ASME BPE (American Society of Mechanical Engineers: Bioprocessing Equipment)

material of construction	*aka* MOC, materials of construction. (The closely related terms construction materials and building materials tend to be associated with civil and building works, rather than process equipment) 1. Generically, what something is made of 2. In engineering contexts usually a metal, and most often of all some kind of steel
material receiving report	see **materials receiving report**
material safety data sheet	*aka* MSDS, product safety data sheet, SDS. A document presenting a fairly standardized form of collected information on scientific, health, safety and environmental properties of a material to aid in risk assessment, undertaken under COSHH (qv) and required by REACH (qv) in the EU and UK
material takeoff	*aka* construction takeoff, takeoff. A listing of specifications and quantities of materials required to produce an item shown on a drawing; used in cost estimation
material test report	*aka* MTR, mill sheet, mill test report. A document containing results of materials testing. Defined formally in ASME BPE (American Society of Mechanical Engineers: Bioprocessing Equipment)
material unaccounted for	*aka* MUF, inventory difference. A euphemism for material 'lost' from a material balance or materials accounting, especially fissile material (qv)
materials receiving report	*aka* MRR, material receiving report. A document confirming receipt of material
materials resource planning	*aka* MRP. A management system designed to improve productivity for businesses by estimating quantities of required raw materials and scheduling deliveries as required
mathematical optimization	see **mathematical programming**
mathematical programming	*aka* mathematical optimization. Mathematical techniques for selecting the best available value of a variable
mating ring	*aka* stationary seat. An axially constrained, static pump seal component against which a wearing seal component runs. Defined formally in API RP 682 - Pump Seals
matrix	Setting aside the mathematical definition, most commonly used in engineering practice to mean a material (most commonly a solid or semisolid) in which a dispersed material is embedded
matte	Mixed molten base metal sulfides insoluble in either metal or slag (qv), produced during certain beneficiation (qv) processes

maturation	Time-related development. May refer to increases in effective filtration in a filter bed, or desirable flavors and textures in food and drink products
maturity level	The degree of development of documents towards readiness for production. Defined formally in BS EN ISO 10209:2012 Technical product documentation – Vocabulary – Terms relating to technical drawings, product definition and related documentation
MAWP	see **maximum allowable working pressure**
maximax	The largest of a range of maxima, or the 'best of the best'. Defined formally in BS ISO 2041:2018 Mechanical vibration, shock and condition monitoring. Vocabulary
maximin	The largest of a range of minima, or the 'best of the worst'
maximum achievable control technology	*aka* MACT. A term used in the US Clean Air Act for evaluating environmental pollution abatement technology, similar to the EU's BAT (qv)
maximum allowable concentration	*aka* MAC. In environmental regulation, the maximum permitted concentration of something in a discharge to environment
maximum allowable continuous speed	*aka* nmax; see **maximum continuous speed**. This particular term is defined formally in BS EN ISO 17769-1:2012 Liquid pumps and installation. General terms, definitions, quantities, letter symbols and units. Liquid pumps; cf minimum allowable continuous speed
maximum allowable operating pressure	*aka* MAOP. The maximum internal operating pressure of a natural gas pipeline segment; cf maximum allowable working pressure, **Box M1**.
maximum allowable quantity	*aka* MAQ. The largest quantity of flammable liquid allowed in a given area; defined in NFPA 30: Flammable and Combustible Liquids Code, 2018 Edition
maximum allowable speed	The highest design speed for continuous operation. Defined formally in EN ISO 10437 Petroleum, petrochemical and natural gas industries - Steam turbines - Special-purpose applications, BS EN ISO 13705:2012/ISO 13705:2012(E) Petroleum, petrochemical and natural gas industries. Fired heaters for general refinery service and EN ISO 13709:2003 Centrifugal pumps for petroleum, petrochemical and natural gas industries
maximum allowable stress value	The maximum permitted unit stress in tank mechanical design calculations. Defined formally in API STD 620 - Low Pressure Storage Tanks

maximum allowable temperature	The highest design temperature for continuous operation. Defined formally in EN ISO 10437 Petroleum, petrochemical and natural gas industries - Steam turbines - Special-purpose applications, BS EN ISO 13705:2012/ISO 13705:2012(E) Petroleum, petrochemical and natural gas industries. Fired heaters for general refinery service, EN ISO 13709:2003 Centrifugal pumps for petroleum, petrochemical and natural gas industries and API RP 682 - Pump Seals
maximum allowable working pressure	*aka* MAWP, working pressure. A contested term - see **Box M1**; cf maximum allowable operating pressure
maximum continuous operating speed	*aka* nmax or nmax,ad. The maximum specified operating speed. Defined formally in BS ISO 14661:2000 Thermal turbines for industrial applications (steam turbines, gas expansion turbines) — General requirements; cf maximum continuous speed

---

Box M1. Maximum Allowable Working Pressure

In summary, and setting aside a couple of significant differences, maximum allowable working pressure (MAWP) means the highest continuous operation design pressure.

MAWP may be calculated as a measure of fitness for service, based on measurements of an item in service, discounting corrosion allowances, at design temperature. This is the approach taken in API RP 579- Fitness for Service. The same approach is taken in API RP 576 - Inspection of Pressure Relieving Devices, API RP 682 - Pump Seals and EN ISO 13709:2003 Centrifugal pumps for petroleum, petrochemical and natural gas industries.

There is a slightly different definition in API RP 520 P1 7th Edition, January 2000 Sizing, Selection, and Installation of Pressure-Relieving Devices in Refineries; Part I - Sizing and Selection, though it still appears compatible with API RP 579. API Standard 521. Pressure-relieving and Depressuring Systems. Sixth Edition | January 2014 differs from both RP579 and RP220, but is again compatible with RP579.

However, MAWP can, alternatively, mean the manufacturer's original design pressure at maximum temperature. This is the definition given in EN ISO 10437 Petroleum, petrochemical and natural gas industries -- Steam turbines -- Special-purpose applications. In all cases, MAWP should not be confused with the similar-sounding maximum allowable operating pressure (MAOP) which is entirely different as defined in NFPA 58: Liquefied Petroleum Gas Code, 2017 Edition.

Related Terms

maximum allowable operating temperature, maximum working pressure, maximum operating pressure

maximum continuous rating	*aka* MCR. The capability of a steam boiler for continuous production of a specified quantity of steam
maximum continuous speed	*aka* maximum allowable continuous speed. There is a range of definitions for this term, which tend to differ from maximum allowable speed (qv) in that they refer to a tested, as-built condition, rather than design figures. Defined formally in ISO 3977 Gas turbines - Procurement - Part 3: Design requirements (a standard which has two different context specific definitions), EN ISO 10437 Petroleum, petrochemical and natural gas industries - Steam turbines - Special-purpose applications and EN ISO 13709:2003 Centrifugal pumps for petroleum, petrochemical and natural gas industries
maximum credible accident	see **design basis event**
maximum discharge pressure	The specified maximum suction pressure plus maximum differential pressure at design conditions. Defined formally in EN ISO 13709:2003 Centrifugal pumps for petroleum, petrochemical and natural gas industries
maximum drying rate	Maximum moisture removal rate; cf constant rate drying
maximum dynamic sealing pressure	*aka* MDSP. The highest pump seal pressure expected under specified operating conditions. Defined formally in EN ISO 13709:2003 Centrifugal pumps for petroleum, petrochemical and natural gas industries and API RP 682 - Pump Seals; cf maximum static sealing pressure
maximum exhaust casing pressure	The highest instantaneous exhaust steam design pressure at maximum design temperature for a steam turbine. Defined formally in EN ISO 10437 Petroleum, petrochemical and natural gas industries - Steam turbines - Special-purpose applications
maximum exhaust pressure	The highest continuous exhaust steam design pressure for a steam turbine. Defined formally in EN ISO 10437 Petroleum, petrochemical and natural gas industries-- Steam turbines - Special-purpose applications
maximum expected inlet temperature	The normal operating inlet temperature plus an added margin for foreseeable process upsets. Defined formally in API STD 560 - Fired Heaters for General Refinery Service

maximum experimental safe gap	*aka* MESG. The gap between parts of a flame propagation test chamber; a key parameter in classifying classification of liquids by their potential for creating flammable atmospheres. Defined formally in API RP 505 2nd Edition, August 2018 Recommended Practice for Classification of Locations for Electrical Installations at Petroleum Facilities Classified as Class I, Zone 0, Zone 1, and Zone 2
maximum exposure limit	*aka* MEL. A collective term for short term exposure limit (qv) and long term exposure limit (qv); acceptable weighted average concentrations of certain hazardous substances in workplace air according to UK health and safety regulation
maximum fill height	*aka* MFH. The maximum permitted height of a liquid of a specified density in a vessel; defined formally in API RP 579 - Fitness for Service
maximum firing rate of burners	The maximum operating firing rate of a water tube boiler. Defined formally in BS EN 12952-8:2002 Water-tube boilers and auxiliary installations. Requirements for firing systems for liquid and gaseous fuels for the boiler
maximum heat flux density	Fired heater coil section heat transfer rate. Defined formally in BS EN ISO 13705:2012/ISO 13705:2012(E) Petroleum, petrochemical and natural gas industries. Fired heaters for general refinery service, API STD 530 - Calculation of Tube Heater Thickness and API STD 560 - Fired Heaters for General Refinery Service
maximum heat input of the firing system	The maximum safe operating heat input for a water tube boiler. Defined formally in BS EN 12952-8:2002 Water-tube boilers and auxiliary installations. Requirements for firing systems for liquid and gaseous fuels for the boiler
maximum inlet pressure and temperature	The highest instantaneous inlet steam operating pressure at maximum operating temperature for a steam turbine. Defined formally in EN ISO 10437 Petroleum, petrochemical and natural gas industries - Steam turbines - Special-purpose applications
maximum operating pressure	*aka* MOP. Commonly defined to mean maximum internal pressure during normal operation. Defined formally in API RP 520 P1 7th Edition, January 2000 Sizing, Selection, and Installation of Pressure-Relieving Devices in Refineries; Part I - Sizing and Selection and API RP 576 - Inspection of Pressure Relieving Devices

maximum operating steam or gas conditions	Highest steam or gas conditions (qv) for continuous turbine operation. Defined formally in BS ISO 14661:2000 Thermal turbines for industrial applications (steam turbines, gas expansion turbines) – General requirements; cf minimum operating steam or gas conditions
maximum operating temperature	Instantaneous maximum pump seal operating temperature. Defined formally in API RP 682 - Pump Seals
maximum power output	*aka* Pmax. A turbine manufacturer's stated maximum available mechanical or electrical power output. Defined formally in BS ISO 14661:2000 Thermal turbines for industrial applications (steam turbines, gas expansion turbines) – General requirements
maximum pump brake horsepower	A shop-tested maximum pump input power requirement at rated speed, suction and delivery conditions. Defined formally in NFPA 20: Standard for the Installation of Stationary Pumps for Fire Protection, 2019 Edition
maximum sealing pressure	The highest operating pressure which turbine seals must effectively seal against. Defined formally in EN ISO 10437 Petroleum, petrochemical and natural gas industries - Steam turbines - Special-purpose applications
maximum specific growth rate	(*symbol* μ_{max}) A term with at least two different definitions, the commonest of which is the maximum growth rate in the Monod equation (qv). Care must be taken to ensure which definition is being used
maximum static sealing pressure	*aka* MSSP. The highest pump seal pressure expected under all conditions other than hydrostatic testing. Defined formally in EN ISO 13709:2003 Centrifugal pumps for petroleum, petrochemical and natural gas industries and API RP 682 - Pump Seals; cf maximum dynamic sealing pressure
maximum steam or gas conditions	The most severe (as opposed to maximum) steam or gas conditions (qv) for continuous turbine operation. Defined formally in BS ISO 14661:2000 Thermal turbines for industrial applications (steam turbines, gas expansion turbines) –General requirements; cf maximum operating steam or gas conditions
maximum suction pressure	Highest operational pump suction pressure. Defined formally in EN ISO 13709:2003 Centrifugal pumps for petroleum, petrochemical and natural gas industries
maximum working pressure	Maximum sustained operating pressure. Defined formally in ASME BPE (American Society of Mechanical Engineers: Bioprocessing Equipment)

maximum working temperature	The maximum sustained operating temperature at maximum working pressure (qv) with a given fluid. Defined formally in ASME BPE (American Society of Mechanical Engineers: Bioprocessing Equipment)
Maxwell-Bonnel method	A method used in oil refining to estimate the vapor pressure of fractions
MB	*(drawing notation)* Machine bolt
MBBR	see **moving bed biofilm reactor**
MBOED	see **million barrels of oil equivalent per day**
MBR	1. see **master batch record** 2. see **membrane bioreactor**
MBWA	see **management by walking around**
McAvoy correlation	*aka* Jafarey, Douglas and McAvoy correlation. The basis of a short-cut method for distillation column design
MCC	1. see **motor control center** 2. see **mechanical completion certificate**
McCabe–Thiele	An almost obsolete graphical method formerly used to design distillation columns, still universally taught in academia despite being professionally superseded by software long ago
MCR	see **maximum continuous rating**
MCRT	see **mean cell residence time**
MDEA	see **methyl diethanolamine**
MDMT	see **minimum design metal temperature**
MDO	see **marine diesel oil**
MDoF system	see **multiple degrees of freedom system**
MDPE	Medium density polyethylene; cf polyethylene
MDSP	see **maximum dynamic sealing pressure**
ME	Microscopic examination
MEA	see **monoethanolamine**
MeABP	see **mean average boiling point**
mean average boiling point	*aka* MeABP. One of five ways used in oil refining to express average boiling point of a mixture of hydrocarbons (the others being molal, volume, weight, and cubic)
mean cell residence time	*aka* MCRT. A measure of solids retention in an activated sludge plant, which is different from solids retention time (qv) and sludge age (qv), even though this term is often informally used interchangeably with the other two. It is not strictly the mean residence time (qv) of cells. MCRT in days is properly calculated as mass of mixed liquor suspended solids (qv) in system/ (mass/day suspended solids wasted + mass/day suspended solids in treated effluent)

mean hydraulic depth	*aka* hydraulic depth, hydraulic mean depth. In open channel hydraulics, the cross sectional area for flow divided by the wetted perimeter
mean hydraulic diameter	*aka* Dh, hydraulic diameter, hydraulic mean diameter. Four times the hydraulic radius (qv)[71]
mean residence time	The average residence time (qv) in a reactor of particles or fluid parcels
mean time between failures	*aka* MTBF. A measure of the reliability of a system, equal to average operating time of equipment between failures, as calculated on a statistical basis from the known failure rates of various components of the system
mean time to failure	*aka* MTTF. A measure of reliability, giving the average time before the first failure
mean time to repair	*aka* MTTR. A measure of reliability of a piece of repairable equipment, giving the average time between repairs
mean velocity	see **average velocity**
meandering	A term for a weld bead (qv) which wanders from a straight track along the joint; defined formally in ASME BPE (American Society of Mechanical Engineers: Bioprocessing Equipment)
measurand	A quantity intended to be measured
measured value	*aka* MV. In process control, an older alternative term for a process variable (qv) which is compared to the set value (qv) to give controller error (E=SV-MV but now E=PV-SV)
measurement	Using equipment to determine the value of a quantity. Defined formally in BS EN ISO 9000 Quality management systems Fundamentals and vocabulary
MEC	see **minimum explosible concentration**
mechanical backing pump	see **holding pump**
mechanical cleaning	"Cleaning solely by circulating and/or flowing chemical detergent solutions and water rinses onto and over the surfaces to be cleaned, by mechanical means" according to the EHEDG Glossary
mechanical commissioning	Somewhat contested, but in chemical engineering contexts, the stage of plant erection, construction (and arguably commissioning) which results in mechanical completion (qv). Mechanical engineers would tend to view this as a type of commissioning (as the name implies), but some process engineers consider only process commissioning (qv) to be commissioning; see **Box C3**.

---

[71] Not two as you might have guessed

mechanical completion	The state of a plant or any part of the plant once it has been erected in accordance with drawings, specifications and applicable codes and precommissioning activities have been completed to the extent necessary to permit the client to accept the plant and begin commissioning activities - i.e., it is ready for commissioning (qv)
mechanical completion certificate	*aka* MCC. A completion certificate (qv) confirming that mechanical completion (qv) has been achieved, usually associated with a stage payment
mechanical polishing	Reduction of surface roughness with an abrasive medium. Defined formally in ASME BPE (American Society of Mechanical Engineers: Bioprocessing Equipment)
mechanical seal	A prefabricated assembly used to seal the drive shaft of rotating equipment against fluid pressure. Defined formally in ASME BPE (American Society of Mechanical Engineers: Bioprocessing Equipment)
mechanical variator	see **variator**
MED	see **multiple effect distillation**
media	The Latin plural of medium[72]. Media is commonly used as a very broad term encompassing various context-specific means to an end, such as growth media (qv), the packing in a packed bed (qv) or the material added to water to increase its density in dense medium cyclone (qv) separation. Some (but not all) of these means to an end might also be called 'medium' (qv)
media blasting	A method of removing contaminants and debris from the surface of metals using the impact of an entrained stream of hard particles, such as sand blasting, grit blasting and bead blasting (qv)
Medicines & Healthcare Products Regulatory Agency	*aka* MHRA. The UK pharmaceutical regulatory agency; the equivalent of the US FDA (qv) and the EMEA (qv) in the EU

---

[72] The other plural form, 'mediums' tends not to be needed in engineering!

medium	1. *aka* growth medium. Most commonly in biochemical engineering, a liquid or semisolid matrix on or in which cells are grown 2. An information storage device, as defined in **BS EN ISO 10209:2012** Technical product documentation — Vocabulary — Terms relating to technical drawings, product definition and related documentation. N.B. Meanings 1 and 2 might also be referred to as media (cf) under certain circumstances, but this singular form tends to predominate in these contexts 3. Middling; an amount between high and low
medium and high expansion foam concentrate	A foam concentrate with hydrocarbon surfactants used to produce firefighting foam; defined formally in NFPA 11: Standard for Low-, Medium-, and High-Expansion Foam, 2016 Edition
medium care areas	see **zoning**
mega-	(*symbol* M-) The SI unit prefix denoting a factor of 10^6
Megger	Proprietary eponym for an insulation resistance tester
MEL	see **maximum exposure limit**
Mellapak	Proprietary eponym for packing
melt	1. To turn from solid to liquid 2. A molten substance or mix of substances
melt crystallization	Purification by crystallization from a molten mix of substances known as a melt (qv)
melting point	The temperature at which a solid substance changes to the liquid state
membrane	"A sheet of porous material that is permeable to a liquid. Example: a reverse osmosis membrane for water treatment" according to the EHEDG Glossary; though we might enhance and broaden this definition by replacing 'liquid' with 'fluid' and also adding the word 'thin' before 'sheet'
membrane bioreactor	Most commonly, a bioreactor which uses membranes for biomass retention
membrane distillation	A novel and presently somewhat unreliable separation process, combining the principles of distillation with membranes of various kinds
membrane filtration	*aka* membrane separation. The separation of suspended or dissolved solids from fluids using membranes; cf **Box L1**.
membrane permeation flux	Permeate volume passing through a membrane in a given time over a given area of surface
membrane plugging	see **plugging**
membrane separation	see **membrane filtration**

membrane stress	Average normal cross-sectional stress; defined formally in API RP 579 - Fitness for Service
MEMU	Mobile explosives manufacturing unit
mEq	see **milliequivalent**
mercantile occupancy	A building or part of one used as a shop or market; defined formally in NFPA 30: Flammable and Combustible Liquids Code, 2018 Edition
Merox process	Short for mercaptan oxidation, a proprietary catalytic process for conversion of mercaptans in hydrocarbon fuels to disulfides by air oxidation
MES	see **manufacturing execution system**
MESG	see **maximum experimental safe gap**
mesh	*aka* mesh number. The number of openings per inch in a screen, used to specify its fineness; also used by extension to classify powder size, thus a '-8' mesh powder passes through an 8-mesh screen, and a '+10' powder is retained by a 10-mesh screen
MESH equations	An acronym for the material balances, equilibrium relationships, summation equations and heat balances which are the chemical engineer's stock in trade, though the acronym itself is used almost entirely within academia
mesh number	see **mesh**
mesophilic bacteria	Bacteria with an optimum temperature between those of cryophilic bacteria (qv) and thermophilic bacteria (qv)
metabolism	Broadly, the biochemical reactions of a living cell
metabolite	1. A low-molecular-weight biological compound, usually enzymically synthesized 2. A compound essential for a metabolic process 3. A substance synthesized by an organism, or taken in from the environment
Metaflex	Proprietary eponym for wire-wound gaskets
metal fatigue	A loss of strength in metals due to the accumulation of small cracks caused by repeated stress
metal fiber reinforcement	The addition of stainless steel reinforcing needles to castable refractories for fired heaters. Defined formally in BS EN ISO 13705:2012/ISO 13705:2012(E) Petroleum, petrochemical and natural gas industries. Fired heaters for general refinery service, API STD 530 - Calculation of Tube Heater Thickness and API STD 560 - Fired Heaters for General Refinery Service

metal inert gas welding	*aka* MIG. An arc welding (qv) process that uses a continuous solid wire electrode, heated and fed into the weld pool from a welding gun which feeds an inert gas alongside the electrode, helping protect the weld pool from air and airborne contaminants
metallic protected flexible hose connector	An LPG hose with a metallic casing for mechanical protection; defined formally in NFPA 58: Liquefied Petroleum Gas Code, 2017 Edition
metering	see **dosing**
metering and regulating station	*aka* M&R station. Installations containing equipment to measure the amount of gas entering or leaving a pipeline system and, sometimes, to regulate gas pressure
metering pump	see **dosing pump**
meters water gauge	*aka* mWG, metres water gauge. A common measure of pressure expressed as a head (qv) of water. 1mWG is the pressure exerted by a 1m column of water under standard conditions
methanation	The catalytic hydrogenation of COx (qv) to methane
methane clathrate	*aka* fire ice, gas hydrate, hydromethane, methane hydrate, methane ice, natural gas hydrate. A naturally occurring ice-like solid, comprising large quantities of methane trapped in the crystal structure of ice, which can be an economically important source of natural gas
methane hydrate	see **methane clathrate**
methane ice	see **methane clathrate**
methanogenesis	The stage of anaerobic digestion where methane is formed
methanol to gasoline	*aka* MTG. A generic term for the processes used in the oil and gas industry with the aim of making this conversion
method statement	*aka* MS, work method statement, safe work method statement (SWMS). A document which explicitly states how a planned operation will be performed; often associated with a complementary risk assessment
methyl orange alkalinity	*aka* M-Alk. Alkalinity as measured by titration with methyl orange indicator; cf P-Alk
methyl diethanolamine	*aka* MDEA. A chemical used in the amine gas treating process (qv) to remove acid gas (qv) from hydrocarbons
metres water gauge	see **meters water gauge**
metric system	A measurement system, not synonymous with SI units[73] (qv); see **Box B1**.
metric ton	see **tonne**

---

[73] ...and the reason why 'Europeans' have the Royale with Cheese!

metric ton unit	see **metric tonne unit**
metric tonne unit	*aka* mtu, metric ton unit. A term used in metal trading for a variable, but specified small quantity of high value metal, based on an assumed yield per tonne of ore
metrological characteristic	Something capable of influencing measurement results; defined formally in BS EN ISO 9000 Quality management systems Fundamentals and vocabulary
metrological confirmation	Operations which ensure that measuring equipment meets requirements; defined formally in BS EN ISO 9000 Quality management systems Fundamentals and vocabulary; cf **Box A1**.
metrological traceability	The traceability (qv) of a measurement result; defined formally in ISO/IEC GUIDE 99:2007 International vocabulary of metrology. Basic and general concepts and associated terms (VIM)
metrology	The study of measurement
mezzanine gate	A gate providing permanent loading/unloading access to a mezzanine floor. Defined formally in BS EN ISO 14122 Safety of machinery. Permanent means of access to machinery. Working platforms and walkways
MF	see **microfiltration**
MFD	1. see **mass flow diagram** 2. see **multistage flash distillation**
MFH	see **maximum fill height**
MFL	see **magnetic flux leakage**
MFL-ILI	Magnetic flux leakage (qv) inline inspection; cf smart pig
MH	*(drawing notation)* Manhole
mho	*(symbol* ℧*)* A deprecated (but still current in some contexts) term for the unit of conductance now known as siemens (qv) in SI units (qv)[74]
MHRA	see **Medicines & Healthcare Products Regulatory Agency**
MI	*(drawing notation)* Malleable iron
MIC	see **minimum ignition current**
MIC ratio	see **minimum igniting current ratio**
Michaelis–Menten equation	*aka* Michaelis–Menten kinetics. A theoretically-derived equation modelling the relationship between reaction rate and substrate concentration for enzyme catalyzed reactions
Michaelis–Menten kinetics	see **Michaelis–Menten equation**
micro-	*(symbol* μ-*)* The SI unit prefix denoting a factor of 10^{-6}

---

[74] Mho being the reciprocal of ohm is why its name is 'ohm' spelled backwards, and it has an inverted omega as its symbol, a couple of little scientist jokes courtesy of Lord Kelvin

microbial impermeability	*aka* microbial tightness. "The ability of material or equipment to prevent the ingress of bacteria, yeasts and moulds from the outside (environment) to the inside (product area)" according to the EHEDG Glossary
microbial tightness	see **microbial impermeability**
microcarbon residue	see **Conradson carbon residue**
microcarrier	A microscopic particle (often, a 200μm polymer bead) that supports cell attachment and growth in suspension culture
microfiche	A near-obsolete information storage medium based on very much reduced photographic copies on transparent film. Defined formally in BS EN 82079-1:2020 Preparation of information for use (instructions for use) of products - Principles and general requirements
microfiltration	*aka* MF. A somewhat vague and contested term used for a membrane filtration process removing very fine particles, even bacteria and viruses; cf ultrafiltration, **Box L1**.
micrometer	*aka* micrometre. 10^{-6} m; N.B. not be confused with the shortened form of micrometer gauge (qv)
micrometer gauge	A type of gauge for the accurate measurement of small distances; cf micrometer
micrometre	see **micrometer**
micron	A deprecated term for micrometer (qv), (the distance unit) still however defined in ASME BPE (American Society of Mechanical Engineers: Bioprocessing Equipment)
Micronair	Proprietary eponym for atomizer
micronizing	1. Most commonly, turning solids into a very fine powder 2. Infrared radiation cooking of cereal, grains and pulses
microorganism	"Microorganisms are living organisms that can be seen only with the aid of a microscope. Note 1: Microorganisms include bacteria, archaea, viruses and certain protozoa, algae and fungi. Note 2: Most microorganisms are unicellular" according to the EHEDG Glossary
microwaves	Electromagnetic waves shorter in wavelength than radio waves; longer than infrared
midstream	Transport and marketing of bulk crude or refined oil and gas; cf downstream, upstream
MIE	see **minimum ignition energy**
MIG	see **metal inert gas welding**
mild steel	Low carbon steel; cf wrought iron
milk separator	see **centrifugal separator**

milk stone	*aka* milkstone. A type of heat exchanger scale unique to the dairy industry, a complex mixture of inorganic hardness and organics such as denatured protein
milkstone	see **milk stone**
mill	1. Generically, in everyday language a mill is any factory, hence windmill, steel mill, textile mill etc. 2. In engineering, it almost always refers to one of many varieties of milling (qv) or grinding (qv) machine. In process engineering, milling is usually particle size reduction; in mechanical engineering it is often to do with surface treatment
mill finish	A surface finish produced on sheet and plate; characteristic of the ground finish used on the rolls produced by a steel mill
mill sheet	see **material test report**
mill test report	see **material test report**
milli-	(*symbol* m-) The SI unit prefix denoting a factor of 10^{-3}
milliequivalent	*aka* mEq. A 1000th of an equivalent (qv) cf equivalents per million
millimeters of mercury	*aka* mmHg, millimetres of mercury. Pressure expressed as a head (qv) of mercury at 0°C. To save you a sum, 760 mm Hg = 133 Pa = 1 Atm
millimetres of mercury	see **millimeters of mercury**
milling	In common with the term grinding (qv), this has non-identical meanings in mechanical and chemical engineering. The mechanical engineering senses are associated with removal of metal from a workpiece by two different mechanisms. The differences between milling and grinding in chemical engineering processes (both always meaning a size reduction process) are to some extent dependent on industry sector
million barrels of oil equivalent per day	*aka* MMboed, MMBOED, MBOED, Mmboepd. An oil industry measurement of daily production volumes expressed as energy equivalent; cf barrel of oil equivalent
millions of standard cubic feet per day	*aka* MMSCFPD. A US customary units (qv) measure of volumetric gas flow, corrected to standard conditions; cf standard cubic foot
mils	A near-obsolete unit of measurement: thousandths of an inch
mineral acid	A generic term for common inorganic acids e.g.; hydrochloric, nitric, sulfuric
mineral beneficiation	see **mineral dressing**

mineral dressing	*aka* ore dressing, mineral/ore beneficiation, mineral upgradation, beneficiation (qv), mineral extraction, mineral processing, ore processing, extraction. Most commonly used to mean preliminary mineral processing in the extractive industries. Some of the akas above may not therefore always be considered to have identical meanings with either this term or each other
mineral extraction	1. Usually, the primary collection of minerals from the environment (mining (qv)) 2. Less commonly, the preliminary treatment of extracted material, more widely known as mineral dressing (qv)
mineral oil	*aka* liquid paraffin, white oil, petroleum oil. A somewhat vague term with a range of related meanings; generally, refers to a hydrocarbon petroleum distillate (as opposed to a triglyceride vegetable oil) and/or a low volatility liquid mixture of alkanes distilled from petroleum; cf mineral spirits
mineral processing	*aka* mineral dressing, beneficiation. Improving ore by removing unwanted minerals known as gangue (qv), producing a concentrate product and a tailings (qv) waste stream
mineral spirits	*aka* mineral turpentine, odorless mineral spirits, solvent naphtha, Stoddard solvent, paint thinners, petroleum spirits, turpentine substitute, white spirit. A common organic solvent; N.B. some of the alternative terms given may not always be thought to have identical meanings with either this term or each other
mineral turpentine	see **mineral spirits**
mineral upgradation	see **mineral dressing**
mingling	A sugar refining term for mixing
minimal medium	Biological growth media with the minimum simple nutrients required for growth of organisms; cf complete medium
minimum allowable continuous speed	*aka* nmin, ad; see **minimum allowable speed.** This particular term is defined formally in BS EN ISO 17769-1:2012 Liquid pumps and installation. General terms, definitions, quantities, letter symbols and units. Liquid pumps
minimum allowable speed	*aka* minimum allowable continuous speed. The lowest design pump RPM for continuous operation; defined formally in EN ISO 13709:2003 Centrifugal pumps for petroleum, petrochemical and natural gas industries
minimum allowable temperature	*aka* MAT. The lowest permitted operating temperature for a given thickness of a given metal to avoid brittle fracture, or a range of temperatures at given pressures. Defined formally in API RP 579 - Fitness for Service

minimum allowable thickness	In the context of vessel design, the metal thickness required for each element of the vessel. Defined formally in API RP 579 - Fitness for Service
minimum continuous operating speed	*aka* nmin. The lowest specified speed, in the context of turbines. Defined formally in BS ISO 14661:2000 Thermal turbines for industrial applications (steam turbines, gas expansion turbines) — General requirements
minimum continuous stable flow	The lowest pump operating flow, considering vibration limits. Defined formally in EN ISO 13709:2003 Centrifugal pumps for petroleum, petrochemical and natural gas industries
minimum continuous thermal flow	The lowest pump operating flow considering the effect of temperature rise on pump operation. Defined formally in EN ISO 13709:2003 Centrifugal pumps for petroleum, petrochemical and natural gas industries
minimum design metal temperature	*aka* MDMT. There are at least two quite different and non-obvious definitions of this term, even within the oil and gas industry, in EN ISO 13709:2003 Centrifugal pumps for petroleum, petrochemical and natural gas industries and API RP 579 - Fitness for Service
minimum exhaust pressure	The lowest exhaust steam pressure for continuous turbine operation. Defined formally in EN ISO 10437 Petroleum, petrochemical and natural gas industries - Steam turbines - Special-purpose applications
minimum explosible concentration	*aka* MEC. The minimum air concentration of combustible material (w/v) allowing deflagration (qv). Defined formally in the context of dust in NFPA 654: Standard for the Prevention of Fire and Dust Explosions from the Manufacturing, Processing, and Handling of Combustible Particulate Solids, 2017 Edition
minimum hot gas ignition temperature	see **hot gas ignition temperature**
minimum igniting current ratio	*aka* MIC ratio. The minimum ignition current (qv) for a gas divided by the minimum ignition current for methane. Defined formally in API RP 505 2nd Edition, August 2018 Recommended Practice for Classification of Locations for Electrical Installations at Petroleum Facilities Classified as Class I, Zone 0, Zone 1, and Zone 2 and NFPA 497
minimum ignition current	*aka* MIC. The minimum spark current capable of igniting the most ignitable mixture of air and a given dust or gas under specified test conditions. Defined formally in API RP 505 2nd Edition, August 2018 Recommended Practice for Classification of Locations for Electrical Installations at Petroleum Facilities Classified as Class I, Zone 0, Zone 1, and Zone 2

minimum ignition energy	*aka* MIE. The minimum spark energy required to ignite the most ignitable mixture of air and a given dust or vapor under specified test conditions. Defined formally in API RP 505 2nd Edition, August 2018 Recommended Practice for Classification of Locations for Electrical Installations at Petroleum Facilities Classified as Class I, Zone 0, Zone 1, and Zone 2 and NFPA 654: Standard for the Prevention of Fire and Dust Explosions from the Manufacturing, Processing, and Handling of Combustible Particulate Solids, 2017 Edition
minimum ignition temperature	The minimum hot surface temperature required to ignite a cloud of dust in air
minimum inlet pressure and temperature	The lowest instantaneous inlet steam pressure/temperature for turbine operation. Defined formally in EN ISO 10437 Petroleum, petrochemical and natural gas industries - Steam turbines - Special-purpose applications
minimum net flow area	Term used for flow area of burst rupture disc. Defined formally in API RP 520 P1 7th Edition, January 2000 Sizing, Selection, and Installation of Pressure-Relieving Devices in Refineries; Part I - Sizing and Selection
minimum operating steam or gas conditions	The least severe steam or gas conditions (qv) (as opposed to 'lowest' steam or gas conditions) for continuous turbine operation. Defined formally in BS ISO 14661:2000 Thermal turbines for industrial applications (steam turbines, gas expansion turbines) — General requirements; cf maximum operating steam or gas conditions
minimum pressurizing temperature	The lowest temperature at which a vessel should be subjected to greater than 40% of MAWP (qv). Defined formally in API RP 2510A - Fire Protection Considerations for the Design and Operation of Liquefied Petroleum Gas (LPG) Storage Facilities
minimum reflux ratio	*aka* Rm, Rmin. The lowest distillation column reflux ratio (qv) at which component separation is theoretically achievable
minimum surface ignition temperature	see **hot surface ignition temperature**
minimum thickess	A term used in heat exchanger tube design for minimum wall thickness in a new condition, including all allowances. Defined formally in API STD 530 - Calculation of Tube Heater Thickness
minimum water supply	The minimum required quantity of firefighting water. Defined formally in NFPA 1142: Standard on Water Supplies for Suburban and Rural Fire Fighting, 2017 Edition
mining	*aka* mineral extraction. The primary collection of minerals from the environment

misalignment	*aka* mismatch. An axial joint offset; defined formally in ASME BPE (American Society of Mechanical Engineers: Bioprocessing Equipment)
miscible	(Usually of two liquids) capable of being mixed to stable homogeneity; cf immiscible
mismatch	see **misalignment**
mission	Senior management's formally stated view of the purpose of an organization. Defined formally in BS EN ISO 9000 Quality management systems Fundamentals and vocabulary
mist eliminator	*aka* demister. A device which coalesces and collects small solid particles and liquid droplets from a gas stream, see **Figure M.1**. Defined formally in API RP 535 - Burners for Fired Heaters at Refineries; cf inertial mist eliminator, vane type mist eliminator, wire mesh type mist eliminator; separator
mist flow	1. A somewhat contested term, most commonly used for a two-phase flow regime with fine liquid droplets entrained in a continuous high velocity gas phase (low gas velocities leave an annulus of liquid at the wall, producing a regime called annular mist flow). Confusingly, the term is also used in a well completion context for gas bubbles in a continuous liquid phase. The term is also used in a combination of the two senses, as a regime with a mist of liquid in the gas stream, and bubbles in the liquid 2. *aka* mist lift. see **mist lift pump**

Figure M.1 Installation of a Mist Eliminator (Courtesy: Sulzer)

mist flow pump	see **mist lift pump**
mist lift pump	*aka* mist flow pump, steam lift pump. A type of simple low head pump, using compressed steam to reduce the bulk density of water to cause it to rise up a tube as a multiphase flow. A type of gas lift (qv) pump
miter	*aka* mitre. A type of pipe bend made by cutting pipe at an angle and welding the cut pipe ends together. Defined formally in ASME BPE (American Society of Mechanical Engineers: Bioprocessing Equipment)
mitigation	The reduction of undesirable consequences or, sometimes, the measures taken to achieve this
mitre	see **miter**
mixed airflow room	A room, typically in a pharmaceutical manufacturing facility, supplied with air by conventional 'turbulent' means, such as a diffuser or terminal HEPA filter, but which also includes a unidirectional flow zone (such as a hood over a critical area); required total air changes for the room are greatly reduced by the inclusion of the hood
mixed bed	A column of mixed cationic and anionic ion exchange (qv) resins
mixed flow	In the context of pumps, a flow regime offering a compromise between radial and axial flow characteristics
mixed liquor suspended solids	*aka* MLSS. A measure of suspended solids concentration in a biological aeration basin used in activated sludge plant design and operation
mixed pressure turbine	A turbine having multiple working fluid inlets at different pressures. Defined formally in BS ISO 14661:2000 Thermal turbines for industrial applications (steam turbines, gas expansion turbines) – General requirements
mixer	A device for homogenizing a heterogeneous mixture (qv); cf homogenizer
mixer/settler	A mineral processing solvent extraction vessel with sequential mixing and settling processes, commonly combined in a countercurrent series
mixing	The unit operation of homogenization (qv) of a heterogeneous mixture(qv)
mixing blower	A blower mixing and compressing air–fuel gas mixtures supplied to fired heaters. Defined formally in NFPA 86: Standard for Ovens and Furnaces, 2019 Edition
mixing machine	Any machine mixing and compressing air–fuel gas mixtures to fired heaters. Defined formally in NFPA 86: Standard for Ovens and Furnaces, 2019 Edition

mixing rules	*aka* van der Waals mixing rules, random mixing rules. A class of empirically-derived rules applied to equations of state to estimate properties of mixtures
mixproof	The automatic prevention by design of the mixing of two incompatible streams, usually having a drainable, cleanable neutral zone between streams, especially clean and dirty streams in hygienic contexts; see **Box M2**.
mixproof valve	A valve designed to be mixproof (qv); cf zero deadleg valve, **Box M2**.
mixture	The product of mixing (qv), or any heterogeneous combination of substances or phases
MJ	*(drawing notation)* Mechanical joint
ML	Material list
MLSS	see **mixed liquor suspended solids**
MMboed	see **million barrels of oil equivalent per day**
MMBOED	see **million barrels of oil equivalent per day**
Mmboepd	see **million barrels of oil equivalent per day**
MMF	Master manufacturing formula, see master batch record
mmHg	see **millimeters of mercury**
MMI	Man-machine interface, see human machine interface
MMS	see **maintenance management system**
MMSCFPD	see **millions of standard cubic feet per day**
MMVF	see **man made vitreous fiber**
MNPT	Male national pipe thread, see **national pipe thread**
MOAb	see **monoclonal antibody**
mobile container	In the context of LPG storage, any vehicle-mounted container (qv) other than its fuel source. Defined formally in NFPA 58: Liquefied Petroleum Gas Code, 2017 Edition
mobile equipment	Process equipment that is not fixed in place and is capable of being moved by a process operator(s) within a process train and/or for cleaning or other operations
mobile phase	In the context of chromatography, the materials which move through or past the stationary phase (qv)
mobile system	A term used for any wheeled firefighting foam-producing unit. Defined formally in NFPA 11: Standard for Low-, Medium-, and High-Expansion Foam, 2016 Edition
mobile water supply apparatus	Any vehicle-mounted firewater delivery system. Defined formally in NFPA 1142: Standard on Water Supplies for Suburban and Rural Fire Fighting, 2017 Edition
mobile water supply tanker	see **mobile water supply apparatus**

Box M2. Mixproof

In my professional practice I have seen failures to segregate clean in place (CIP) fluids from product in the food and drink sector which led to an expensive product contamination issue. Such incidents can be avoided through good 'mixproof' design, though there are essentially two mutually incompatible schools of thought on how to proceed.

In the pharmaceutical industry, an arrangement of zero deadleg valve (ZDL) types is used to produce a mixproof situation for both CIP and sterilization in place (SIP) scenarios. The food and drink industry, on the other hand, uses 'double seat mixproof valves', since ZDL valve arrangements are considered insufficiently mixproof. They tend not to use SIP, so the shortcomings of their favored valve in SIP situations do not trouble them.

Both industries are effectively regulated worldwide by the US FDA, but they differ markedly in their interpretation of its requirements with respect to product contamination, hygiene and cleanliness. Essentially, pharma designers consider their ZDL arrangement always superior to the food industry's double-seat mixproof valve, (and vice versa), on what looks like a version of not invented here syndrome.

But what about other valve types? If we accept that the essence of 'mixproofness' is a double block and bleed arrangement with an ability to see if one of the valves is passing, can this not be achieved with something cheaper and easier to control than a ZDL diaphragm valve? What about the water engineer's favourite, the (not always trusty) wafer pattern butterfly valve?

Whilst you can buy something called a 'hygienic butterfly valve' or even a so-called dedicated 'sanitary' double disc /double seal butterfly valve, these items might not fully live up to their description. They may be made from polished stainless steel, with sanitary connections, but when they are described as 'FDA approved', this usually refers only to the seal elastomers. Butterfly valves are mentioned in EHEDG guidance but they are not presently 3-A approved, nor mentioned by ASME BPE.

So, whether they are hygienic enough as a standalone item is questionable, if you agree there are degrees of hygiene (and not everyone in the hygienic industries does – many will insist that hygiene is absolute, rather like 'absolute safety' and 'zero environmental impact'). They are not perfectly hygienic, but then, nothing is, in practice. Putting a few of these valves together to make a double block and bleed might arguably multiply hygienic problems, but even if the more favoured ZDL-based mixproof arrangements or double seat mixproof valves are perfectly hygienic when first installed, they will not be later on, even when maintained in accordance with manufacturer's instructions. It's all a question of degree.

Related Terms
hygiene, clean

MOC	1. see **management of change**
	2. see **material of construction**
modal analysis	A way of characterizing a complex linear system of mechanical vibration by modes of vibration. Defined formally in BS ISO 2041:2018 Mechanical vibration, shock and condition monitoring. Vocabulary
modal stiffness	The stiffness element of a given mode of vibration; defined formally in BS ISO 2041:2018 Mechanical vibration, shock and condition monitoring. Vocabulary
Modbus	Proprietary eponym for a data communications protocol
model	Most commonly used nowadays to refer to a computer model, though can still sometimes mean a physical representation, as defined formally in BS EN ISO 10209:2012 Technical product documentation — Vocabulary — Terms relating to technical drawings, product definition and related documentation
model identification	The identification of the best mathematical model to quantify process dynamics
modeling	see **process modeling**
moderator	A material which slows fast neutrons in a nuclear reactor
modification	Altering a product for uses outside its original specification. Defined formally in BS EN 82079-1:2012 Preparation of instructions for use. Structuring, content and presentation. General principles and detailed requirements
modular	Term for plant section constructed offsite as a module (qv) in a yard or factory and transported to site; cf stick built
modular construction	A system where each section of a plant, or module (qv), is factory-fabricated such that site works consist only of linking these modules together. Modules are usually not capable of standalone operation, unlike skid (qv) mounted packages
modular process skid	A type of skid (qv) rather than a module (qv)
modulating control	Control of an actuator across a wide range of positions; cf on/off control
modulating valve	A valve equipped for modulating control (qv)
module	*aka* pre-assembled unit (PAU). A section of a plant preassembled offsite and shipped as a substantially complete assembly, allowing rapid plant construction
Mogden formula	A formula used in the UK to calculate trade effluent charges based on effluent suspended solids, COD (qv) concentration and daily volume
Mohs scale of hardness	(N.B. not 'Moh's' scale) An ordinal mineral hardness scale based on a ranking of 'what can scratch what'

moiety	One of the portions into which something is divided; a component, part, or fraction. In chemistry, a specific section of a molecule, usually complex, that has a characteristic chemical effect or pharmacological property
Moineau pump	see **progressing cavity pump**
moisture	A somewhat vague term for small quantities of a room-temperature liquid (especially water) as a component of a mixed vapor, or within the body of and/or on the surface of a solid; cf humidity
moisture content	see **wet basis**
Mokveld valve	A type of non-slam check valve; cf one way valve
molal average boiling point	*aka* MABP. One of five ways used in oil refining to express average boiling point of a mixture of hydrocarbons (the others being volume, weight, cubic and mean)
molal volume	The volume occupied by one mole of an ideal gas at arbitrarily selected 'standard conditions'
molality	Concentration expressed as moles of solute per kilogram of solvent
molar density	Most commonly, the number of moles of something in a unit volume; sometimes used for mass of a species per unit volume
molar gas constant	see **gas constant**
molar units	There is a suite of units starting with 'molar' or 'mole', such as molar flow rate, equivalent to their corresponding 'mass' unit term e.g.; molar flowrate is moles per unit of time, rather than mass per unit of time; (N.B. these have not been listed separately in this dictionary)
mold flash	*aka* flash. The excess material which escapes through the seal of a mold and sets, and must subsequently be removed. Defined formally in ASME BPE (American Society of Mechanical Engineers: Bioprocessing Equipment)
molded seal	A seal manufactured by molding. Defined formally in ASME BPE (American Society of Mechanical Engineers: Bioprocessing Equipment)
mole balance	A mass balance (qv) where masses of substances are expressed as moles
mole fraction	The number of moles of a component in question divided by the total moles of all components present
mole plough	Equipment used to place small bore pipes and cables directly in the ground; not strictly considered trenchless technology (qv)
mole ratio	The ratio in moles of any two compounds involved in a chemical reaction
mole sieve	*(informal)* see **molecular sieve**

molecular biology	The study and understanding of biological activity and processes at the molecular level; and especially, the molecular basis of gene function and protein synthesis
molecular diffusion	see **diffusion**
molecular distillation	A short-path high-vacuum distillation process used for high-value heat labile (qv) substances such as fragrance molecules
molecular sieve	*aka* mole sieve. Crystalline metal aluminosilicates (qv zeolite), treated so as to have adsorbent cavities, uniformly sized to match specific molecules
Mollier chart	see **Mollier diagram**
Mollier diagram	*aka* Enthalpy-Entropy chart, H-S chart, Mollier chart. A plot of fluid enthalpy (qv), H against entropy (qv), S; cf psychrometric chart (a different plot of the same data)
molten salt bath furnace	A process heater producing molten salts used for high temperature process heating or metal treatment. Defined formally in NFPA 86: Standard for Ovens and Furnaces, 2019 Edition
Molykote	Proprietary eponym for molybdenum disulfide lubricants
moment	see **torque**
moment of a force	see **torque**
moment of force	see **torque**
Mond index	A hazard index
Mond nickel process	An obsolete nickel purification process
monel	A strong nickel-copper alloy with excellent resistance to chloride corrosion
monitor	1. A visual display unit (qv) or human machine interface (qv) 2. A limit transducer, as defined formally in BS EN 12952-8:2002 Water-tube boilers and auxiliary installations. Requirements for firing systems for liquid and gaseous fuels for the boiler 3. A firefighting foam turret, capable of directing and controlling a flow of foam
monitoring	"Conducting a planned sequence of observations or measurements to assess whether control measures are operating as intended. Note: Monitoring is done while the control measures are operating" according to the EHEDG Glossary. An alternative definition in a quality context may be found in BS EN ISO 9000 Quality management systems Fundamentals and vocabulary

monitoring pressure regulator	A standby pressure regulator placed in series with a duty pressure regulator, to control overpressure in the event of failure of the duty regulator. Defined formally in NFPA 86: Standard for Ovens and Furnaces, 2019 Edition
monkey metal	*(informal) aka* pot metal. Low grade, low melting metal alloy, especially that used for castings; cf chinesium
monkey wrench	*aka* pipe wrench; cf crescent wrench, Stillson
monkeyboard	In oil drilling, a platform above the oil rig floor where a derrickman (qv) works
Mono pump	Proprietary eponym synonymous with progressing cavity pump (qv) in water treatment[75]; cf screw pump
monoclonal antibody	*aka* mAb, MAB, MoAb. Antibodies derived from a single clone of cells that recognize only one type of antigen
Monod equation	*aka* Monod kinetics. The empirical version of the Michaelis-Menten equation (qv), describing the relationship between reaction rate and substrate concentration for enzyme catalyzed reactions or bacterial cultures
Monod kinetics	see **Monod equation**
monoethanolamine	*aka* MEA. A chemical used in the amine gas treating process (qv) to remove acid gas (qv) from hydrocarbons
monolithic lining	A single layer, single component refractory lining. Defined formally in BS EN ISO 13705:2012/ISO 13705:2012(E) Petroleum, petrochemical and natural gas industries. Fired heaters for general refinery service, API STD 530 - Calculation of Tube Heater Thickness and API STD 560 - Fired Heaters for General Refinery Service
monomer	The basic subunit from which, by repetition of a single reaction, a polymer (qv) is made
monomeric silica	*aka* dissolved- or reactive- silica. The simplest form of silica (SiO_2)
monosaccharides	The building blocks of carbohydrates, hence known as 'simple sugars'; classified by the number of carbon atoms in the molecule: pentoses have five and hexoses six
Monte Carlo method	see **Monte Carlo simulation**
Monte Carlo simulation	*aka* Monte Carlo method, probability simulation. A way of modelling the probability of a number of different outcomes in a complex system using various computational algorithms with many practical applications

---

[75] Even though Mono manufactures various other types of pumps, and most 'mono pumps' aren't actually made by Mono

montejus	1. Strictly, a cylindrical vessel steel similarly operated to an acid egg (qv) 2. Sometimes used as a synonym for acid egg
Moody chart	see **Moody plot**
Moody diagram	see **Moody plot**
Moody plot	*aka* Moody chart, Moody diagram. A summary of empirical experiments as a dimensionless plot of friction factors[76]
moon pool	*aka* wet porch. An opening underneath a drillship (qv) or drilling platform into a pressurized chamber, used to pass drilling tools, divers and ROVs into the water
MOP	see **maximum operating pressure**
mortar	In the context of fired heaters, a refractory cement used with refractory bricks. Defined formally in BS EN ISO 13705:2012/ISO 13705:2012(E) Petroleum, petrochemical and natural gas industries. Fired heaters for general refinery service, API STD 530 - Calculation of Tube Heater Thickness and API STD 560 - Fired Heaters for General Refinery Service
mothballing	Placing a plant in a safe state ready for a planned extended period of disuse
mother liquor	The residual liquid that remains after crystallization or isolation processes which may contain unreacted materials, intermediates, product and/or impurities. It may be used for further processing
motionless mixer	see **inline mixer**
motor control center	*aka* motor control centre, MCC, controlgear. A term favored in Europe for a cabinet downstream of switchgear (qv) which controls motors on a plant, containing motor starters, instrumentation, power incomers and possibly a PLC (qv)
motor control centre	see **motor control center**
motor operated valve	*aka* MOV; see **actuated valve**
motor speed	Speed on a motor nameplate. Defined formally in NFPA 20: Standard for the Installation of Stationary Pumps for Fire Protection, 2019 Edition
mounded container	An ASME container (qv) designed to be covered in a mound of earth; defined formally in NFPA 58: Liquefied Petroleum Gas Code, 2017 Edition
mounting bolts	see **hold down bolts**

---

[76] Essentially an admission of defeat on the part of the mathematicians responsible for fluid mechanics—they couldn't make their sums work without these fiddle factors taken from experimental data). Nowadays, superseded by the use of curve-fitting equations in MS Excel (the output of which is probably closer to the true experimental value than can be read from a printed Moody chart)

mouse hole	In oil drilling, a hole in the rig floor where a piece of drilling string (qv) or coring tool (qv) can be temporarily stored
MOV	see **motor operated valve**
moveable fuel storage tender	*aka* farm cart. A wheeled movable cargo tank used to transport fuel (but not LPG). Defined formally in NFPA 58: Liquefied Petroleum Gas Code, 2017 Edition
moveable guard	A guard (qv) which is openable without a tool. Defined formally in BS EN ISO 12100:2010 Safety of machinery. General principles for design. Risk assessment and risk reduction
moveable rest landing	An area in an access ladder system allowing a user to rest but not interchange. Defined formally in BS EN ISO 14122 Safety of machinery. Permanent means of access to machinery. Working platforms and walkways
moving bed biofilm reactor	*aka* MBBR. A bioreactor used in effluent treatment with growth attached to small neutrally buoyant moving plastic carriers
Moyno pump	see **progressing cavity pump**
MPa	Absolute pressure (qv) in megapascals
MPag	Gauge pressure (qv) in megapascals
MPC	see **multivariable predictive control**
MPEG	Methoxy polyethylene glycol
MPR	Master production record, see **master batch record**
MPT	Male pipe thread, see **national pipe thread**
MPY	Mils per year
MRP	see **materials resource planning**
MRR	see **materials receiving report**
MS	1. see **mass spectrometry** 2. see **method statement**
MSDS	see **material safety data sheet**
MSF	see **multistage flash distillation**
MSL	Mean sea level
MSR	Maximum speed rise
MSSP	see **maximum static sealing pressure**
MSW	see **municipal solid waste**
MT	1. Magnetic particle examination, according to API RP 579 - Fitness for Service 2. Magnetic particle testing, according to API RP 571 - Damage Mechanism Affecting Fixed Refinery Equipment
MTBF	see **mean time between failures**
MTG	see **methanol to gasoline**
MTO	Material take-off
MTR	see **material test report**
MTTF	see **mean time to failure**

MTTR	see **mean time to repair**
MTU	see **metric tonne unit**
Mucon	Proprietary eponym for an iris diaphragm valve
Mudan model	A model used to calculate heat flux from flames
mudballing	The formation of balls of mud within a depth filter (qv) bed which do not break up on backwashing, caused by insufficient air scour (qv)
MUF	see **material unaccounted for**
muffle	An arrangement to separate the heat source from the working chamber of a furnace. Defined formally in NFPA 86: Standard for Ovens and Furnaces, 2019 Edition; cf muffler
muffle block	see **tile**
muffler	1. Generally, US English for engine 'silencer' 2. Specifically, a device to reduce the propagation of combustion noise back through fired heater burner. Defined formally in API RP 537 Flare Details for Petroleum, Petrochemical, and Natural Gas Industries and API RP 535 - Burners for Fired Heaters at Refineries
muffler block	see **tile**
Müller–Kühne process	The coproduction of sulfuric acid and Portland cement by thermal decomposition of gypsum
multiburner flare	A flare with multiple (usually staged) burners; defined formally in API RP 537 Flare Details for Petroleum, Petrochemical, and Natural Gas Industries
multicomponent distillation	Distilling mixtures with more than two components
multicomponent lining	A layered system of multiple types of refractories. Defined formally in BS EN ISO 13705:2012/ISO 13705:2012(E) Petroleum, petrochemical and natural gas industries. Fired heaters for general refinery service, API STD 530 - Calculation of Tube Heater Thickness and API STD 560 - Fired Heaters for General Refinery Service; cf multilayer lining
multicomponent separation	Separating mixtures with more than two components
multicore cable	A cable with multiple conductors, insulated from each other within a common outer insulation; cf stranded conductors
multi-effect distillation	see **multiple effect distillation**
multifamily dwelling	A building comprising at least three dwelling units; defined formally in NFPA 30: Flammable and Combustible Liquids Code, 2018 Edition

multifuel burner	A burner which can burn more than one fuel separately or simultaneously. Defined formally in BS EN 12952-8:2002 Water-tube boilers and auxiliary installations. Requirements for firing systems for liquid and gaseous fuels for the boiler; cf dual fuel burners
multifuel firing system	A firing system with a common combustion chamber which can burn more than one fuel separately or simultaneously. Defined formally in BS EN 12952-8:2002 Water-tube boilers and auxiliary installations. Requirements for firing systems for liquid and gaseous fuels for the boiler
multilayer lining	A layered system of a single type of refractory. Defined formally in BS EN ISO 13705:2012/ISO 13705:2012(E) Petroleum, petrochemical and natural gas industries. Fired heaters for general refinery service, API STD 530 - Calculation of Tube Heater Thickness and API STD 560 - Fired Heaters for General Refinery Service; cf multicomponent lining
multipass	A type of system which passes material through a process several times
multiphase	Having multiple phases
multiple degrees of freedom system	*aka* MDoF system. A system whose state at a given instant cannot be defined completely by a single coordinate; qv degrees of freedom; cf SDoF system
multiple effect	A separation process with multiple linked (usually countercurrent) stages
multiple effect distillation	*aka* MED, multi-effect distillation, multistage flash distillation. A system consisting of a series of stages or 'effects', each comprising a consecutive condensation/evaporation process occurring in a decreasing order of temperatures and pressures and typically consisting of a series of shell and tube heat exchangers (qv)
multiple port burner	A burner with multiple discharge openings (ports). Defined formally in NFPA 86: Standard for Ovens and Furnaces, 2019 Edition
multiple turn valve	*aka* multiturn valve. A valve requiring multiple 360-degree rotations to go from fully open to fully closed position, such as a globe or needle valve; cf quarter turn valve
multipoint flare	A flare comprising multiple burners. Defined formally in API RP 537 Flare Details for Petroleum, Petrochemical, and Natural Gas Industries
multiproduct facility	*aka* multiuse facility. A facility that supports production of two or more products, either in a campaigned or concurrent manner

multistage	see **multiple effect**
multistage flash distillation	see **multiple effect distillation**
multistage multiport pump	A pump having several stages, and several ports. Defined formally in NFPA 20: Standard for the Installation of Stationary Pumps for Fire Protection, 2019 Edition
multiturn valve	see **multiple turn valve**
multiuse facility	see **multiproduct facility**
multivariable model predictive control	see **multivariable predictive control**
multivariable predictive control	*aka* multivariable model predictive control, MPC. An approach to control of processes with multiple non-independent inputs and outputs
municipal solid waste	*aka* MSW, garbage, rubbish, trash. Solid wastes arising from households
municipal type water system	Most commonly, a system similar to that used for municipal water supply. In the context of firewater, a system of pipes and hydrants designed to supply at least 950 l/min of firewater at least 138 kPa for two hours. Defined formally in NFPA 1142: Standard on Water Supplies for Suburban and Rural Fire Fighting, 2017 Edition
Murphree plate efficiency	*aka* Murphree tray efficiency, Murphree vapor efficiency. A fiddle factor (qv) used alongside McCabe-Thiele (qv)
Murphree tray efficiency	see **Murphree plate efficiency**
Murphree vapor efficiency	see **Murphree plate efficiency**
muster area	A place where site occupants are required to gather in an emergency
mutual aid agreement	*aka* mutual assistance agreement. A formal arrangement between parties to share emergency response resources. Defined formally in NFPA 1142: Standard on Water Supplies for Suburban and Rural Fire Fighting, 2017 Edition
mutual assistance agreement	see **mutual aid agreement**
MV	1. see **manipulated variable** 2. see **measured value** 3. see **manual valve**
MW	*(drawing notation)* Manway
mWG	see **meters water gauge**

N

NAAQs	see **National Ambient Air Quality Standards**
NAC	Naphthenic acid corrosion
nano-	(*symbol* n-) The SI unit prefix denoting a factor of 10^{-9}
nanofiltration	Membrane filtration processes which overlap with the coarse end of reverse osmosis (qv) and the fine end of ultrafiltration (qv); see **Box L1**.
nanomaterials	Materials with a particle size most conveniently measured in nanometers
naphtha	A naphthenic petrochemical, and an obsolete name for crude oil (qv)
naphthalene	Specifically, the simplest polynuclear aromatic hydrocarbon; cf naphthenes
naphthenes	*aka* cycloparaffins. An old term for cycloalkanes, still common in oil and gas
naphthenic	An oil and gas industry term for a hydrocarbon mixture containing a high proportion of hydrogenated benzenes and naphthenes
NAPLs	see **non-aqueous phase liquids**
nappe	*aka* vain. The sheet of water which flows over a weir (qv)
naphthenic acids	An oil and gas industry term for a mixture of cycloalkane carboxylic acids
National Ambient Air Quality Standards	*aka* NAAQs. US air quality standards under the Clean Air Act
national pipe taper	see **national pipe thread**
national pipe tapered thread	see **national pipe thread**
national pipe thread taper	see **national pipe thread**
national pipe thread	*aka* American national standard pipe thread, NPT, national pipe taper, national pipe thread taper, national pipe tapered thread, sealing thread, tightly threaded connection. A USA thread standard, measured in US customary units (qv), which may be tapered or straight. When externally threaded, may be known as male pipe thread (MPT), MNPT, or NPT(M). When internally threaded, may be known as female pipe thread (FPT), FNPT, or NPT(F). There are many other variants and associated abbreviations.

national pipe thread water oil gas	*aka* NPT WOG. A commonly used NPT threaded valve type acceptable for use with water, oil or gas, not meeting ASME B31.34
National Pollutant Discharge Elimination System	*aka* NPDES. US water quality standards under the Clean Water Act
Natta catalyst	see **Ziegler-Natta catalyst**
natural convection	*aka* free convection. Heat transfer by convection driven only by reduced density of heated fluid; cf forced convection
natural draft	In the context of fired heaters, the pressure difference across the system, i.e.; draft (qv) driven only by stack effect (qv) (reduced density of heated gases); defined formally in API RP 535 - Burners for Fired Heaters at Refineries; cf induced draft
natural draft heater	see **natural draught heater**
natural draught heater	*aka* natural draft heater. A fired heater with combustion air induced (and flue gases removed) by stack effect (qv) (reduced density of heated gases) alone; defined formally in BS EN ISO 13705:2012/ISO 13705:2012(E) Petroleum, petrochemical and natural gas industries. Fired heaters for general refinery service, API STD 530 - Calculation of Tube Heater Thickness and API STD 560 - Fired Heaters for General Refinery Service; cf induced draft heater
natural frequency	The free undamped linear vibration frequency of a mechanical system; defined formally in BS ISO 2041:2018 Mechanical vibration, shock and condition monitoring. Vocabulary
natural gas	A methane based fossil fuel gas
natural gas condensate	see **natural gas liquids**
natural gas hydrate	see **methane clathrate**
natural gas liquids	*aka* NGL, natural gas condensate, gas condensate, condensate, condy, drip gas. Liquids that drop out of natural gas during conditioning and processing at a gas plant, (or in the case of drip gas, in a pipeline) typically comprising C2 (qv) hydrocarbons such as ethane and ethylene; C3 (qv), C4 (qv) and some heavier hydrocarbons; cf liquefied natural gas
natural mode of vibration	The vibration mode of a freely vibrating system; defined formally in BS ISO 2041:2018 Mechanical vibration, shock and condition monitoring. Vocabulary
natural science	The activity of trying to understand natural phenomena; cf engineering science
Nauta	Proprietary eponym for batch conical screw mixer used for free-flowing powders and pastes, especially in the pharmaceutical industry

naval brass	see **marine brass**
Navier–Stokes equations	A mathematical model of viscous incompressible flow
NB	see **nominal bore**
NC	1. see **nominal capacity**
	2. *(drawing notation)* Normally closed
NDE	Non-destructive examination
NDT	see **nondestructive testing**
needle coke	*aka* acicular coke. A high grade of petroleum coke (qv)
negative list	see **exclusion list**
negative peak value	see **peak value**
negative pressure	*aka* vacuum pressure. Pressure lower than atmospheric pressure
negative pressure pneumatic conveying system	A pneumatic conveying system (qv) driven by gas under negative pressure (qv); defined formally in NFPA 654: Standard for the Prevention of Fire and Dust Explosions from the Manufacturing, Processing, and Handling of Combustible Particulate Solids, 2017 Edition
negative recruitment period	*aka* layoff. A management speak (qv) euphemism for firing workers
Nelson–Farrar cost index	*aka* NFCI. A venerable oil and gas industry cost index (qv).
neoprene	A polychloroprene elastomer
nephelometric turbidity unit	*aka* NTU. A unit of turbidity (qv), most often used when referencing the USEPA Method 180.1 or Standard Methods For the Examination of Water and Wastewater
net heating value	see **lower heating value**
net positive suction head	*aka* NPSH. A way to check that pump suction pressure does not exceed vapor pressure under prevailing conditions, to avoid cavitation (qv); defined formally in NFPA 20: Standard for the Installation of Stationary Pumps for Fire Protection, 2019 Edition, API RP 2510A - Fire Protection Considerations for the Design and Operation of Liquefied Petroleum Gas (LPG) Storage Facilities and EN ISO 13709:2003 Centrifugal pumps for petroleum, petrochemical and natural gas industries
net positive suction head available	*aka* NPSHa, available net positive suction head (qv). NPSH calculated for a system; this particular term defined formally in EN ISO 13709:2003 Centrifugal pumps for petroleum, petrochemical and natural gas industries
net positive suction head required	*aka* NPSHr. A manufacturer's specified minimum NPSH for a given pump; defined formally in EN ISO 13709:2003 Centrifugal pumps for petroleum, petrochemical and natural gas industries

net present value	*aka* NPV. One of a number of accounting techniques which discount future costs and benefits of a project
net pressure	The pressure difference between pump suction and discharge flanges; defined formally in the context of firewater pumps in NFPA 20: Standard for the Installation of Stationary Pumps for Fire Protection, 2019 Edition; cf differential pressure
net rate	In the context of fire protection spray systems, the rate of water discharge, ignoring any water that does not contribute to fire protection; defined formally in NFPA 15 Standard for Water Spray Fixed Systems for Fire Protection
net specific energy	The minimum energy per unit of mass of a specified fuel, ignoring the latent heat of condensation of water; defined formally in ISO 3977 Gas turbines - Procurement - Part 3: Design requirements cf LHV
net working capital	*aka* NWC; see **working capital**
net zero	see **Box N1**.

Box N1. Net Zero

The concept of net zero was originally closely related to the older one of 'carbon neutrality', a general idea that the carbon dioxide releases associated with an activity could be offset or undone by undertaking (or paying others to undertake on your behalf) carbon-absorbing activities. Some still use this definition, but other seek to take net zero further, taking non-carbon greenhouse gases into consideration. Some go further still, rejecting science-based approaches entirely.

On the other hand, one of my panel of reviewers suggested net zero was no more than 'a marketing-related calculation arrived at by performing a very naive mass balance over a small, arbitrarily defined part of a much larger system', or to put it another way, greenwash.

However far people go, net zero remains controversial amongst some environmentalists as it is perceived as not going far enough, whilst some trade unions have sought to harness the idea to their own political agendas by including for example a requirement for a 'just transition', favoring their members in the interim.

Perhaps a more reasonable and less self-interested criticism of net zero is it can be rather like net present value, kicking the can of reining in carbon emissions into the future, whilst we continue business as usual in the present.

Whatever your view of these issues, it is clear that net zero can mean a range of things, most notably in respect of what should add up to zero, the period over which the calculation is done, and whether or not to include the various unrelated issues which stakeholders may seek to smuggle in alongside the actual reduction in greenhouse gas emissions. In this way it is rather like sustainability.

network diagram	A diagram illustrating interconnections between e.g.; the stages of a project, or an electrical or electronic network; defined formally in only the second of these senses in ISO 10209:2012 Technical product documentation – Vocabulary – Terms relating to technical drawings, product definition and related documentation
network map	A diagram illustrating interconnections of an electrical or electronic network, or showing such interconnections on a physical map; defined formally in only the second of these senses in ISO 10209:2012 Technical product documentation – Vocabulary – Terms relating to technical drawings, product definition and related documentation
neutral spirit	*aka* ethyl alcohol of agricultural origin, distillate of agricultural origin, rectified alcohol, rectified spirit. Distilled potable alcohol of 55-97% ethanol ABV (qv)
neutralisation number	see **neutralization number**
neutralization number	*aka* neutralisation number; see **acid number**
neutron absorber	see **absorber**
new product development	*aka* NPD. The complete process from inception to marketing of a new product, including the ill-defined start and end phases; cf new product introduction
new product introduction	*aka* NPD. The later and more formal part of the product development process from around design completion to marketing of a new product; cf new product development
new source performance standards	*aka* NSPS. USA pollution control standards for major new pollution sources under the Clean Air and Clean Water Acts
new source review	*aka* NSR. A USA permitting program for major new pollution sources under the Clean Air Act
newton	(*symbol* N) The SI unit of force
newton meter	(*symbol* Nm) The SI unit of torque
Newton number	see **power number**
newton per meter	(*symbol* N/m) The SI unit of surface tension
Newton's law of cooling	An empirical 'law' (at best approximately true in the real world) which states that the rate of cooling of an object is proportional to the temperature difference between the object and its surroundings
Newtonian fluid	A fluid with a viscosity which does not vary with shear stress; cf non Newtonian fluid
Newton–Raphson method	*aka* Newton's method. A mathematical procedure for finding approximate solutions to equations by means of iteration
NFCI	see **Nelson-Farrar cost index**
NFPA	National Fire Protection Association (USA)

NGL	see **natural gas liquids**
NH3-N	The most common abbreviation of ammoniacal nitrogen (qv)
NH4HS	Ammonium bisulfide
NH4-N	Less common abbreviation of ammoniacal nitrogen
nibbler	1. A tool for cutting metal plate
	2. A type of mill (best described as a 'giant parmesan grater') used to break up chocolate crumb into small enough pieces for the next stage of its refining process
Nichols plot	*aka* Nichol plot, Nichol's plot[77]. A graph used in control engineering, similar to a Nyquist plot (qv), but with cartesian rather than polar coordinates
nick	A surface dent or cut, especially one with a 'vee' cross section. A compatible formal definition may be found in ASME BPE (American Society of Mechanical Engineers: Bioprocessing Equipment)
nickel base	Alloys with more than 30% nickel content; defined formally in API RP 571 - Damage Mechanism Affecting Fixed Refinery Equipment
NIH	see **not invented here**
nil ductility temperature	*aka* NDT. The temperature at which metal fracture mode changes from ductile to brittle; defined formally in API RP 579 - Fitness for Service
NIOSH	National Institute for Occupational Safety and Health (USA)
nip	*(drawing notation)* Nipple (qv); see **Figure N1**
nipple	A pipe connection; a short pipe with male threads on both ends; see **Figure N1**

Figure N1 Example of a nipple pipe connection

---

[77] Incorrectly, in both cases, since the inventor was Nichols, not Nichol

Nippolet	A proprietary eponym, an abbreviated form of NIPple outLET. One of various olet (qv) types
nitriding	A form of case hardening (qv) of steel using nitrogen; cf carburization
nitrification	Biological oxidation of ammonia in effluent treatment; cf denitrification
nmax	see **maximum continuous operating speed**
nmax	see **maximum continuous operating speed**
nmin	see **minimum continuous operating speed**
nmin	see **minimum continuous operating speed**
NMR	Nuclear magnetic resonance
NO	*(drawing notation)* Normally open
no flow	*aka* churn, shutoff. Generically, the condition in which a pump runs with a closed outlet. Defined formally, in the case of firepumps, in NFPA 20: Standard for the Installation of Stationary Pumps for Fire Protection, 2019 Edition
noble gases	*aka* inert gases, aerogens, rare gases. The group 18 elements (e.g.; helium, neon, argon, krypton, xenon, radon)
n-octanol-water partition coefficient	see **octanol/water partition coefficient**
NOD	see **nominal outside diameter**
nodding donkey	*aka* beam pump, big Texan pump, dinosaur pump, donkey pumper, grasshopper pump, horsehead pump, nodding donkey pump, pumpjack, pumping unit, rocking horse pump, sucker rod pump, thirsty bird pump, walking beam pump. An aboveground driver for an underground piston pump in an onshore oil well
nodding donkey pump	see **nodding donkey**
noise	1. An undesirable, random, disruptive sound 2. By analogy, an undesirable, random, disruptive signal; defined formally in BS ISO 2041:2018 Mechanical vibration, shock and condition monitoring. Vocabulary
Nomex	Proprietary eponym for clothing material used in fire retardant PPE
nominal	Whilst the word nominal is to do with 'name', that 'name' in engineering is most commonly a number, and most often refers to an integer approximating a specified value (e.g.; nominal pipe size)
nominal bore	*aka* NB. In Europe, a metric pipe size specification synonymous with DN (qv); in the US, synonymous with nominal pipe size (qv) (NPS) measured in 'British Units'

nominal capacity	The nominal (qv) volumetric liquid capacity of a tank; defined formally in EN 12566 - Small wastewater treatment systems for up to 50 PT Part 1: Prefabricated septic tanks and API STD 620 - Low Pressure Storage Tanks
nominal filter	A filter which removes a stated minimum percentage of hard spherical particles greater than its rated pore size under constant low pressure in a laboratory; this measure should not however be assumed to predict real world behavior; cf absolute filter
nominal filter rating	The approximate minimum percentage removal of test material by a filter of test particles greater than a specified size
nominal liquid capacity	see **nominal capacity**
nominal outside diameter	*aka* NOD. The nominal (qv) outside diameter of a pipe
nominal pipe size	*aka* NPS. US standard pipe sizes by nominal bore (qv) in inches
nominal wall thickness	Pipe nominal (qv) wall thickness, especially in the case of NPS pipes
nomogram	*aka* nomograph. A graph which allows you to look up an approximate value for something unknown based on known values
nomograph	see **nomogram**
non absorbant material	see **non absorbent material**
non absorbent material	"Materials which, under the intended conditions of use, do not internally retain substances with which they come into contact" according to the EHEDG Glossary. Very similar definition in BS EN 1672-2:2005+A1:2009 Food processing machinery. Basic concepts. Hygiene requirements. In addition, BS EN ISO 14159:2008 Safety of machinery. Hygiene requirements for the design of machinery has a similar definition, but it spells absorbent as 'absorbant'
non air aspirating discharge device	A class of devices which make and discharge firefighting foam; defined formally in NFPA 11: Standard for Low-, Medium-, and High-Expansion Foam, 2016 Edition; cf air aspirating discharge device
non Fickian diffusion	Diffusion not obeying Fick's laws of diffusion (qv)
non food area	Not a food area (qv)
non polar covalent	see **covalent**
non product contact surface	Not a product contact surface (cf)
non pusher seal	A pump seal with a static secondary seal; defined formally in API RP 682 - Pump Seals

nonabsorbing ground	Ground with low liquid permeability or absorbency; defined formally in NFPA 15 Standard for Water Spray Fixed Systems for Fire Protection
non-aqueous phase liquids	*aka* NAPLs. An environmental engineering term for a liquid with a different density from, immiscible with, and insoluble in, water; qv dense non-aqueous phase liquid, light non-aqueous phase liquid
non-capital costs	Project costs which do not meet the definition of capital costs, such as inventory; cf operating costs
noncombustible	Not combustible (qv). However, qv noncombustible material
noncombustible material	1. Generally used to describe any material which will not burn under foreseeable conditions of use 2. Could however be used to mean material which does not ignite, burn, or release combustible vapor when heated, as per formal definition in NFPA 654: Standard for the Prevention of Fire and Dust Explosions from the Manufacturing, Processing, and Handling of Combustible Particulate Solids, 2017 Edition; cf **Box F1**.
noncompetitive inhibition	Enzyme inhibition caused by something which binds somewhere other than the active site
noncondensable gas	Generically, a gas or vapor which does not condense under process conditions, the opposite of condensable gas (qv). Slightly different definition of this term in API Standard 521 Pressure-relieving and Depressuring Systems. Sixth Edition | January 2014
nonconformities per million opportunities	*aka* NPMO; see **defects per million opportunities**
nonconformity	The opposite of conformity (qv)
noncontacting seal	A seal other than a contacting seal (qv)
non-destructive testing	see **nondestructive testing**
nondestructive testing	*aka* NDT, non-destructive testing. A material testing method which does not damage test materials
nonelevated flare	see **ground flare**
non-enzymatic browning	In the context of foods, the Maillard reaction (qv) and caramelization (qv)
non-feedback control	see **open loop control**
nonferrous metal	A metal other than a ferrous metal (cf)
nonflammable	Not flammable (cf)
nonflammable special atmosphere	In the context of fired heaters, a special atmosphere of nonflammable gases; defined formally in NFPA 86: Standard for Ovens and Furnaces, 2019 Edition

nonflashing hydrocarbon	A hydrocarbon other than a flashing hydrocarbon (cf)
nonfragmenting rupture disk	A rupture disk which does not impair operation of downstream NRVs after rupture; defined formally in API RP 520 P1 7th Edition, January 2000 Sizing, Selection, and Installation of Pressure-Relieving Devices in Refineries; Part I - Sizing and Selection
nonhazardous location	Not a hazardous (classified) location (cf)
nonhydrocarbon service	In the context of pumps in the oil and gas industry, generally speaking, pumping aqueous solutions, defined formally in API RP 682 - Pump Seals
nonideal flow	Real world flow, rather than theoretical
non-ideal gas	see **real gas**
nonideal mixture	Real world mixture, rather than theoretical
nonlinear damping	Damping other than linear damping (cf)
nonlinear vibration	see **rectilinear vibration**
nonmetallic container	In the context of flammable liquid storage, a container (qv) which is not made of metal
nonmetallic intermediate bulk container	In the context of flammable liquid storage, an intermediate bulk container (qv) which is not made of metal
nonmetallic portable tank	In the context of flammable liquid storage, a portable tank (qv) which is not made of metal
non-Newtonian fluid	A fluid with a viscosity which varies with stress; cf Newtonian fluid
non-nuclear testing phase	*aka* preoperational stage. A term used in the nuclear industry to describe the stage of commissioning which covers all the activities prior to fuel loading, such as general checks, monitoring, modifications, component settings, component tests and functional testing of individual systems; see **Box C4**.
nonparametric statistics	Statistics other than parametric statistics (cf)
nonpolar compound	A compound which is not polar (qv)
nonrandom two liquid model	*aka* NRTL, NRTL model, NRTL activity coefficient mode. A thermodynamic model used for phase equilibrium calculations
nonreclosing pressure relief device	A PRD (qv) which does not reclose once operated, therefore requiring manual reset; defined formally in API RP 520 P1 7th Edition, January 2000 Sizing, Selection, and Installation of Pressure-Relieving Devices in Refineries; Part I - Sizing and Selection and API RP 576 - Inspection of Pressure Relieving Devices
nonreturn valve	*aka* NRV; see **one way valve**

nonsliding seal	A seal which does not move relative to mating surface; cf sliding seal
nontoxic construction materials	"Materials that, under the conditions of intended use, do not release any substance in amounts that would be harmful to the consumer", according to the EHEDG Glossary
nontoxic material	Synonym of nontoxic construction materials; this particular term defined formally in BS EN 1672-2:2005+A1:2009 Food processing machinery. Basic concepts. Hygiene requirements and BS EN ISO 14159:2008 Safety of machinery. Hygiene requirements for the design of machinery
nontrained user	As well as the obvious generic meaning, the term may be used in the context of fall arresters to describe someone without experience of their use (rather than training); defined formally in this sense in BS EN ISO 14122 Safety of machinery. Permanent means of access to machinery. Working platforms and walkways
nonuniform mechanical polishing marks	A patch of surface polishing noticeably different from the surrounding area; defined formally in ASME BPE (American Society of Mechanical Engineers: Bioprocessing Equipment)
normal	A term with a range of potential meanings, not all of which are synonymous with the everyday meaning. For example, a 1 molar solution can be called 'normal' (and a 0.1M solution decinormal), and a 'normal distribution' is a statistical model underlying parametric statistics. qv normal heat release, normal operating point, normal stress, normal temperature and pressure and normal working conditions for various other meanings
normal heat release	In the context of fired heaters, the ratio of design heat absorption to calculated fuel efficiency; defined formally in BS EN ISO 13705:2012/ISO 13705:2012(E) Petroleum, petrochemical and natural gas industries. Fired heaters for general refinery service and API STD 560 - Fired Heaters for General Refinery Service
normal operating conditions	see **normal operating point**

normal operating point	Conditions under which equipment is mostly expected to operate, and therefore at which maximum efficiency is desired; defined formally in EN ISO 10437 Petroleum, petrochemical and natural gas industries - Steam turbines - Special-purpose applications, BS EN ISO 13705:2012/ISO 13705:2012(E) Petroleum, petrochemical and natural gas industries. Fired heaters for general refinery service, EN ISO 13709:2003 Centrifugal pumps for petroleum, petrochemical and natural gas industries and ISO14661 Thermal turbines for industrial applications (steam turbines, gas expansion turbines) - General requirements
normal operation	Either synonymous with normal operating point (qv), or operation within design envelope, as per formal definition in API RP 505 2nd Edition, August 2018 Recommended Practice for Classification of Locations for Electrical Installations at Petroleum Facilities Classified as Class I, Zone 0, Zone 1, and Zone 2
normal stress	Component of stress perpendicular to material cross section; cf shear stress
normal temperature and pressure	*aka* NTP. Most commonly (but not always), NIST's equivalent of standard temperature and pressure (qv)
normal vent	A mechanism for excess pressure difference equalization during normal operation; defined formally in NFPA 30: Flammable and Combustible Liquids Code, 2018 Edition
normal wear part	*aka* wearing part. A machine part requiring replacement at every service interval; defined formally in EN ISO 13709:2003 Centrifugal pumps for petroleum, petrochemical and natural gas industries
normal working conditions	Agreed specified working conditions; defined formally in ISO 1819 Continuous mechanical handling equipment — Safety code and BS 5667-1: 1979— General rules and Continuous mechanical handling equipment —Safety requirements— Part 1: General
normality	see **normal**
nosing	A step or landing's front top edge; defined formally in BS EN ISO 14122 Safety of machinery. Permanent means of access to machinery. Working platforms and walkways
not invented here	*aka* NIH. A strong bias against ideas coming from outside a group
notch	1. A nick (qv) in a metal surface 2. A shape cut out from a weir plate, such as a vee notch (qv)
notch sensitivity	Susceptibility to fracture under stress of a notched material

notch toughness	The ability to resist fracture under stress of a notched material; defined formally in API RP 579 - Fitness for Service
noteworthy industry practices	Best practice (qv) in the context of pipeline integrity management; cf recognized and generally accepted good engineering practice(s)
NOx	Generally, oxides of nitrogen, especially in an environmental context. However, NO and NO_2 are frequently (but not universally) understood to be the oxides in question, excluding dinitrogen dioxide (N_2O_2), nitrous oxide (N_2O), dinitrogen trioxide (N_2O_3), dinitrogen tetroxide (N2O4) and dinitrogen pentoxide (N_2O_5) cf NOy and NOz
NOy	All reactive nitrogen oxides, including NOx and other oxidized nitrogen species like HNO_3 and organic nitrates
NOz	The difference between NOx and NOy
nozzle	1. A fluid transfer connection on a pressure vessel 2. see **spray device**
nozzle area	see **bore area**
nozzle meter	A type of flowmeter similar to venturi and orifice meters
nozzle mixing burner	In the context of fired heaters, a burner with separate fuel and air introduction; defined formally in NFPA 86: Standard for Ovens and Furnaces, 2019 Edition
nozzle throat area	see **bore area**
NPD	see **new product development**
NPDES	see **National Pollutant Discharge Elimination System**
NPGA	National Propane Gas Association (USA)
NPI	see **new product introduction**
NPMO	see **nonconformities per million opportunities**
NPS	see **nominal pipe size**
NPSH	see **net positive suction head**
NPSHa	see **net positive suction head available**
NPSHr	see **net positive suction head required**
NPT	see **national pipe thread**
NPT WOG	see **national pipe thread water oil gas**
NPT(F)	National pipe thread (female), see **national pipe thread**
NPT(M)	National pipe thread (male), see **national pipe thread**
NPV	see **net present value**
NRTL	see **nonrandom two liquid model**
NRTL model	see **nonrandom two liquid model**
NRTL model activity coefficient model	see **nonrandom two liquid model**
NRV	see **nonreturn valve**
ns	see **specific speed**; cf characteristic coefficient

NSF International	A USA-based international testing and standards organization in the area of food safety; see **Box C2**.
NSPS	see **new source performance standards**
NSR	see **new source review**
NTP	see **normal temperature and pressure**
NTS	*(drawing notation)* Not to scale
NTU	1. see **nephelometric turbidity unit**
	2. see **number of transfer units**
NU	*(drawing notation)* Non-upset [ends]
Nu	see **Nusselt number**
nubbin	A ridge machined onto a flange (especially a heat exchanger flange) which reduces the required bolt load to seal a gasket; defined formally in API 660 - Shell-and-Tube Heat Exchangers and API RP 661 - Heat Exchangers
nuclear fuel reprocessing	Recovering fissionable material from spent nuclear fuel
nuclear meltdown	A very serious safety incident, in which there is thermal damage to (though not necessarily melting of) the core of a nuclear reactor
nuclear testing phase	*aka* operational phase. A term used in the nuclear industry to describe the stage of commissioning which encompasses fuel loading, pre-critical tests, criticality and power escalation tests in steps up to nominal power. The nuclear testing phase usually begins by validating core physics parameters, calibrating nuclear instrumentation, and functional testing of conventional island (qv) equipment and systems
nuclear waste	see **radioactive waste**
nucleate boiling	Boiling which takes place at a surface not much hotter than a liquid's boiling point, such that produced vapor bubbles are carried away, maintaining heat transfer efficiency; cf film boiling
nucleation	1. Generically, the initiation of crystal formation around microscopic particles
	2. In the sugar industry, the controlled production of the small sugar crystals which will form the nuclei of larger ones
nuisance	1. An activity or situation that causes offence or detracts from an environment; a term used in some jurisdictions used to cover light, odor, smoke and noise emissions which, whilst not physically damaging, "substantially interfere with the use or enjoyment of a home or other premises" (legal definition of a statutory nuisance in England & Wales)
	2. In process control, nuisance alarms are low priority alarms that distract board operators from critical tasks

null hypothesis	*aka* H0. The statistical hypothesis that there is no real difference between two populations. If the hypothesis is proved false, then the opposite is true: there is a difference
number of transfer units	*aka* NTU. The required number of theoretical successive steps towards a final result in a heat exchanger or distillation column
nurdles	*(informal)* A term for small plastic pellets used for the production of molded plastic items; their escape during transport presents a significant cause of oceanic pollution
Nusselt number	*(symbol* Nu) A dimensionless group, the ratio of convective to conductive heat transfer
nutrient removal	The removal of nitrate and phosphate from treated effluent
Nutsche	Proprietary eponym for a type of filter-dryer[78]
NWC	Net working capital, see **working capital**
nylon	Properly, certain polyamide thermoplastics; though the term may also be used informally for similar materials
Nyquist frequency	*aka* folding frequency. 50% of sampling rate; defined formally in BS ISO 2041:2018 Mechanical vibration, shock and condition monitoring. Vocabulary
Nyquist plot	A type of frequency response plot used in control engineering; defined formally in BS ISO 2041:2018 Mechanical vibration, shock and condition monitoring. Vocabulary; cf Nichols plot

---

[78] Remember Buchner funnels from your chem lab courses? Just think the same, but bigger!

O

O&M	Operating and maintenance
O&M manual	see **operating and maintenance manual**
O'Shaughnessy equation	A way to calculate degree of volatile matter removal in anaerobic digestion (qv)
OARS	Occupational Alliance for Risk Science
object	Literally anything you can think of, according to BS EN ISO 9000 Quality management systems Fundamentals and vocabulary. More specific defined formally in ISO 10209:2012 Technical product documentation – Vocabulary – Terms relating to technical drawings, product definition and related documentation
objective	1. A desired result 2. Based in facts, rather than opinions BS EN ISO 9000 Quality management systems Fundamentals and vocabulary defines the first of these formally, but also uses the term in the second sense without defining it
objective evidence	Factual support for a belief. BS EN ISO 9000 Quality management systems Fundamentals and vocabulary uses the term in this sense, despite defining objective (qv) differently
obligate anaerobe	An organism which cannot reproduce in the presence of oxygen
observation door	see **peephole**
observed	In the context of testing, a term which does not necessarily mean that an event was actually observed by the purchaser, merely that they were notified and invited to observe. Defined formally in EN ISO 10437 Petroleum, petrochemical and natural gas industries - Steam turbines - Special-purpose applications and EN ISO 13709:2003 Centrifugal pumps for petroleum, petrochemical and natural gas industries; cf witnessed
observed inspection	In the context of testing, a term which does not necessarily mean that an inspection was actually observed by the purchaser; merely that they were notified and invited to observe. Defined formally in ISO14661 Thermal turbines for industrial applications (steam turbines, gas expansion turbines) - General requirements; cf witnessed inspection

observed test	In the context of testing, a term which does not necessarily mean that a test was actually observed by the purchaser; merely that they were notified and invited to observe. Defined formally in API RP 682 - Pump Seals
observer	In the case of quality audit, someone accompanying an audit team who does not audit. Defined formally in BS EN ISO 9000 Quality management systems Fundamentals and vocabulary
OC	*(drawing notation)* On center
occupancy	1. A term used extensively in NFPA literature to mean the purpose a structure is intended to be used for, shorthand for occupancy classification (qv); defined formally in NFPA 30: Flammable and Combustible Liquids Code, 2018 Edition) 2. The ratio of the number of occupants to the theoretical maximum number of occupants
occupancy classification	The NFPA system for characterizing the primary use of a building from a fire safety perspective; defined formally in NFPA 30: Flammable and Combustible Liquids Code, 2018 Edition
occupancy hazard classification number	NFPA's method for estimating total firewater requirements based on occupancy (qv) classification. Defined formally in NFPA 1142: Standard on Water Supplies for Suburban and Rural Fire Fighting, 2017 Edition
occupational safety	*aka* personal safety. Safety issues which tend to affect single individuals, such as slips, trips and falls; cf process safety
occupational safety management	*aka* personal safety management. Formal systems used to avoid personal injury, other than as a result of failures of process safety management e.g.; slips, trips and falls
OCGT	see **open cycle gas turbine**
octane number	*aka* octane rating. The measure of a fuel's resistance to premature ignition on compression, qv knocking
octane rating	see **octane number**
octanizing	A proprietary catalytic reforming process with a platinum catalyst
octanol/water partition coefficient	*aka* KOW, n-octanol-water partition coefficient. The equilibrium ratio of concentrations of a substance in water and octanol
OD	Outside diameter (of pipes etc.)
Oddo-Tomson index	An index with facility for correction for two or three fluid phases; used in the oil and gas industry to provide guidance on whether a water is likely to be scaling or aggressive. As with the Ryznar stability index (qv) and Langelier saturation index (qv) it is not an exact science, and experts may differ on its validity

odorisation	see **stenching**
odorising	see **stenching**
odorization	see **stenching**
odorizing	see **stenching**
odorless mineral spirits	see **mineral spirits**
odourising	see **stenching**
ODS	see **oxydesulfurization process**
OEM	Original equipment manufacturer
OF	Organic fouling
off angle	The deviation from squareness of two faces of a component; used only in the hygienic industries. Defined formally in ASME BPE (American Society of Mechanical Engineers: Bioprocessing Equipment)
off plane	The offset between centerlines or planes of a component; used only in the hygienic industries. Defined formally in ASME BPE (American Society of Mechanical Engineers: Bioprocessing Equipment)
offgas	A gaseous byproduct, typically consumed within plant fuel gas systems or flared. Alternatively, sometimes a synonym of outgassing (qv)
offline	1. A term for measurements made on discrete samples in a laboratory some distance from the sample collection point; cf online, atline, inline 2. see **out of service**
offline compressor washing	Washing a gas turbine compressor offline by slow rotation through a cleaning solution inlet with the gas turbine not operating. Defined formally in ISO 3977 Gas turbines - Procurement - Part 3: Design requirements
offset	In process control, a sustained deviation (qv) from setpoint
offshore drilling rig	see **production platform**
offshore platform	see **production platform**
offsite spacing	1. The spacing between offsite units 2. The spacing between process plant and certain types of equipment (offsites (qv)) not normally placed inside a process plant, e.g.; flares, LPG storage vessels and petroleum tanks
offsites	*aka* outside battery limits (OSBL), balance of plant (BOP). Supporting facilities which are neither a direct part of the process reaction train nor utilities (e.g.; transport pipelines, tank farms, flares, effluent treatment facilities, etc.)
OFG	Oils, fats and greases, see **fats oils and greases**
OGLV	see **operating gas lift valve**
ohm	(*symbol* Ω) The SI unit of electrical resistance; cf siemens, mho

Ohmart	Proprietary eponym for gamma-ray based level detection systems
oil	A liquid hydrocarbon or lipid; cf fat, grease, wax
oil and gas	An industry sector encompassing the exploration, recovery and processing of crude oil and natural gas. Also sometimes used in a broader sense to encompass the refining and petrochemicals industry[79]
oil and gas reservoir	see **petroleum reservoir**
oil gasification	Producing gaseous fuel from any type of oil (qv)
oil mist lubrication	A type of centrifugal pump bearing housing lubrication system. Defined formally in EN ISO 13709:2003 Centrifugal pumps for petroleum, petrochemical and natural gas industries; cf pure oil mist lubrication, purge oil mist lubrication
oil platform	see **production platform**
oil refinery	*aka* petroleum refinery. A process plant where commercial products are produced from any type of oil (qv)
oil separator	*aka* catch can. A simple inertial oil mist removal device used in compressed air lines and vacuum pump discharge; cf mist eliminator
oil shale	Shale containing significant quantities of a solid organic material, kerogen (qv), which can be used to produce a liquid shale oil (qv)
oil tanker	An informal term for a ship used for bulk transport of oil; more properly known as a very large crude carrier (qv) or an ultra large crude carrier (qv)
oil sweetening	Removing hydrogen sulfide from oil
oil water separator	*aka* OWS, oily water separator. A device for gravity separation of a light non-aqueous phase liquid and contaminated water based on their different densities which may also separate solids and dense non-aqueous phase liquids. Commonly designed in accordance with API 421, though the term may also be applied to other gravity settlement devices
oil whip	Self-excited rotor vibration caused by a sleeve bearing's oil film. Defined formally in BS ISO 2041:2018 Mechanical vibration, shock and condition monitoring. Vocabulary
oilfield rotary slips	see **slips**
Oldshue–Rushton column	A type of liquid–liquid extraction column
olefin	A traditional name for alkenes, still current outside academia

---

[79] Much to the consternation of some working in the narrower definition!

olet	*aka* -Olet, O-let. A collective term for a class of weld-on pipe outlet fitting including Elbolet (qv), Latrolet, Nippolet (qv), Weldolet, sockolet, Sweepolet (qv) and threadolet (qv)
O-let	see **olet**
-Olet	see **olet**
olfactorithmetic	The ability to detect an implausible numerical answer by its bad 'smell'
Olin–Raschig process	see **Raschig process**
OM	see **organic matter**
OMM	see **operating and maintenance manual**
on stream inspection	*aka* OSI. The non-destructive testing of pressurized components other than during a plant shutdown; defined formally in API RP 579 - Fitness for Service
on/off control	*aka* two-position control, three position control, floating control. Control of an actuator allowing only two points to be set. Two- and three-position control differ in how this is achieved; cf modulating control
one family dwelling	A single dwelling unit building; defined formally in NFPA 30: Flammable and Combustible Liquids Code, 2018 Edition
one line diagram	see **single line diagram**
one pot synthesis	Multistage synthesis in a single batch reactor
one way valve	*aka* check valve, clack valve, nonreturn valve (NRV), reflux valve. A device which allows fluid flow in a pipe in only one direction; defined formally in NFPA 24: Standard for the Installation of Private Fire Service Mains and Their Appurtenances, 2019 Edition
online	1. Strictly, analysis using an instrument mounted in a sampling loop, but may be used as a synonym of inline; cf inline, atline, offline 2. Operational, or in service
online compressor washing	Washing a gas turbine compressor online by injecting cleaning fluid into its inlet with the gas turbine operating. Defined formally in ISO 3977 Gas turbines - Procurement - Part 3: Design requirements
onsite power production facility	Onsite main electric power generation (as opposed to alternate/standby provision). Defined formally in NFPA 20: Standard for the Installation of Stationary Pumps for Fire Protection, 2019 Edition; cf onsite standby generator
onsite spacing	Spacing between equipment within the same process or utility unit; cf offsite spacing

onsite standby generator	Onsite alternate/standby electric power generation (as opposed to main). Defined formally in NFPA 20: Standard for the Installation of Stationary Pumps for Fire Protection, 2019 Edition; cf onsite power production facility
OOS	see **out of service**
opacity	Lack of transparency, almost always referring to visible light, though radiopacity is the rarer equivalent of non-visible radiation. Closely related to turbidity (qv)
OPD	see **overfilling prevention device**
OPEC	Organisation of the Petroleum Exporting Countries
open book	A form of contract whereby the purchaser is party to the costs to the vendor of materials, labor etc., and pays these costs plus an agreed markup
open channel	An open topped fluid conduit
open cooling system	A fired heater cooling system with operator-visible sight drains. Defined formally in NFPA 86: Standard for Ovens and Furnaces, 2019 Edition
open cycle gas turbine	*aka* OCGT. A gas turbine without energy recovery; cf closed cycle gas turbine
open head	The open type of orbital TIG welding (qv) head; defined formally in ASME BPE (American Society of Mechanical Engineers: Bioprocessing Equipment)
open hearth furnace	An almost entirely obsolete batch steelmaking process
open liquid or salt media quench-type tank	An open tank of quenchant used to rapidly cool furnace workpieces; defined formally in NFPA 86: Standard for Ovens and Furnaces, 2019 Edition; cf integral liquid or salt media quench-type tank
open loop control	*aka* non-feedback control. cf closed loop control 1. Simple control without feedback from measurement of the controlled variable 2. A deprecated term for manual control (qv)
open motor	A motor ventilated through unrestricted openings (ingress protection rating (qv) IP00 to IP23). Defined formally in NFPA 20: Standard for the Installation of Stationary Pumps for Fire Protection, 2019 Edition
open system	1. Generally, a system which has interactions with the wider environment 2. In thermodynamics, a system in which mass, energy or both cross the system boundary (like all real systems); cf closed system
open water spray nozzle	An open static spray device (qv); defined formally in NFPA 15 Standard for Water Spray Fixed Systems for Fire Protection

opening pressure	The PRV inlet pressure which is required to lift its disk. Defined formally in API RP 520 P1 7th Edition, January 2000 Sizing, Selection, and Installation of Pressure-Relieving Devices in Refineries; Part I - Sizing and Selection and API RP 576 - Inspection of Pressure Relieving Devices
operating and maintenance manual	aka O&M manual, OMM, operational manual. The manual for a plant including a general description, standard operating procedures, troubleshooting guidance, drawings, spares lists, etc.
operating capacity	1. In a management context, how much can be produced in a given time 2. In the context of equipment, concepts such as how much weight can safely be lifted without a vehicle tipping over; or the practical capacity of an ion exchange resin to remove ions under specified conditions
operating company	A company whose main activity is managing and operating process plants
operating conditions	The physical or operational conditions to which equipment is exposed, including but not limited to temperature, pressure, time in service and mode of operation
operating costs	aka opex. The non-capital costs (qv) of a project; cf capex
operating cycle	A complete batch process
operating envelope	aka safe operating envelope. Equipment's design operating conditions. Defined formally in one context in API RP 932B Corrosion Air Coolers
operating gas lift valve	aka OGLV. A term used in oil and gas for the valve which feeds the lowest point of gas introduction on a gas lift well (qv)
operating instructions	1. An O&M manual (qv) 2. A standard operating procedure (qv)
operating manual	see **operating and maintenance manual**
operating pressure	The normal range of process system pressure; defined formally in API Standard 521. Pressure-relieving and Depressuring Systems. Sixth Edition | January 2014
operating qualification	aka OQ; see **V model**
operating ratio of a pressure relief valve	The ratio of maximum operating pressure to PRV set pressure. Defined formally in API RP 520 P1 7th Edition, January 2000 Sizing, Selection, and Installation of Pressure-Relieving Devices in Refineries; Part I - Sizing and Selection
operating ratio of a rupture disk	The ratio of maximum operating pressure to rupture disk burst pressure. Defined formally in API RP 520 P1 7th Edition, January 2000 Sizing, Selection, and Installation of Pressure-Relieving Devices in Refineries; Part I - Sizing and Selection

operating region	The range of pump flows and heads in operation; defined formally in EN ISO 13709:2003 Centrifugal pumps for petroleum, petrochemical and natural gas industries; see also ANSI/HI 9.6.3- 2017 Rotodynamic Pumps – Guideline for Operating Regions; cf preferred operating region, allowable operating region
operating speed range	The design range of continuous speeds; defined formally in ISO 3977 Gas turbines - Procurement - Part 3: Design requirements
operating temperature	The normal range of process system temperatures
operating unit	see **operating vessel**
operating vessel	*aka* operating unit. Containment for a unit operation (qv); defined formally in NFPA 30: Flammable and Combustible Liquids Code, 2018 Edition
operational availability	see **availability**
operational cycle	Going from an initial set of process conditions to a new set, and back to the initial condition; covering startup/shutdown, normal operation and upset conditions; defined formally in API RP 579 - Fitness for Service
operational manual	A less used term for operating and maintenance manual (qv); this particular term is however defined formally in ISO 10209:2012 Technical product documentation — Vocabulary — Terms relating to technical drawings, product definition and related documentation
operational phase	see **nuclear testing phase**
operational prerequisite program	*aka* OPRP. A "program identified by the hazard analysis as essential in order to control the likelihood of introducing food safety hazards to and/or the contamination or proliferation of food safety hazards in the product(s) or in the processing environment" according to the EHEDG Glossary
operational readiness review	see **pre start up safety review**
operations	1. Usually, a term for a manufacturing process or its management 2. In the context of flammable liquid safety, handling flammable liquids; defined formally in NFPA 30: Flammable and Combustible Liquids Code, 2018 Edition
operator	1. Most commonly, someone who operates process equipment; defined formally in the case of fired heaters in NFPA 86: Standard for Ovens and Furnaces, 2019 Edition 2. A shorthand for operating company (qv)

operator access	The space required between items of equipment to permit safe walking, operating valves, viewing instruments, climbing ladders or stairs and safe emergency exit
operator competency	*aka* operator competence. The competence (qv) of a process operator
operator training simulator	*aka* OTS. A virtual reality plant used to train operators
operatory competence	see **operator competency**
opex	Operating expenditure; see **operating costs**
opportune shutdown	Unplanned maintenance activity, for which workpacks and materials are kept ready to do a job, instead of taking a shutdown, or waiting for a TAR/Planned shutdown
OPRP	see **operational prerequisite program**
optical density	*aka* absorbance. A dimensionless logarithmic measure of attenuation of visible light; related to turbidity (qv), opacity (qv) and color (qv)
optimization	Improving process design or operation by balancing a number of variables against cost, safety and robustness
optimum reflux ratio	*aka* economical reflux ratio. The most cost effective reflux ratio for a distillation column
OQ	see **operating qualification**
OR	Outside radius
orange hydrogen	1. *aka* biogenic hydrogen. A German term for hydrogen which is made directly or indirectly from biomass; cf hydrogen colors 2. A Dutch term for hydrogen produced in the Netherlands
orange peel	A type of surface finish defect, either (a) a roughening of a metal surface, with the appearance of orange peel, defined formally in ASME BPE (American Society of Mechanical Engineers: Bioprocessing Equipment); or (b) a sprayed coating with the appearance of orange peel
orbital welding	A technique where a welding head is rotated mechanically around a static workpiece to produce a highly consistent weld; used to join and seal pipes in the hygienic industries
order of reaction	The exponent applied to a reactant concentration in a rate equation
ordinal data	Data which can be ranked in order, but for which the gaps between ranked items are unknown
ore	A natural mineral aggregate from which something (commonly a metal) is refined
ore beneficiation	see **mineral dressing**
ore dressing	see **mineral dressing**
ore flotation	see **froth flotation**

ore processing	see **beneficiation**
ORFF	see **orifice flange**
organic fouling	Fouling of membranes or ion exchange media by adherent organic matter
organic loading	In environmental applications, the mass (usually in kg) of COD (qv) or BOD (qv) applied per unit (usually m^2) of media surface area or per unit volume (usually m^3) of media per unit time (usually h)
organic matter	*aka* OM, TOC. Raw water contaminants, mostly naturally occurring substances such as humic acids (qv) and fulvic acids (qv)
organic scavenger	*aka* OS. An additive or unit operation which removes organic matter (qv)
organic vapor respirator	see **acid gas respirator**
organization	People and facilities formally organized with respect to authority, relationships and responsibility. Defined formally in BS EN ISO 9000 Quality management systems Fundamentals and vocabulary and BS EN 82079-1:2012 Preparation of instructions for use. Structuring, content and presentation. General principles and detailed requirements
organizational silo	see **silo**
orifice	An aperture, especially one through which fluids can pass
orifice flange	*aka* ORFF. A set of flanges and ancillaries holding an orifice plate which together form part of an orifice plate meter (qv), as specified in ASME B16.36 Orifice Flanges
orifice nipple	A pipe nipple with regulating orifice; defined formally in API RP 682 - Pump Seals
orifice plate	*aka* restriction orifice. A plate with an aperture (orifice) of controlled size running through it, commonly fitted between two flanges to restrict flow, or to provide a pressure drop proportional to flow to allow flow to be calculated from differential pressure (qv orifice plate meter)
orifice plate meter	A device which uses headloss through an orifice to measure flow
orifice seal	see **velocity seal**
O-ring	A preformed ring of elastomer (qv) with a circular cross section used as a seal. Defined formally in API RP 682 - Pump Seals and ASME BPE (American Society of Mechanical Engineers: Bioprocessing Equipment)
orsat gas analyzer	Old fashioned lab equipment still sometimes used for flue gas analysis
orthogonally	Arranged at right angles only

orthogonal axonometry	A single plane orthogonal projection, though often used as synonymous with orthogonal projection (qv)
orthogonal projection	A parallel projection where projection lines intersect the projection plane orthogonally
OS	see **organic scavenger**
OS&D	see **over, short and damaged**
OS&Y	see **outside screw and yoke**
OSBL	Outside battery limits, see **offsites**
oscillation	1. A regular variation in magnitude of some variable with respect to a central value. Defined formally in BS ISO 2041:2018 Mechanical vibration, shock and condition monitoring. Vocabulary 2. A term used in distillation for a wave type motion of the liquid on the tray, perpendicular to the normal direction
OSHA	Occupational Safety & Health Administration (USA)
OSI	see **on stream inspection**
osmosis	A net movement of molecules across a semipermeable membrane by diffusion which tends to equalize concentrations; cf reverse osmosis, forward osmosis
osmotic shock	A rapid change of size due to change in osmotic pressure, especially when leading to physical degradation
osmotic stability	The ability to resist osmotic shock
Ostwald coefficient	The gas solubility coefficient (qv) at stated temperature and standard pressure
Ostwald process	The catalytic oxidation of ammonia with air to produce nitric acid
Ostwald–de Waele equation	The power law describing non-Newtonian fluids (qv)
other synthetic foam concentrate	A firefighting foam concentrate using hydrocarbon based surfactants. Defined formally in NFPA 11: Standard for Low-, Medium-, and High-Expansion Foam, 2016 Edition
OTS	see **operator training simulator**
OUR	see **oxygen uptake rate**
out of service	*aka* OOS, offline. Currently unavailable. A common design case, requiring a plant to continue functioning without a redundant parallel stream, or streams, to accommodate it becoming unserviceable or requiring periodic downtime e.g.; offline compressor washing (qv)

outage	A period of unavailability of plant and equipment. Sometimes applied to all periods in which a plant is out of service, including unscheduled temporary unavailability, but most commonly applied to deliberate, scheduled periods such as a turnaround (qv)
outboard	A term used to describe components further from the drive coupling of a pump or motor; cf inboard
outboard seal	A shaft seal without process fluid contact, used in hygienic and highly corrosive service; defined formally in ASME BPE (American Society of Mechanical Engineers: Bioprocessing Equipment)
outdoor occupancy classification	A fire safety classification applied to operations outside of buildings; defined formally in NFPA 30: Flammable and Combustible Liquids Code, 2018 Edition
outer seal	In an arrangement 2 seal (qv) or an arrangement 3 seal (qv), the seal furthest away from the impeller; defined formally in API RP 682 - Pump Seals
outgassing	*aka* offgassing. The release of gases from (most commonly a solid) material exposed to heat or vacuum; defined formally in NFPA 86: Standard for Ovens and Furnaces, 2019 Edition; cf offgas
outlet	*aka* drawoff. A pipe or opening through which gas or liquid may escape; cf inlet
outlet size	In the context of a PRV (qv), outlet bore expressed as NPS; defined formally in API RP 520 P1 7th Edition, January 2000 Sizing, Selection, and Installation of Pressure-Relieving Devices in Refineries; Part I - Sizing and Selection and API RP 576 - Inspection of Pressure Relieving Devices
outline drawing	A dimensioned drawing of the outside of an object; defined formally in ISO 10209:2012 Technical product documentation — Vocabulary — Terms relating to technical drawings, product definition and related documentation
outloading	Discharging
output	The mass or energy leaving a system; defined formally in BS EN ISO 9000 Quality management systems Fundamentals and vocabulary and EN ISO 14040:2006 Environmental management - Life cycle assessment - Principles and framework
outside battery limits	*aka* OSBL; see **offsites**
outside mounted seal	see **externally mounted seal**

outside screw and yoke	*aka* OS&Y, outside stem and yoke, indicating gate valve, indicating valve (qv). A gate valve (qv) with a visual indication of whether it is open or closed, commonly found in fire protection systems.
outside stem and yoke	see **outside screw and yoke**
outsource	To engage a third party to carry out part of an organization's work. Defined formally in BS EN ISO 9000 Quality management systems Fundamentals and vocabulary
ovality	A deviation from circularity of a pipe's bore
oven	Often taken as synonymous with furnace (qv), but sometimes considered to be any heater operating below 1000° F (538° C), whilst furnace refers to heaters operating above this temperature. Defined formally in NFPA 86: Standard for Ovens and Furnaces, 2019 Edition
over, short and damage	see **over, short and damaged**
over, short and damaged	*aka* OS&D, over, short and damage. An over, short and damaged report is used for a receiving party to report on any over-shipment, short-shipment or damage to ordered goods, and to request compensation where appropriate
overall collapse	The collapse of a stiffened shell section of a pressure vessel. Defined formally in PD 5500:2018+A1:2018 Specification for unfired fusion welded pressure vessels
overall heat transfer coefficient	(*symbol* U) The sum of individual heat transfer coefficients
overall mass transfer coefficient	(*symbols* k_G *and* k_L) *aka* total transfer velocity. The sum of individual mass transfer coefficients
overall plate efficiency	*aka* overall tray efficiency, E_O. A fiddle factor (qv) for correcting theoretical models with respect to the real separation efficiency of a distillation column. The ratio of the number of plates actually needed to those theoretically needed; cf plate efficiency
overall selectivity	Selectivity (qv) expressed as the molar ratio of desired product to undesired product for a chemical reaction
overall tray efficiency	see **overall plate efficiency**
overburden	*aka* spoil, waste. Material overlying economically exploitable ore reserves; cf gangue, tailings
overfilling prevention device	A device which automatically prevents an LPG container from overfilling. Defined formally in NFPA 58: Liquefied Petroleum Gas Code, 2017 Edition
overhead product	Product leaving the top of a distillation column
overhead vapor	Uncondensed vapor leaving the top of a distillation column

overhung impeller pump	see **overhung pump**
overhung pump	*aka* overhung impeller pump. A pump with a single impeller which is hung from a single set of bearings; defined formally in EN ISO 13709:2003 Centrifugal pumps for petroleum, petrochemical and natural gas industries; cf between bearings pump
overlap	How much something protrudes over something else, e.g. (a) between parts of an access stair, as in BS EN ISO 14122 Safety of machinery. Permanent means of access to machinery. Working platforms and walkways, or (b) between weld metal and a weld as in ASME BPE (American Society of Mechanical Engineers: Bioprocessing Equipment)
overpressure	1. In the context of a PRV (qv), pressure compared with a specified pressure, especially **PRV set pressure**. Three different formal definitions in API RP 520 P1 7th Edition, January 2000 Sizing, Selection, and Installation of Pressure-Relieving Devices in Refineries; Part I - Sizing and Selection, API Standard 521. Pressure-relieving and Depressuring Systems. Sixth Edition | January 2014 and API RP 576 - Inspection of Pressure Relieving Devices 2. In the context of explosions, a sudden wave of increased pressure or shock wave (qv)
overpressure shutoff device	A device which shuts off LPG vapor flow at a set regulator outlet pressure; defined formally in NFPA 58: Liquefied Petroleum Gas Code, 2017 Edition
overpumping	1. *aka* bypass pumping. Most commonly, using temporary pumps to bypass a water or sewage treatment facility 2. Pumping beyond some limit, such as e.g.; causing saline intrusion by abstracting too much groundwater
overshoot	A response exceeding target, in the context of process control or vibration. Formal definition of the latter in BS ISO 2041:2018 Mechanical vibration, shock and condition monitoring. Vocabulary; cf undershoot
overview diagram	A comprehensive diagram showing limited detail. Defined formally in ISO 10209:2012 Technical product documentation — Vocabulary — Terms relating to technical drawings, product definition and related documentation

owner	The entity destined to possess purchased equipment, as distinct from a purchaser (qv) or an operating company (qv). Defined formally in EN ISO 10437 Petroleum, petrochemical and natural gas industries - Steam turbines - Special-purpose applications
owner/operator	The formal definition of this term in NFPA 654: Standard for the Prevention of Fire and Dust Explosions from the Manufacturing, Processing, and Handling of Combustible Particulate Solids, 2017 Edition is synonymous with operator (qv), and makes no mention of ownership
owner/user	The formal definition of this term in ASME BPE (American Society of Mechanical Engineers: Bioprocessing Equipment) combines owner (qv) and user (qv)
owner's inspector	see inspector
OWS	see oil water separator
oxic	see aerobic
oxidation	A chemical reaction in which something gains oxygen[80]; defined formally in ASME BPE (American Society of Mechanical Engineers: Bioprocessing Equipment); cf oxygenation, reduction
oxidative coupling	Coupling molecules through oxidation, commonly involving a transition metal catalyst
oxidative desulfurization	see oxydesulfurization process
oxide island	A type of nonmetallic inclusion within a weld; defined formally in ASME BPE (American Society of Mechanical Engineers: Bioprocessing Equipment)
oxide layer	An oxidized area of a weld, most commonly found in the HAZ (qv). Defined formally in ASME BPE (American Society of Mechanical Engineers: Bioprocessing Equipment)
oxo process	see hydroformylation
oxo synthesis	see hydroformylation
oxy-acetylene cutting	see oxygen cutting
oxy-acetylene welding	*aka* oxy-fuel welding, gas welding. Welding using a fuel gas (usually acetylene) burning in pure oxygen; cf oxygen cutting
oxychlorination process	Generically, a process in which HCl is used to create C-Cl bonds in organic chemicals
oxydesulfurization process	*aka* ODS, oxidative desulfurization. Generically, a process in which sulfur is removed by oxidation, as in the case of upgrading high sulfur coal
oxy-fuel cutting	see oxygen cutting

---

[80] Or loses electrons or hydrogen, if you ask a chemist

oxy-fuel welding	see **oxy-acetylene welding**
oxygen cutting	*aka* gas cutting, oxy-fuel cutting, oxy-acetylene cutting. Cutting metals using a fuel gas (usually acetylene) burning in pure oxygen
oxygen uptake rate	*aka* OUR. The rate of change of dissolved oxygen concentration in a biological culture; often expressed as SOUR, the specific rate per gramme of biomass (or per gram of MLSS (qv) in sewage treatment)
oxygenates	Oxygen containing organic compounds such as alcohols and ethers used as blendstock (qv)
oxygenation	Physically adding oxygen to a mixture; cf oxidation
ozonation	The addition of ozone (O_3); a less-used water disinfection process
ozonolysis	Breaking down by reaction with ozone (O_3)

P

P&ID	see **piping and instrumentation diagram**; N.B. not 'process and instrumentation diagram'
P:E	Price to earnings ratio
P+ID	see **piping and instrumentation diagram**; N.B. not 'process and instrumentation diagram'
P2	1. Usually pollution (qv) prevention 2. In management contexts, product partnerships 3. In scaleup, polymers and processing
PAC	1. Programmable automation controller 2. Provisional acceptance certificate 3. see **powdered activated carbon** 4. see **polyaluminum chloride**
package plant	see **packaged domestic wastewater treatment plant**
packaged domestic wastewater treatment plant	A factory built/modular domestic effluent treatment plant; defined formally in EN 12566 - Small wastewater treatment systems for up to 50 PT Part 3: Packaged and/or site assembled domestic wastewater treatment plants
packaged fire pump assembly	A skid (qv) mounted fire pump and driver with associated controller and accessories; defined formally in NFPA 20: Standard for the Installation of Stationary Pumps for Fire Protection, 2019 Edition
packager	In the context of gas turbines, a supplier who engineers a packaged gas turbine system. Defined formally in ISO 3977 Gas turbines - Procurement - Part 3: Design requirements
packed bed	A vessel filled with random packing (qv) or structured packing (qv)
packed bed reactor	*aka* PBR. A plug flow reactor filled with packing, often carrying a catalyst
packed column	A packed bed (qv) in the form of a column
packed gland seal	A shaft seal using a mechanically compressed sealing material
packing	1. A solid structure or structures used to enhance mass transfer or reaction 2. A type of shaft sealing material found in a packed gland seal (qv)
pad	International English equivalent of plinth (qv)
paddle	1. A blind plate between flanges 2. A type of mixer (qv)
paint thinners	see **mineral spirits**

Paisley screwdriver	(*jocular*) see **Manchester screwdriver**
P-Alk	see **phenolphthalein alkalinity**
Pall ring	Proprietary eponym for a type of column random packing (qv)
pallet tank	see **intermediate bulk container**
pan amalgamation process	An obsolete process for the extraction of silver from ore
panel	An enclosure for (usually electrical/electronic) instrumentation, switchgear (qv) etc. Defined formally in EN ISO 10437 Petroleum, petrochemical and natural gas industries - Steam turbines-- Special-purpose applications
panel man	see **control room operator**
panel mounted instrument	An instrument mounted as part of an operator accessible group (as the name implies, in a common enclosure). Defined formally in BS1646-1 Symbolic Representation for Process Measurement Control Functions and Instrumentation Part 1: Basic Requirements and BS 1646-3:1984 Symbolic representation for process measurement control functions and instrumentation. Specification for detailed symbols for instrument interconnection diagrams; cf field mounted instrument
panel operator	see **control room operator**
PAR	Pre-assembled rack; see **piperack**
paraffin	An old-fashioned name for an alkane (qv), still used in the oil and gas industry, or (in the UK) a particular light hydrocarbon liquid blend known as kerosene in the US
paraffinic	Similar to or predominantly paraffin (qv) in the alkane sense
parallel	*aka* in parallel. A term for an arrangement of equipment with identical units mounted between equivalent branches of a system, rather than one feeding the other (series (qv) installation)
parallel plate interceptor	*aka* PPI. The lamella clarifier (qv) type of API oil separator (qv)
parallel projection	*aka* axonometric projection/axonometric representation (qv). A projection of an image/object in three dimensions onto a fixed plane (two dimensions). Defined formally in BS EN ISO 10209:2012 Technical product documentation - Vocabulary - Part 1: Terms relating to technical drawings: general and types of drawings
parametric statistics	The commonplace kind of statistics, the valid applicability of which is limited to cases where sample data can be modeled by a probability distribution defined by fixed parameters.[81]

---

[81] This circumstance is less common than might be imagined, so parametric statistics are commonly abused.

Pareto analysis	A statistical technique in decision making that is used for analysis, identification and selection of the limited number of tasks that produce significant overall effect. The premise is that 80% of problems are produced by a few critical causes (20%)
Parker fitting	Proprietary eponym for high pressure/temperature pipe connectors similar to Swagelok (qv)
Parkes process	A pyrometallurgic liquid-liquid extraction process which removes lead from silver using liquid zinc
part 70 Federal operating permit	see **title V operating permit**
part A (1) installation	UK terminology for facilities which carry out the most polluting industrial processes, e.g.; refineries, food and drink factories and intensive farming activities, which are regulated under this classification by the UK Environment Agency
part A (2) installation	UK terminology for facilities which carry out less polluting industrial processes than a part A (1) installation (qv), e.g.; refining gas, metalwork, melting, casting and surface treating metals, grinding cement clinker or metallurgical slag, glass or ceramic product manufacturing, non-hazardous and animal waste incineration, disposal or recycling, which are regulated under this classification by UK Local Authorities
part B installation	UK terminology for the least polluting facilities, i.e.; considered neither a part A (1) installation (qv) nor a part A (2) installation (qv); those which cause only emissions to air, and are regulated under this classification by UK Local Authorities
part drawing	A drawing of a component part of a product. Defined formally in BS EN ISO 10209:2012 Technical product documentation - Vocabulary - Part 1: Terms relating to technical drawings: general and types of drawings
partial arrangement drawing	A general arrangement drawing (qv) of a limited section of an assembly. Defined formally in BS EN ISO 10209:2012 Technical product documentation - Vocabulary - Part 1: Terms relating to technical drawings: general and types of drawings
partial condenser	A distillation column condenser which only condenses some of the incoming vapor, allowing economical light distillate recovery; cf total condenser
partial design	An academic approximation of the easy parts of the design process which falls far short of total design (qv)
partial gas volume	see **partial volume**

partial safety factor	*aka* PSF. The application of engineering judgement-based factors to input variables in a structural design equation, to give a desired reliability (avoiding probabilistic calculations). Defined formally in API RP 579 - Fitness for Service and BS 7910 Guide to Methods for Assessing the Acceptability of Flaws in Metallic Structures
partial volume	*aka* partial gas volume. The volume which would be filled by a component in a mixture of gases if all other gases present in the mixture were absent at system pressure; cf Amagat's law
partially miscible liquids	Liquids which are only miscible (qv) to a certain degree and/or at a certain temperature
particle diffusion	In the context of ion exchange, ion movement within a resin particle towards the exchange site
particle filter	1. A unit operation which removes larger particles from a fluid; see **Box L1**. 2. *aka* sequential Monte Carlo method (qv). A mathematical technique
particle filtration	see **Box L1**.
particle size distribution	The percentage of particles in a sample of mixed particle sizes which fall within a set of size ranges (sometimes confusingly applied to particle weight distributions)
particles	Very small bits of matter, (opinions differ about how just small 'very small' is), most commonly solid matter; cf particulates
particulate matter	see **particulates**
particulates	*aka* atmospheric aerosol particles, atmospheric particulate matter, particulate matter, PMx, SPM, suspended particulate matter. A form of atmospheric pollution comprising very fine solids or liquids suspended in air
partition coefficient	The equilibrium ratio of concentrations of a substance in two immiscible solvents, such as KOW (qv). Usually *aka* distribution coefficient (qv), but to a chemist, that term may be thought distinct from partition coefficient in as much as it takes into consideration ionized and unionized forms of a molecule and is therefore pH-dependent
pascal	(*symbol* Pa) The SI unit of pressure
pascal second	(*symbol* Pa.s) The SI unit of dynamic viscosity
Pascal's law	*aka* Pascal's pressure law. The law which states that the pressure applied to a confined incompressible fluid is transmitted equally to all points in the fluid
pascalization	see **high pressure processing**
Pascal's pressure law	see **Pascal's law**

pass	1. In the context of fired heaters, a heat exchange flow circuit, as defined identically in BS EN ISO 13705:2012/ISO 13705:2012(E) Petroleum, petrochemical and natural gas industries. Fired heaters for general refinery service, API STD 530 - Calculation of Tube Heater Thickness and API STD 560 - Fired Heaters for General Refinery Service; 2. *aka* stream. The number of transits through a unit operation
pass partition plate	A horizontal heat exchanger head cover divider
passing	1. Most commonly, the leaking of a valve 2. A type of deviation from the ideal in the case of drawings, defined formally in BS EN ISO 10209:2012 Technical product documentation - Vocabulary - Part 1: Terms relating to technical drawings: general and types of drawings
passivated	see **passive**
passivation	Usually, reducing corrosion potential of a stainless-steel surface by chemical removal of free iron, and subsequent formation of a protective oxide layer. Defined formally in ASME BPE (American Society of Mechanical Engineers: Bioprocessing Equipment); but the term can also be applied to processes reducing corrosion potential of other materials e.g.; aluminum, titanium and silicon
passive	1. The condition of a metal surface which has been subjected to passivation (qv) 2. Can be applied to building design, as in passive fire protection (qv) or passive HVAC
passive fire protection	*aka* PFP. Those features of a building which resist the spread of fire and smoke
passive layer	The protective oxide layer resulting from passivation. Defined formally in ASME BPE (American Society of Mechanical Engineers: Bioprocessing Equipment)
passivity	The corrosion-resistant state of a passivated surface. Defined formally in ASME BPE (American Society of Mechanical Engineers: Bioprocessing Equipment)
pasteurizable	Designed to be capable of being pasteurized; defined formally in BS EN ISO 14159:2008 Safety of machinery. Hygiene requirements for the design of machinery

pasteurization	1. "A microbiocidal heat treatment aimed at reducing the number of any harmful microorganisms, if present, to a level at which they do not constitute a significant health hazard", according to the EHEDG. Another formal definition may be found in BS EN ISO 14159:2008 Safety of machinery. Hygiene requirements for the design of machinery; sterilization 2. see **pasteurizing section**
pasteurization unit	*aka* PU. A measure of the degree of heat treatment, where 1 PU is achieved by holding at 60° C for 1 minute. Equivalent time/temperature combinations in minutes can be calculated using PU = 1.393(T-60)
pasteurizing section	Part of a distillation column which removes desired distillate as a side product with lighter impurities being taken from the top of the section
path function	A thermodynamic function dependent on the path taken to the point of measurement e.g.; heat, work etc.; cf state function
pathogen	see **pathogenic microorganisms**
pathogenic microorganisms	Microorganisms that can cause adverse health effects; cf indicator microorganisms, relevant microorganisms
PAU	Pre-assembled unit; see **module**
PAUT	see **phased array ultrasonic testing**
pavement	see **tarmac**
PBE	*(drawing notation)* Plain both ends
PBR	1. see **packed bed reactor** 2. see **pulsed baffle reactor**
PC	1. see **personal computer** 2. see **prime cost**
PC sums	see **prime cost**
PCB	1. see **polychlorinated biphenyl** 2. Printed circuit board
PCM	see **precision control mixing**
PCS	Process control system, see **process control**
PCT	see **process centric team**
PCV	see **pressure control valve**
PD	*(drawing notation)* Pressure drain
PDA	1. see **personal digital assistant** 2. see **preliminary design analysis**
PDG	see **permanent downhole gauge**
PDP	see **process design package**

PE	1. *(drawing notation)* Plane end
	2. see **Professional Engineer**
	3. see **population equivalent**
	4. see **polyethylene**
peak load	1. Generally, the maximum electrical load an electrical system can carry
	2. In the context of fire protection pump acceptance testing, the maximum power requirement across the complete range of flows up to 150% of rated flow; defined formally in NFPA 20: Standard for the Installation of Stationary Pumps for Fire Protection, 2019 Edition
peak magnitude	see **peak value**
peak stress	A largely unobjectionable stress that does not cause noticeable distortion, according to API RP 579 - Fitness for Service, though other meanings are possible
peak to peak value	In the context of vibration monitoring, the difference between maximum and minimum values of a vibration. Defined formally in BS ISO 2041:2018 Mechanical vibration, shock and condition monitoring. Vocabulary
peak value	*aka* peak magnitude, positive peak value. In the context of vibration monitoring, maximum value of a vibration in a time period; defined formally in BS ISO 2041:2018 Mechanical vibration, shock and condition monitoring. Vocabulary
peaking factor	A multiplier used in municipal biological effluent treatment plant design to account for diurnal variation (qv)
Péclet number	(*symbol* Pe) A class of dimensionless groups, being the ratio of advective to diffusive transport rates in heat and mass transfer contexts
PED	Pressure Equipment Directive 97/23/EC (EU)
pedestal	see **plinth**
peephole	*aka* inspection door, observation door. A small hole or door allowing the observation of fired heater burners or pyrometallurgic/hydrometallurgic reaction vessels in action; cf sight glass
Pekilo process	The Finnish equivalent of Pruteen (qv)
pelican hook	A quick- release securing device for wires and lanyards
penetrant testing	see **liquid penetrant indication**
penetration	see **depth of fusion**
Peng-Robinson EOS	see **Peng-Robinson equation of state**

Peng–Robinson equation of state	*aka* PR equation of state, Peng-Robinson EOS. A popular two-constant equation of state (qv) used for "predicting the vapor pressure and volumetric behavior of single-component systems, and the phase behavior and volumetric behavior of binary, ternary, and multicomponent systems", according to its eponymous developers
penstock	*aka* sluice gate. Most commonly used to mean a gate in a channel used to control water flow, though can also be used for a hydroelectric turbine feed pipe
PERC	see **powered emergency release coupling**
percent dry solids	*aka* %DS. Dry solids (qv) content, expressed as w/w percent of total sample mass
percolating filter	see **trickling filter**
percussive maintenance	*(jocular) aka* shift man's persuasion, technical tap. Attempting to correct equipment malfunction by hitting with a with a hammer, known in this context as an adjuster or calibrator
perfluoroalkoxy	see **perfluoroalkoxyalkane**
perfluoroalkoxyalkane	*aka* PFA. Specifically in a hygienic setting, known as 'perfluoroalkoxy', a copolymer of tetrafluoroethylene and perfluorovinylether. Defined formally in ASME BPE (American Society of Mechanical Engineers: Bioprocessing Equipment).
performance	Defined formally in BS EN ISO 9000 Quality management systems Fundamentals and vocabulary as a "measurable result", though often used in other senses
performance coefficient	(*symbol* λ) A dimensionless coefficient derived from the affinity laws (qv)
performance qualification	*aka* PQ. The final stage of validation in the V model (qv)
performance specification	A document which defines required performance. Defined formally in BS EN ISO 10209:2012 Technical product documentation - Vocabulary - Part 1: Terms relating to technical drawings: general and types of drawings
performance trial	A formal test to establish that commissioned equipment meets a predetermined performance under defined conditions for a specified period; cf commissioning and qualification, trial
periodic vibration	A vibration which cycles through values in a fixed time period; defined formally in BS ISO 2041:2018 Mechanical vibration, shock and condition monitoring. Vocabulary; cf aperiodic vibration

peristaltic pump	A positive displacement pump (qv) in which a wave of compression of a flexible hose (most commonly by a roller in a circular casing) pushes out the pumped fluid
permanent downhole gauge	*aka* PDG; see **down hole pressure and temperature transducer**
permanent gas	An archaic term for gases once thought to be impossible to liquefy, such as air or its main components. Nowadays, any substance with a critical temperature (qv) substantially below room temperature
permanent hardness	Hardness (qv) not removed by boiling
permanent installation	see **stationary installation**
permanganate value	*aka* PV. A quick site test which allows COD (qv) to be estimated
permeability	The degree to which a substance (especially one in the form of a semipermeable membrane) allows the passage of some form of matter
permeate	Most commonly in engineering, a fluid which has passed through a membrane, though the word also has the common meaning of penetrate; cf retentate
permit to work	A documented procedure that authorizes certain people to carry out specific work within a specified time frame, according to the UK Health and Safety Executive. Also used for the associated, usually hardcopy authorization document; cf work permit
permitted	1. Usually, having had a permit (such as a permit to work) issued 2. Sometimes used to refer to activity allowable without any permit being issued, as defined formally in NFPA 58: Liquefied Petroleum Gas Code, 2017 Edition. N.B.: the two senses of this term are pronounced with different emphasis, but are written identically
persistent organic pollutant	*aka* POP. A toxic organic substance which resists environmental degradation and tends to bioaccumulate
personal care products	Toiletries; defined formally in ASME BPE (American Society of Mechanical Engineers: Bioprocessing Equipment)
personal computer	*aka* PC. A genericized trademark from IBM. A computer similar to a commonplace desktop device which is used to run control software, as well as high level systems such as SCADA (qv) and DCS (qv)
personal digital assistant	A near-obsolete term, like the equipment it refers to, an early example of business electronics

personal protective equipment	*aka* PPE. Equipment used to protect an individual against a hazard; includes items such as safety spectacles, helmets, ear protectors, gloves, shoes/boots. Defined formally in BS EN 82079-1:2012 Preparation of instructions for use. Structuring, content and presentation. General principles and detailed requirements
personal safety	see **occupational safety**
personal safety management	see **occupational safety management**
personnel	Employees, especially operational staff. Defined formally in ISO 1819 Continuous mechanical handling equipment — Safety code — General rules
PERT	see **program evaluation and review technique**
pervaporation	An experimental membrane technology with liquid inside a membrane and vapor outside, derived from PERmeation/eVAPORATION
pest	An animal which can affect food product quality; defined formally in BS EN ISO 14159:2008 Safety of machinery. Hygiene requirements for the design of machinery. Plural is pests or vermin
PET	1. Generically polyester, or more specifically polyethylene terephthalate (qv) 2. Less commonly, see **piezoelectric transducer**
peta-	(*symbol* P-) The SI unit prefix denoting a factor of 10^{15}
petcoke	see **petroleum coke**
petrochemical feedstocks	Feedstocks used in petrochemicals (qv) production
petrochemicals	Strictly, chemicals derived from crude oil, though sometimes expanded to cover similar chemicals made from natural gas and coal, or even biomass feedstocks
petrol water separator	see **interceptor**
petroleum	Properly, crude oil, but also commonly used for certain petrochemicals, most notably in the UK for what is known in the US as 'gasoline', in which context it is shortened to 'petrol'.[82]
petroleum coke	*aka* petcoke. Coke produced from petroleum processing; cf honeycomb coke, needle coke, shotcoke, sponge coke
petroleum ether	Not an ether at all, confusingly, but a light hydrocarbon mixture used as a solvent
petroleum feedstock	A deprecated term for petrochemical feedstocks (qv)
petroleum oil	see **mineral oil**

---

[82] Both gasoline and petrol are actually genericized trademarks

petroleum refinery	*aka* oil refinery (qv). Generically, a plant where liquid and gaseous products are produced from petrochemical feedstocks. A safety focused formal definition is given in API RP 505 2nd Edition, August 2018 Recommended Practice for Classification of Locations for Electrical Installations at Petroleum Facilities Classified as Class I, Zone 0, Zone 1, and Zone 2
petroleum reservoir	*aka* oil and gas reservoir. An underground pocket of hydrocarbons
petroleum spirits	see **mineral spirits**
PFA	1. see **perfluoroalkoxyalkane** 2. see **pulverized flue ash**
Pfaudler	Proprietary eponym which is used by batch organic or pharma engineers to refer to any glass-lined reactor
PFD	see **process flow diagram**
PFMEA	see **process failure mode and effect analysis**
PFP	see **passive fire protection**
PFR	see **plug flow reactor**
PG	see **produced gas**
pH	Theoretically, the negative logarithm of water's hydrogen ion concentration (though only approximately this in practice)
PHA	see **process hazard analysis**
pharmaceutical	Related to medications; defined formally in ASME BPE (American Society of Mechanical Engineers: Bioprocessing Equipment)
Pharmacovigilance Risk Assessment Committee	*aka* PRAC. An EU committee which assesses and monitors the safety of human medicines
phase	Most commonly in chemical engineering either: 1. A sequential stage of a project 2. One of the three common forms of matter, (solid, liquid or gas) 3. Something associated with recurring cycles, often referring to the two types of AC electrical supply (single or three phase power (qv))
phase change	see **change of phase**
phase change diagram	see **pressure/temperature diagram**
phase diagram	A graphical representation in two or three dimensions of relationships between the three phases of matter under different physical conditions, usually for a pure substance
phase gate process	see **stage gate model**
phase rule	see **Gibbs phase rule**
phase separation	Separation into two or more phases of matter
phase state diagram	see **pressure/temperature diagram**

phase transition	see **change of phase**
phased array ultrasonic testing	*aka PAUT.* A type of ultrasonic testing (qv) which uses an array of synchronized transducers to produce a focused beam which can scan through a test sample without moving the array
phenolphthalein alkalinity	*aka* P-Alk. Alkalinity as measured by titration with phenolphthalein indicator; cf methyl orange alkalinity
Phillips catalyst	A chromium oxide catalyst used in the Phillips process (qv)
Phillips process	Most commonly an HDPE (qv) production process, but also used for the freeze concentration of beverages and a butane to butadiene process
phlegmatized	In the context of explosives, desensitized or stabilized, like dynamite (qv)
photolysis	see **photolytic reaction**
photolytic reaction	*aka* photolysis. Splitting something into subunits using light energy
physical explosion	An explosion not caused by a chemical or nuclear reaction, such as a boiler explosion
physical solvent	A solvent such as Selexol (qv) which relies purely on a solute's solubility for dissolution, rather than any chemical reaction
physical stability	The ability to resist physical impacts
PI	1. Professional indemnity, see **professional indemnity insurance** 2. *(drawing notation)* Pressure indicator
pickling	Chemical cleaning and descaling prior to passivation (qv); defined formally in ASME BPE (American Society of Mechanical Engineers: Bioprocessing Equipment)
pickling lime	see **slaked lime**
pico-	*(symbol* p-) The SI unit prefix denoting a factor of 10^{-12}
PID control	Proportional, integral, derivative control
PID controller	*aka* three term controller. A process controller capable of PID control (qv)
Pidgeon process	The most common magnesium metal production process
piece	*(drawing notation)* Spool pieces
pier	A vertical structural support, such as a 'pier wall' or a structure projecting from a shore into navigable water with vertical supports. Formal definition of the second meaning in NFPA 30: Flammable and Combustible Liquids Code, 2018 Edition
piezoelectric transducer	*aka* PET. A type of pressure transducer

pig	*(potentially offensive) aka* pipeline scraper. A device inserted into a pipeline for a number of purposes. Often a simple sphere used to clear out liquids, but can also be an abrasive plug to scrape out deposits (a pipeline scraper (qv) in the true sense), or even a sophisticated internal inspection device; e.g.; intelligent pig (qv)
PIG	A backronym for pig (qv), supposedly short for either 'pipeline inspection gauge', or 'pipeline intervention gadget'
pig iron	*aka* crude iron. The impure iron product of a blast furnace
pig launcher	The device which launches a pig (qv) into a pipeline, usually from a pressurized container
pig livers	*(informal)* An electrical lineman's term for special yokes used on extra-high voltage lines
pig trap	A piece of piping on one end of a pipeline to catch a pig (qv)
pigging	Using a pig (qv)
pig's dick	*(informal, potentially offensive)* A progressive cavity pump rotor, so-called due to its corkscrew shape
pigtail	1. A short wire from the underside of a surface buoy to the first pennant
	2. A coil of wire made by an electrical installer at the point of connection to facilitate maintenance; see **Figure P1**.
PII	see **professional indemnity insurance**

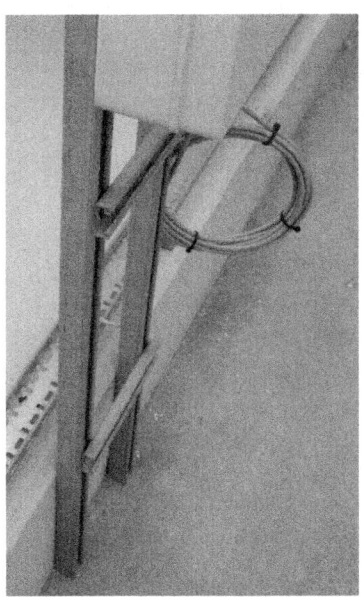

Figure P1. A pigtail wiring arrangement

pilot	*aka* pilot burner. A small flame which stays continuously alight in order to light (or relight to stabilize) the main burner. Defined formally in BS EN ISO 13705:2012/ISO 13705:2012(E) Petroleum, petrochemical and natural gas industries. Fired heaters for general refinery service, BS EN 12952-8:2002 Water-tube boilers and auxiliary installations. Requirements for firing systems for liquid and gaseous fuels for the boiler, API Standard 521. Pressure-relieving and Depressuring Systems. Sixth Edition | January 2014, API RP 535 - Burners for Fired Heaters at Refineries, API RP 537 Flare Details for Petroleum, Petrochemical, and Natural Gas Industries, API STD 530 - Calculation of Tube Heater Thickness, API STD 560 - Fired Heaters for General Refinery Service and NFPA 86: Standard for Ovens and Furnaces, 2019 Edition
pilot burner	see **pilot**
pilot operated modulating pressure safety valve	*aka* pilot operated modulating pressure relief valve (POPRV). A PSV (qv) according to API 520 RP 520 P1 7th Edition, January 2000 Sizing, Selection, and Installation of Pressure-Relieving Devices in Refineries; Part I - Sizing and Selection, even though it modulates, but API 520 also defines "safety valve" as "a spring-loaded pressure relief valve actuated by the static pressure upstream of the valve and characterized by rapid opening or pop action". This is why a pilot operated modulating pressure safety valve is more commonly called a POPRV (pilot-operated modulating pressure relief valve). Even if API 520 had been consistent, it would have differed from everyone else's definition
pilot operated modulating pressure relief valve	*aka* POPRV. A more preferred term for a pilot operated modulating pressure safety valve (qv)
pilot operated pressure relief valve	*aka* pilot operated safety relief valve. A PRV (qv) controlled by a second, self-actuated valve. Defined formally in API RP 520 P1 7th Edition, January 2000 Sizing, Selection, and Installation of Pressure-Relieving Devices in Refineries; Part I - Sizing and Selection, API Standard 521. Pressure-relieving and Depressuring Systems. Sixth Edition | January 2014 and EN ISO 4126; cf spring loaded PRV
pilot operated safety relief valve	see **pilot operated pressure relief valve**
pilot plant	Equipment used to generate design data for process scaleup

pilot sprinkler	A sprinkler used as a detector to initialize the release of a system actuation valve (qv). Defined formally in NFPA 15 Standard for Water Spray Fixed Systems for Fire Protection
pin actuated device	*aka* pin device. A non-reclosing pressure relief device (qv) whose function is based on a pin bending or breaking. Defined formally in API RP 520 P1 7th Edition, January 2000 Sizing, Selection, and Installation of Pressure-Relieving Devices in Refineries; Part I - Sizing and Selection, API RP 537 Flare Details for Petroleum, Petrochemical, and Natural Gas Industries, API RP 576 - Inspection of Pressure Relieving Devices and API Standard 521. Pressure-relieving and Depressuring Systems. Sixth Edition	January 2014
pin device	see **pin actuated device**
pinch analysis	Correctly, a formal exercise to minimize (rather than optimize) usage of certain resources, with some limited professional utility in energy intensive industries; now unfortunately used informally as a term for all energy recovery improvements; cf heat integration
pinch point	1. In an engineering context, a place in or near a machine where part of a person may become trapped 2. In academia, something to do with pinch analysis (qv) - of limited professional relevance
pinch temperature	1. In an engineering context, the difference between a heat recovery steam generator's heating medium's exit temperature and steam saturation temperature, as defined formally in API RP 534 – Heat Recovery Steam Generators 2. In academia, something to do with pinch analysis (qv) - of limited professional relevance
pink diesel	A USA alternative term for red diesel (qv)
pink hydrogen	Hydrogen produced by nuclear powered electrolysis; cf hydrogen colors
pipe	A sectional cylindrical vessel used for fluid conveyance, especially those of smaller diameter, shorter or entirely in-plant, see **Box P1**; cf pipeline
pipe anchor	*aka* thrust block, pipework anchor. A fixed point of attachment used to control pipework movement due to thrust arising from thermal expansion or frictional forces
pipe anchor load	*aka* thrust block load. The force calculated to resist loading (excluding thermal expansion loads) in anchored pipe systems

Box P1. Pipe/Tube/Tubing/Piping/Hose

I give clear definitions between these terms in this book, but discussions with fellow professionals show that there is some confusion, and some strongly held minority opinions.

LinkedIn is full of people sharing a ChatGPT generated graphic about this, adding further to confusion by claiming (incorrectly) that the only difference between pipe and tube is that pipe is measured by internal diameter, and tube by external diameter.

My definitions are as follows:

Pipe: A sectional cylindrical vessel used for fluid conveyance, especially those of smaller diameter, shorter or entirely in-plant; cf pipeline

Pipeline: A large diameter pipe (qv) used for long distance fluid transport, or by analogy, a means of conveyance of something (not necessarily something tangible) in management speak (qv)

Tube *aka* tubing: A tube differs from a pipe (qv) in that it can have a noncircular cross section, and is sized by nominal outside (rather than inside) diameter. Defined formally in ASME BPE (American Society of Mechanical Engineers: Bioprocessing Equipment)

Discussions however show that there is a minority view that "tube" and especially "tubing" is a smaller bore and possibly thinner walled type of pipe/piping, especially if it is flexible (normally called a hose). I have added a new definition in this edition to address this confusion

Hose: A flexible tube or pipe used to convey liquids

Someone also said that they had heard the term "standard tubing" used, though all other commentators said this was meaningless to them. It is therefore strongly deprecated. There is no 'standard' standard.

pipe bent	*aka* bent, piperack bent, rack bent. A frame consisting of vertical and horizontal steel or concrete members which carries pipework (usually above head height) within a pipe rack. The most crowded bent sets the width of the whole pipe rack; N.B. not to be confused with 'bent pipe'
pipe bridge	see **pipebridge**
pipe burner	A simple perforated tube burner. Defined formally in NFPA 86: Standard for Ovens and Furnaces, 2019 Edition
pipe chase	see **chase**
pipe class	see **piping class**
pipe drivehead	see **drivehead**
pipe load	The weight of all piping, including contents, valves, fitting and insulation
pipe rack	see **piperack**

pipe schedule	see **schedule number**
pipe supports	see **supports**
pipe wrench	A Stillson (qv) wrench
pipeband	see **piperack**
pipebridge	*aka* pipe bridge. A specially designed and constructed bridge which carries pipes over a road or other area which needs to be free of support columns at approximately 6-7m above grade. Sometimes confusingly used synonymously with pipe rack (qv)
pipeline	A large diameter pipe (qv) used for long distance fluid transport, or by analogy, a means of conveyance of something (not necessarily something tangible) in management speak (qv); see **Box P1**.
pipeline inspection gauge	see **PIG**
pipeline intervention gadget	see **PIG**
pipeline scraper	Properly, a form of pig (qv) which scrapes deposits for the inside of a pipeline. In countries where any mention of a pig may be considered offensive, a term applied to all forms of pig
pipeline transportation facility	In an oil and gas context, an area associated with the handling, storage and delivery of flammable hydrocarbons via pipeline. Defined formally in API RP 505 2nd Edition, August 2018 Recommended Practice for Classification of Locations for Electrical Installations at Petroleum Facilities Classified as Class I, Zone 0, Zone 1, and Zone 2
piper	*(informal)* see **piping engineer**
piperack	*aka* pipeband, pipetrack, pipeway, pre-assembled rack (PAR). A structure intended primarily to support pipework, but which commonly also carries any traywork (qv), lighting, etc. which cannot pass through adjacent areas around the plant, at 4.4-7m above grade
piperack bent	see **pipe bent**
pipestill	see **crude distillation unit**
pipetrack	Usually, synonymous with piperack (qv), though some define pipetrack as being at ground level and piperack as being at an elevation of 4.4-7m
pipeway	see **piperack**
pipework	A system of pipes, fittings and valves
pipework anchor	see **pipe anchor**
piping	Synonymous with pipework (qv), also the discipline followed by a piping engineer (qv); see **Box P1**

piping and instrumentation diagram	*aka* P&ID, P+ID. The process engineer's signature drawing. A drawing setting out the physical and logical interrelationships between process plant components in the form of a topologically correct symbolic drawing which shows the unit operations, piping, and instrumentation of a process plant. Defined formally in BS EN ISO 10209:2012 Technical product documentation - Vocabulary-- Part 1: Terms relating to technical drawings: general and types of drawings; cf process and instrumentation diagram
piping class	A classification of temperature and pressure rating of pipe flanges (rather than piping per se), based on ASME/ANSI B16.5 Pipe Flanges and Flanged Fittings: NPS 1/2 through NPS 24, Metric/Inch Standard 2020
piping designer	see **piping engineer**
piping engineer	*aka* piper, piping designer. A specialist in piping and sometimes plant layout used in some industries and countries to lay out plant and piping
piping layout	The layout of piping and associated support systems, usually undertaken by piping engineers; a subset of site layout (qv), plant layout (qv) or plot layout (qv)
piping spacer	see **spacer ring**
piping study	The detailed design of piping systems undertaken from detailed design (qv) stage onwards
piping system	1. Generally, synonymous with piping (qv), making the word 'system' superfluous 2. In an LPG context, refers only to LPG conveyance, as defined formally in NFPA 58: Liquefied Petroleum Gas Code, 2017 Edition
piston	A moving component in a reciprocating pump cylinder which has a circumferential moving high-pressure seal; cf plunger
piston diaphragm pump	A type of high precision/high pressure reciprocating positive displacement pump (qv), in which a hydraulic fluid driven by a piston in turn drives a diaphragm which pumps the process fluid
piston plunger pump	A deprecated term for a plunger pump (qv)
piston pump	A positive displacement reciprocating pump using a piston to pump the process fluid; cf plunger pump
pit	An element of localized corrosion in the form of a void, the width and depth of which are small compared with plate thickness. Slightly different formal definition in a hygienic context in ASME BPE (American Society of Mechanical Engineers: Bioprocessing Equipment)

pitch	1. Generally, the angle at which something is set relative to something else, as defined formally in ASME BPE (American Society of Mechanical Engineers: Bioprocessing Equipment) 2. In the context of heat exchangers, the distance between tube centerlines expressed in multiples of tube outside diameter 3 A synonym for tar (qv) 4. The frequency of sound produced by an operating piece of mechanical equipment such as a pump
Pitot tube	*aka* pitot probe. One of a number of devices used to convert the flow rate of a fluid into a pressure difference, in this case to allow its measurement in a pitot tube flow meter
pitting	Corrosion in the form of pit(s) (qv). Slightly different formal definition in an oil and gas context in API RP 579 - Fitness for Service
pitting resistance equivalent number	*aka* PREN. A prediction of resistance to pitting of stainless steels based on their composition
pka	A measure of acid strength, the negative log of an acid's dissociation constant (qv). Strong acids have a negative pka
PL	1. see **planar longitudinal** 2. see **public liability insurance**
placing on the market	In the EU, this term has a very specific legal meaning, which in summary means that all requirements of CE marking (qv) must be met
plait point	The point on a ternary phase diagram (qv) where raffinate (qv) and extract (qv) phases have the same composition
plan	1. Most commonly in engineering, a drawing showing a top-down view; defined formally in BS EN ISO 10209:2012 Technical product documentation - Vocabulary - Part 1: Terms relating to technical drawings: general and types of drawings; cf elevation 2. Rarely, synonymous with program (qv)
planar longitudinal	*aka* PL. A term for a wave whose motion is in the plane of the direction of travel of the wave
planar surface	*aka* PS. A mathematical plane, a flat or level surface
planar transverse	*aka* PT. A term for a wave whose motion is in a plane at right angles to the direction of travel of the wave
planning consent	see **planning permission**
planning permission	*aka* planning consent. A UK requirement prior to building on or changing the use of land. The process required to obtain permission is analogous to the requirements of land use and zoning regulations in the USA

planometric axonometry	A type of oblique axonometry; defined formally in BS EN ISO 10209:2012 Technical product documentation - Vocabulary - Part 1: Terms relating to technical drawings: general and types of drawings
plant	The complete set of process units and direct supporting infrastructure required to produce a product or products from raw or part-processed materials. The term may also encompass other elements, e.g.; buildings housing process plant, warehousing, research, quality control, operational control and administration functions. Shorter but vaguer formal definitions may be found in BS EN 82079-1:2012 Preparation of instructions for use. Structuring, content and presentation. General principles and detailed requirements or NFPA 59: Utility LP-Gas Plant Code, 2018 Edition
plant emergency escape route	Operator egress and emergency escape route on a plant
plant floor	see **interstitial space**
plant layout	The spatial arrangement of process equipment and its interconnections
plant section	An independently operable part of a process plant. Defined formally in BS EN ISO 10209:2012 Technical product documentation - Vocabulary - Part 1: Terms relating to technical drawings: general and types of drawings
plant trial	see **trial**
plasma arc	An electrical current creating and passing through an ionized gas plasma. Defined formally in NFPA 86: Standard for Ovens and Furnaces, 2019 Edition
plasma arc furnace	A furnace which heats work by means of a plasma arc (qv). Defined formally in NFPA 86: Standard for Ovens and Furnaces, 2019 Edition
plastic	Used to describe a type of polymer (properly called a thermoplastic); or the physical property of plasticity (qv)
plastic deformation	see **plasticity**
plasticity	*aka* plastic deformation. The ability to permanently change shape in response to applied forces; defined formally in API RP 579 - Fitness for Service

plate	1. Strictly, a rectangular flat metal stock, usually >6mm thick and thicker than sheet (qv) 2. Also used informally as synonymous with sheet 3. *aka* tray. A name for a number of thinnish, flattish items manufactured from such stock, used in distillation columns, heat exchangers etc. 4. A verb referring to coating base metal with another metal, especially by electrolysis
plate and fin heat exchanger	An expensive but compact heat exchanger, prone to fouling and hard to clean
plate and frame filter press	Traditionally, a stack of plates with coarse filter elements clamped in a frame operating batchwise by cake filtration (qv), with manual solids discharge. Modern designs may however use finer membrane filtration media, and feature automated cake discharge
plate and frame heat exchanger	*aka* plate heat exchanger. An expensive but compact heat exchanger, less prone to fouling than others, and designed for thorough cleaning; popular in hygienic applications
plate count	The number of bacterial colonies on a petri dish
plate efficiency	A fiddle factor (qv) for correcting theoretical models with respect to the real separation efficiency of a distillation column plate; cf Baur efficiency, overall plate efficiency, Murphree plate efficiency
plate heat exchanger	see **plate and frame heat exchanger**
plate settler	see **lamella clarifier**
platforming process	Catalytic reforming (qv) using a platinum catalyst
platinum cobalt scale	*aka* Pt Co scale; see **color**
platinum-cobalt color	see **color**
Platts	Shorthand for S&P Platts, the commercial website used by energy and commodities traders for price data. In insurance, Platts data is used in the evaluation of business interruption insurance (qv) claims from refineries and petrochemical plants
PLC	see **programmable logic controller**
plc	Public limited company (UK)
PLE	*(drawing notation)* Plain large end

plenum	1. Generically, a chamber or space in which a fluid is at higher than atmospheric pressure 2. In the context of fired heaters, the plenum feeds the burners; defined formally in this sense in BS EN ISO 13705:2012/ISO 13705:2012(E) Petroleum, petrochemical and natural gas industries. Fired heaters for general refinery service, API RP 535 - Burners for Fired Heaters at Refineries, API STD 530 - Calculation of Tube Heater Thickness and API STD 560 - Fired Heaters for General Refinery Service
plenum chamber	see **plenum**
plenum space	The plenum (qv) between a building's floor and ceiling used for ductless HVAC circulation; cf interstitial space
Plinke	Proprietary eponym for an oil and gas industry acid recycling plant
plinth	UK English for the raised pad of concrete slab upon with an item of equipment is mounted, internationally known as a footing, pad, pedestal, base etc.
plot	An area of a site most commonly defined as being bounded by the road system, although it may be single-side accessed or be directly adjacent to another plant taking a feed or feeds from that location
plot layout	Layout at a plot (qv) level: the consideration of process units in relation to each other's disposition within a plot
plot plan	see **general arrangement drawing**
plot roads	Roads which bound (and therefore often define) individual plots
plough	UK English for plow (qv)
plow	*aka* plough. A structure which diverts bulk materials from a conveyor
plug flow	1. Generally, having a constant velocity profile perpendicular to flow direction 2. Far less commonly, a type of two-phase flow regime
plug flow reactor	*aka* PFR. A theoretical reactor model, with plug flow and - by implication - no axial mixing, but with perfect radial mixing; or the real-world item which the model approximates; cf continuous stirred tank reactor
plug header	A cast heat exchanger return bend with inspection and cleaning ports. Defined formally in BS EN ISO 13705:2012/ISO 13705:2012(E) Petroleum, petrochemical and natural gas industries. Fired heaters for general refinery service, API STD 530 - Calculation of Tube Heater Thickness and API STD 560 - Fired Heaters for General Refinery Service

plug number	A number you plug into a formula; cf SWAG
plug valve	A valve controlling flow by means of an internal cylindrical plug with a hole through it, rotating through one quarter of a turn from open to close; cf ball valve, globe valve
plugging	*aka* pore plugging, membrane plugging. The blocking of membrane pores by solid particles
plume	A steady release of gas or aerosol, or the cloud produced by such a release; cf puff
plunger	Similar to a piston (qv), except that the high-pressure seal is stationary, and the reciprocating element travels past it
plunger pump	*aka* ram pump. A positive displacement reciprocating pump using a plunger rather than a piston, and consequently capable of generating higher delivery pressures than a piston pump (qv)
Plus Gas™	Alternative proprietary eponym for products similar to WD40 (qv)
Pmax	see **maximum power output**
PMI	see **positive materials identification**
PMP	see **project management professional**
PMx	A size range of atmospheric particulate matter (qv), where x stands for a number, such as for example PM10, which is the fraction of particulate matter with aerodynamic diameter smaller than 10 microns
pneumatic control valve	A pneumatically actuated control valve
pneumatic conveying	The controlled transfer of a particulate solid by fluidization (qv) in a gas stream
pneumatic conveying system	Everything required for pneumatic conveying of solids including solid and gas introduction equipment, ductwork, and gas-solid separation; defined formally in NFPA 654: Standard for the Prevention of Fire and Dust Explosions from the Manufacturing, Processing, and Handling of Combustible Particulate Solids, 2017 Edition
pneumatic ejector	A gas-driven ejector (qv)
pneumatic mixer	*aka* air-mix mixer, air driven mixer. A mixer which uses an air stream to mix fine powders or liquids
pneumatics	Systems driven by compressed air, or the engineering science of such systems
pneumercator	A rather complex and old-fashioned type of liquid level sensor
PO	1. see **produced oil** 2. *(drawing notation)* Pump outs 3. see **purchase order**

POA	1. Plan of action
	2. Price on application
	3. Proof of acceptance
POC	Point of connection
POC switch	see **proof of closure switch**
POD	1. see **point of departure**
	2. see **probability of detection**
	3. see **programmable optimal density**
POE	*(drawing notation)* Plain one end
point of departure	*aka* POD. A starting point
point of measurement	The point where a measurement can be made. Defined formally in BS1646-1 Symbolic Representation for Process Measurement Control Functions and Instrumentation Part 1: Basic Requirements and BS 1646-3:1984 Symbolic representation for process measurement control functions and instrumentation. Specification for detailed symbols for instrument interconnection diagrams
point of tangency	see **end of curve**
point of transfer	An LPG equipment connection or venting point. Defined formally in NFPA 58: Liquefied Petroleum Gas Code, 2017 Edition and NFPA 59: Utility LP-Gas Plant Code, 2018 Edition
point of use	*aka* POU. A term used in water supply and hygienic service for the end point of a distribution system (qv)
point source method	A statistical method based on the analysis of a single identifiable localized source of something (e.g.; pollution, explosion)
poise	(*symbol* P) A non-SI metric unit of dynamic viscosity
Poiseuille equation	see **Hagen–Poiseuille equation**
Poiseuille law	see **Hagen–Poiseuille equation**
poison	1. In a biological context, a potential toxin (there are only potential toxins as the dose makes the poison)
	2. In the context of catalysts, something which potentially chemically deactivates a catalyst, more properly called a catalyst poison (qv)
Poisson's ratio	The ratio of lateral to longitudinal strain
poka yoke	Japanese for 'mistake-proofing' (previously known as the less polite b*aka* yoke (qv) or 'idiot-proofing'); techniques to avoid human error in manufacturing industries by preventing, correcting, or drawing attention to human errors as they occur

polar	1. In chemistry, short for 'polar covalent', a form of chemical bonding where unequal sharing of electrons produces a molecule with electrical charge separation between ends (poles); cf **covalent** 2. In a drawing context, see **polar coordinate system**
polar coordinate system	*aka* polar. A system of coordinates relative to a horizontal straight line; defined formally in BS EN ISO 10209:2012 Technical product documentation - Vocabulary - Part 1: Terms relating to technical drawings: general and types of drawings
polar covalent	see **polar**
polarimeter	A device which measures the rotation of plane polarized light, a property which is influenced by concentration of solutes such as sugar
Polarite	1. A proprietary eponym for granular filter media which enhance autocatalytic manganese removal in drinking water treatment 2. A palladium containing mineral
polarity	see **polar**
polisher	In the context of ion exchange, the unit responsible for final removal of trace contaminants
polishing	1. Reducing the surface roughness of metals 2. The final removal of trace contaminants
pollutant	A detrimental environmental contaminant (can be physical as well as chemical)
pollution	The release of a pollutant (qv), and/or the consequences of such a release
Pollution Prevention and Control Act 1999	*aka* PPC Act. UK environmental legislation which enacted the European IPPC Directive (qv)
poly	see **polyelectrolyte**
polyaluminum chloride	*aka* PAC. A water treatment coagulant
polychlorinated biphenyl	*aka* PCB. A class of persistent organic pollutants
polycondensation	see **condensation polymerization**
polydispersed resin	An ion exchange resin with a wide particle size range
polyelectrolyte	*aka* polymer, poly. An organic flocculant, (and sometimes coagulant) used in water treatment
polyethene	see **polyethylene**
polyethylene	*aka* polythene, polyethene, PE, HDPE, LDPE, MDPE. A polymer of ethene (*aka* ethylene)
polyethylene terephthalate	*aka* PET. Polyester used for liquid containers and textiles

polymer	1. A long molecule made of a string of similar (but not necessarily identical) subunits known as monomers (qv) 2. see **polymeric material**; defined formally in ASME BPE (American Society of Mechanical Engineers: Bioprocessing Equipment 3. An informal term used in the water industry for polyelectrolyte (qv)
polymeric material	A macroscopic material made from polymer molecules. Defined formally in ASME BPE (American Society of Mechanical Engineers: Bioprocessing Equipment)
polymeric silica	A term used in water treatment for dissolved silica species, other than the monomeric form (SiO_2)
polymerization	Chemically combining monomers to make a polymer (qv)
polymerizing fluid	A fluid capable of polymerization (qv) under process conditions; defined formally in API RP 682 - Pump Seals
polypropylene	*aka* PP. A polymer of propene (known as propylene in the past/in the petrochemicals industry)
polythene	see **polyethylene**
polythionic acid stress corrosion cracking	*aka* PTASCC. A form of stress corrosion cracking (qv) of stainless steels in the presence of certain oxoacids, potentially important when processing high sulfur hydrocarbons. Defined formally in API RP 932B Corrosion Air Coolers
polytropic	A term applied to a thermodynamic process where energy may leave or enter the system, but PV^n is constant, n being the polytropic index; cf adiabatic, isentropic, isobaric
polytropic gas	*aka* ideal polytropic gas. A gas with polytropic (qv) expansion and compression
Ponchon–Savarit	A more precise, more time-consuming, though equally obsolete alternative to McCabe-Thiele (qv)
pony collar	A short piece of drill collar (qv) that can be used to adjust the total length of a string of pipe
pool boiling	Boiling below the surface of a static fluid reservoir; cf flow boiling, nucleate boiling, film boiling
pool fire	A burning pool of liquid
POP	see **persistent organic pollutant**
pop	*aka* popping. Although API use the term in their PSV definition, "popping" is commonly applied to all PRVs as a synonym of "lifting", with an implication of rapid opening (but not necessarily complete opening)
POPRV	see **pilot operated modulating pressure relief valve**
pop pressure	*aka* popping pressure. The inlet pressure required to lift a PRV disk; not necessarily the same as the set pressure (qv)

pop test	A practical test to compare the pop pressure (qv) with set pressure (qv) of a PRV
Popping	see **pop**
popping pressure	see **pop pressure**
population equivalent	*aka* PE, equivalent person (EP), unit per capita loading. In effluent treatment plant design, an amount of BOD (qv) nominally equivalent to that of a single person using a sewage system
pore plugging	see **plugging**
porosity	1. *aka* void fraction, voidage. Quantitatively, the fractional volume of a bulk material which is not occupied by that material 2. Qualitatively, the property of having any such discontinuities, as defined formally in API STD 620 - Low Pressure Storage Tanks and ASME BPE (American Society of Mechanical Engineers: Bioprocessing Equipment)
port	An access point, whether applied to vessel penetrations, or a place for ships to sail to
portable cannon	see **portable monitor**
portable container	An LPG transport container which can be moved whilst full; defined formally in NFPA 58: Liquefied Petroleum Gas Code, 2017 Edition; cf portable storage container
portable monitor	*aka* portable cannon. A firefighting foam monitor (qv) which can be readily transported to a fire. Defined formally in NFPA 11: Standard for Low-, Medium-, and High-Expansion Foam, 2016 Edition
portable storage container	An LPG container used for storage but (confusingly) not for the transport of LPG, because it is only designed to be moveable whilst empty. Defined formally in NFPA 58: Liquefied Petroleum Gas Code, 2017 Edition cf portable container
portable system	Firefighting foam system components which can be readily hand-transported to a fire. Defined formally in NFPA 11: Standard for Low-, Medium-, and High-Expansion Foam, 2016 Edition

portable tank	Has a specialist meaning in the context of transportation of flammable fluids, for which the NFPA offers two different definitions: 1. A transportable tank of more than 230 L capacity for use with flammable and combustible liquids, defined formally in NFPA 30: Flammable and Combustible Liquids Code, 2018 Edition 2. A far larger transportable tank for LPG, defined formally in NFPA 58: Liquefied Petroleum Gas Code, 2017 Edition
Portakabin	The UK proprietary eponym for temporary accommodation; called a 'trailer' in the USA
portal frame	A form of continuous frame structure common in industrial buildings which provides a clear span unobstructed by bracing
Portaloo	A UK proprietary eponym for a Portapotty (qv)
portapac	A portable storage container (qv) in the form of an ASME container (qv) with retractable wheels. Defined formally in NFPA 58: Liquefied Petroleum Gas Code, 2017 Edition
Portapotty	A US proprietary eponym for a transportable toilet; cf Portaloo
positive displacement compressor	One of a number of types of compressor which generate a flow of gas mechanically, by filling a cavity at one end with gas, then emptying that cavity at the other end. There are reciprocating and rotary subtypes
positive displacement pump	One of a number of types of pump which generate a flow of fluid mechanically, by filling a cavity at one end with fluid, then emptying that cavity at the other end. There are reciprocating and rotary subtypes
positive list	see **mandatory list**
positive materials identification	*aka* PMI. Various NDT (qv) methods which allow a material's composition to be positively identified
positive peak value	see **peak value**
positive pressure pneumatic conveying system	A pneumatic conveying system (qv) using a gas above atmospheric pressure. Defined formally in NFPA 654: Standard for the Prevention of Fire and Dust Explosions from the Manufacturing, Processing, and Handling of Combustible Particulate Solids, 2017 Edition
positive shutoff valve	A shutoff valve which, when closed, prevents flow of fluid in both directions; defined formally in NFPA 58: Liquefied Petroleum Gas Code, 2017 Edition; cf one way valve
POSRV	Pilot operated safety relief valve; see **pilot operated pressure relief valve**

possum belly	*aka* distribution box, flowline trap. A vessel on a drilling rig which receives drilling fluid (qv) from the end of the flow line
post construction design	The stages of process design in which the for construction (qv) design has to be modified to match real world conditions, and post-handover optimization occurs
post weld heat treatment	The controlled heating and cooling of a weldment (qv) to greatly reduce residual weld stresses. Defined formally in API RP 579 - Fitness for Service
pot metal	see **monkey metal**
pot still	A simple, inefficient batch distillation rig used to manufacture some alcoholic spirits
potable water	*aka* PW. Water fit for human consumption (which has usually been filtered, chemically conditioned, and disinfected) usually implying that it meets WHO and national standards
potential alkalinity	In the context of a bicarbonate buffered water, a measure of all the various carbon dioxide derived species in the system[83]
potential maximum power	A term used in the context of turbines, defined formally (and differently) in ISO 3977 Gas turbines - Procurement - Part 3: Design requirements and EN ISO 10437 Petroleum, petrochemical and natural gas industries - Steam turbines - Special-purpose applications
POU	see **point of use**
pound	1. (*symbol* £) *aka* pound sterling. A UK currency unit 2. (*symbol* lb.) *aka* international pound, avoirdupois pound. A US customary unit (qv) of mass 3. (*symbol* #) *aka* pound sign (US only), hash tag etc. A rarely used alternative abbreviation for a pound in weight (avoirdupois pound)
poundal	An obsolete Imperial unit of force
pounds per square inch	*aka* psi. A US customary unit of pressure; cf psia, psig
pour point	*aka* pourpoint. The lowest temperature at which a sample can be poured from a beaker; used as a heuristic for estimating a crude oil's paraffinicity/aromaticity
Pourbaix diagram	A plot of equilibrium potential against pH for various oxidation states of a metal, useful in predicting corrosion
pourpoint	see **pour point**
powder	A bulk solid comprised of dry fine granules produced by a size reduction process

---

[83] Covered in more detail in Moran, S. (2018) *An Applied Guide to Water and Effluent Treatment Plant Design*, Oxford: Elsevier

powdered activated carbon	*aka* PAC. Activated carbon (qv) particles too small to be retained by a 50-mesh sieve; cf granular activated carbon
power actuated safety relief valve	A pressure relief valve (qv) with an externally powered controller; cf spring loaded PRV, pilot operated pressure relief valve
power number	(*symbol* Np *or* Po) *aka* Newton number. A dimensionless group representing the ratio of drag and inertial forces used in the analysis of mixing
power supply bushing	A fitting allowing electrical power to be carried to and from the inside of a pressure vessel shell. Defined formally in NFPA 58: Liquefied Petroleum Gas Code, 2017 Edition
Powerforming	A proprietary catalytic reforming process with a platinum catalyst
powered emergency release coupling	*aka* PERC. A safety measure on an LNG tanker
PP	see **polypropylene**
ppb	Parts per billion; a unit of concentration
PPC Act	see **Pollution Prevention and Control Act 1999**
PPE	see **personal protective equipment**
PPI	1. see **parallel plate interceptor**
	2. see **producer price indexes**
ppm	Parts per million; a unit of concentration; often confused with mg/l (which it is equivalent to under certain circumstances), so sometimes clarified with v/v or m/v
PPMVD	Parts per million by volume (dry)
PPPPPP	*aka* the six Ps. Short for perfect planning prevents piss poor performance (or pretty poor performance for some)
PQ	see **performance qualification**
PR equation of state	see **Peng–Robinson equation of state**
PRA	1. Preliminary risk assessment
	2. Probabilistic risk analysis; see quantitative risk analysis
PRAC	see **Pharmacovigilance Risk Assessment Committee**
practical test	A documented evaluation procedure, defined very formally in BS EN ISO 14159:2008 Safety of machinery. Hygiene requirements for the design of machinery
Prandtl number	(*symbol* Pr) A dimensionless group; the ratio of kinematic viscosity to thermal diffusivity; useful in heat transfer analysis
Prandtl's one seventh power law	An empirically determined predictor of velocity profile for turbulent flow in pipes
PRD	see **pressure relief device**
pre assembled rack	*aka* PAR; see **piperack**

pre start up safety review	*aka* operational readiness review. A process safety review (qv) carried out just prior to commissioning as part of process safety management
pre-assembled unit	*aka* PAU; see **module**
precip.	see **electrostatic precipitator**
precipitation	The formation of particles from solution by exceeding the solubility of an ionic solid
precipitation crystallization	see **salting out**
precipitation potential	1. The probability of atmospheric precipitation (such as rain) 2. The amount of a substance which must be precipitated from a solution to reach the concentration in equilibrium with that material as a solid; cf calcium carbonate precipitation potential
precision	Whether an instrument will give the same reading against the same true value the next time it is tested (though commonly used in other senses); cf accuracy; see **Box A1**.
precision continuous mixer	see **precision control mixing**
precision control mixing	*aka* PCM, precision continuous mixer. A type of hydration unit used to make up frac fluid (qv)
precoat filter	A relatively coarse filter designed to be impregnated with a cake of fine particles (which serve as the effective filtration medium) prior to use
precommissioning	*aka* pre-commissioning. The non-operating work responsibilities such as adjustments, cold alignment checks, etc. which are performed by the contractor prior to the ready for commissioning (qv) or mechanical completion (qv) stage; see **Box C3**.
pre-commissioning	see **precommissioning**
Pred	Reduced pressure, see **vented explosion pressure**
prefabricated septic tank	A unitary factory-made septic tank. Defined formally in EN 12566 - Small wastewater treatment systems for up to 50 PT Part 1: Prefabricated septic tanks
preferential weld corrosion	*aka* PWC, preferential weldment corrosion, grooving corrosion, knife-line attack, trench-like corrosion. Rapid corrosion of the weld bead and/or heat affected zone (qv) in certain carbon steels exposed to corrosive environments such as high CO_2 water or seawater
preferential weldment corrosion	see **preferential weld corrosion**

preferred operating region	1. Most commonly defined as operation at 70-120% of flow at BEP (qv), as per API 610 Centrifugal Pumps for Petroleum, Petrochemical, and Natural Gas Industries 2. Also defined with respect to vibration compared with base limit as per EN ISO 13709:2003 Centrifugal pumps for petroleum, petrochemical and natural gas industries. N.B. care must be taken, as these standards specifically cover the same equipment, but define this term differently
preheated air	In the context of fired heaters, combustion air (qv) heated prior to use; defined formally in API RP 535 - Burners for Fired Heaters at Refineries
preliminary design analysis	*aka* PDA. A deprecated vague term for one of the early stages of design, either conceptual design (qv) or FEED (qv)
preliminary drawing	see **draft drawing**
preliminary treatment	In the context of sewage treatment, properly used to describe the removal of the most gross contaminants by screening (or possibly comminution) and degritting prior to primary treatment (qv)
premix burner	In the context of fired heaters, a burner fed with premixed fuel and air. Two different formal definitions may be found in API RP 535 - Burners for Fired Heaters at Refineries and NFPA 86: Standard for Ovens and Furnaces, 2019 Edition; cf raw gas burner
premixed foam solution	A batch of firefighting foam solution produced by mixing a measured amount of concentrate with a measured amount of water. Defined formally in NFPA 11: Standard for Low-, Medium-, and High-Expansion Foam, 2016 Edition
PREN	see **pitting resistance equivalent number**
preoperational stage	see **non-nuclear testing phase**
prerequisite program	*aka* PRP. "Food safety basic conditions and activities that are necessary to maintain a hygienic environment throughout the food chain suitable for the production, handling and provision of safe end products and safe food for human consumption", according to the EHEDG Glossary
preservation	A chemical or physical treatment which extends the useful life of food or drink
pressing	The dewatering of slurry by mechanical compression
pressure	An amount of force applied per unit area
pressure atomizing burner	In the context of fired heaters, a burner atomizing a liquid fuel using pressure to force it through small orifices. Defined formally in NFPA 86: Standard for Ovens and Furnaces, 2019 Edition

pressure casing	The stationary pressure-containing parts of a piece of rotodynamic equipment. Defined formally in EN ISO 10437 Petroleum, petrochemical and natural gas industries - Steam turbines - Special-purpose applications, EN ISO 13709:2003 Centrifugal pumps for petroleum, petrochemical and natural gas industries and API RP 682 - Pump Seals
pressure control valve	*aka* PCV. Has a highly specific meaning in the context of fire pumps, defined formally in NFPA 20: Standard for the Installation of Stationary Pumps for Fire Protection, 2019 Edition
pressure design code	A recognized, purchaser-specified pressure vessel equipment design standard. Defined formally in BS EN ISO 13705:2012/ISO 13705:2012(E) Petroleum, petrochemical and natural gas industries. Fired heaters for general refinery service, API 660 - Shell-and-Tube Heat Exchangers, API RP 537 Flare Details for Petroleum, Petrochemical, and Natural Gas Industries, API RP 661 - Heat Exchangers, API STD 560 - Fired Heaters for General Refinery Service and API Standard 521. Pressure-relieving and Depressuring Systems. Sixth Edition|January 2014
pressure drop	The difference in pressure between two points in a flowing fluid as a result of friction, akin to headloss (qv). Defined formally in BS EN ISO 13705:2012/ISO 13705:2012(E) Petroleum, petrochemical and natural gas industries. Fired heaters for general refinery service, API STD 530 - Calculation of Tube Heater Thickness and API STD 560 - Fired Heaters for General Refinery Service, though these are rather confusing
pressure drop multiplier	(*symbol* R, $\Phi L2$, $\Phi G2$, $\Phi LO2$, $\Phi GO2$) *aka* two phase multiplier. One of a number of factors applied to a single-phase frictional pressure drop to account for two phase flow
pressure filter	A filter enclosed in a pressure vessel, especially a depth filter (qv)
pressure foam maker (high back pressure or forcing type)	A venturi-based firefighting foam maker; defined formally in NFPA 11: Standard for Low-, Medium-, and High-Expansion Foam, 2016 Edition
pressure gauge	A field mounted mechanical pressure indicator, most commonly a Bourdon gauge (qv)
pressure maintenance pump	*aka* jockey pump, make up pump. In the context of firewater systems, a pump which maintains acceptable system pressure when not in use for firefighting. Defined formally in NFPA 20: Standard for the Installation of Stationary Pumps for Fire Protection, 2019 Edition

pressure proportioning tank	A bladderless tank which passively doses foam concentrate into a water line proportionally to its flow. Defined formally in NFPA 11: Standard for Low-, Medium-, and High-Expansion Foam, 2016 Edition
pressure rating	Broadly synonymous with design pressure (qv); defined formally in one context in ASME BPE (American Society of Mechanical Engineers: Bioprocessing Equipment)
pressure reduced	*aka* Pred; see **vented explosion pressure**
pressure reducing regulator	see **pressure reducing valve**
pressure reducing valve	*aka* pressure reducing regulator. A valve which maintains a set downstream pressure
pressure regulating device	A water pressure regulating device. Defined formally in NFPA 20: Standard for the Installation of Stationary Pumps for Fire Protection, 2019 Edition and NFPA 24: Standard for the Installation of Private Fire Service Mains and Their Appurtenances, 2019 Edition
pressure regulating valve	see **pressure sustaining valve**
pressure regulator	A device which controls pressure downstream of the device; defined formally in one context in NFPA 86: Standard for Ovens and Furnaces, 2019 Edition
pressure relief device	*aka* PRD. A valve or bursting disk which opens a relief vent in response to high inlet pressure; used to protect pressurized systems. Defined formally in NFPA 58: Liquefied Petroleum Gas Code, 2017 Edition and API RP 520 P1 7th Edition, January 2000 Sizing, Selection, and Installation of Pressure-Relieving Devices in Refineries; Part I - Sizing and Selection
pressure relief valve	*aka* PRV. A pressure relief device (qv) opening to a degree proportional to excess pressure, and capable of reclosing when inlet pressure has fallen; defined formally in NFPA 58: Liquefied Petroleum Gas Code, 2017 Edition, NFPA 86: Standard for Ovens and Furnaces, 2019 Edition, API RP 520 P1 7th Edition, January 2000 Sizing, Selection, and Installation of Pressure-Relieving Devices in Refineries; Part I - Sizing and Selection and API Standard 521. Pressure-relieving and Depressuring Systems. Sixth Edition|January 2014
pressure safety valve	*aka* safety relief valve, PSV. A PRV (qv) which opens fully (rather than proportionally to excess pressure) as soon as a set inlet pressure is reached, usually found in gas service
pressure sustaining valve	*aka* PSV, pressure regulating valve (PrV). A valve which maintains a set upstream pressure

pressure swing adsorption	*aka* PSA. A process which allows separation of gas mixture components by selective adsorption at high pressure, followed by desorption at low pressure
Pressure Systems Safety Regulations 2000	*aka* PSSR. The UK primary regulations governing the design and inspection criteria for pressure systems
pressure tank	A tank using pressurized gas to passively empty itself of liquid. Defined formally in NFPA 22: Standard for Water Tanks for Private Fire Protection, 2018 Edition; cf vacuum tank
pressure test	The static pressure testing of a system of piping and vessels, commonly synonymous with a hydrostatic test (qv); a late stage in mechanical commissioning (qv). Defined formally in one context in NFPA 58: Liquefied Petroleum Gas Code, 2017 Edition
pressure vessel	A closed fluid container designed to withstand internal or external pressure. A common design is cylindrical with end caps called heads, which are usually either hemispherical (qv) or torispherical (qv) in shape. A US-centric formal definition may be found in NFPA 30: Flammable and Combustible Liquids Code, 2018 Edition
pressure/temperature diagram	*aka* phase state diagram, phase change diagram, PT diagram, P-T diagram. A type of thermodynamic phase diagram showing the relationship between pressure and temperature in a closed system
pressure/volume diagram	*aka* indicator diagram, PV diagram, p-V diagram. P-V diagram. A type of thermodynamic property diagram showing the relationship between pressure and volume in a closed system; cf TS diagram
pressurised water reactor	see **pressurized water reactor**
pressurized water reactor	*aka* pressurised water reactor, PWR. A nuclear reactor whose core is cooled by water which does not boil in the process, due to being pressurized; cf boiling water reactor
prestart purge mode	The purging of fuel lines of vapor with fuel prior to an engine start. Defined formally in NFPA 58: Liquefied Petroleum Gas Code, 2017 Edition
pretreatment	In the context of water treatment, initial treatment processes removing certain contaminants prior to processes which might be irrevocably damaged by such contaminants such as ion exchange (qv) or reverse osmosis (qv)
preventative maintenance	*aka* preventive maintenance, proactive maintenance. Keeping machinery in a safely operable state using routine, specified procedures. Can be time-based or usage based

prevention of significant deterioration program	*aka* Clean Air Act Program. A US EPA program to prevent air quality deteriorating significantly as a result of new large facilities
preventive action	Acting to eliminate at source a potentially undesirable consequence. Defined formally in BS EN ISO 9000 Quality management systems Fundamentals and vocabulary
preventive maintenance	see **preventative maintenance**
prill	A small, approximately spherical pellet produced by rapidly solidifying a sprayed melt (qv); cf granule
prilling	The production of prills (qv) from a melt
primary air	In the context of fired heaters, all air mixed with fuel prior to the burner. Four identical formal definitions in BS EN ISO 13705:2012/ISO 13705:2012(E) Petroleum, petrochemical and natural gas industries. Fired heaters for general refinery service, API STD 530 - Calculation of Tube Heater Thickness, API RP 535 - Burners for Fired Heaters at Refineries and API STD 560 - Fired Heaters for General Refinery Service; a slightly different one in NFPA 86: Standard for Ovens and Furnaces, 2019 Edition; cf secondary air
primary containment	In the context of loss prevention, process equipment including pipes, vessels etc.; cf secondary containment, tertiary containment
primary explosion	The explosion of a pre-existing combustible dust cloud; cf dust explosion, secondary explosion
primary grade of release	A concept in hazardous area classification. The condition in which a release of hazardous material - whilst not continuous - is to be expected to occur in normal operation. Defined formally in API RP 505 2nd Edition, August 2018 Recommended Practice for Classification of Locations for Electrical Installations at Petroleum Facilities Classified as Class I, Zone 0, Zone 1, and Zone 2 and IEC 61892-7, Mobile and fixed offshore units – Electrical installations – Part 7: Hazardous areas; cf continuous grade of release, secondary grade of release
primary sedimentation	see **primary settlement**
primary settlement	*aka* primary sedimentation, primary treatment. In sewage treatment, the solids settling process between preliminary treatment (qv) and secondary treatment (qv)
primary settlement tank	see **primary settling tank**
primary settling tank	*aka* primary settlement tank. In sewage treatment, a tank in which primary settlement (qv) happens
primary stress	Stress imposed directly by mechanical loading which is not self-limiting. Defined formally in API RP 579 - Fitness for Service; cf secondary stress

primary treatment	see **primary settlement**
prime cost	*aka* PC, PC sum. The total direct costs of an item; cf provisional cost
priming	1. In the context of pumps, filling with process liquid 2. More broadly, by analogy, a way of getting something ready to start
PRINCE2	PRojects IN Controlled Environments; a project management technique
private fire hydrant	A fire hydrant on private property. Defined formally in NFPA 24: Standard for the Installation of Private Fire Service Mains and Their Appurtenances, 2019 Edition; cf public hydrant, which is not as you might perhaps expect the opposite of this term, and seems to allow for the possibility of a private fire hydrant being also a public hydrant
private fire service main	A firewater main on private property; defined formally in NFPA 24: Standard for the Installation of Private Fire Service Mains and Their Appurtenances, 2019 Edition
proactive maintenance	see **preventative maintenance**
probabilistic risk analysis	*aka* PRA; see **quantitative risk analysis**
probability	The likelihood of occurrence of an event
probability of detection	*aka* POD. The probability of consistent detection of a given component flaw by standard nondestructive testing (qv); defined formally in API RP 579 - Fitness for Service
probability of failure on demand	A safety function reliability metric; cf safety integrity level
probability simulation	see **Monte Carlo simulation**
probits	Probability units; a concept used in toxicity modelling to relate what percentage of a population will be killed by a given dose of a toxin
problem	A design problem will require the use of engineering judgement and imagination to solve, since data and/or design methodologies are lacking; cf task
procedure	A specified methodology; defined formally in BS EN ISO 9000 Quality management systems Fundamentals and vocabulary

process	A system of integrated operations which transforms inputs into outputs in a controlled manner. Formal definitions vary according to the context and may be found in BS EN ISO 9000 Quality management systems Fundamentals and vocabulary, EN ISO 14040:2006 Environmental management - Life cycle assessment - Principles and framework, API RP 2001 - Fire Protection at Refineries, BS EN ISO 10209:2012 Technical product documentation - Vocabulary - Part 1: Terms relating to technical drawings: general and types of drawings and NFPA 30: Flammable and Combustible Liquids Code, 2018 Edition
process and instrumentation diagram	A strongly deprecated PFD (qv)/ P&ID (qv) hybrid
process behavior chart	see **control chart**
process block diagram	see **block flow diagram**
process capability	A concept used in six-sigma (qv). The ability of a process to produce output to specification
process capability index	*aka* process capability ratio. A statistical measure of process capability (qv). The term Cp is used when the process metric mean is centered between the lower and upper specification limits. Cpk is used when the process metric mean is not centered, i.e.; closer to the lower or upper limit. The lowest of the two values is then chosen as the Cpk
process capability ratio	see **process capability index**
process centric team	A term referring to a business value creation process, rather than process engineering
process commissioning	Somewhat contested, but generally, once all equipment has been connected, and mechanically and electrically tested, a process commissioning engineer makes all the equipment work together in order to reliably meet the performance specification. This is called process commissioning; see **Box C3**.
process component	In a hygienic context, any component potentially in contact with product. Defined formally in ASME BPE (American Society of Mechanical Engineers: Bioprocessing Equipment); N.B. this term has a different meaning to software engineers (and possibly to other chemical engineers)

process computer	A programmable online process monitoring and control device, such as a PLC (qv). Defined formally in BS EN ISO 10209:2012 Technical product documentation - Vocabulary - Part 1: Terms relating to technical drawings: general and types of drawings and BS 1646-4:1984 Symbolic representation for process measurement control functions and instrumentation. Specification for basic symbols for process computer, interface and shared display/control functions
process contact surface	In a hygienic context, any surface that has the potential to affect product quality by indirect contact. Defined formally in ASME BPE (American Society of Mechanical Engineers: Bioprocessing Equipment); cf product contact surface
process control	The control of process variability by the monitoring of process parameters, and the control of those parameters via signals to actuating element(s) (qv)
process control air	In the context of fired heaters, air added to control carbon potential (qv) or oxygen level. Defined formally in NFPA 86: Standard for Ovens and Furnaces, 2019 Edition; cf instrument air
process controller	Generally, a standalone device which provides process control (qv) of part of a process. A specific formal definition in the context of gas turbines may be found in ISO 3977 Gas turbines - Procurement - Part 3: Design requirements
process costing	In product manufacture, dividing total direct and indirect production costs by number of items produced; cf job costing
process design	1. In the profession, the design of a process plant; see **Box D1**. 2. In academia, the abstract conceptual 'design' of a chemical process (rather than a process plant) omitting proper consideration of cost, safety or robustness, yet frequently including things which fall within the remit of other professional disciplines
process design house	An entity offering specialist design services[84]
process design package	*aka* PDP. The set of deliverables (qv) issued by a process licensor to a licensee, intended to be sufficient to allow the licensee to reliably produce a working plant based on the licensed process
process dynamics	The interactions between elements of a complex system, in the context of process control
process economics	The financial costs and benefits of building and operating process plants

---

[84] Hopefully in the design of process plant, rather than academic process design (qv)

process energy	The energy input to a process, measured at the site boundary. Defined formally in EN ISO 14040:2006 Environmental management - Life cycle assessment - Principles and framework
process engineer	The most common job title of a chemical engineering practitioner; see **Box C1**.
process engineering	What a process engineer (qv) does
process failure mode and effect analysis	*aka* PFMEA. A subtype of failure mode and effects analysis (qv), as applied to processes
process flow diagram	*aka* PFD. A diagram which shows in outline the main unit operations, piped interconnections and mass flows of a process plant. It represents the mass balance, and resembles in many ways a simplified P&ID (qv). Defined formally in BS EN ISO 10209:2012 Technical product documentation - Vocabulary - Part 1: Terms relating to technical drawings: general and types of drawings
process fluid	Usually, main process stream fluid (as opposed to a utility, for example). However, the formal definition in API RP 534 – Heat Recovery Steam Generators uses this term for a particular type of heat transfer fluid
process gain	see **gain**
process guarantee	*aka* warranty. A process guarantee may be offered by a designer, setting out a guaranteed plant performance usually as an amount of product produced to a given specification under given conditions in a performance trial. Such guarantees are usually backed by agreed penalties or liquidated damages (qv) for non-compliance
process hazard analysis	*aka* PHA, hazard analysis (qv). A term defined by OSHA as "systematic effort designed to identify and analyze hazards associated with the processing or handling of highly hazardous materials"; and "a method to provide information which will help workers and employers in making decisions that will improve safety"
process integration	A much taught, rarely-practised academic approach to process design (qv)
process intensification	1. A largely academic concept of combining unit operations 2. Trevor Kletz's term for what is now usually called 'minimization' in inherent safety (qv) techniques 3. A term now commonly applied to any process efficiency improvement
process modeling	*aka* modeling, process modelling. Producing a simplified mathematical model of the important characteristics of a process, especially when done with specialist modeling software

process modelling	see **process modeling**
process parameter	see **process variable**
process plant	A plant in which a process takes place. Defined formally in BS EN ISO 10209:2012 Technical product documentation - Vocabulary - Part 1: Terms relating to technical drawings: general and types of drawings
process pressure vessel	In the context of LPG there is a formal definition for this term in NFPA 59: Utility LP-Gas Plant Code, 2018 Edition which includes most pressure vessels within LPG production or storage facilities
process reaction curve	A curve showing a controlled system's reaction to a disturbance, used in control loop (qv) tuning
process safety	*aka* loss prevention. Avoidance of loss of containment of hazardous processes; cf occupational safety management
process safety information	*aka* PSI. A package of information about equipment, hazards and process used as the basis of process safety management (qv)
process safety management	*aka* PSM. Formal systems used to avoid loss of containment of hazardous processes; cf occupational safety management
process safety review	*aka* PSR. A plant inspection to review process safety hazards and mitigation measures as part of process safety management (qv)
process safety time	see **safety time**
process sewer	A sewer which drains process vessels or an area potentially containing process fluids; cf foul sewer, drainage
process shutdown	see **shutdown**
process simulation	Using a simplified mathematical model of a process to attempt to predict its real-world performance to aid decision-making; cf process modeling
process step	A key stage in a process, perhaps not a single unit operation. Defined formally in BS EN ISO 10209:2012 Technical product documentation - Vocabulary - Part 1: Terms relating to technical drawings: general and types of drawings
process tank	*aka* process vessel. Generally, any tank in a process plant, though this term might be restricted to unpressurized tanks or main process tanks depending on context. Formal definition of this term in API Standard 521. Pressure-relieving and Depressuring Systems. Sixth Edition | January 2014
process train	*aka* train, stream. A group of process units operated in series

process unit	Synonymous with unit operation, i.e.; a single item of equipment or unit operation (often a set of vessels and equipment) that provides one functional operation within the whole. (There are however exceptions: in a refinery, a crude distillation unit (qv) is a process unit with a number of unit operations); cf operating vessel
process upset	*aka* upset. A departure from intended process conditions
process validation	see **validation**
process value	*aka* PV; see **process variable**
process variable	*aka* PV, process value, process parameter. In the context of process control, the measured value of the property being controlled
process vessel	see **process tank**
process waste	see **waste**
process water	Usually, potable water which has passed through a site break tank, though it is sometimes produced on site from natural sources by similar processes to that used by municipal treatment works
processing	see **process**
produced fluids	see **well fluids**
produced gas	*aka* PG. Gas recovered from well fluids
produced oil	*aka* PO. Generally, crude oil (qv) which has been separated from produced water (qv) and produced gas (qv)
produced water	Water recovered from well fluids
producer gas	*aka* suction gas, town gas, coal gas, wood gas etc. A predominantly nitrogen/hydrogen mixture made by partial combustion in air of feedstocks such as coal; cf coal gas, water gas, syngas
producer price indexes	*aka* PPI. Indexes of the prices of subcomponents of process plants, used to generate indexes such as the CEPCI (qv) and its subindexes
producing platform	see **production platform**

product	Generically, the output of a production process. Formal definitions in a number of specific cases can be found in BS EN 82079-1:2012 Preparation of instructions for use. Structuring, content and presentation. General principles and detailed requirements, BS EN ISO 9000 Quality management systems Fundamentals and vocabulary, EN ISO 14040:2006 Environmental management - Life cycle assessment - Principles and framework, BS EN ISO 14159:2008 Safety of machinery. Hygiene requirements for the design of machinery, BS EN ISO 10209:2012 Technical product documentation - Vocabulary - Part 1: Terms relating to technical drawings: general and types of drawings and the EHEDG Glossary
product contact surface	"Surfaces which are exposed intentionally or unintentionally to the product and surfaces from which splashed product, condensate, liquids or material may drain, drop, diffuse or be drawn into the product or onto product contact surfaces or surfaces that come into contact with product contact surfaces of packaging materials. Note: product contact surfaces may contribute to cross-contamination, and must therefore be included in the hazard analysis" according to the EHEDG Glossary. Very similar formal definitions in BS EN ISO 14159:2008 Safety of machinery. Hygiene requirements for the design of machinery and ASME BPE (American Society of Mechanical Engineers: Bioprocessing Equipment); cf process contact surface, splash area
product factor	A fiddle factor (q_v) used in liquid evaporative loss characteristics calculations. Defined formally in API MPMS 19.1 Manual of Petroleum Measurement Standards Chapter 19.1 Evaporative Loss from Fixed-roof Tanks
product flow	In an environmental context, the flow of products between systems. Defined formally in EN ISO 14040:2006 Environmental management - Life cycle assessment - Principles and framework
product life cycle	The time from initial conception to final disposal of product. Two slightly different formal definitions may be found in BS EN ISO 10209:2012 Technical product documentation - Vocabulary - Part 1: Terms relating to technical drawings: general and types of drawings

product recovery	1. Most commonly in chemical engineering, the process of obtaining a saleable product from a mixture, especially a high value product 2. Processes used to enhance yield of product, or to restore value to discarded products; cf recovery
product requirement specification	*aka* PRS, product requirements document. A document containing an early stage, high level description of the business requirements and regulatory constraints of a proposed product; cf user requirement specification, functional requirements specification
product requirements document	see **product requirement specification**
product safety data sheet	*aka* PSDS; see **material safety data sheet**
product safety label	A label informing the reader of potential product hazards. Defined formally in BS EN 82079-1:2012 Preparation of instructions for use. Structuring, content and presentation. General principles and detailed requirements
product slate	The mix of different products of an oil refinery; cf crude slate
product specification sheet	*aka* cut sheet. A document containing the technical specification of a product
product system	A product life cycle process model; defined formally in EN ISO 14040:2006 Environmental management - Life cycle assessment - Principles and framework
product temperature margin	In the context of pump seals, the difference (at seal chamber pressure) between a pumped fluid's temperature and its vaporization temperature; defined formally in API RP 682 - Pump Seals
product water	Term for potable water used in the hygienic industries
production areas	Areas where significant quantities of flammable hydrocarbons are handled, other than for transport. Defined formally in API RP 505 2nd Edition, August 2018 Recommended Practice for Classification of Locations for Electrical Installations at Petroleum Facilities Classified as Class I, Zone 0, Zone 1, and Zone 2
production drawing	Usually, one of the set of drawings which contain all information required to produce a part. Defined formally in BS EN ISO 10209:2012 Technical product documentation - Vocabulary - Part 1: Terms relating to technical drawings: general and types of drawings; cf working drawing
production platform	*aka* offshore drilling rig, offshore platform, oil platform, producing platform. A standalone offshore structure used to extract, store, process and export oil and gas

production separator	*aka* separator. In the oil and gas industry, a pressurized gravity separator, separating well fluids into various streams of oil, water and gas
production wing valve	A valve typically located on the right-hand side of a Christmas tree (qv) that controls the flow of oil and gas from the well to production facilities
Professional Engineer	*aka* PE. An engineer with sufficient and suitable education and experience for professional licensure in the USA[85]; see **Box E1**.
professional indemnity insurance	*aka* errors & omissions insurance, indemnity cover, PI, PII, professional liability insurance. Insurance against costs associated with a claim by a client of the provision of an inadequate professional service
professional liability insurance	see **professional indemnity insurance**
profibus	A fieldbus communication standard; cf profinet
profilometer	A surface roughness measurement instrument. Defined formally in ASME BPE (American Society of Mechanical Engineers: Bioprocessing Equipment)
profinet	An industrial ethernet standard; cf profibus
program	*aka* programme. 1. see **project program** 2. A set of instructions used by a programmable controller (qv)
program evaluation and review technique	*aka* PERT. A project planning methodology, a more sophisticated (some say more pessimistic) variant of critical path analysis (qv)
programmable	Capable of accepting instructions to automatically control a process. Slightly different formal definitions in BS EN ISO 10209:2012 Technical product documentation - Vocabulary - Part 1: Terms relating to technical drawings: general and types of drawings and BS 1646-4:1984 Symbolic representation for process measurement control functions and instrumentation. Specification for basic symbols for process computer, interface and shared display/control functions
programmable controller	Most commonly a programmable (qv) digital device; defined formally in NFPA 86: Standard for Ovens and Furnaces, 2019 Edition
programmable logic controller	*aka* PLC, programmable controller (qv). Industrial computers, capable of reliably controlling industrial processes
programmable optimal density	*aka* POD. A type of blender which blends and pumps frac fluid (qv)

---

[85] Elsewhere, it could be the person who fixes your domestic boiler!

programme	see **program**
progressing cavity pump	*aka* wobble pump, eccentric screw pump, screw pump, cavity pump, Moineau pump, Moyno pump, Mono pump. A rotodynamic positive displacement pump (qv) based upon a helical rotor rotating within a helical elastomer stator
progressive polishing	Mechanical polishing (qv) using successively finer grades of abrasives
project	A sequence of activities intended to achieve a specified aim with certain resources; most commonly refers to the construction phase of a plant. Defined formally in BS EN ISO 9000 Quality management systems Fundamentals and vocabulary
project engineer	Most commonly, an engineer responsible for some aspect of the construction phase of a project involving multi-disciplinary teams
project management	The management of a project (qv)
project management plan	A document setting out a planned sequence of activities intended to achieve a specified aim with certain resources. Defined formally in BS EN ISO 9000 Quality management systems Fundamentals and vocabulary
project management professional	*aka* PMP. The most popular and conventional project management certification; cf certified scrum master; PRINCE2
project manager	The individual responsible for management of a project (qv)
project network techniques	*aka* network analysis methods. Project management techniques such as critical path analysis (qv) and program evaluation and review technique (qv)
project program	*aka* schedule, program, project schedule. A diagram at a minimum showing the times taken and interrelationships between the various discrete tasks which have to be completed to achieve a project
project schedule	see **project program**
projection	see **view**
projection line	*aka* projector. An imaginary line on a drawing projected outwards from another line or point to assist with drafting
projector	see **projection line**
prokaryotic	A classification of 'lower' forms of life, the cells of which do not have a membrane-bound nucleus, amongst other things; cf eukaryotic

proof	To a mathematician, or a philosopher, a proof is a convincing argument that a statement is true, though the nature of evidence and argument may differ in these two disciplines. In engineering, proof tends to be a unit of measurement of alcohol content of potable distillates, or more accurately a number of competing measures. In the USA, proof is straightforwardly twice alcohol by volume (ABV); i.e.; two times % v/v. As with pints, proof is theoretically no longer used in the UK, though in practice it is. UK proof is more like 1.7 times ABV, whilst French proof is equal to ABV
proof of closure switch	*aka* POC switch. A manufacturer-installed switch on the stem of a safety shutoff valve, to positively identify complete valve closure. Defined formally in NFPA 86: Standard for Ovens and Furnaces, 2019 Edition
propane LPG dispenser	An LPG (qv) dispenser for road vehicle fuel tanks. Defined formally in NFPA 58: Liquefied Petroleum Gas Code, 2017 Edition
propellant	A material whose expansion propels something, such as the compressed fluid used to propel the contents of an aerosol dispenser, or the energetic materials in a rocket motor
propeller	*aka* propellor. A bladed device which converts rotation into linear flow, used for agitation; cf impeller
propellor	see **propeller**
property diagram	see **thermodynamic diagram**
property line	A line dividing one lot of land from another, or from a street or other public space
proportional action control	In the context of process control, using controller output proportional to the difference between set value (qv) and measured value (qv). The P of PID control (qv), with a stabilizing action
proportional band	In the context of process control, the input range over which controller output is proportional to input
proportional mixer	A type of venturi mixer used to accurately and safely premix fuel gas and air for a fired heater. Defined formally in NFPA 86: Standard for Ovens and Furnaces, 2019 Edition
proportional safety valve	A safety valve which opens in proportion to an increase in pressure

proportioning	1. Generically, dosing (qv) a specified proportion of additive into something 2. In the context of firefighting foam production, the continuous addition of a recommended proportion of foam concentrate to water; defined formally in NFPA 11: Standard for Low-, Medium-, and High-Expansion Foam, 2016 Edition
proportioning methods for foam systems	In the context of firefighting foam production, methods used for the continuous addition of a recommended proportion of foam concentrate to water. Defined formally in NFPA 11: Standard for Low-, Medium-, and High-Expansion Foam, 2016 Edition
proportioning type balanced pressure pump	In the context of firefighting foam production, a pump used for the continuous addition of a recommended proportion of foam concentrate to water. Defined formally in NFPA 11: Standard for Low-, Medium-, and High-Expansion Foam, 2016 Edition
proposals engineer	A person responsible for producing the commercial and/or technical documentation included in a commercial offer to provide plant or equipment
proprietary eponym	*aka* generic trademark, genericized trademark. A trade name commonly used as a generic name
protected fired vessel	A fired vessel whose air intake and exhaust are designed to prevent them becoming ignition sources. Defined formally in API RP 505 2nd Edition, August 2018 Recommended Practice for Classification of Locations for Electrical Installations at Petroleum Facilities Classified as Class I, Zone 0, Zone 1, and Zone 2
protection for exposures	A private or public fire service able to protect the property adjacent to flammable liquid storage using water jets. Defined formally in NFPA 30: Flammable and Combustible Liquids Code, 2018 Edition
protective coating	A surface coating on metal intended to resist corrosion. Defined formally in BS EN ISO 13705:2012/ISO 13705:2012(E) Petroleum, petrochemical and natural gas industries. Fired heaters for general refinery service, API STD 530 - Calculation of Tube Heater Thickness and API STD 560 - Fired Heaters for General Refinery Service
protective device	A device which is a safeguard (qv), but not a guard (qv). Defined formally in BS EN ISO 12100:2010 Safety of machinery. General principles for design. Risk assessment and risk reduction

protective measure	A risk reduction feature in the design or operation of machinery. Defined formally in BS EN ISO 12100:2010 Safety of machinery. General principles for design. Risk assessment and risk reduction
protein foam concentrate	A hydrolyzed protein based firefighting foam concentrate. Defined formally in NFPA 11: Standard for Low-, Medium-, and High-Expansion Foam, 2016 Edition
protium	(*symbol* H) *aka* hydrogen. A stable, light isotope of hydrogen, the most common form; cf deuterium
proton number	see **atomic number**
proved pilot	A pilot flame, the presence of which is monitored by a flame detection system (qv). Defined formally in NFPA 86: Standard for Ovens and Furnaces, 2019 Edition
provider	see **supplier**
provisional cost	*aka* provisional sum. A stated estimate of cost for something too ill-defined to price accurately in a costing. Not the same as PC sum (qv)
provisional sum	see **provisional cost**
proximate analysis	The coarse determination of composition of a fuel with respect to key parameters such as sulfur, moisture, volatile matter, ash, and fixed carbon; cf ultimate analysis
proximate cause	An event or circumstances immediately responsible for causing something; cf ultimate cause
PRP	see **prerequisite program**
PRR	Pressure reducing regulator, see **pressure reducing valve**
PRS	see **product requirement specification**
Pruteen	An uneconomic single cell protein animal feed production process developed by ICI in the UK; cf Pekilo process
PrV	Pressure regulating valve, see **pressure sustaining valve**
PRV	see **pressure relief valve**
PS	1. *(drawing notation)* Pipe support
	2. see **planar surface**
PSA	see **pressure swing adsorption**
PSD	1. see **particle size distribution**
	2. see **prevention of significant deterioration program**
	3. Process shutdown, see **shutdown**
PSDS	Product safety data sheet; see **material safety data sheet**
PSE	*(drawing notation)* Plain small end
pseudocomponent	*aka* single oil, hypothetical component. A way of simplifying process modelling with software by replacing some components with non-existent substances thought to have similar properties to those omitted

pseudocritical properties	Values of temperature and pressure extrapolated from empirical data to facilitate the practical estimation of the critical point of multicomponent mixtures
pseudo-order	see **pseudoorder**
pseudoorder	*aka* pseudo-order. A concept used to simplify the analysis of chemical reactions, by treating the reaction as being of a lower order than is actually the case, in a situation where it makes little practical difference
pseudoplastic	*aka* shear thinning. A fluid property of apparent viscosity decreasing with rate of shear (a time-independent non-Newtonian behavior); cf dilatant
PSF	see **partial safety factor**
PSI	see **process safety information**
psi	see **pounds per square inch**
psia	Pounds per square inch absolute
psig	Pounds per square inch gauge
PSM	see **process safety management**
PSR	see **process safety review**
PSSR	1. see **Pressure Systems Safety Regulations 2000** 2. see **pre start up safety review**
PST	see **primary settling tank**
PSV	see **pressure safety valve**
psychrometric chart	A graph showing physical and thermodynamic properties of a gas/vapor mixture (most commonly air/water vapor); cf Mollier diagram
psychrophilic bacteria	see **cryophilic bacteria**
PT	1. *(drawing notation)* Pressure transmitter 2. see **planar transverse** 3. Penetrant testing, see **liquid penetrant indication**
Pt Co scale	see **platinum cobalt scale**
PT diagram	see **pressure/temperature diagram**
P-T diagram	see **pressure/temperature diagram**
PTASCC	see **polythionic acid stress corrosion cracking**
PTFE	Polytetrafluoroethylene (qv Teflon)
PU	1. see **pasteurization unit** 2. Polyurethane
public hydrant	A hydrant (qv) on a public water supply. Defined formally in NFPA 24: Standard for the Installation of Private Fire Service Mains and Their Appurtenances, 2019 Edition; cf private fire hydrant, which is not as you might perhaps expect the opposite of this term, and seems to allow for the possibility of a private fire hydrant being also a public hydrant

public liability insurance	*aka* PL. Insurance against the costs associated with a claim by a member of the public for damages associated with the insured's commercial activities
public way	A publicly accessible open-air route at least ten feet wide and high throughout; defined formally in NFPA 101, Life Safety Code, 2021 Edition
puff	*aka* puff release, instantaneous release. A short (or at least of finite duration) release of gas or aerosol, or the cloud produced by such a release; cf plume
puff release	see **puff**
pulsatance	see **angular frequency**
pulsation damper	A device which reduces pressure pulses in flowing fluids, often used in conjunction with positive displacement pumps or compressors
pulsed baffle reactor	*aka* PBR. A class of tubular plug flow reactors with intermittent flow cessation or reversal
pulsed column	*aka* pulsed column extractor. A liquid–liquid extraction column using pulsed flow with intermittent addition of compressed air
pulsed column extractor	see **pulsed column**
pulverised flue ash	see **pulverized flue ash**
pulverized flue ash	*aka* PFA, pulverised flue ash, pulverized fly ash, fly ash, flue ash (FA), coal ash. Fine particulates in flue gases from pulverized coal combustion, often recovered and used for various purposes
pulverized fly ash	see **pulverized flue ash**
pulverizer	*aka* grinder. A device which reduces bulk material to powder ('pulver' being German for powder)
pump	One of many types of fluid moving device; also, the activity of using one of them
pump curve	see **characteristic curves**
pump fluid	*aka* pump oil, working fluid, working medium. The operating fluid of certain types of vacuum pump; defined formally in NFPA 86: Standard for Ovens and Furnaces, 2019 Edition
pump laws	see **affinity laws**
pump manufacturer	The manufacturer (qv) of a pump. This particular term is defined formally in API RP 682 - Pump Seals
pump oil	see **pump fluid**
pump priming	see **priming**
pump proportioner	*aka* around the pump proportioner. A pressure foam maker (qv) in a pump bypass line; defined formally in NFPA 11: Standard for Low-, Medium-, and High-Expansion Foam, 2016 Edition

pump selection chart	see **pump curve**
pumparound	Removing heat from a distillation column by pumping out hot liquid from a lower tray, cooling it and returning it to a higher one
pumpdown factor	In the context of vacuum pumps, an estimate (commonly obtained from a graph) of evacuation time divided by system volume; used to allow for the time required to bring a vessel of a known volume to a given pressure. Defined formally in NFPA 86: Standard for Ovens and Furnaces, 2019 Edition
pumpdown time	see **evacuation time**
pumped liquid level	see **pumping liquid level**
pumper	US English for a fire engine with pumps and hoses
pumper outlet	A hydrant butt (qv) for connection to a pumper (qv). Defined formally in NFPA 24: Standard for the Installation of Private Fire Service Mains and Their Appurtenances, 2019 Edition
pumping liquid level	*aka* pumped liquid level, feed tank level, reservoir level. The level of liquid being pumped, relative to the pump. Formal definition of this term in NFPA 20: Standard for the Installation of Stationary Pumps for Fire Protection, 2019 Edition
pumping ring	see **internal circulating device**
pumping unit	see **nodding donkey**
pumpjack	see **nodding donkey**
punch list	see **snagging list**
pup	Similar to a pony collar (qv), but in the form of a short piece of drill pipe, casing, or tubing
PUP	*(drawing notation)* Pick up point
purchase order	*aka* PO. A formal document issued by a purchaser to a vendor confirming an order for a specified item under specified commercial conditions
purchaser	The entity which orders something. Various related formal definitions may be found in PD 5500:2018+A1:2018 Specification for unfired fusion welded pressure vessels, EN ISO 10437 Petroleum, petrochemical and natural gas industries - Steam turbines - Special-purpose applications, EN ISO 13709:2003 Centrifugal pumps for petroleum, petrochemical and natural gas industries, ISO14661 Thermal turbines for industrial applications (steam turbines, gas expansion turbines) - General requirement, API RP 682 - Pump Seals and API STD 620 - Low Pressure Storage Tanks
pure mass	see **deadweight**

pure oil mist lubrication	A type of dry sump centrifugal pump bearing lubrication system. Defined formally in EN ISO 13709:2003 Centrifugal pumps for petroleum, petrochemical and natural gas industries; cf purge oil mist lubrication
pure steam	Steam whose condensate is of water for injection (qv) quality or better. Defined formally in ASME BPE (American Society of Mechanical Engineers: Bioprocessing Equipment) and 3A Standard 609-02; cf clean steam
PUREX	Plutonium uranium reduction extraction
PUREX process	A pulsed column (qv) plutonium uranium reduction extraction process used in nuclear reprocessing
purge	1. Generically, the complete replacement of something with something else 2. More commonly in engineering, the replacement of a potentially flammable or explosive atmosphere with a non-potentially flammable or explosive one; defined formally in NFPA 86: Standard for Ovens and Furnaces, 2019 Edition
purge gas	Mostly synonymous with inert special atmosphere (qv), but API standards also include fuel gas. Defined formally in API Standard 521. Pressure-relieving and Depressuring Systems. Sixth Edition | January 2014 and API RP 537 Flare Details for Petroleum, Petrochemical, and Natural Gas Industries
purge of burner guns	In the context of fired heaters, purging a burner gun (qv) or guns, commonly with steam. Defined formally in BS EN 12952-8:2002 Water-tube boilers and auxiliary installations. Requirements for firing systems for liquid and gaseous fuels for the boiler
purge oil mist lubrication	A type of wet sump centrifugal pump bearing housing lubrication system. Defined formally in EN ISO 13709:2003 Centrifugal pumps for petroleum, petrochemical and natural gas industries; cf pure oil mist lubrication
purge reduction device	see **air seal**
purge valve	An actuated valve used to purge engine fuel supply and return lines of vapor with fuel, prior to startup. Defined formally in NFPA 58: Liquefied Petroleum Gas Code, 2017 Edition
purged	Having been subjected to a purge (qv)
purged building	see **purged enclosure**

purged enclosure	*aka* purged building. An enclosed space which is purged and maintained under positive pressure by air or inert gas. Defined formally in API RP 505 2nd Edition, August 2018 Recommended Practice for Classification of Locations for Electrical Installations at Petroleum Facilities Classified as Class I, Zone 0, Zone 1, and Zone 2 and Standard for Purged and Pressurized Enclosures for Electrical Equipment, ANSI/NFPA 496
purging of the flue gas passes	In the context of fired heaters, the purging of the combustion chamber, flue gas side of boiler heat exchanger and ducts, using air. Defined formally in BS EN 12952-8:2002 Water-tube boilers and auxiliary installations. Requirements for firing systems for liquid and gaseous fuels for the boiler
purified water	*aka* PW, compendial water. Water used in the pharmaceutical industry, of a quality specified in the relevant pharmacopoeia. Defined formally in ASME BPE (American Society of Mechanical Engineers: Bioprocessing Equipment)
Purisol	A pyrrolidone-based proprietary physical solvent (qv) for gas sweetening (qv); cf Selexol, Rectisol
purple hydrogen	*aka* violet hydrogen. The same as pink hydrogen (qv), but also uses a nuclear reactor's heat and electricity; cf hydrogen colors
pusher seal	A pump seal with a dynamic secondary seal; defined formally in API RP 682 - Pump Seals
PV	1. *(drawing notation)* Plug valve 2. *(drawing notation)* Pressure vessel 3. *(drawing notation)* Photovoltaic 4. see **permanganate value** 5. see **process variable or process value**
PV diagram	see **pressure/volume diagram**
P-V diagram	see **pressure/volume diagram**
p-V diagram	see **pressure/volume diagram**
PVA	Poly vinyl alcohol
PVCc	Chlorinated poly vinyl chloride
PVCu	Unplasticized poly vinyl chloride
PVDF	Polyvinylidene difluoride
PVF	Piper abbreviation for pipes, valves and fittings
PW	1. see **purified water** 2. see **potable water**
PWC	see **preferential weld corrosion**
PWHT	see **post weld heat treatment**
PWR	see **pressurized water reactor**
Pyrex	Proprietary eponym for borosilicate glass (qv)

pyrogen	A fever-inducing substance
pyrolysis	The decomposition of organic feedstock by anaerobic heat into fluids and a solid char (qv); cf hydropyrolysis
pyrophoric	1. Most commonly, a term meaning 'igniting spontaneously on exposure to air' 2. However, it may be used to describe a merely self-heating material, according to the formal definition in API RP 2001 - Fire Protection at Refineries; cf pyrophoric material
pyrophoric iron	Iron sulfide, a pyrophoric (qv) substance which can build up in hydrocarbon processing equipment; formed when rust is exposed to hydrogen sulfide
pyrophoric material	An NFPA-defined term which appears to define pyrophoric (qv) more rigorously than the API, requiring the material to be capable of autoignition (qv) rather than simply self heating (qv) at less than 54.4°C, as per NFPA 654: Standard for the Prevention of Fire and Dust Explosions from the Manufacturing, Processing, and Handling of Combustible Particulate Solids 2017 Edition

Q

Q factor	A dimensionless number used to characterize the resonance of an underdamped harmonic oscillator. Defined formally in BS ISO 2041:2018 Mechanical vibration, shock and condition monitoring. Vocabulary
Q/H curves	see **characteristic curves**
QA	see **quality assurance**
QAC	see **quaternary ammonium compound**
QAD	see **quarter damping**
QAT	see **quaternary ammonium compound**
QC	see **quality control**
QESH	see **quality, environment, safety, health**
QHSE	see **quality, environment, safety, health**
QHSSE	see **quality, environment, safety, health**
Q-line	A line drawn on a graph as part of the now-obsolete McCabe-Thiele (qv) method
QP	see **qualified person**
QRA	1. see **qualitative risk analysis** 2. see **qualitative risk assessment**; N.B. not used to abbreviate quantitative risk analysis (qv) or quantitative risk assessment (qv)
QSL process	see **Queneau-Schuhmann-Lurgi process**
quadruple effect	A four-effect unit operation; particularly used in the context of multiple effect evaporation
qualified	Qualified is a somewhat contested term - the NFPA for example does not even agree with itself on what it means. See formal definitions in NFPA 15 Standard for Water Spray Fixed Systems for Fire Protection, NFPA 20: Standard for the Installation of Stationary Pumps for Fire Protection, 2019 Edition, NFPA 654: Standard for the Prevention of Fire and Dust Explosions from the Manufacturing, Processing, and Handling of Combustible Particulate Solids, 2017 Edition and OSHA 1910.29 - Fall protection systems and falling object protection-criteria and practices; see also **Box Q1**.
qualified person	*aka* QP. A person who is qualified (qv) in a particular context; see **Box Q1**.

Box Q1. Qualified

Qualified, competent, responsible, authorized and terms derived from them have a number of potential meanings, some of which are important from a regulatory point of view. Some of these interact with another contested term, reasonable.

Competence is defined as 'the ability to achieve desired results by the application of knowledge and skill' in BS EN ISO 9000. However, a competent authority is a body responsible for some legal duty, often safety related, and a competent person is one trained in identifying hazards, and authorized to promptly eliminate them.

In the EU, a qualified person (QP) is responsible for ensuring the safety of medicines and will have attained specified academic and professional qualifications, and several years' experience in pharmaceutical manufacturing. (Note that someone equivalent to a QP may be known as a responsible person (RP) or authorized person (AP) in other jurisdictions).

These role descriptions share common features with formal definitions of 'qualified' in other contexts, requiring a combination of academic and practical training which, arguably, should add up to competence. However, competence and qualification are not necessarily the same.

OSHA, for example, defines a qualified person as one who, "by possession of a recognized degree, certificate, or professional standing, or who by extensive knowledge, training and experience, has successfully demonstrated his ability to solve or resolve problems relating to the subject matter, the work, or the project"; whilst defining competence as the ability to spot hazards and correct them. Qualified workers might have more technical expertise than 'competent' ones, but not necessarily expertise in hazard recognition or correction. As well as 'competent' and 'qualified', OSHA also define 'authorized' or 'certified' as different terms with corresponding personal statuses.

Meanwhile, the NPFA defines a qualified person as "one who has demonstrated skills and knowledge related to the construction and operation of electrical equipment and installations, and has received safety training to identify and avoid the hazards involved." NFPA 70E requires that only a qualified person perform work on or near exposed and energized electrical conductors or circuit parts, and the UK's NEC definition is virtually identical.

Risk management appears in all but the EU definition. Is it right that a person responsible for releasing pharmaceuticals on the market is probably not an engineer, and may not have the formal training required to understand the production process?

More broadly, who is a qualified person in engineering? Is an engineer not automatically a qualified person? Do we require licensing, academic qualification, practical experience, and professional training validated by examinations to be 'qualified', or is a relevant degree sufficient? Should a qualified engineer not be able to spot hazards and correct them?

Related Terms
Chartered Engineer, Professional Engineer, skilled person

qualitative risk analysis	*aka* QRA (deprecated, to avoid confusing with quantitative risk assessment -qv). A formal analysis of sequences of events that might plausibly lead a hazard to become an accident; used to identify and assess risks, then rank them according to risk, (commonly using a risk matrix - qv) allowing preventive and mitigative measures to be prioritized. Often considered synonymous with qualitative risk assessment (qv), but some claim that qualitative risk assessment is a broader brush approach, whilst qualitative risk analysis is more of a micro-level investigation. Perhaps confusingly, some forms of qualitative risk analysis may assign numerical values to the likelihood and consequences of risks. This is not however necessarily spurious precision (qv) or inappropriate rigor: all numbers used in all forms of risk assessment are to some extent subjective. The idea that qualitative risk analysis is automatically more objective than quantitative risk analysis may not be well founded. See also **Box R2**
qualitative risk assessment	*aka* QRA (deprecated, to avoid confusing with quantitative risk assessment -qv). See **qualitative risk analysis**; see also **Box R2**
quality	The extent to which an object or process meets stated specifications or quality characteristics. Note that 'quality' is to do with meeting specifications: consistently hitting a low specification is thus arguably a 'high quality' outcome. Defined formally in BS EN ISO 9000 Quality management systems Fundamentals and vocabulary
quality assurance	*aka* QA. That part of quality management concerned with preventing defects in products by controlling the design or production process, with a view to obtaining consistency[86]. ISO 9000 is the de facto international QA standard; the term is also defined formally in BS EN ISO 9000 Quality management systems Fundamentals and vocabulary; cf quality control
quality characteristic	A specification, or a characteristic inherent to an object or process defined in a requirement (qv); defined formally in BS EN ISO 9000 Quality management systems Fundamentals and vocabulary
quality control	*aka* QC. The product inspection aspect of quality management, catching defects after they have happened, rather than preventing them; defined formally in BS EN ISO 9000 Quality management systems Fundamentals and vocabulary; cf quality assurance

---

[86] Even if that means producing a product which is consistently terrible!

quality improvement	The product improvement aspect of quality management. Defined formally in BS EN ISO 9000 Quality management systems Fundamentals and vocabulary
quality management	Managing quality (qv). The longer formal definition in BS EN ISO 9000 Quality management systems Fundamentals and vocabulary is equally obvious
quality management system	A system for quality management (qv). The longer formal definition in BS EN ISO 9000 Quality management systems Fundamentals and vocabulary is equally obvious
quality management system consultant	A person who assists with introducing and/or managing a quality management system (qv). The longer formal definition in BS EN ISO 9000 Quality management systems Fundamentals and vocabulary is equally obvious
quality management system realization	Introducing and/or managing a quality management system (qv). The longer formal definition in BS EN ISO 9000 Quality management systems Fundamentals and vocabulary is equally obvious
quality manual	A quality management system (qv) specification; defined formally in BS EN ISO 9000 Quality management systems Fundamentals and vocabulary
quality objective	Just what it sounds like, as long as you understand what quality (qv) means in this context. Defined formally in BS EN ISO 9000 Quality management systems Fundamentals and vocabulary
quality plan	A document specifying the procedures and resources to be used to achieve quality (qv). Defined formally in BS EN ISO 9000 Quality management systems Fundamentals and vocabulary and ISO 10209-1:1992 Technical product documentation - Vocabulary - Part 1: Terms relating to technical drawings: general and types of drawings; cf quality manual
quality planning	The part of quality management to do with specifying quality objectives (qv) and processes. Whilst producing a quality plan (qv) might form a part of this, quality planning is not the same thing as producing a quality plan. Defined formally in BS EN ISO 9000 Quality management systems Fundamentals and vocabulary
quality policy	Just what it sounds like, as long as you understand what quality (qv) means in this context. Defined formally in BS EN ISO 9000 Quality management systems Fundamentals and vocabulary

quality requirement	Just what it sounds like, as long as you understand what quality (qv) means in this context. Defined formally in BS EN ISO 9000 Quality management systems Fundamentals and vocabulary
quality, environment, safety, health	*aka* QESH, also SHEQ, QHSE. A rolling-together of non-profit-making obligations by those businesses which take a minimalist approach to such matters, including quality, in the case of this specific term. All too often delegated to a particular subset of sidelined staff who may not be taken seriously in the wider organization; cf environment health and safety
quantitative risk analysis	*aka* quantitative risk assessment, probabilistic risk analysis (PRA). Theoretically, a stringent numerical analysis of risks, though all too often the numbers used reflect spurious precision (qv). Commonly used as synonymous with quantitative risk assessment (qv), but some claim that quantitative risk assessment is a broader brush approach, and quantitative risk analysis is more of a micro-level investigation. See also **Box R2**
quantitative risk assessment	*aka* quantitative risk analysis, probabilistic risk analysis (PRA). See **quantitative risk analysis**. See also **Box R2**
quarl	see **tile**
quart	A measure of volume, equivalent to two pints; obsolete everywhere except the USA
quarter amplitude damping	see **quarter damping**
quarter amplitude decay	see **quarter damping**
quarter damping	*aka* quarter amplitude damping, quarter amplitude decay, QAD. A tuning objective used in the (strongly disrecommended) Ziegler-Nichols tuning method (qv)
quarter turn valve	A valve requiring 90 degrees of rotation to go from fully open to fully closed, such as a ball, butterfly or plug valve; cf multiple turn valve
quarter twat	A degree of opening for valves and pumps; cf full twat, half twat
quarternary	see **quaternary**
quat	see **quaternary ammonium compound**
quaternary	*aka* quarternary (commonly and incorrectly). Related to the fourth in a ranked series
quaternary amine	An oil and gas term for quaternary ammonium compound (qv)
quaternary ammonium cation	see **quaternary ammonium compound**

quaternary ammonium compound	*aka* QAC, QAT, quat, quaternary ammonium cation, quaternary amine, quaternary ammonium salt. An amine produced by the fourth and final level of substituting ammonia with organic groups; or the salt of such a compound. cf tertiary ammonium compound
quaternary ammonium salt	see **quaternary ammonium compound**
quench	1. Most commonly, to cool something rapidly, by exposure to cooler fluid; or informally, the fluid used to do so 2. The quenching of a reaction either by rapid cooling as in (1.), by rapidly introducing a significant concentration of reaction product, or otherwise shifting the equilibrium/kinetics of the reaction; cf kill 3. In the context of pump seals, a fluid used on the air side of a seal to prevent solids buildup, as defined formally in API RP 682 - Pump Seals
quench column	*aka* quench tower. A column used to quench (qv), i.e.; rapidly cool, a gas stream
quench medium	see **quenchant**
quench tank	A tank of quenchant (qv)
quench tower	see **quench column**
quench type tank	A term used but not defined in NFPA 86 in statements such as "integral liquid or salt media quench-type tank", the definitions of which make clear that a quench type tank is a tank of quenchant (qv). 'Quench type tank' therefore appears to be synonymous with quench tank (qv)
quenchant	*aka* quench, quench medium. A fluid used to quench (qv), such as quench oil in the steam cracking process
quenching	Cooling a fluid by mixing with colder fluid, according to API Standard 521. Pressure-relieving and Depressuring Systems. Sixth Edition | January 2014; more generally, the gerund of quench (qv)
Queneau-Schuhmann-Lurgi process	*aka* QSL process. A relatively new lead smelting process which claims to have enhanced efficiency and reduced environmental impact
quick acting shutoff device	A safety shut-off device which closes in one second or less; defined formally in BS EN 12952-8:2002 Water-tube boilers and auxiliary installations. Requirements for firing systems for liquid and gaseous fuels for the boiler
quick connectors	Tool-free hose assembly fittings; defined formally in NFPA 58: Liquefied Petroleum Gas Code, 2017 Edition

quick freezing	Sufficiently rapid freezing of foodstuffs as to prevent product degradation by ice crystal formation; cf blast freezing
quickfit	A type of borosilicate glass labware which goes up to pilot scale (QVF -qv), sometimes used for commercial production in pharmaceuticals and fine chemicals production
quickflange	Proprietary eponym for mechanical piping joint processes
quicklime	*aka* burnt lime. Calcium oxide (CaO); the product of calcining (qv) of limestone, which reacts vigorously and very exothermically with water. (Its vigor is why it is called 'quick' lime, in the old sense of 'alive')
quiescence	In the context of biological cells, a reversible state in which they are not dividing
quill shaft	A thin flexible shaft in a gas turbine; defined formally in ISO 3977 Gas turbines - Procurement-- Part 3: Design requirements
qv	Quod vide or 'which see'; a cross-reference to a related entry in the context of this dictionary
QVF	Proprietary eponym for pilot scale borosilicate glassware; cf quickfit

R

R	see **gas constant**
R2P2	Reducing Risk, Protecting People: a commonly referenced UK HSE publication which is the closest approximation made by the UK regulator to outlining acceptable risk criteria
RA	see **risk assessment**
Ra	Roughness average: the log mean surface profile measurement; defined formally in ASME BPE (American Society of Mechanical Engineers: Bioprocessing Equipment)
Ra max	The highest recorded Ra (qv) value; defined formally in ASME BPE (American Society of Mechanical Engineers: Bioprocessing Equipment)
rabbit	In oil and gas drilling, a short cylindrical piece of metal which is dropped through a drill pipe to ensure its interior diameter is fully open, to allow coring and logging tools to pass through
rachet dog	*aka* ratchet dog; see **dogs**
rack	1. *aka* racking. A frame used for storage; defined formally in NFPA 30: Flammable and Combustible Liquids Code, 2018 Edition; cf piperack 2. A fuel tanker loading facility
rack bay	The space between each stanchion (qv) in a rack; defined formally in NFPA 30: Flammable and Combustible Liquids Code, 2018 Edition
rack bent	see **pipe bent**
rack section	An enclosed rack or set of racks; defined formally in NFPA 30: Flammable and Combustible Liquids Code, 2018 Edition
radial	In the direction of a radius; cf axial, centrifugal, transverse
radial dispersion	*aka* transverse dispersion. Dispersion in a radial direction; cf axial dispersion
radial flow fan	see **centrifugal fan**
radial flow fixed bed reactor	*aka* RFBR, radial flow packed bed reactor, FRBR. A low headloss, high volume reactor type
radial flow packed bed reactor	see **radial flow fixed bed reactor**
radial flow pump	see **centrifugal pump**
radial velocity	Velocity in a radial direction

radially split	In the context of rotodynamic equipment, split in a radial direction (perpendicular to shaft centerline); defined formally in EN ISO 10437 Petroleum, petrochemical and natural gas industries -- Steam turbines -- Special-purpose applications and EN ISO 13709:2003 Centrifugal pumps for petroleum, petrochemical and natural gas industries
radian	(*symbol* rad) The SI unit of angle: one radian is the angle drawn on a circle by a section of the circumference which equals the same circle's radius. π times one radian equals 180 degrees
radian per second	(*symbol* rad/s or ω) The SI unit of angular velocity
radian per second squared	(*symbol* rad/s^2) The SI unit of angular acceleration
radiant burner	A fired heater burner mainly transferring heat of combustion by radiation; defined formally in NFPA 86: Standard for Ovens and Furnaces, 2019 Edition
radiant section	The part of a fired heater which contains rows of tubes containing fluid to be heated; defined formally in BS EN ISO 13705:2012/ISO 13705:2012(E) Petroleum, petrochemical and natural gas industries. Fired heaters for general refinery service, API STD 530 - Calculation of Tube Heater Thickness and API STD 560 - Fired Heaters for General Refinery Service
radiant tube	A tube in a fired heater substantially filled with produced flame, so that the tube radiates heat evenly
radiant tube burner	A fired heater burner which substantially fills a tube with produced flame, so that the tube radiates heat evenly; defined formally in NFPA 86: Standard for Ovens and Furnaces, 2019 Edition
radiant tube heating system	A fired heater based on radiant tube burners; defined formally in NFPA 86: Standard for Ovens and Furnaces, 2019 Edition
radiant wall burner	A fired heater radiant burner (qv) mainly transferring heat of combustion by radiation from a radiant wall; defined formally in API RP 535 - Burners for Fired Heaters at Refineries
radiation	Heat transfer by electromagnetic radiation, such as infrared light; cf conduction; convection
radiation intensity	In the context of flare stacks, the grade-level radiant heat transfer rate from the flare flame; defined formally in API Standard 521. Pressure-relieving and Depressuring Systems. Sixth Edition| January 2014 and API RP 537 Flare Details for Petroleum, Petrochemical, and Natural Gas Industries

radiation loss	*aka* setting loss. Generically, heat loss due to radiation. Identical specific definitions in the context of heaters can be found in BS EN ISO 13705:2012/ISO 13705:2012(E) Petroleum, petrochemical and natural gas industries. Fired heaters for general refinery service, API STD 530 - Calculation of Tube Heater Thickness and API STD 560 - Fired Heaters for General Refinery Service
radioactive waste	Waste (qv) which is hazardous because it is radioactive (though it may have other hazardous properties). Different legal jurisdictions use differing definitions of how radioactive the waste must be in order to be classed as radioactive waste. Different levels of activity will determine the grade of radioactive waste and minimum criteria for legal disposal routes qv highly active waste, low level waste, high level waste. Also see **Box W1**.
radiography	A non-destructive testing (qv) technique, akin to an x-ray; most widely used for the identification of sub-surface defects in welded metal process equipment and a codal requirement of a number of pressure vessel codes e.g., PD5500 or ASME VIII
radioisotope	see **radionuclide**
radiological contamination	In the nuclear industry, the deposition of radioactive particles on equipment/persons
radiological controlled area	*aka* RCA. An area designated by an employer to restrict personnel radiation exposure. Everything leaving the RCA, including people, must be monitored for radiological contamination (qv) to prevent the spread of radioactive material beyond the RCA
radiolysis	Breaking down with ionizing radiation
radionuclide	*aka* radioisotope. A radioactive isotope of an element
radiosity	The total radiation from a surface, emitted and reflected; used in some heat transfer calculations
raffinate	The lean stream leaving solvent extraction
rag layer	A mixture of fine solids and an emulsion of water and non-aqueous phase liquid which presents separation problems
RAGAGEP	see **recognized and generally accepted good engineering practice(s)**
raining	A term used in distillation for a condition similar to dumping (qv) in which no liquid flows over the weir. The difference between dumping and raining is that liquid trickles through all holes at a uniform rate in a "raining" condition
ram pump	see **plunger pump**

ramp	1. Usually, a fixed inclined access surface; defined formally in BS EN ISO 14122 Safety of machinery. Permanent means of access to machinery. Working platforms and walkways 2. May also be used in control engineering to describe a change in a variable which looks like a ramp on a graph, or less formally, to increasing the rate of something: "ramp up to X Te/hr"
ramp response	In the context of control engineering, a response which looks like a ramp on a graph
random mixing rules	see **mixing rules**
random noise	Unpredictable noise (qv); defined formally in BS ISO 2041:2018 Mechanical vibration, shock and condition monitoring. Vocabulary
random packing	*aka* unstructured packing. Vessel packing materials supplied in bulk as small shapes; see **Figure R1** for an example of random packing from a MBBR

Figure R1 Random Packing

random vibration	Unpredictable vibration (qv); defined formally in BS ISO 2041:2018 Mechanical vibration, shock and condition monitoring. Vocabulary
Raney nickel	*aka* spongy nickel. A proprietary eponym for a metallic sponge of activated aluminum/nickel alloy used as a catalyst
Rankin	see **Rankine scale**
Rankine cycle	see **steam cycle**

Rankine scale	*aka* Rankin (sic). A Fahrenheit temperature scale whose zero is absolute zero
Rankine steam cycle	see **steam cycle**
Rankine vapor cycle	see **steam cycle**
Raoult's law	The law which states that the partial pressure of any given component above a mixture of liquids is proportional to the mole fraction of that component in the mixture, and its saturated vapor pressure under those conditions
rapid evacuation ejector	see **hogger jet**
rapid gravity filter	*aka* RGF. An open topped depth filter (qv) used in water treatment; cf pressure filter, slow sand filter
rapid hygiene methods	A term used in the hygienic industries for atline (qv) process monitoring methods which are sufficiently rapid to allow the results to be used for process control
rare earth elements	*aka* rare earth metals. A collective term for the lanthanides, scandium and yttrium, (rather than their oxides, as the term is sometimes mistakenly defined: those are properly rare earth oxides (qv))
rare earth metals	see **rare earth elements**
rare earth oxides	The proper term for the oxides of rare earth elements (qv)
rare gases	see **noble gases**
rarefaction	The opposite of compression for gases
RAS	see **return activated sludge**
Raschig hydroxylamine process	see **Raschig process**
Raschig process	There are three Raschig processes: The Raschig hydroxylamine process and the Olin-Raschig process are used to manufacture hydroxylamine, whilst the Raschig–Hooker process is used to manufacture phenol
Raschig ring	A type of random packing (qv)
Raschig-Hooker process	see **Raschig process**
rat hole	1. see **ratholing** 2. In oil and gas drilling, a hole whose bottom is deeper than the length of the casing string entering the ground
ratchet dog	*aka* rachet dog; see **dogs**
ratcheting	1. The progressive stepwise irreversible deformation of a component; defined formally in API RP 579 - Fitness for Service 2. In bulk solids handling, a mechanism by which larger sized particulates tend to move to the top of an agitated bed

rate constant	(*symbol* k) *aka* reaction rate constant, reaction rate coefficient, specific rate constant. The relationship between concentration of reactants and rate of reaction
rate determining step	*aka* RDS, rate limiting step. The slowest step of a reaction
rate limiting step	see **rate determining step**
rated burst pressure	see **marked burst pressure**
rated capacity	In the context of fire hydrants, available flow at rated pressure; defined formally in NFPA 24: Standard for the Installation of Private Fire Service Mains and Their Appurtenances, 2019 Edition; cf rated flow
rated coefficient of discharge	In the context of a PRV (qv), the coefficient of discharge (qv) used to calculate rated flow capacity; defined formally in API RP 520 P1 7th Edition, January 2000 Sizing, Selection, and Installation of Pressure-Relieving Devices in Refineries; Part I - Sizing and Selection
rated flow	In the context of fire pumps, maximum flow/pressure as per manufacturer's nameplate; defined formally in NFPA 20: Standard for the Installation of Stationary Pumps for Fire Protection, 2019 Edition
rated flow capacity	A term used but not defined by API in API RP 520 P1 7th Edition, January 2000 Sizing, Selection, and Installation of Pressure-Relieving Devices in Refineries; Part I - Sizing and Selection. N.B. Care should be taken not to mix API and ASME standards in determining the value of rated flow capacity
rated operating point	Vendor certified pump performance; defined formally in EN ISO 13709:2003 Centrifugal pumps for petroleum, petrochemical and natural gas industries
rated point	At a given turbine speed, the point where maximum power is produced; defined formally in ISO14661 Thermal turbines for industrial applications (steam turbines, gas expansion turbines) - General requirements
rated power	The greatest specified turbine power production/speed; defined formally in EN ISO 10437 Petroleum, petrochemical and natural gas industries - Steam turbines - Special-purpose applications; cf rated point, rated power output
rated power output	(*symbol* Pr) The maximum specified turbine power output, as measured at generator terminals; defined formally in ISO14661 Thermal turbines for industrial applications (steam turbines, gas expansion turbines) -- General requirements; cf rated power

rated pressure	In the context of fire pumps, differential pressure across a pump at rated flow/speed as per the manufacturer's nameplate; defined formally in NFPA 20: Standard for the Installation of Stationary Pumps for Fire Protection, 2019 Edition
rated relieving capacity	A manufacturer's rating of PRV (qv) capacity; defined formally in API RP 520 P1 7th Edition, January 2000 Sizing, Selection, and Installation of Pressure-Relieving Devices in Refineries; Part I - Sizing and Selection, API Standard 521. Pressure-relieving and Depressuring Systems. Sixth Edition	January 2014 and API RP 576 - Inspection of Pressure Relieving Devices; cf rated flow capacity
rated speed	(*symbol* N_r) *aka* 100% speed. In the context of rotodynamic equipment, a rotational speed: either the manufacturer's speed rating on pump nameplate (as defined more formally in NFPA 20: Standard for the Installation of Stationary Pumps for Fire Protection, 2019 Edition); or the highest specified speed of a turbine, as defined more formally in EN ISO 10437 Petroleum, petrochemical and natural gas industries - Steam turbines - Special-purpose applications or slightly differently in ISO 14661 Thermal turbines for industrial applications (steam turbines, gas expansion turbines) - General requirements
ratholing	1. The undesirable effect of pulling material preferentially from the center of a silo, or more generally from directly above a tank outlet 2. *(informal)* Channeling within a packed bed, or the formation of holes through refractory and other coatings
rattle gun	A pneumatic wrench
raw gas burner	A gas burner which injects fuel gas alone into the combustion zone, where it mixes with air and burns; defined formally in API RP 535 - Burners for Fired Heaters at Refineries; cf premix burner
raw material	see **feedstock**. This term defined formally in EN ISO 14040:2006 Environmental management - Life cycle assessment - Principles and framework
raw water	Untreated water from a natural source
Rayleigh equation	Most commonly in engineering contexts, an equation relating quantity to concentration of the more volatile component used to analyze simple distillation (though there are other Rayleigh equations); cf Rayleigh's equation
Rayleigh stability equation	see **Rayleigh's equation**
Rayleigh wave	see **surface wave**

Rayleigh's dimensional analysis method	The basis of the Buckingham π theorem (qv)
Rayleigh's equation	*aka* Rayleigh stability equation. An equation used to study hydrodynamic stability in fluid dynamics
Raymond mill	Proprietary eponym for a vertical roller mill
rayon	see **viscose**
RBC	see **rotating biological contactor**
RBM	see **risk based maintenance**
RBOB	see **reformulated blendstock for oxygenate blending**
RC	Reinforced concrete
RCA	1. see **radiological controlled area**
	2. see **root cause analysis**
RCDS	Reactant control and dilution system
RCM	see **reliability centered maintenance**
RCO	see **regenerative catalytic oxidizer**
RCRA	see **Resource Conservation and Recovery Act**
RDS	1. see **rate determining step**
	2. see **refractometric dry solids**
RE estimate	*(informal, offensive) aka* REE. RE stands for 'rectally extracted'[87]; cf SWAG
REAC	see **reactor effluent air cooler**
REACH	Registration, evaluation, authorization and restriction of chemicals: an EU regulation
reactant	see **reagent**
reactant ammonia	Ammonia used in post combustion SNCR (qv) NOx reduction plant; defined formally in API RP 536 - NOx control on Fired Heaters at Refineries
reactant flow control unit	A complete system responsible for the controlled introduction of reactant urea (qv) or reactant ammonia (qv) for post combustion SNCR (qv) NOx reduction
reactant injection system	*aka* RIS. An injection system used for controlled introduction of reactant urea (qv) or reactant ammonia (qv) for post combustion SNCR (qv) NOx reduction
reactant urea	Urea used in post combustion SNCR (qv) NOx reduction plant; defined formally in API RP 536 - NOx control on Fired Heaters at Refineries
reaction air	Air used to oxidize reaction gas (qv) to produce a special atmosphere (qv); defined formally in NFPA 86: Standard for Ovens and Furnaces, 2019 Edition

---

[87] i.e.; a number 'pulled out of your ass'!

reaction gas	Gas oxidized by reaction air (qv) to produce a special atmosphere (qv); defined formally in NFPA 86: Standard for Ovens and Furnaces, 2019 Edition
reaction rate coefficient	see **rate constant**
reaction rate constant	see **rate constant**
reactive distillation	A form of process intensification (qv) combining distillation with reaction
reactive maintenance	*aka* run to failure maintenance. At best, restoring machinery to a safely operable state only once it has failed. All too often, restoring machinery to a barely operable state only after it has failed; cf routine maintenance, preventative maintenance
reactive metals	In the context of equipment fabrication, the family of materials including aluminum, titanium, tantalum and zirconium which have desirable mechanical properties but can be notoriously difficult to work with
reactive silica	see **monomeric silica**
reactor	A vessel designed to contain a reaction. A narrow formal definition in one context may be found in API RP 536 - NOx control on Fired Heaters at Refineries
reactor effluent air cooler	*aka* REAC. A unit downstream of a hydroprocessing (qv) unit which provides a final stage of cooling prior to recycle gas separation; both process-critical and prone to corrosion, hence API produced RP 932-B Design, Materials, Fabrication, Operation, and Inspection Guidelines for Corrosion Control in Hydroprocessing Reactor Effluent Air Cooler (REAC) Systems
readily accessible	Capable of rapid, safe emergency access; defined formally in BS EN ISO 14159:2008 Safety of machinery. Hygiene requirements for the design of machinery and NFPA 86: Standard for Ovens and Furnaces, 2019 Edition
readily removable	see **easily or readily removable**. This particular alternative term is defined formally in BS EN ISO 14159:2008 Safety of machinery. Hygiene requirements for the design of machinery
ready for commissioning	A term which usually means that mechanical completion (qv) has been achieved
reagent	*aka* reactant. A substance which reacts
real gas	*aka* non-ideal gas. Confusingly, not a term which means a non-theoretical gas, or to put it another way 'all gases under all circumstances'; rather it is a term used for a gas under circumstances where its properties differ sufficiently from that of an ideal gas (qv) that a non-ideal model must be used
real time control	*aka* RTC. Time-critical closed-loop process control
reasonable	A somewhat contested term; see **Box R1**.

Box R1. Reasonable

The word reasonable is baked into many concepts that chemical engineers need to understand, such as ALARP (as low as reasonably practicable), SFAIRP (so far as is reasonably practicable), reasonably competent, reasonable skill and care, etc. Of these, the first two are the standards for managing risk in the UK, whilst the second two are the standards we must work to when discharging our duties as professional engineers. In all four cases, these are the standards we will be held to in court if things go wrong, as are related terms in other jurisdictions such as 'reasonable foreseeability'. The exercise of reason is required of us as professionals everywhere.

What does 'reasonable' mean in these contexts? Ultimately, it is a matter of legal interpretation, but as far as the first two are concerned, the HSE provides guidance: "Reasonably practicable' is a narrower term than 'physically possible' ... a computation must be made by the owner in which the quantum of risk is placed on one scale and the sacrifice involved in the measures necessary for averting the risk (whether in money, time or trouble) is placed in the other, and that, if it be shown that there is a gross disproportion between them – the risk being insignificant in relation to the sacrifice – the defendants discharge the onus on them".

In summary, we are required to exercise professional judgement in determining what constitutes ALARP in our situation, in the context of 'good practice', and with due consideration of the costs of risk reduction. So, any 'safety experts' insisting that no cost is too high for any given risk reduction are working to a different standard. We are required to work to good (rather than best) practice, and we should therefore consider whether the cost of risk reduction is disproportionate. Working to the 'physically possible' standard is not good engineering practice.

Reasonable competence is a term in English law used in the assessment of liability for negligence involving a breach of duty. Its tests are rather like those for ALARP above. It is allowable to exercise judgement in reducing foreseeable risks, the costs of mitigation measures can be considered, and the required degree of skill is that which is professionally commonplace, rather than extraordinary.

Reasonable skill and care is a contractual duty analogous to 'reasonably competent'. Its evil twin is fit for purpose, a term which, when used in a contract, prevents the use of all the defences implicit in those terms containing the word reasonable. We might therefore consider it to require us to be unreasonably competent, and to exercise unreasonable skill and care. Which hardly seems reasonable.

Related Terms
reasonably foreseeable, fit for purpose

reasonable competence	A term in English law used in the assessment of liability for negligence involving a breach of duty; the standard of competence to which a professional is held; cf **Box R1**.
reasonable skill and care	A term used in contracts to define the required standard of diligence; cf fit for purpose, **Box R1**.
reasonably foreseeable misuse	Using something in a way it was not intended to be used (as people reliably do); defined formally in BS EN 82079-1:2012 Preparation of instructions for use. Structuring, content and presentation. General principles and detailed requirements and BS EN ISO 12100:2010 Safety of machinery. General principles for design. Risk assessment and risk reduction
rebar	see **reinforcing bar**
reboiler	A heat exchanger used to power distillation by evaporating liquid (using a utility or hot process stream) usually located at the bottom of a distillation column. Also used in amine scrubbing service to provide the heat input needed to remove absorbed gases (hydrogen sulfide or carbon dioxide) from rich amine
rebuilding drawing	A type of general arrangement drawing (qv) specifying rebuilding works; defined formally in BS EN ISO 10209:2012 Technical product documentation — Vocabulary — Terms relating to technical drawings, product definition and related documentation, as a drawing of construction works
recip.	see **reciprocating compressor**
reciprocating compressor	*aka* recip. A piston or diaphragm positive displacement gas compressor
reciprocating pump	A piston, plunger, bellows or diaphragm positive displacement pump (qv)
reclaiming	Recovering bulk material from a stockpile (qv)
recognized and generally accepted good engineering practice(s)	*aka* RAGAGEP. A term commonly used in the USA for engineering, operation, or maintenance activities based on established codes, standards, published technical reports, recommended practices or similar documents; very similar to good engineering practice(s) (qv)
recognized code or standard	A code or standard with local legal standing; defined formally in API RP 579 - Fitness for Service
record	Documentary evidence of activity; defined formally in BS EN ISO 9000 Quality management systems Fundamentals and vocabulary
record drawing	see **as-built drawing**
recovery	The fraction of what one puts into a process that one gets out

recrystallization	The purification of a solute by repeated crystallization and redissolution
rectangular coordinate system	A system of coordinates relative to three mutually orthogonal axes; defined formally in BS EN ISO 10209:2012 Technical product documentation - Vocabulary - Part 1: Terms relating to technical drawings: general and types of drawings
rectification	1. The general, mathematical, electrical and legal meanings for the term all essentially mean 'putting right' 2. In engineering contexts, almost always means repeated distillation to produce a high purity product
rectified alcohol	see **neutral spirit**
rectified spirit	see **neutral spirit**
rectifying section	*aka* enriching section. A distillation column section above the feed section where enriching predominates; cf stripping section
rectilinear transducer	A term used in BS ISO 2041:2018 Mechanical vibration, shock and condition monitoring. Vocabulary to mean a linear motion sensor (rather than a linear transducer (qv)) to differentiate it from one which detects rotational motion, despite this standard listing 'linear' and 'rectilinear' as synonyms elsewhere
rectilinear vibration	A straight-line vibration; defined formally in BS ISO 2041:2018 Mechanical vibration, shock and condition monitoring. Vocabulary
Rectisol	A methanol-based proprietary physical solvent for gas sweetening (qv); cf Selexol, Purisol
recuperative catalytic oxidizer	A recuperative thermal oxidizer (qv) which uses a catalyst; defined formally in NFPA 86: Standard for Ovens and Furnaces, 2019 Edition
recuperative thermal oxidizer	*aka* thermal recuperative oxidizer (TRO). A pollution control device, a thermal oxidizer (qv) which uses a recuperator (qv) to heat incoming feed; defined formally in NFPA 86: Standard for Ovens and Furnaces, 2019 Edition; cf regenerative thermal oxidizer
recuperator	A heat exchanger which heats a feed stream with a hot waste stream
recycle	Adding a proportion of the output from a process to its feed stream
recycle ratio	The proportion of recycled process output to fresh feed going into a process
red diesel	*aka* dyed diesel, pink diesel. In the UK and USA, an untaxed gas oil (qv) fuel, dyed red and used for heating and certain off-road vehicles

red hydrogen	Hydrogen made using nuclear heat only, i.e.; without electrolysis; cf pink hydrogen, hydrogen colors
red list	1. One of various lists of substances prohibited on environmental grounds, mostly of limited relevance in engineering contexts
	2. A deprecated use in the context of aqueous pollutants, gray list (qv) and black list (qv) being the correct terminology
red rubber	*(informal) aka* RR. A colloquial name for styrene butadiene (Buna-S) rubber gasket material
red tide	see **algal bloom**
REDCF	*(drawing notation)* Reducing flange
Redlich-Kwong equation of state	*aka* RK equation of state. One of a number of empirical equations relating temperature, pressure and volume; used to allow unknown properties to be estimated based on known quantities; cf Soave-Redlich-Kwong equation of state
redox reaction	A reaction involving simultaneous mutual oxidation and reduction
reduced crude	see **atmospheric residue**
reduced pressure	*(symbol* p_r*)* Actual pressure divided by critical pressure
reduced properties	A general term for state properties (pressure and temperature) divided by their corresponding critical value
reduced temperature	*(symbol* T_r*)* Actual temperature divided by critical temperature
reducer	A pipe fitting which reduces pipe bore, allowing pipes of different sizes to be joined. May be concentric (qv) or eccentric (qv); defined formally in NFPA 1142: Standard on Water Supplies for Suburban and Rural Fire Fighting, 2017 Edition
reduction	A chemical reaction in which something loses oxygen (or gains electrons or hydrogen if you ask a chemist); cf oxidation
reduction efficiency	In the context of environmental controls on fired heaters, the percentage of flue gas NOx removal by reduction; defined formally in API RP 536 - NOx control on Fired Heaters at Refineries
reduction scale	1. *aka* reduction scale factor. A drawing scale factor less than 1:1; defined formally in ISO 10209:2012 Technical product documentation — Vocabulary — Terms relating to technical drawings, product definition and related documentation
	2. The ratio of a dimension on a drawing to its larger real world equivalent
	3. A calibrated ruler used to read real world dimensions directly from a drawing; cf enlargement scale
reduction scale factor	see **reduction scale**

reductive thermal oxidizer	A commonplace incorrect understanding of the meaning of RTO (properly regenerative thermal oxidizer (qv))
redundancy	see **redundant**
redundant	In engineering, normally refers to duplicate equipment, rather than equipment surplus to requirements. Some differentiate between active and standby redundancy, active redundancy being an always-online assist (qv) duty, and standby redundancy being a standby (qv) unit, brought online either as required, and/or periodically switching places with the duty unit. Others still insist that a redundant unit is only the assist unit
REE	see **RE estimate**
reeving	Fastening ropes
reference arrow layout	A drawing with informally positioned views and sections; defined formally in ISO 10209:2012 Technical product documentation — Vocabulary — Terms relating to technical drawings, product definition and related documentation
reference designation	A tag number (qv), a unique code for something used on a drawing such as a P&ID, and ideally used consistently throughout other deliverables. Two different definitions compatible with this are given in ISO 10209:2012 Technical product documentation — Vocabulary — Terms relating to technical drawings, product definition and related documentation
reference point	1. A coordinate system origin; defined formally in ISO 10209:2012 Technical product documentation — Vocabulary — Terms relating to technical drawings, product definition and related documentation 2. In chemistry, a type of datum, e.g.; the freezing and boiling points of water are used as reference points in temperature measurement
reference stress	A factor used in the failure assessment diagram (qv) method for the evaluation of crack-like flaws in components, based on primary stress, component geometry and crack dimensions; defined formally in API RP 579 - Fitness for Service
refine	To purify
refinery	Generically, a process plant where something is purified by separation or chemical processes. Commonly used as a synonym of petroleum refinery (qv), as per the formal definition in NFPA 30: Flammable and Combustible Liquids Code, 2018 Edition
refinery fuel gas	see **fuel gas**
refinery gas	see **still gas**

refining	Purifying
reflow	Remelting a weld bead to correct structural or aesthetic defects; defined formally in ASME BPE (American Society of Mechanical Engineers: Bioprocessing Equipment)
reflux	1. The return of condensed vapor to a distillation column 2. A mode of operation where condensed vapor is returned
reflux classifier	A proprietary fluidized bed classifier used in mineral processing; cf lamella plate
reflux drum	*aka* accumulator. A small vessel used as a buffer tank for condensed liquid at the top of a distillation column
reflux ratio	The ratio of boilup rate (qv) to takeoff rate (qv); cf boilup ratio
reflux valve	see **one way valve**
reformates	The higher-octane products produced from naphthas by catalytic reforming
reformer furnaces	Essentially, heated catalytic reactors which produce hydrogen; used to make low octane distillates into higher octane ones plus hydrogen, or to turn a mix of natural gas and steam into syngas (qv)
reforming	see **catalytic reforming, steam methane reforming.**
reformulated blendstock for oxygenate blending	*aka* RBOB, reformulated gasoline blendstock for oxygenate blending. A blendstock (qv) which is blended with oxygenates to make reformulated gasoline (qv)
reformulated gasoline	*aka* RFG. Gasoline reformulated to produce fewer air pollutants according to 1990 Clean Air Act amendments, which must be used in the USA in urban areas with poor air quality; cf conventional gasoline
reformulated gasoline blendstock for oxygenate blending	see **reformulated blendstock for oxygenate blending**
refractometric dry solids	*aka* RDS. A measure of total dissolved solids in a sugar liquor using a refractometer
refractory	1. A term generally meaning resistant 2. Insulating bricks which can resist high temperatures 3. In effluent treatment, applied to contaminants which resist treatment, especially biological treatment
refractory design temperature	The hot face temperature which a refractory lining is designed for; defined formally in API RP 534 – Heat Recovery Steam Generators
refractory rating temperature	The maximum rated continuous operating temperature of a refractory; defined formally in API RP 534 – Heat Recovery Steam Generators

refractory service temperature	The maximum temperature a refractory is suitable for, based on shrinkage; defined formally in API RP 534 - Heat Recovery Steam Generators
refrigerant	A fluid with a boiling point close to ambient temperatures used to refrigerate something, transferring heat by phase change
refrigerated LP gas	LPG which is sufficiently cooled to have a vapor pressure of no more than 1 atm; defined formally in NFPA 58: Liquefied Petroleum Gas Code, 2017 Edition
refrigeration	*(informal) aka* fridge. Cooling with a refrigerant
refrigeration ton	*aka* RT; see **tons refrigeration**
regas sendout rate	The rate of export of regasified LNG (qv)
regenerant	A chemical used to make exhausted ion exchange resin available for reuse by replacement of ions adsorbed in service
regeneration	1. Reactivating a catalyst or other heterogenous process support 2. In oil refining, restoring catalyst properties by combustion of coke laid down on its surface 3. Making exhausted ion exchange resin available for reuse by replacement of ions adsorbed in service
regeneration efficiency	The ratio of ion exchange resin capacity after regeneration to new resin capacity
regeneration level	The amount of regenerant (qv) used in an ion exchange regeneration cycle (commonly in Lb./ft3 of resin)
regenerative blower	see **side channel blower**
regenerative catalytic oxidizer	*aka* RCO. A regenerative thermal oxidizer (qv) which uses a catalyst; defined formally in NFPA 86: Standard for Ovens and Furnaces, 2019 Edition
regenerative heat exchanger	*aka* regenerator. A heat exchanger in which heat is transferred from a fluid to a solid material, and subsequently from the solid material to a second fluid
regenerative thermal oxidizer	*aka* RTO. A pollution control device, a thermal oxidizer (qv) which uses a catalytic bed preheated by a previous exhaust cycle to heat reactant gases; defined formally in NFPA 86: Standard for Ovens and Furnaces, 2019 Edition; cf recuperative thermal oxidizer
regenerator	1. Shorthand for regenerative heat exchanger (qv) 2. The section of a fluid catalytic cracker (qv) where coke is removed from the catalyst by combustion
regrade	Modification of nonconforming product or service to meet a grade (qv) or specification other than originally intended; defined formally in BS EN ISO 9000 Quality management systems Fundamentals and vocabulary; cf rework

regulating authority	*aka* regulator. A legal enforcement authority; defined formally in the context of pressure vessels in PD 5500:2018+A1:2018 Specification for unfired fusion welded pressure vessels
regulating valve	*aka* regulator. A control valve (qv). Although a valve actually controls flow (qv flow regulating valve), this can be used indirectly to control pressure (qv pressure regulating valve), or temperature (qv temperature regulating valve)
regulator	1. see **regulating authority** 2. see **regulating valve**
regulatory requirement	A legally mandated requirement; defined formally in BS EN ISO 9000 Quality management systems Fundamentals and vocabulary
Reid vapor pressure	*aka* RVP. A measure of vapor pressure commonly used in the oil and gas industry under test conditions set out in ASTM-D-323, mainly used for petrol/gasoline specifications for storage. Too high a value indicates an unacceptable level of LPG (C3/C4) components
reinforcement	1. In the context of welds, a synonym for convexity (qv); defined formally in ASME BPE (American Society of Mechanical Engineers: Bioprocessing Equipment) 2. Also sometimes used to mean rebar (qv)
reinforcement drawing	A drawing illustrating required rebar (qv) positions; defined formally in ISO 10209:2012 Technical product documentation — Vocabulary — Terms relating to technical drawings, product definition and related documentation
reinforcement of weld	*aka* reinforcement. A synonym of weld convexity; however, this alternative term is the one defined in API STD 620 - Low Pressure Storage Tanks
reinforcement steel	see **reinforcing bar**
reinforcing bar	*aka* rebar, reinforcing steel, reinforcement steel. The steel reinforcement of reinforced concrete
reinforcing steel	see **reinforcing bar**
rejection coefficient	The effectiveness of a membrane in excluding a given solute, expressed as a percentage (100 times the difference between its feed and permeate concentrations, divided by its feed concentration)
rejuvenation	An attempt to restore some catalyst activity while on-stream, such as removing some/all feed and purging with H_2 or steam[88]

---

[88] Which usually fails

relative atomic mass	(*symbol* Ar) *aka* atomic weight. The average mass of an atom in a sample of an element, compared with 1/12th the mass of a carbon-12 atom
relative density	Synonymous with specific gravity (qv). This alternative term is however used in the formal definitions in BS EN 12952-8:2002 Water-tube boilers and auxiliary installations. Requirements for firing systems for liquid and gaseous fuels for the boiler, EN ISO 13709:2003 Centrifugal pumps for petroleum, petrochemical and natural gas industries and API RP 505 2nd Edition, August 2018 Recommended Practice for Classification of Locations for Electrical Installations at Petroleum Facilities Classified as Class I, Zone 0, Zone 1, and Zone 2
relative humidity	The ratio of absolute humidity to maximum possible humidity at measurement temperature, usually expressed as a percentage
relative roughness	(*symbol* ε/D) In pipes, the ratio of the absolute pipe wall roughness to the inside diameter; used in hydraulic calculations; N.B. it is not the same as hydraulic roughness (qv)
relative volatility	(*symbol* A) The ratio of Henry's law constants(qv) for components used in distillation calculations
relaxation in the bath	The realignment of polymer molecules when fibers are washed in water
release	1. In a documentation context, a formal permission to issue a document or proceed to the next stage of a process, as per formal definitions in BS EN ISO 9000 Quality management systems Fundamentals and vocabulary, ISO 10209:2012 Technical product documentation – Vocabulary – Terms relating to technical drawings, product definition and related documentation and ISO 10209:2012 Technical product documentation – Vocabulary – Terms relating to technical drawings, product definition and related documentation 2. In the context of a loss of containment, see **releases** 3. Occasionally used in the context of unsticking things, as in release agent (qv)
release agent	A coating which prevents a molding sticking to a mold
release phase	Document release (qv) stage; defined formally in ISO 10209:2012 Technical product documentation – Vocabulary – Terms relating to technical drawings, product definition and related documentation

release rate	In the context of releases (qv) of flammable gases, the quantity per unit time; defined formally in API RP 505 2nd Edition, August 2018 Recommended Practice for Classification of Locations for Electrical Installations at Petroleum Facilities Classified as Class I, Zone 0, Zone 1, and Zone 2
releases	Fugitive emissions, or losses of containment; defined formally in EN ISO 14040:2006 Environmental management - Life cycle assessment - Principles and framework
relevant hazard	A hazard associated with a given machine; defined formally in BS EN ISO 12100:2010 Safety of machinery. General principles for design. Risk assessment and risk reduction
relevant microorganisms	"Microorganisms able to contaminate, multiply or survive in the product and be harmful to the consumer or product quality", according to the EHEDG glossary; a very similar definition may also be found in BS EN ISO 14159:2008 Safety of machinery. Hygiene requirements for the design of machinery
reliability	Somewhat contested, see **Box A2**, but: 1. Most commonly, the ability to perform required function for a specified time under specified conditions 2. Also used to refer to a probabilistic measure of this ability; defined formally in IEC 60050-192:2015 International electrotechnical vocabulary. Dependability and BS EN ISO 12100:2010 Safety of machinery. General principles for design. Risk assessment and risk reduction
reliability centered maintenance	*aka* RCM. A technique for planning preventive maintenance to optimize reliability (qv); defined formally in SAE JA 1011-2009 Evaluation Criteria For Reliability-Centered Maintenance (RCM) Processes; cf risk based maintenance
reliability engineering	A somewhat contested term, but essentially, the application of engineering principles to optimization of reliability through all stages of a product/project lifecycle
reliability process	The process through which reliability engineering (qv) contributes to design, construction, and assurance
relief gas	*aka* flared gas, waste gas, waste vapor. Gas relieved into a flare header (qv). Defined formally in API Standard 521. Pressure-relieving and Depressuring Systems. Sixth Edition	January 2014 and API RP 537 Flare Details for Petroleum, Petrochemical, and Natural Gas Industries

relief valve	*aka* RV; see **spring-loaded PRV**. However, this alternative term is defined formally in API RP 520 P1 7th Edition, January 2000 Sizing, Selection, and Installation of Pressure-Relieving Devices in Refineries; Part I - Sizing and Selection and API Standard 521. Pressure-relieving and Depressuring Systems. Sixth Edition| January 2014
relief valve set pressure	The pressure at which a PRV (qv) is set to lift. Not necessarily the same as pop pressure (qv), though the formal definition in EN ISO 10437 Petroleum, petrochemical and natural gas industries - Steam turbines - Special-purpose applications seems to describe pop pressure
relieving conditions	The inlet pressure and temperature of a PRV (qv) whilst relieving; defined formally in API RP 520 P1 7th Edition, January 2000 Sizing, Selection, and Installation of Pressure-Relieving Devices in Refineries; Part I - Sizing and Selection and API Standard 521. Pressure-relieving and Depressuring Systems. Sixth Edition	January 2014
remaining strength factor	*aka* RSF. The ratio of estimated collapse pressure of a damaged component to original collapse pressure; used to determine whether it is acceptable for continued service. Defined formally with detailed guidance in API RP 579 - Fitness for Service
remanence	see **residual magnetism**
remanent magnetism	see **residual magnetism**
remote	In the context of instrumentation, distant from monitored equipment; defined formally in EN ISO 10437 Petroleum, petrochemical and natural gas industries - Steam turbines - Special-purpose applications; cf local
remote field testing	*aka* RFT; see **remote field eddy current testing**
remote field eddy current testing	*aka* remote field testing (RFT), RFEC, RFET, and (incorrectly) remote field electromagnetic technique. A non-destructive testing (qv) technique used to detect flaws in conductive materials
remote field electromagnetic technique	see **remote field eddy current testing**
remote location	In the context of fire protection, at least 4000 feet from a reference point; defined formally in API RP 2510A - Fire Protection Considerations for the Design and Operation of Liquefied Petroleum Gas (LPG) Storage Facilities
remote monitoring	see **distance monitoring**
remote operating center	*aka* ROC. A central control room (qv) not colocated on site

remotely operated emergency isolation valve	*aka* ROEIV; see **remotely operated shutoff valve**
remotely operated isolation valve	*aka* ROIV; see **remotely operated shutoff valve**
remotely operated operation valve	An actuated valve which permits remote operation
remotely operated shutoff valve	*aka* ROSoV, ROSOFF (sic), remotely operated emergency isolation valve (ROEIV), remotely operated isolation valve (ROIV), emergency shutdown valve (ESV); see **emergency shutdown valve**
remotely operated underwater vehicle	*aka* ROUV; see **remotely operated vehicle**
remotely operated vehicle	*aka* ROV, remotely operated underwater vehicle (ROUV). An underwater robot, or underwater autonomous vehicle (qv), usually operated via a tethered connection from the surface
removable	see **easily or readily removable**
renaturation	The reversal of denaturation (qv)
renewable diesel	Unmodified biologically derived oils and fats, as opposed to biodiesel (qv) which is produced by esterification and trans-esterification of such fats and oils
repair	Making unserviceable equipment suitable for its intended purpose; defined formally in BS EN 82079-1:2012 Preparation of instructions for use. Structuring, content and presentation. General principles and detailed requirements, BS EN ISO 9000 Quality management systems Fundamentals and vocabulary and API RP 579 - Fitness for Service
repeatability	Precision (qv) under tightly controlled conditions over a short time period
replacement in kind	Replacement with a component which meets the design specification of the item replaced. Within the context of management of change (qv) procedures, a replacement of equipment not requiring full assessment as a modification; defined formally in NFPA 654: Standard for the Prevention of Fire and Dust Explosions from the Manufacturing, Processing, and Handling of Combustible Particulate Solids, 2017 Edition
replica fidelity	The degree of accuracy and completeness of a copying process; defined formally in ISO 10209:2012 Technical product documentation — Vocabulary — Terms relating to technical drawings, product definition and related documentation

Reppe chemistry	*aka* Reppe processes. A largely obsolete set of high pressure catalytic reactions involving acetylene, especially those yielding vinyl compounds
Reppe processes	see **Reppe chemistry**
reproducibility	Consistency (qv) on repeated measurement
required impact energy	The Charpy impact test (qv) value required to demonstrate compliance of an item with PD 5500:2018+A1:2018 Specification for unfired fusion welded pressure vessels
required impact test temperature	*aka* RITT. The specified temperature for demonstration of required impact energy (qv); defined formally in PD 5500:2018+A1:2018 Specification for unfired fusion welded pressure vessels
requirement	A stated expectation or obligation. A slightly broader formal definition may be found in BS EN ISO 9000 Quality management systems Fundamentals and vocabulary
requirement specification	*aka* specification of requirements (SOR). A generic term, see user requirement specification (qv), functional requirements specification (qv) and product requirement specification (qv) for specific instances. This particular term is defined in ISO 10209:2012 Technical product documentation — Vocabulary — Terms relating to technical drawings, product definition and related documentation
reradiating wall	see **target wall**
rerating	A reduced maximum acceptable working pressure (qv) or temperature rating of a vessel, based on insufficient remaining strength factor (qv); defined formally in API RP 579 - Fitness for Service
rescue hoist	see **rescue winch**
rescue winch	*aka* rescue hoist. A winch used solely for emergency personnel egress at height or from below grade confined spaces; cf man riding winch
reservoir level	see **pumping liquid level**
reset	1. To manually reinitialize a machine, control system, computer, or other electronic system; cf reset control, restart 2. Commonly used to describe clearing a fault in a control system requiring manual intervention, especially releasing an EMS button; defined formally in this last sense in ISO 3977 Gas turbines - Procurement - Part 3: Design requirements
reset control	see **integral action control**
resid.	see **residue**

residence time	The amount of time something (particles or fluid parcels, commonly) spends in a reactor, or in contact with the catalyst; cf space time, retention time
residence time distribution	*aka* RTD. The frequency distribution of residence times (qv); cf mean residence time
residential board and care occupancy	A building or part of one used for the board, lodging and personal care of at least four residents who are unrelated to building operators; defined formally in NFPA 30: Flammable and Combustible Liquids Code, 2018 Edition
residential occupancy	A building or part of one provided with sleeping accommodation (other than a health care occupancy (qv) or detention and correctional occupancy (qv)); defined formally in NFPA 30: Flammable and Combustible Liquids Code, 2018 Edition
residual fuel oil	see **bunker oil**
residual hydrant	The hydrant where pressure is measured during a firemain flow test; defined formally in NFPA 24: Standard for the Installation of Private Fire Service Mains and Their Appurtenances, 2019 Edition
residual magnetism	*aka* remanence, remanent magnetization, retentivity. Magnetism in a ferromagnetic material exposed to an external magnetic field, after that field is removed; defined formally in ISO 3977 Gas turbines - Procurement - Part 3: Design requirements
residual pressure	Pressure at the residual hydrant (qv) at the point of taking flow readings during a firemain flow test; defined formally NFPA 24: Standard for the Installation of Private Fire Service Mains and Their Appurtenances, 2019 Edition
residual risk	Any risk remaining after risk mitigation; defined formally in BS EN 1050:1997 Safety of machinery. Principles for risk assessment and BS EN ISO 12100:2010 Safety of machinery. General principles for design. Risk assessment and risk reduction
residue	1. Generically, that which remains 2. *aka* resid., residuum. Commonly used in oil and gas to describe that which remains after distillation, such as atmospheric residue, vacuum residue (qv) 3. Used to describe traces of contaminants such as pesticides in environmental and food contexts
residuum	see **residue**
resistance	An extrinsic property; the measure of the opposition to electrical, hydraulic or thermal conduction (qv) of a material; the opposite of conductance (qv)

resistance coefficient	see **Darcy friction factor**
resistance heating system	Heating by means of electrical resistance; defined formally in NFPA 86: Standard for Ovens and Furnaces, 2019 Edition
resistance temperature detector	*aka* RTD, resistance thermometer. A thermometer based on a variation in resistance proportional to temperature
resistance thermometer	see **resistance temperature detector**
resistivity	An intrinsic property, the measure of the opposition to electrical, hydraulic or thermal conduction (qv) of an item; the opposite of conductivity (qv)
Resistoflex	Proprietary eponym for a fluoropolymer pipe lining material
resolution	In control engineering, the smallest change detectable by an instrument; cf accuracy, precision
resonance speed	see **critical speed**
Resource Conservation and Recovery Act	*aka* RCRA. USA environmental legislation covering the disposal of solid/hazardous waste
resource depletion	see **depletion**
resource efficiency	*aka* waste minimization. Reducing a process's inputs and outputs of materials and energy per unit of useful product
respiration	1. An energy yielding redox reaction (qv) involving either oxygen (aerobic respiration) or other electron acceptors (anaerobic respiration) 2. May also be used incorrectly to describe the exchange by animals of oxygen and carbon dioxide with their surroundings (breathing)
respiratory quotient	*aka* RQ. RQ = mass of CO_2 eliminated/mass of O_2 consumed during respiration
response	In the context of process control, the output of a control system; defined formally in BS ISO 2041:2018 Mechanical vibration, shock and condition monitoring. Vocabulary
rest platform	A horizontal platform, large enough for more than one person to rest, spaced every 6-9m on a single ladder flight; defined formally in BS EN ISO 14122 Safety of machinery. Permanent means of access to machinery. Working platforms and walkways
restart	To automatically reinitialize a machine, control system, computer, or other electronic system (or the automatic reinitialization thereof); defined formally in the first sense in the context of fired heaters in BS EN 12952-8:2002 Water-tube boilers and auxiliary installations. Requirements for firing systems for liquid and gaseous fuels for the boiler; cf reset

restoring force	A force or forces tending to return something to equilibrium/original position, especially those arising from material deformation; defined formally in BS ISO 2041:2018 Mechanical vibration, shock and condition monitoring. Vocabulary
restriction orifice	*aka* restriction orifice plate. Used in some industries to refer to an orifice plate (qv)
restriction orifice plate	*aka* ROP; see **restriction orifice**
retentate	That part of a feed fluid which does not pass through a membrane; cf permeate
retention time	*aka* RT. A common rule of thumb, nominally the time that something spends in a unit operation, usually (but not always) calculated as a mean value; cf hydraulic retention time, solids retention time
retentivity	see **residual magnetism**
retrofit	see **revamp**
return	An inlet
return activated sludge	*aka* RAS. That part of activated sludge recirculated from a clarifier (qv) to an aeration basin (qv) in an activated sludge process
return flow atomizer	In the context of fired heaters, oil burners where the fuel oil flow is controlled by an actuated valve in a bypass back to the oil storage tank; defined formally in BS EN 12952-8:2002 Water-tube boilers and auxiliary installations. Requirements for firing systems for liquid and gaseous fuels for the boiler
return on capital employed	*aka* ROCE. An accounting ratio used to assess how well a business's capital is generating profit
reuse dismantling drawing	A drawing showing how something (especially a building) can be dismantled for reuse; defined formally in ISO 10209:2012 Technical product documentation — Vocabulary — Terms relating to technical drawings, product definition and related documentation
revamp	*aka* retrofit. Upgrading existing processes or technology, especially on an oil refinery; cf debottlenecking, turnaround
reverse acting controller	A controller with output inversely proportional to input; cf direct acting controller
reverse osmosis	*aka* RO. The net movement of solvent molecules across a semipermeable membrane driven by physical pressure which leads to increased solute concentration on the side of the membrane to which pressure is applied; cf osmosis, forward osmosis; **Box L1**.

reversible	1. Generally, capable of being reversed, especially in the context of reaction; cf irreversible 2. In an ion exchange context, the ability of a mixture of resin types to be regenerated by variations in component concentrations or selectivity. Perhaps confusingly, all ion exchange processes are reversible in the first sense
revision notice	A listing of revisions made to a drawing, commonly part of the relevant drawing; defined formally in ISO 10209:2012 Technical product documentation – Vocabulary – Terms relating to technical drawings, product definition and related documentation
revision phase	A stage where revisions to documents are allowed; defined formally in ISO 10209:2012 Technical product documentation – Vocabulary – Terms relating to technical drawings, product definition and related documentation
rework	The modification of nonconforming product or service to meet original specification; defined formally in BS EN ISO 9000 Quality management systems Fundamentals and vocabulary; cf regrade
Reynolds number	*(symbol* Re) A dimensionless group, being the ratio of inertial and viscous forces, very commonly used in fluid dynamics
RF	*(drawing notation)* Raised face
RFBR	see **radial flow fixed bed reactor**
RFEC	see **remote field eddy current testing**
RFET	Remote field eddy testing; see **remote field eddy current testing**
RFG	1. Refinery fuel gas, see **fuel gas** 2. see **reformulated gasoline**
RFSU	Ready for start up
RFT	Remote field testing, see **remote field eddy current testing**
RGF	see **rapid gravity filter**
Rheniforming	A proprietary catalytic reforming process with a rhenium catalyst
rheogram	A graph of rheology (qv), most commonly plotting a fluid's shear stress against its shear rate
rheology	The science of deformation of matter; most commonly, in a chemical engineering context, the flow of fluids
rheopectic fluid	*aka* antithixotropic fluid (qv). A fluid which exhibits rheopecty (qv)
rheopecty	*aka* rheopexy, antithixotropy, inverse thixotropy. The opposite of thixotropy (qv) i.e.; time-dependent shear thickening, a behavior of some non-Newtonian fluids
rheopexy	see **rheopecty**

ribbon cable	see **ribbon cable wiring**
ribbon cable wiring	*aka* ribbon cable. A flat common insulator containing and separating multiple parallel conductors; defined formally in ISO 3977 Gas turbines - Procurement - Part 3: Design requirements
RIBC	see **rigid intermediate bulk container**
rich	Generically, a term applied to a process stream with a higher concentration of a component; cf lean
rider bands	*aka* rider rings, guide rings. Components which guide a piston in a compressor cylinder bore, preventing the piston from contacting the liner
rider rings	see **rider bands**
rig	1. As a noun, most commonly an oil production platform (qv) 2. May also (less commonly) be applied to a test rig or pilot plant (qv) used to generate design data 3. *(informal)* As a verb, to rig something up means to set it up or put it together, especially something temporary 4. To prepare for lifting (qv rigger)
rig floor	see **drill floor**
rig table	see **rotary table**
rigger	*aka* banksman/slinger. A term associated with facilitating safe lifting. The operation of cranes or derricks is not rigging, whereas attaching the load to such lifting equipment with tackle, strops, shackles etc. is.
right of way	*aka* ROW. 1. Generically, a defined, linear piece of land which there is a right to access 2. Specifically applied to land which a pipeline operator has a right to use for a pipeline
rigid acceptance criteria	A term used in commissioning to describe acceptance criteria which are critical to safety or operation and which cannot be resolved with design modifications; cf soft acceptance criteria
rigid frame	see **braced frame**
rigid intermediate bulk container	*aka* RIBC, IBC, intermediate bulk container. An IBC (qv) in a rigid frame; defined formally in NFPA 654: Standard for the Prevention of Fire and Dust Explosions from the Manufacturing, Processing, and Handling of Combustible Particulate Solids, 2017 Edition; cf flexible intermediate bulk container
rigorous multicomponent distillation methods	A plate-by-plate calculation from first principles of mass flows across a distillation column; rarely done in practice except to the extent that it is built into programs such as Hysys (qv)

rim seal	Most commonly, a flexible seal on a floating roof hydrocarbon storage tank; defined formally in API MPMS 19.1 Manual of Petroleum Measurement Standards Chapter 19.1 Evaporative Loss from Fixed-roof Tanks
Ringelmann number	A visual measure of smoke opacity (qv); defined formally in API RP 537 Flare Details for Petroleum, Petrochemical, and Natural Gas Industries
rinse	Flushing out of residual material, usually with flowing water; commonly applied to fixed beds after backwash or regeneration, or to surface cleaning in a hygienic context
rinsing	1. Generically, see **flushing** 2. In a hygienic context, "The removal of product, dirt, chemicals, cleaning residues or any objectionable matter by flowing potable water" according to the EHEDG Glossary
RIS	see **reactant injection system**
rise	The difference in tread surface heights between consecutive steps on an access stairway; defined formally in BS EN ISO 14122 Safety of machinery. Permanent means of access to machinery. Working platforms and walkways
rise rate	see **surface loading**
riser	1. Most commonly, a pipe which carries something upwards (usually therefore at least partially vertical), e.g.; the inlet section of a fluid catalytic cracker (qv); defined formally in two different contexts in API RP 537 Flare Details for Petroleum, Petrochemical, and Natural Gas Industries and API RP 534 - Heat Recovery Steam Generators 2. An access stair upright as defined in OSHA 1910.29 - Fall protection systems and falling object protection-criteria and practices
rising film evaporator	*aka* climbing film evaporator, vertical tube evaporator, vertical long tube evaporator. An evaporator (qv) in which a continuous film of liquid to be evaporated flows upwards under the influence of rising vapor
rising main	1. Most commonly, a pipe containing pumped flows, usually of sewage or less commonly potable water, almost always with the pump lower than the point of discharge 2. Less frequently, the point where such a pipe, (or by extension other vertical utilities such as electricity) come out of the ground within a building within a building

risk Most commonly held to be the probability of an adverse effect or hazard occurring, multiplied by the severity of outcome if it occurs. It is however a highly contested term: see **Box R2**. A range of formal definitions can be found in BS EN ISO 9000 Quality management systems Fundamentals and vocabulary, BS EN 82079-1:2012 Preparation of instructions for use. Structuring, content and presentation. General principles and detailed requirements; Regulation (EC) No 178/2002 of the European Parliament and of the Council of 28 January 2002, API RP 2001 - Fire Protection at Refineries, BS EN ISO 12100:2010 Safety of machinery. General principles for design. Risk assessment and risk reduction, BS EN ISO 17776:2016 Petroleum and natural gas industries. Offshore production installations. Major Accident hazard management during the design of new installations, BS ISO 31000:2018 Risk management Guidelines and the EHEDG glossary. See also Reducing Risk, Protecting People by UK HSE for extensive definitions used by UK competent authorities

risk analysis *aka* risk based analysis. One of several formal processes for analyzing risk (qv), which may be qualitative or quantitative, and are often more rigorous and at a more detailed level than risk assessment (qv). Different formal definitions can be found in Regulation (EC) No 178/2002 of the European Parliament and of the Council of 28 January 2002, BS EN ISO 12100:2010 Safety of machinery. General principles for design. Risk assessment and risk reduction and the EHEDG glossary; cf quantitative risk analysis, qualitative risk analysis

R2. Risk

Risk is a strongly contested term. At its broadest, it is used to mean the effect of uncertainty, and is most commonly held to mean "the probability of an adverse effect or hazard occurring". However, it is also used to mean the possibility of an effect (not quantified as with a probability), the severity of the effect, or the product of the probability and the severity. I would differentiate risk from hazard in that where a hazard is a potential harm, risk is the likelihood of it happening, but, whilst this is a commonly held view, it clashes with other formal definitions of both hazard and risk.

In fact, virtually every part of my "most commonly held" definition is disputed: whether it involves possibility or probability of an effect, whether it only refers to adverse effects, whether the effect constitutes a hazard are all up for grabs, even in formal definitions from reputable international standards organizations.

I have been informed that ISO alone – which makes its definitions publicly available – sets out at least forty different definitions of risk, which use a carousel of different keyword permutations. Most significantly, the ISO definitions cannot agree about what a risk is; what or who it affects; whether it is always adverse; whether it is the probability of something, the outcome of something, the product of these two, the sum of them, both together or each separately; or simply the effect of uncertainty.

A lexicographer's additional quibble would be that the keywords probability and outcome in the above sentence are replaced in many of ISO's definitions with synonyms or near-synonyms which adds to potential confusion.

Technical commentators have been scathing about this state of affairs, but it seems unlikely that ISO is any less rigorous than other standards organizations, or professionals working in the area. On the contrary, my discussions about risk with fellow practitioners have shown me that even the most highly qualified risk specialists do not agree with each other about the definition of the subject of their specialism.

So, a term which is arguably key to process safety is highly ambiguous in meaning. As with all such contested terms, engineers need to be sure that they all mean the same thing when discussing risk.

Related Terms
risk analysis, risk assessment, risk matrix

risk assessment	*aka* RA, risk estimation. One of a number of formal processes for assessing risk (qv), often at a less detailed level than risk analysis (qv). A term commonly used for a tick-sheet type document used as an aide-memoire to ensure that common site work hazards have been addressed, used alongside a method statement (qv). Risk assessment may however incorporate a risk analysis and risk evaluation (qv). Formal definitions may be found in Regulation (EC) No 178/2002 of the European Parliament and of the Council of 28 January 2002, API RP 2001 - Fire Protection at Refineries, BS EN ISO 12100:2010 Safety of machinery. General principles for design. Risk assessment and risk reduction, NFPA 654: Standard for the Prevention of Fire and Dust Explosions from the Manufacturing, Processing, and Handling of Combustible Particulate Solids, 2017 Edition and the EHEDG glossary; cf quantitative risk assessment, qualitative risk assessment
risk based analysis	see risk analysis. This alternative term is defined in API RP 2001 - Fire Protection at Refineries
risk based maintenance	*aka* RBM, risk centered maintenance. A technique for planning preventive maintenance to minimize risk; cf reliability centered maintenance
risk centered maintenance	see risk based maintenance
risk communication	The interactive exchange of information about risk (qv); defined formally in Regulation (EC) No 178/2002 of the European Parliament and of the Council of 28 January 2002 and the EHEDG glossary
risk estimation	see **risk assessment**. This alternative term is defined formally in BS EN ISO 12100:2010 Safety of machinery. General principles for design. Risk assessment and risk reduction
risk evaluation	Judging, based on post mitigation risk analysis, whether a residual risk (qv) is tolerable; defined formally in BS EN ISO 12100:2010 Safety of machinery. General principles for design. Risk assessment and risk reduction
risk management	The "process of identifying, evaluating and prioritizing risks to eliminate them, to mitigate their likelihood or severity or to exploit opportunities", according to the EHEDG glossary and Regulation (EC) No 178/2002 of the European Parliament and of the Council of 28 January 2002
risk matrix	A commonly used graphical illustration of qualitatively estimated probability and severity of adverse consequences
risk zone	see **zoning** (hygienic sense)

RITT	see **required impact test temperature**
RK equation of state	see **Redlich–Kwong equation of state**
Rm	see **minimum reflux ratio**
RMC	Ready mixed concrete
Rmin	see **minimum reflux ratio**
RMS	Root mean square
RO	see **reverse osmosis**
road safety barriers	see **vehicle safety barriers**
road tanker	A mobile tank for bulk liquid transport by road
roasting	Heating sulfide ores in air, most commonly with a view to oxidizing the sulfides to SOx
robustness	One of the three key metrics of good design (the other two being cost and safety); a robust design is proof against all foreseeable disturbances
ROC	see **remote operating center**
ROCE	see **return on capital employed**
rock	Naturally occurring part of the ground (qv) which is harder than soil (qv); defined formally in BS 5930:2015 Code of practice for ground investigations
rocking horse pump	see **nodding donkey**
Rockwool	Proprietary eponym for a mineral fiber insulating material
rodding eye	see **rodding point**
rodding point	*aka* sewer cleanout, drainage cleanout, rodding eye. An access point used to insert a rodding device to clear a sewer blockage
ROEIV	Remotely operated emergency isolation valve; see **remotely operated shutoff valve**
roentgen	(*symbol* R) An obsolete unit of ionizing radiation exposure, superseded by coulombs/kg in SI
ROI	Return on investment
ROIV	Remotely operated isolation valve, see **remotely operated shutoff valve**
rolled asphalt	see **tarmac**
roman candle effect	*aka* burning rain, golden rain. The (highly hazardous) effect produced when liquid goes up a flare stack
Romankov number	(*symbol* Ro) A dimensionless group, being the ratio of exhaust gas temperature to dried solids temperature; used for analysis of drying
room relation drawing	A drawing of a building, showing the position of rooms relative to each other; defined formally in ISO 10209:2012 Technical product documentation – Vocabulary – Terms relating to technical drawings, product definition and related documentation

root cause	see **ultimate cause**
root cause analysis	A technique for determining the ultimate cause (qv) of an occurrence
root valve	*aka* isolation valve. A valve used to isolate equipment; this particular term is defined formally in API RP 2510A - Fire Protection Considerations for the Design and Operation of Liquefied Petroleum Gas (LPG) Storage Facilities
Roots blower	Proprietary eponym for a type of low pressure positive displacement compressor (qv) (though some dispute whether a blower is a compressor)
ROP	see **restriction orifice plate**
Rose equation	*aka* Rose's equation. An equation used to estimate the head loss through a depth filter in the clean condition
rose tinted hydrogen	(jocular) A hydrogen type only ever seen with subsidy spectacles (qv hydrogen colors)
Rosenmund	Proprietary eponym for an agitated Nutsche (qv) filter-dryer
Rose's equation	see **Rose equation**
ROSOFF	A common misspelling of ROSoV (qv)
ROSoV	see **remotely operated shutoff valve**
Rotameter	Proprietary eponym for a variable area flowmeter (qv)
rotary airlock	see **rotary valve**
rotary air-to-air enthalpy wheel	see **rotary heat exchanger**
rotary atomizing burner	A burner which atomizes oil with centrifugal force imparted by a rotating element; defined formally in NFPA 86: Standard for Ovens and Furnaces, 2019 Edition
rotary blower pump	Can be used to mean any rotary positive displacement compressor (qv), such as a Roots blower (qv), especially in the context of a vacuum system (in which case also called a mechanical booster pump if used with a holding pump (qv)); defined formally in this sense in NFPA 86: Standard for Ovens and Furnaces, 2019 Edition
rotary damper	*aka* rotational damper. A device which dampens rotary movement; cf linear damper
rotary drum filter	see **drum filter**
rotary feeder	see **rotary valve**
rotary furnace	see **rotary kiln**
rotary gauge	In the context of an ASME container (qv) for LPG, a specific type of liquid level gauge with a dial, defined formally in NFPA 58: Liquefied Petroleum Gas Code, 2017 Edition

rotary heat exchanger *aka* rotary recovery wheel, thermal wheel, rotary air-to-air enthalpy wheel, energy recovery wheel, heat recovery wheel, enthalpy wheel, desiccant wheel, Kyoto wheel. A rotating cylindrical device in which a matrix heated by an outgoing stream is continuously rotated into a position to heat an incoming stream. As many of its alternative names imply, it is often used for waste heat recovery; see **Figure R2**.

rotary kiln *aka* calciner, rotary furnace. A rotating cylindrical furnace commonly used for calcining (qv) materials

rotary lobe pump *aka* lobe pump. A type of positive displacement pump (qv) with two or more multilobed rotating elements; defined formally in NFPA 20: Standard for the Installation of Stationary Pumps for Fire Protection, 2019 Edition

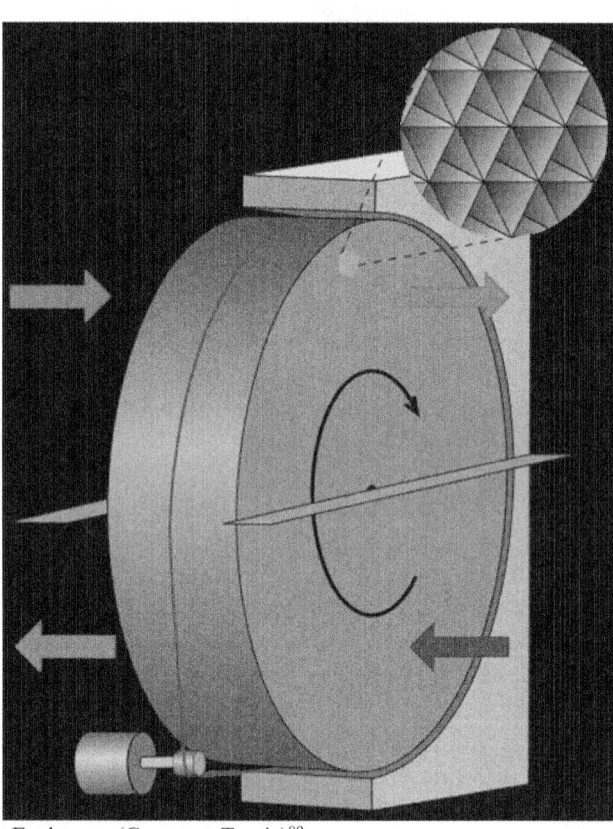

Figure R2 Rotary Heat Exchanger (Courtesy: Tomia)[89]

---

[89] Reproduced under CC BY 2.5 https://creativecommons.org/licenses/by/2.5/

rotary pump	One of a number of types of positive displacement pump (qv) with a rotating prime mover; cf rotodynamic pump
rotary recovery wheel	see **rotary heat exchanger**
rotary screw compressor	see **screw compressor**
rotary slips	see **slips**
rotary switch	see **rotational switch**
rotary table	*aka* rig table. In oil and gas drilling, the rotating part of the drill floor which drives a drilling string (qv)
rotary vacuum filter	see **vacuum filter**
rotary valve	1. A valve with a rotating plug or ball containing a passage for fluid 2. *aka* diverter, rotary feeder, rotary airlock. A metering valve for particulate solids
rotary vane pump	A type of positive displacement pump (qv) with a single multivaned rotating element; defined formally in NFPA 20: Standard for the Installation of Stationary Pumps for Fire Protection, 2019 Edition
rotating biological contactor	*aka* RBC. A type of simple attached growth (qv) biological sewage treatment device, commonly supplied as a packaged domestic wastewater treatment plant (qv), though may less commonly be used at large municipal scale
rotation test	*aka* bump test. A part of commissioning in which the direction of rotation of three-phase motors is established by a brief energization
rotational damper	see **rotary damper**
rotational force	see **torque**
rotational switch	*aka* rotary switch. 1. Generically, any switch operated by rotation 2. In the particular case of fans, a switch driven by the fan with a contact which closes at a set fan speed. Defined formally in NFPA 86: Standard for Ovens and Furnaces, 2019 Edition
rotodynamic pump	see **centrifugal pump**
rotor	A complete rotating assembly, as found e.g.; in a centrifugal pump (qv); defined formally in this case in EN ISO 13709:2003 Centrifugal pumps for petroleum, petrochemical and natural gas industries
rotor blade	*aka* bucket. In the context of gas turbines, a blade forming part of the rotor (qv); defined formally in ISO 3977 Gas turbines - Procurement - Part 3: Design requirements
rotor dynamics	The analysis of lateral and torsional perturbations in the motion of a turbine rotor; defined formally in ISO 3977 Gas turbines - Procurement - Part 3: Design requirements

rouge	A term used in the hygienic industries for certain types of surface discoloration of stainless steel; defined formally in ASME BPE (American Society of Mechanical Engineers: Bioprocessing Equipment)
rough order of estimate	see **gross order of magnitude estimate**
roughing	A preliminary pressure reduction prior to the use of a vacuum pump
roughing line	A line from a roughing pump to a vacuum system used for roughing (qv); defined formally in NFPA 86: Standard for Ovens and Furnaces, 2019 Edition
roughing pump	A pump used for roughing (qv); defined formally in NFPA 86: Standard for Ovens and Furnaces, 2019 Edition
roughing time	The time taken for roughing (qv) from atmospheric pressure to that required for the startup of a vacuum pump; defined formally in NFPA 86: Standard for Ovens and Furnaces, 2019 Edition
roughneck	*aka* floorhand. A supervisor of roustabouts (qv) on an oil rig, akin to a chargehand (qv)
roughness	A term with a range of meanings, which might be unhelpfully confused: 1. *aka* surface roughness. A deviation from smoothness (qv) of a surface, usually expressed as some function of surface height, such as Ra (qv) 2. *aka* hydraulic roughness. A measure of resistance to flow in channels, affected by many factors other the surface roughness of the channel wall; cf absolute roughness, relative roughness
rouging	Development of rouge (qv), especially on stainless steel
roundness	The sharpness of a particle's corners; N.B. not synonymous with sphericity (qv), though sometimes used in error as if it was
roustabout	An unskilled laborer on an oil rig
routine maintenance	*aka* scheduled maintenance. Keeping machinery in a safely operable state using routine, specified procedures; cf reactive maintenance
ROUV	Remotely operated underwater vehicle, a synonym of remotely operated vehicle (qv)
ROV	see **remotely operated vehicle**
ROW	see **right of way**
Royal Commission Standard	A venerable standard for effluent quality in the UK
RQ	see **respiratory quotient**
RR	see **red rubber**
RS	*(drawing notation)* Rising stem [valve]

RSF	see **remaining strength factor**
RSI	see **Ryznar stability index**
RT	1. Radiographic testing, according to API RP 571 - Damage Mechanism Affecting Fixed Refinery Equipment; N.B. API RP 579 - Fitness for Service uses RT as an abbreviation for radiographic examination
	2. Refrigeration ton, see **tons refrigeration**
	3. see **retention time**
RTC	see **real time control**
RTD	1. see **residence time distribution**
	2. see **resistance temperature detector**
RTE	*(drawing notation)* Reducing tee
RTFM	*(informal, offensive)* Read the fucking manual - a dismissive reply to a lazy query
RTJ(F)	*(drawing notation)* Ring type joint [flange]
RTO	see **regenerative thermal oxidizer**
rubber bible	see **rubber book**
rubber book	*(informal)* What older chemical engineers call the "CRC Handbook of Chemistry and Physics", a key reference source in the olden days before the internet
rubber duck	A UK site term for a wheeled (as opposed to tracked) crane, or other site vehicle
rubbish	see **municipal solid waste**
rule of thumb	Approximate guidance founded in practice
Rulon	Proprietary eponym for a metal filled gasket
run time	Operating time, especially the effective part of cycle time for a batch process
run to failure maintenance	see **reactive maintenance**
runaway	*aka* thermal runaway. An uncontrolled acceleration of the rate of a chemical reaction, especially an exothermic reaction promoted by increased temperature
rundown	Liquid (especially water) running down a surface due to gravity, momentum, or both; defined formally in NFPA 15 Standard for Water Spray Fixed Systems for Fire Protection
runout	*aka* circular runout. A deviation from circularity of a rotating part's cross section; defined formally in BS ISO 2041:2018 Mechanical vibration, shock and condition monitoring. Vocabulary

rupture allowable stress	(*symbol* sr) The stress allowed in a design to avoid failure by creep rupture; defined formally in API STD 530 - Calculation of Tube Heater Thickness and BS EN ISO 13704:2007 Petroleum, petrochemical and natural gas industries — Calculation of heater-tube thickness in petroleum refineries
rupture design pressure	The design maximum sustained operating pressure of a tube heater coil section; defined formally in API STD 530 - Calculation of Tube Heater Thickness and BS EN ISO 13704:2007 Petroleum, petrochemical and natural gas industries —Calculation of heater-tube thickness in petroleum refineries
rupture disc	see **rupture disk**
rupture disk	*aka* rupture disc, bursting disc, bursting disk. A sheet of (usually metallic) material, precision engineered to reliably and irreversibly rupture at a given differential pressure between faces, used as a pressure relief device (qv); defined formally in API RP 520 P1 7th Edition, January 2000 Sizing, Selection, and Installation of Pressure-Relieving Devices in Refineries; Part I - Sizing and Selection
rupture disk device	A non-reclosing PRD (qv) based on a rupture disk (qv); defined formally in API RP 520 P1 7th Edition, January 2000 Sizing, Selection, and Installation of Pressure-Relieving Devices in Refineries; Part I - Sizing and Selection and API Standard 521. Pressure-relieving and Depressuring Systems. Sixth Edition | January 2014
rupture disk holder	Just what it sounds like, but a formal definition is offered in API RP 520 P1 7th Edition, January 2000 Sizing, Selection, and Installation of Pressure-Relieving Devices in Refineries; Part I - Sizing and Selection
rupture exponent	A parameter used in tube heater design to avoid failure by creep rupture; defined formally in API STD 530 - Calculation of Tube Heater Thickness and BS EN ISO 13704:2007 Petroleum, petrochemical and natural gas industries — Calculation of heater-tube thickness in petroleum refineries
rupture pressure	see **elastic design pressure**
rural	A term used by the USA NFPA for inhabitable, settled areas with population densities lower than $500/mile^2$; defined formally in NFPA 1142: Standard on Water Supplies for Suburban and Rural Fire Fighting, 2017 Edition
Rushton turbine	A simple radial flow impeller
rusting	The wet oxidation of ferrous metal to ferric oxides
rutherford	(*symbol* Rd) A non-SI unit of radioactive decay

rutile	1. Strictly, a mineral comprised of a crystalline form of titanium dioxide
	2. Informally, all high titanium oxide content (>90%) ores
RV	see **relief valve**
RVP	see **Reid vapor pressure**
Ryznar stability index	*aka* RSI. An index of water 'aggressiveness', similar to the Langelier saturation index (qv)

S

S wave	see **shear wave**
S&T	Shell and tube, see **shell and tube heat exchanger**
S/x	Schedule x (of pipe or fittings), where x is either a number from 5-160, XXH (qv) or XXS (qv); see **schedule**
Sabatier process	A process for the catalytic production of methane from hydrogen and carbon dioxide
Sabatier-Senderens process	A process for the catalytic hydrogenation of organic compounds
SAC	Strong acid cation (in the context of ion exchange (qv) resin)
SAC exchange resin	see **strong acid cation resin**
SAC resin	see **strong acid cation resin**
saccharimeter	A polarimeter (qv) calibrated to measure the strength of sugar solutions
Sachse process	The production of ethyne from methane (some sources say LPG)
sacrificial protection	see **cathodic protection**
SADT	see **self accelerating decomposition temperature**
SAF	see **sustainable aviation fuel**
safe area	An area not classified as hazardous from the point of view of explosive atmospheres. Defined formally in ISO 3977 Gas turbines - Procurement - Part 3: Design requirements
safe design limit	see **upper safe design limit**
Safe Drinking Water Act	*aka* SDWA. A USA federal law under which the EPA (qv) is empowered to set drinking water quality standards
safe operating envelope	see **operating envelope**
safe start check	An inbuilt safeguard test preventing fired heater start-up in the apparent presence of a flame; defined formally in NFPA 86: Standard for Ovens and Furnaces, 2019 Edition
safe work method statement	*aka* SWMS; see **method statement**
safeguard	A device used for the control of hazards of equipment which cannot be made sufficiently inherently safe. The term is sometimes also used for procedures rather than devices with the same purpose. Defined formally in BS EN ISO 12100:2010 Safety of machinery. General principles for design. Risk assessment and risk reduction

safeguarding	A system of devices used for the control of hazards of equipment which cannot be made sufficiently inherently safe. The term is sometimes also used for procedures rather than devices with the same purpose. Defined formally in BS EN ISO 12100:2010 Safety of machinery. General principles for design. Risk assessment and risk reduction cf inherently safe design measure
safety	see **Box S1**.
safety audit	*aka* health and safety audit. A formal investigation of effectiveness of safety measures on a plant, especially procedural ones
safety blowout	In the context of fired heaters, a system of devices which shuts the system down and relieves pressure in the event of a flashback in the fuel/air mixture piping. Defined formally in NFPA 86: Standard for Ovens and Furnaces, 2019 Edition
safety cage	A permanent cage fixed to an access ladder to protect against the risk of falling. Defined formally in BS EN ISO 14122 Safety of machinery. Permanent means of access to machinery. Working platforms and walkways
safety can	A container holding no more than 20 l of flammable liquids; defined formally in NFPA 30: Flammable and Combustible Liquids Code, 2018 Edition, which also specifies appropriate fire safety measures
safety data sheet	see **material safety data sheet**
safety device	In the context of fired heaters, a term essentially identical in meaning to safeguard (qv). Defined formally in NFPA 86: Standard for Ovens and Furnaces, 2019 Edition
safety factor	*aka* factor of safety. According to DOE/NV/25946-1891, the ratio of ultimate load (or stress) to allowable load (or stress). A term commonly confused with the entirely different safety margin/ margin of safety (qv)
safety function	Any machine function leading to immediately increased risk of failure. Defined formally in BS EN ISO 12100:2010 Safety of machinery. General principles for design. Risk assessment and risk reduction; cf safety instrumented function
safety instrumented function	*aka* SIF. A safeguard (qv), or layer of protection, especially one which involves an electronic control loop. A term associated with IEC 61511 and 61508. Whether each SIF has an individual safety integrity level (qv) is sometimes a matter of contention

Box S1. Safety

In the first edition of this book, I identified "Risk" as a highly contested term, but that was just the beginning of a rabbithole.

> *"Many of the central terms in safety research are in need of clarification, such as "risk"..."precaution"..."safety factor"..."inherent safety"..."substitution" and "(safety) barrier".*
>
> ***Sven Ove Hansson, Safety is an inherently inconsistent concept (2012)***

It seems implausible that the profession simply just hasn't got around to defining its most fundamental terms yet. Most scholars think that "safety engineering" started with the Industrial Revolution, which was a while back now. Something is stopping them from agreeing on definitions.

My primary suspect is that it is something to do with the fact that "safety specialists" come from a wide range of backgrounds. There are graduate engineers, scientists, psychologists, an assortment of management/business studies types, even social workers. There are also quite a lot of non-graduates, often time-served, who may have moved into safety via NEBOSH-type qualifications.

This means that "safety specialists" are not one thing at all. They have different levels of education, ranging from around A level to PhD. Their disciplines have different rules for determining the truth of an argument, and different degrees of mathematical and linguistic skill are required of practitioners. Some of them love a bit of beard-stroking philosophising, some of them want to make the world a nicer, fairer place, but many just want to get the boxes ticked on the form so that the job can go ahead.

So, it would appear that the reason why there is no agreed lexicon of "safety" is that there is no such thing as a safety specialist. Each discipline within the field (especially the more numerate ones) could probably agree a lexicon if left to themselves, but the existence of the other - seemingly equivalent - disciplines muddies the waters. The least numerate disciplines generate junk research which makes it possible to cite "scientific references" for wrong-headed views. Some from the more numerate disciplines might then latch on to this, both because verbal reasoning is not their strength, and because they have a personal preference for whatever it is that the humanities / business / management studies types are peddling.

I can understand the basic feeling of resentment that many working in safety have. All too often, no-one likes them, or takes them seriously. They are widely seen as at best a necessary evil, as the annoying little externally imposed jobsworth with a clipboard, getting in the way of the job getting done. It must be awful being that person, but don't squabble amongst yourselves on LinkedIn about stuff no-one else cares about. It's just playing to the stereotype.

As for me, I'm going to define these terms when I use them as chemical engineers do, as this is a dictionary of chemical engineering. Just be aware that you may come across people who don't share your definitions.

safety instrumented system	*aka* SIS, safety interlock system, emergency shutdown system, ESD, safety shutdown system, SSD. This particular term is associated with IEC 61511 and 61508 and comprises a number of safety instrumented functions (qv)
safety integrity level	*aka* SIL. The integrity level of a safety instrumented function (qv), on a scale of 1-4, as determined by a methodology from IEC 61508/61511. This particular term is also defined formally in API Standard 521. Pressure-relieving and Depressuring Systems. Sixth Edition	January 2014
safety interlock	An interlocking device (qv) which prevents unsafe operation of equipment; defined formally in NFPA 86: Standard for Ovens and Furnaces, 2019 Edition
safety interlock system	A system of sensors, controllers and actuators which automatically takes a process to a safe state; defined formally in API Standard 521. Pressure-relieving and Depressuring Systems. Sixth Edition	January 2014
safety management system	*aka* SMS. A business administration system which controls occupational health and safety such as BS OHSAS 18001 or ISO 45001
safety margin	see **margin of safety**
safety measure	A means to reduce risk; defined formally in BS EN 1050:1997 Safety of machinery. Principles for risk assessment
safety note	A document (or part thereof) containing safety information; defined formally in BS EN 82079-1:2012 Preparation of instructions for use. Structuring, content and presentation. General principles and detailed requirements
safety relay	A relay with special features to increase its reliability appropriately for a safety duty; defined formally in NFPA 86: Standard for Ovens and Furnaces, 2019 Edition
safety relief valve	*aka* SRV. A spring loaded PRV (qv), though there is a formal definition of this particular term in API RP 520 P1 7th Edition, January 2000 Sizing, Selection, and Installation of Pressure-Relieving Devices in Refineries; Part I - Sizing and Selection
safety shutdown	In the context of fired heaters, an automatic shutdown initiated by safety controls and requiring manual intervention to restart; defined formally in NFPA 86: Standard for Ovens and Furnaces, 2019 Edition
safety shutdown system	*aka* SSD. See **safety instrumented system**

safety shutoff device	*aka* safety trip valve. In the context of fired heaters, an automatic fuel supply cutoff valve; defined formally in BS EN 12952-8:2002 Water-tube boilers and auxiliary installations. Requirements for firing systems for liquid and gaseous fuels for the boiler; cf safety shutoff valve
safety shutoff valve	In the context of fired heaters, a normally closed automatic valve in fuel, air or oxygen supply piping, used to shut down a fired heater; defined formally in NFPA 86: Standard for Ovens and Furnaces, 2019 Edition; cf safety shutoff device
safety square	see **fire diamond**
safety time	*aka* process safety time. In a process control context, the delay between a failure and an associated hazardous event. Process safety time is defined formally in BS EN 12952-8:2002 Water-tube boilers and auxiliary installations. Requirements for firing systems for liquid and gaseous fuels for the boiler and IEC 61508-2
safety trip valve	see **safety shutoff device**
safety valve	*aka* safety relief valve. A spring loaded PRV (qv); though this particular term is defined formally in API RP 520 P1 7th Edition, January 2000 Sizing, Selection, and Installation of Pressure-Relieving Devices in Refineries; Part I - Sizing and Selection and API Standard 521. Pressure-relieving and Depressuring Systems. Sixth Edition | January 2014
safety ventilation	Ventilation which dilutes a class A furnace (qv) atmosphere below the permitted fraction of the LFL (qv); defined formally in NFPA 86: Standard for Ovens and Furnaces, 2019 Edition
safety, health, environment, quality	*aka* SHEQ; see **quality, environment, safety, health**
SAFT	see **statistical associating fluid theory**
SAG	see **semi-autogenous grinding**
SAG mill	see **semi-autogenous grinding mill**
sailaway	The setting to sea of a constructed oil or gas rig
salt fractionation	see **salting out**
salt splitting	The decomposition of carboxylic acid salts into corresponding acids and bases via ion exchange (qv) or electrolysis (qv)
saltation velocity	The minimum superficial gas velocity in a horizontal gas-solid flow required to keep particles in suspension
salt-induced precipitation	see **salting out**

salting out	*aka* anti-solvent crystallization, drowning out, precipitation crystallization, salt fractionation, salt-induced precipitation. The precipitation of specific macromolecules such as DNA and protein in an undenatured form from mixtures using high ionic strength solutions
sample	A part of something intended to represent the whole, or the action of obtaining this part
sample point	A place where a sample can be collected
sampling interval	The units between successive samples. Where units are time, synonymous with sampling period (qv). Defined formally in BS ISO 2041:2018 Mechanical vibration, shock and condition monitoring. Vocabulary
sampling period	*aka* sampling period. The time between successive samples; defined formally in BS ISO 2041:2018 Mechanical vibration, shock and condition monitoring. Vocabulary
sanction	1. Permission to proceed to the next stage of design, usually with a formal form of contract in place and accompanying promises of payment 2. (especially in the UK) Preventing something from happening (more or less the opposite of meaning 1.)
sand filtration	The filtration, usually batchwise, of suspended solids from water by a bed of sand; a subset of depth filtration; cf depth filter
Sandpiper pump	Proprietary eponym for an air operated double diaphragm pump; cf Wilden pump
Sandvik 2205	A grade of duplex stainless steel with very high mechanical strength, good weldability and high resistance to many forms of corrosion and erosion
Sandvik 254SMO	A grade of high-alloy austenitic stainless steel with high mechanical strength, good weldability and excellent resistance to many forms of corrosion and erosion, especially from chloride
sandwich molding	see co-**injection molding**
sanitary	Synonymous with hygienic (qv), according to Americans in general and ASME BPE (American Society of Mechanical Engineers: Bioprocessing Equipment) in particular. However, see **Box C2**.
sanitary sewer	see **foul sewer**
sanitary weld	*aka* hygienic weld. A weld without any defects which might interfere with maintenance of a clean condition. Prescriptive formal definition in ASME BPE (American Society of Mechanical Engineers: Bioprocessing Equipment)

sanitation	"Cleaning, disinfection if necessary, pest control and waste management. Sometimes used in place of hygiene" (qv), according to the EHEDG Glossary. However, see **Box C2**.
sanitisation	see **sanitizing**
sanitization	see **sanitizing**
sanitizer	"A substance that reduces the microbial contaminants on inanimate surfaces to levels that are considered safe for public health. According to the official food contact surface sanitizer test, a sanitizer is a chemical that reduces the microbial contamination of two standard organisms, Staphylococcus aureus and Escherichia coli, by 99.999% or 5 logs in 30 seconds, at 25° C. Non-food contact sanitizers must reduce contamination by 99.9% or 3 logs in 5 minutes", according to EHEDG Glossary. To put it another way, almost identical to disinfectant (qv)
sanitizing	*aka* sanitising, sanitizing. "A process applied to a cleaned surface capable of reducing the numbers of the most resistant human pathogens by at least 5 log 10" according to EHEDG Glossary. Or to put it another way, disinfection (qv); see **Box C2**.
saponification	The hydrolysis of triglycerides, as in soapmaking
SARA	see **saturates, aromatics, resins and asphaltenes**
SAS	Surplus activated sludge, see **waste activated sludge**
SAT	see **site acceptance test**
SATP	see **standard ambient temperature and pressure**
saturated	1. A term applied to the maximum amount of a solute which can be dissolved in a solvent under given conditions (saturated solution) 2. The amount of multiple bonds in a hydrocarbon chain (saturated/unsaturated hydrocarbon) 3. *(informal)* the maximum loading on an adsorbent, e.g.; in gas separation 4. A term applied to the point at which gas-phase vapor pressure of a substance is equal to that at equilibrium with a flat surface of the liquid substance at a given temperature 5. By extension of the last meaning, saturated steam (qv) is water vapor in equilibrium with liquid water
saturated steam	Steam in equilibrium with liquid water, commonly called dry steam (qv); cf superheated steam
saturated vapor pressure	Pressure exerted by the vapor of a substance in equilibrium with its liquid form

saturates, aromatics, resins and asphaltenes	*aka* SARA. The characterization of crude oil (qv) by fractionation (qv) and quantification of components based on polarizability and polarity
saturation coefficient	1. In civil engineering contexts, the amount of water a brick absorbs
	2. In the sugar beet industry, synonymous with what the rest of the sugar industry calls the solubility coefficient (qv)
saturation level	The saturation concentration of a solute, above which it comes out of solution
saturation temperature	A liquid's boiling point (qv)
Saunders valve	Proprietary eponym for diaphragm valve (qv)
Sauter mean diameter	A method used to characterize average particle size
SAW	Submerged arc weld
Saybolt color	A color (qv) measurement scale used to characterize petroleum fuels; defined formally in ASTM D156 Standard Test Method for Saybolt Color of Petroleum Products (Saybolt Chromometer Method)
Saybolt universal second	An obsolete unit of kinematic viscosity (qv)
SBA	Strong base anion (in the context of ion exchange resin)
SBA exchange resin	see **strong base anion resin**
SBA resin	see **strong base anion resin**
SBR	1. see **sequencing batch reactor**
	2. see **styrene butadiene rubber**
SBS	Sodium bisulfite
SC	*(drawing notation)* Sample connections
Sc	see **Schmidt number**
SCADA	see **supervisory control and data acquisition**
scale	1. Mineral deposits built up on a surface
	2. *aka* scale factor, scaling factor. The ratio of linear dimension on a drawing to its real world equivalent; this meaning defined formally in BS EN ISO 10209:2012 Technical product documentation - Vocabulary - Part 1: Terms relating to technical drawings: general and types of drawings.
scale drawing	A drawing, the dimensions of which are consistently in some ratio of the size of their corresponding real-world counterparts
scale factor	1. Synonymous with scale (qv)
	2. In the case of instrumentation, the sensitivity in a specified range of frequency, as defined formally in BS ISO 2041:2018 Mechanical vibration, shock and condition monitoring. Vocabulary
scaled distance	A technique to predict blast effects based on size of explosive and distance from explosion

scaleout	Making a unit operation or process larger by increasing the number of units
scaleup	Making a unit operation or process larger by increasing the size of units
scaling	1. An autoantonym, used to mean both the buildup of mineral deposits or corrosion (scale) on a surface and the removal of those deposits; cf **descaling** 2. The process of translating from a scale drawing to real world dimensions 3. see **scaleup**
scaling factor	*aka* scale. The ratio of linear dimension on a drawing to its real world equivalent. Defined formally in BS EN ISO 10209:2012 Technical product documentation - Vocabulary - Part 1: Terms relating to technical drawings: general and types of drawings.
scalping	The removal of oversize particles
Scanima	Proprietary eponym for a vacuum high shear mixer used in the food industry, widely known as a Scanima after the original brand name. The technology's new owner tried rebranding it to 'Almix' but it didn't stick; now officially called a high shear vacuum mixer
scavenger	A unit operation or material used to specifically remove certain types of contaminants, especially organics or oxygen
SCBA	see **self-contained breathing apparatus**
SCC	see **stress corrosion cracking**
SCEs	Safety critical elements
SCF	see **standard cubic foot**
SCFM	see **standard cubic feet per minute**
schedule	1. see **project program** 2. *aka* pipe schedule, schedule number. A dimensional pipe standard; as defined formally in ASME/ANSI B 36.10 Welded and Seamless Wrought Steel Pipe, ASME/ANSI B36.19 Stainless Steel Pipe and ASME BPE (American Society of Mechanical Engineers: Bioprocessing Equipment)
schedule number	see **schedule**
scheduled maintenance	see **routine maintenance**
Schmidt number	*aka* Sc. A dimensionless group, the ratio of kinematic viscosity to mass diffusivity, used in analysis of convective mass transfer

schmutzdecke	The thin layer on top of a slow sand filter (qv) which is the effective filter medium; see **Figure S1**
Schoepentoeter	Proprietary eponym for Shell's patented feed inlet vane device; see **Figure S2**
scientific wild ass guess	*aka* SWAG. Any experience-based plug number (qv) used in the absence of real data when estimating costs
scooter	A kind of cursor on a SCADA (qv) trend screen which allows values to be tracked
scoping	Identifying the required content and extent of a further study
scoping study	A design/project planning stage that typically delivers a project capital estimate, typically AACE (qv) class 4 ~ +/- 30%
SCR	see **selective catalytic reduction**
scram	Emergency shutdown of a nuclear reactor; not, as some think, an acronym for "safety control rod axe man". It means "scram" as in "run away" and is thus a backronym, not an acronym
scrap	A nonconforming product or service, or the action of discarding or removing these from service (scrapping). Also used for materials from manufacturing destined for reprocessing (or possibly rework (qv)), especially metals

Figure S1 Schmutzdecke (courtesy: Olov Eriksson)[90]

---

[90] Reproduced under Creative Commons CC BY-SA 4.0, https://creativecommons.org/licenses/by-sa/4.0/

Figure S2 Schoepentoeter (courtesy: Sulzer)

scraped surface heat exchanger	*aka* SSHE. A double pipe heat exchanger with a scraped heat exchange surface used for viscous or crystallizing fluids
scraper bridge	see **bridge scraper**
scratch	An elongated surface incision. Defined formally in ASME BPE (American Society of Mechanical Engineers: Bioprocessing Equipment)
screen dump	An image from a complete computer screen, commonly used to produce a hard copy (qv)
screening	Filtering out solids from a fluid with a mesh or grating, often using successively finer filter media to produce multiple size graded outputs; or by analogy, filtering out options in an analysis
screw	1. *aka* scroll, worm. A helical device within a cylindrical housing (such as the bowl of a decanter centrifuge) which rotates to drive solids to a discharge point 2. An externally threaded fastener, defined formally in ASME B18.2.1 - Square and Hex Bolts and Screws, Inch Series
screw compressor	*aka* rotary screw compressor, scroll compressor. A rotary positive displacement gas compressor
screw conveyer	*aka* auger. A device that uses a rotating screw to move solids
screw pump	Used variously to mean a progressing cavity pump (qv) or Archimedean screw pump (qv)
scroll	see **screw**

scroll compressor	see **screw compressor**
scrubber	*aka* scrubbing column. A gas absorption column
scrubbing	The removal of an undesirable gas or dust component from a gas stream with a scrubber (qv)
scrubbing column	see **scrubber**
scum	Floating impurities in water treatment, often containing fats oils and greases (qv)
SCWO	see **supercritical water oxidation**
SD&P	see **simultaneous drilling and production**
SDoF system	see **single degree of freedom system**
SDR	see **standard dimension ratio**
SDS	Safety data sheet; see **material safety data sheet**
SDWA	see **Safe Drinking Water Act**
seal	1. The set of components used to close an aperture, as for example defined formally in API RP 682 - Pump Seals 2. Alternatively, to close an aperture against the passage of contamination or loss of fluid with a seal, as defined formally in BS EN 1672-2:2005+A1:2009 Food processing machinery. Basic concepts. Hygiene requirements and BS EN ISO 14159:2008 Safety of machinery. Hygiene requirements for the design of machinery
seal balance ratio	For pump seals, a function of the ratio of pressurized face area to total seal area. A balanced seal (qv) has a ratio of 1:1. Full formal definition in API RP 682 - Pump Seals
seal chamber	A stuffing box or other seal component which lies between a rotating component such as a pump shaft and the casing. Defined formally in API RP 682 - Pump Seals and ASME BPE (American Society of Mechanical Engineers: Bioprocessing Equipment)
seal face	A seal or mating ring surface which forms a sealing surface; defined formally in API RP 682 - Pump Seals and ASME BPE (American Society of Mechanical Engineers: Bioprocessing Equipment)
seal oil consoles	Compressor ancillaries which provide a supply of oil to the hydraulic seals at the ends of a compressor shaft
seal point	1. The point at which contact between components creates a boundary between process fluid and surroundings. Defined formally in ASME BPE (American Society of Mechanical Engineers: Bioprocessing Equipment) 2. A term used in distillation for the point at which "weeping" changes into "raining"

seal pot	*aka* condensate pots. A liquid reservoir used for various purposes in pump seals, instrumentation and for the collection of condensed vapors
seal ring	A flexibly mounted counterpart to the mating ring (qv) in a pump seal. Defined formally in API RP 682 - Pump Seals
seal weld	A weld which seals in a fluid. Defined formally in ASME BPE (American Society of Mechanical Engineers: Bioprocessing Equipment)
seal welded	In the context of heat exchangers, sealed by means of a tube to tubesheet seal weld (qv). Defined formally in API 660 - Shell-and-Tube Heat Exchangers and API RP 661 - Heat Exchangers
sealing thread	see **national pipe thread**
seat leakage	A quantity of fluid passing through a closed valve under test conditions. Defined formally in ASME BPE (American Society of Mechanical Engineers: Bioprocessing Equipment)
SEC	see **specific energy consumption**
second	1. The base SI unit of time, or one of a number of obsolete units of viscosity such as the Saybolt universal second (qv) 2. 1/60th of a degree of rotation
second generation biofuel	*aka* advanced biofuel. A biofuel (qv) manufactured from non-food biomass (qv)
second law of thermodynamics	The second law of thermodynamics is that entropy never decreases in an isolated system[91]; cf first law of thermodynamics, third law of thermodynamics
second stage regulator	A vapor service LPG pressure regulator in between the first stage regulator (qv) outlet and the delivery point. Defined formally in NFPA 58: Liquefied Petroleum Gas Code, 2017 Edition
secondary air	In the context of fired heaters, combustion air supplied in addition to primary air (qv). Defined formally in BS EN ISO 13705:2012/ISO 13705:2012(E) Petroleum, petrochemical and natural gas industries. Fired heaters for general refinery service, API STD 530 - Calculation of Tube Heater Thickness, API STD 560 - Fired Heaters for General Refinery Service, NFPA 86: Standard for Ovens and Furnaces, 2019 Edition and API RP 535 - Burners for Fired Heaters at Refineries
secondary containment	In the context of loss prevention, a layer of containment around primary containment, including bunding, drip trays, toe walls etc.; cf primary containment, tertiary containment

---

[91] Or, to put it another way, 'the most you can accomplish by work is to break even'

secondary containment tank	A double walled tank, ideally with interstitial space leak monitoring; defined formally in NFPA 30: Flammable and Combustible Liquids Code, 2018 Edition
secondary explosion	An explosion of a dust cloud created when accumulated dust is disturbed by a primary explosion; cf primary explosion, dust explosion
secondary fuel	Additional fuel injected downstream of the burner block (qv) in a staged fuel burner (qv). Defined formally in API RP 535 - Burners for Fired Heaters at Refineries
secondary grade of release	A concept in hazardous area classification: a condition in which release of hazardous material is not to be expected in normal operation. Defined formally in API RP 505 2nd Edition, August 2018 Recommended Practice for Classification of Locations for Electrical Installations at Petroleum Facilities Classified as Class I, Zone 0, Zone 1, and Zone 2 and IEC 61892-7, Mobile and fixed offshore units – Electrical installations – Part 7: Hazardous areas; cf continuous grade of release, primary grade of release
secondary oil	see **slop oil**
secondary seal	A seal around another seal. Defined formally in API RP 682 - Pump Seals
secondary settlement	*aka* final settlement. In secondary/biological sewage treatment, especially activated sludge treatment, the settlement of solids from a bioreactor
secondary stress	A classification of self-limiting stress, often as developed by constraint. Defined formally in API RP 579 - Fitness for Service; cf primary stress
secondary treatment	The biological treatment of sewage and any associated secondary settlement (qv)
secondary wave	see **shear wave**
SECp	Specific energy consumption (qv) based on primary energy input
section	1. Part of a plant which can be operated independently (plant section) 2. see **sectional view** Both senses defined formally in BS EN ISO 10209:2012 Technical product documentation - Vocabulary - Part 1: Terms relating to technical drawings: general and types of drawings
sectional view	*aka* section. A drawing of an object as if cut through; defined formally in BS EN ISO 10209:2012 Technical product documentation - Vocabulary - Part 1: Terms relating to technical drawings: general and types of drawings

sectionalizing block valve	see **mainline valve**
sedimentation	*aka* gravity separation. The settlement of solids in a fluid under gravity or centrifugal force
sedimentation basin	*aka* sedimentation tank; see **settling tank**
sedimentation tank	see **settling tank**
seed crystals	Small crystals of a solute which act as nucleation sites for the formation of crystals from supersaturated solution
seed train	A set of small fermentation vessels used to grow a biological culture from small storage volumes to sufficient quantity to inoculate an industrial scale bioreactor
seeding	The act of adding an initiator (qv) to start a change of phase (qv) - usually in crystallization, but also for condensation. Seeding may be accomplished with impurities or with the material undergoing phase change
segregation	Managing hazards by using physical barriers between hazards and the protected area. Defined formally in NFPA 654: Standard for the Prevention of Fire and Dust Explosions from the Manufacturing, Processing, and Handling of Combustible Particulate Solids, 2017 Edition
segregation coefficient	*aka* equilibrium segregation coefficient. The ratio of the concentration of an impurity in a solid to that in a liquid with which it is in equilibrium
Seider-Tate equation	An equation used for estimating the heat transfer coefficients for flow in tubes
selective catalytic reduction	*aka* SCR. A flue gas NOx removal process which uses a catalyst and a dosed additive such as ammonia or urea; cf selective non-catalytic reduction
selective medium	A microbial medium which discourages or prevents the growth of undesired microorganisms, whilst encouraging the growth of desired microorganisms
selective non-catalytic reduction	*aka* SNCR. A flue gas NOx removal process which uses a dosed additive such as ammonia or urea, but no catalyst; cf selective catalytic reduction
selectivity	1. In the context of ion exchange, a difference in attraction for one ion over another 2. More generally, to chemists, overall selectivity (qv) is most commonly the molar ratio of desired product to undesired product for a chemical reaction. They however also use selectivity to mean other things, most unhelpfully as a synonym for yield (qv)

selector zone	A zone used to selectively culture denitrifying organisms in an activated sludge (qv) plant
Selexol	A proprietary DEPG (qv) mixture used as a physical solvent for gas sweetening (qv); cf Rectisol
self accelerating decomposition temperature	*aka* SADT. The lowest temperature at which the temperature-dependent decomposition of a substance (such as an organic peroxide) held in a container exceeds the heat lost to surroundings, increasing the substance's temperature and hence rate of decomposition
self-closing gate	A pivoting part of a guard-rail used for access. Defined formally in BS EN ISO 14122 Safety of machinery. Permanent means of access to machinery. Working platforms and walkways
self-contained breathing apparatus	*aka* SCBA, compressed air breathing apparatus (CABA), breathing apparatus set (BA set). A device worn by rescue workers, firefighters, and others to provide breathable air for a duration greater than 30 minutes in an immediately dangerous to life or health (IDLH) atmosphere
selfdraining	Designed and constructed in such a way as to passively eliminate all liquid from a system, or that property of a system so designed and constructed. Three different formal definitions may be found in BS EN 1672-2:2005+A1:2009 Food processing machinery. Basic concepts. Hygiene requirements, ASME BPE (American Society of Mechanical Engineers: Bioprocessing Equipment) and BS EN ISO 14159:2008 Safety of machinery. Hygiene requirements for the design of machinery cf gravity flow
selfeducting nozzle	A nozzle incorporating an eductor (qv) to feed it with foam concentrate from supply. Defined formally in NFPA 11: Standard for Low-, Medium-, and High-Expansion Foam, 2016 Edition
self-elevating unit	see **jack up rig**
selfexcited vibration	see **selfinduced vibration**
selfextinguishing	Incapable of sustaining combustion after an initiating flame is removed
selfheating	see **spontaneous heating**
selfignition	Sometimes defined as ignition without external heat energy supply, and distinct from (cf) autoignition and (cf) spontaneous ignition
selfinduced vibration	System vibration resulting from conversion of non-oscillatory energy to an oscillatory form. Formal definition in BS ISO 2041:2018 Mechanical vibration, shock and condition monitoring. Vocabulary

selfpiloted burner	A fired heater burner with integrated pilot (qv). Defined formally in NFPA 86: Standard for Ovens and Furnaces, 2019 Edition
selfregulating control	The speed and power controller of a selfregulating variable speed fire pump unit (qv). Defined formally in NFPA 20: Standard for the Installation of Stationary Pumps for Fire Protection, 2019 Edition
selfregulating variable speed fire pump unit	A packaged unit of fire pump, motor, and selfregulating control (qv). Defined formally in NFPA 20: Standard for the Installation of Stationary Pumps for Fire Protection, 2019 Edition
selfregulation	1. The regulation of engineers by professional engineering bodies[92] 2. One of a number of types of closed loop control (qv) of pump output, most commonly used to automatically prevent cavitation (qv)
selfsustaining speed	The minimum normal operational steady state gas turbine rotor speed. Defined formally in ISO 3977 Gas turbines - Procurement - Part 3: Design requirements
selfvaporizing liquid burner	see **vaporizing burner**
SEM	Scanning electron microscope
semiannular flow	see **churn flow**
semi-autogenous grinding	*aka* SAG. A primary stage of ore preparation in a semi-autogenous grinding mill (qv)
semi-autogenous grinding mill	*aka* SAG mill. An ore mill containing balls, like a ball mill (qv), but with the ore as a continuous phase with interstitial balls, and a significant amount of grinding being done by the ore being ground; used for preliminary grinding of minerals
semiautomatic arc welding	Arc welding (qv) with automatic control of filler metal feed only. Defined formally in API STD 620 - Low Pressure Storage Tanks and ASME BPE (American Society of Mechanical Engineers: Bioprocessing Equipment)
semiautomatic burners	Manually initiated burners with automatic ignition, safety monitoring and control devices. Defined formally in BS EN 12952-8:2002 Water-tube boilers and auxiliary installations. Requirements for firing systems for liquid and gaseous fuels for the boiler

---

[92] Professional engineering bodies may not in practice themselves be controlled by engineers, arguably making the 'self' part of the term selfregulation questionable)

semiautomatic shutdown	A partly manually controlled or initiated shutdown. Defined formally in the context of gas turbines in ISO 3977 Gas turbines - Procurement - Part 3: Design requirements
semibatch process	A batch process in which further reactants are added after an initial charge, and product may also be withdrawn periodically; cf fed batch process
semicombined CO_2	*aka* half-bound CO_2. In the context of carbonate buffering, a term for CO_2 bound in hydrogen carbonates/bicarbonates; cf free CO_2, combined CO_2
semicontinuous process	A process with a discrete period of continuous feed introduction and product discharge, popular in pharmaceutical production; cf semibatch process, campaign manufacture
semifixed system	A firefighting foam system with fixed discharge outlets adjacent to a hazard, piped to a connection point a safe distance away. Defined formally in NFPA 11: Standard for Low-, Medium-, and High-Expansion Foam, 2016 Edition
semi-infinite	Infinite only in certain dimensions
semipermeable membrane	A membrane with selective permeability, such as a reverse osmosis (qv) membrane
semisubsurface foam injection	The introduction of firefighting foam onto the surface of a storage tank via a floating hose connected to an external foam supply. Defined formally in NFPA 11: Standard for Low-, Medium-, and High-Expansion Foam, 2016 Edition
sensible heat	Added heat which changes temperature, rather than phase; cf latent heat
sensing element	*aka* sensor. A device in 'contact' with a process that generates a signal of some sort for control and/or information purposes. Defined formally in BS ISO 2041:2018 Mechanical vibration, shock and condition monitoring. Vocabulary and BS 1646-2:1983 Symbolic representation for process measurement control functions and instrumentation. Specification for additional basic requirements
sensitive axis	The direction in which a rectilinear transducer (qv) has maximum sensitivity. Defined formally in BS ISO 2041:2018 Mechanical vibration, shock and condition monitoring. Vocabulary
sensitive protective equipment	*aka* SPE, electrosensitive protective equipment (ESPE). Safety equipment which detects the presence of a person in a hazardous area; defined formally in BS EN ISO 12100:2010 Safety of machinery. General principles for design. Risk assessment and risk reduction

sensitivity	The ratio of output to input quantities. Defined formally in BS ISO 2041:2018 Mechanical vibration, shock and condition monitoring. Vocabulary
sensitivity analysis	A systematic procedure of varying inputs to a model to see how large an effect it has on the outputs, for the purposes of establishing, e.g.; robustness of a design. Defined formally in two different contexts in EN ISO 14040:2006 Environmental management - Life cycle assessment - Principles and framework and API RP 579 - Fitness for Service
sensitivity check	Verification that sensitivity analysis output is relevant in a particular context. Defined formally in EN ISO 14040:2006 Environmental management - Life cycle assessment - Principles and framework
sensitization	A type of damage leading to a reduction in robustness. Most commonly applied to human skin which has been made susceptible to contact dermatitis, but also to alloys which have been made susceptible to stress corrosion cracking as in API RP 932B Corrosion Air Coolers; cf sensitivity
sensor	see **sensing element**. However, this particular term is defined formally in BS EN ISO 14159:2008 Safety of machinery. Hygiene requirements for the design of machinery
separation	1. The partitioning of mixtures of substances into at least two streams, at least one of which is richer in a desired product 2. Allowing sufficient space in plant layout to avoid a domino effect (qv), as defined formally in NFPA 654: Standard for the Prevention of Fire and Dust Explosions from the Manufacturing, Processing, and Handling of Combustible Particulate Solids, 2017 Edition
separation margin	The separation between critical speed (qv) and the closest operating speed. Defined formally in EN ISO 10437 Petroleum, petrochemical and natural gas industries - Steam turbines - Special-purpose applications

separator	1. Generically, a unit operation which takes in a stream of mixed components and produces at least one 'rich' stream enriched in at least one component, and at least one 'lean' stream with less of that component, especially in the case where gravity or centrifugal forces drive separation 2. Sometimes used for some kinds of demister (qv) 3. In the oil and gas industry, applied only to a number of types of pressurized gravity separator, separating well fluids into various streams of oil, water and gas, including but not limited to the test separator (qv); cf production separator 4. In the dairy industry, shorthand for a centrifugal separator (qv)
separator drum	see **knockout drum**
sequencing batch reactor	*aka* SBR. A semi-batch variant on the activated sludge process (qv)
sequestration	1. The removal of something from a system, for example carbon sequestration 2. May also be used (especially by chemists) as a synonym of chelation
serial production	The manufacture, in a specified way, of a series of identical items. Defined formally in the context of pressure vessels in PD 5500:2018+A1:2018 Specification for unfired fusion welded pressure vessels
series	*aka* in series. A term for an arrangement of equipment with units mounted such that one feeds the other; cf parallel
series fire pump unit	Fire pump units within one building and arranged directly in series (qv). Defined formally in NFPA 20: Standard for the Installation of Stationary Pumps for Fire Protection, 2019 Edition
series pressure regulator	One of a number of pressure regulators arranged in series (qv). Defined formally in the context of fired heaters in NFPA 86: Standard for Ovens and Furnaces, 2019 Edition
serum free medium	see **defined medium**
service	1. An activity carried out on behalf of another, (usually on a commercial basis); defined formally in BS EN 82079-1:2012 Preparation of instructions for use. Structuring, content and presentation. General principles and detailed requirements and BS EN ISO 9000 Quality management systems Fundamentals and vocabulary 2. A process utility; defined formally in NFPA 20: Standard for the Installation of Stationary Pumps for Fire Protection, 2019 Edition

service condition	In the context of pump seals, the range of service temperatures and pressures. Defined formally in API RP 682 - Pump Seals; though the definition may be much wider in other contexts
service equipment	In the context of fire pumps, electrical switchgear. Defined formally in NFPA 20: Standard for the Installation of Stationary Pumps for Fire Protection, 2019 Edition
service factor	The ratio of the permissible maximum power loading for an AC motor to the rated power under specified conditions. Defined formally in NFPA 20: Standard for the Installation of Stationary Pumps for Fire Protection, 2019 Edition
service floor	see **interstitial space**
service head adapter	A type of steel-to-plastic transition fitting used in hydrocarbon gas service. Defined formally in NFPA 58: Liquefied Petroleum Gas Code, 2017 Edition
service life	The time or number of cycles for which a component will maintain specified performance in service. Defined formally in ISO 3977 Gas turbines - Procurement - Part 3: Design requirements
service pressure regulator	A gas supplier's pressure regulator, limiting a service line to a specified delivery pressure. Defined formally in NFPA 86: Standard for Ovens and Furnaces, 2019 Edition
services	see **utilities**
set point	*aka* setpoint; see **set value**
set pressure	1. Generically, any set value (qv) for pressure, defined identically in API RP 520 P1 7th Edition, January 2000 Sizing, Selection, and Installation of Pressure-Relieving Devices in Refineries; Part I - Sizing and Selection, API Standard 521. Pressure-relieving and Depressuring Systems. Sixth Edition | January 2014 and API RP 576 - Inspection of Pressure Relieving Devices 2. The inlet gauge pressure at which a PRV (qv) is designed to lift; cf pop pressure 3. The setpoint for a variable speed pressure limiting control (qv) system; defined formally in NFPA 20: Standard for the Installation of Stationary Pumps for Fire Protection, 2019 Edition

set value	*aka* setpoint, set point. The value of a controlled parameter which the control system is set to achieve. Defined formally in BS1646-1 Symbolic Representation for Process Measurement Control Functions and Instrumentation Part 1: Basic Requirements and BS 1646-3:1984 Symbolic representation for process measurement control functions and instrumentation. Specification for detailed symbols for instrument interconnection diagrams
set-in branch	see **stub-in tee**
set-in tee	A misnomer for a stub-in tee (qv). There is however a set-in branch, corresponding to a stub-in tee
set-on branch	see **stub-on tee**
set-on tee	A misnomer for a stub-on tee (qv). There is however a set-on branch, corresponding to a stub-on tee
setpoint	see **set value**
setting	In the context of fired heaters, the refractory, insulating and structural containment. Defined formally in BS EN ISO 13705:2012/ISO 13705:2012(E) Petroleum, petrochemical and natural gas industries. Fired heaters for general refinery service, API STD 530 - Calculation of Tube Heater Thickness and API STD 560 - Fired Heaters for General Refinery Service
setting loss	see **radiation loss**
setting out drawing	A drawing annotated to define the position and level of construction elements for field use, especially by civil engineers. Defined formally in BS EN ISO 10209:2012 Technical product documentation - Vocabulary - Part 1: Terms relating to technical drawings: general and types of drawings
settlement tank	see **settling tank**
settler	see **settling tank**
settling tank	*aka* settlement tank, settler, sedimentation tank, sedimentation basin, clarifier. A tank in which suspended solids are settled from liquid (almost always water) under gravity
settling time	The interval following a stimulus for a control system's output to settle within an error band, or that for solids to settle from suspension (amongst other things)
settling zone	The largest of the four theoretical zones of a settling tank, where low velocity allows solids settlement
severity	'Cracking severity' is a term for the temperature used for thermal cracking (qv)
Seveso Directive	An EU directive responding to the Seveso incident (qv), regulating major accident hazards in the chemical industry; now the Seveso II Directive

Seveso disaster	see **Seveso incident**
Seveso incident	*aka* Seveso disaster. A major industrial 'accident' in 1976[93]
sewage	Liquid wastes arising from toilets, bathing etc. in human habitations; cf effluent
sewage sludge gas	see **biogas**
sewer	A conduit for sewage (qv), in some cases mixed with industrial effluent (qv), and/or storm water (qv), in which case it is known as a combined sewer (qv)
sewer box	A manhole (qv) or rodding point (qv)
sewer cleanout	see **rodding point**
sewer hole	see **manhole**
sewer main	*aka* main sewer. A primary sewer line usually separated into sections by manholes or sewer boxes
sewerage	Sewage (qv) infrastructure, mostly comprising sewers
SFAIRP	So far as is reasonably practicable
SFAS	Statements of financial accounting standards
SFE	see **supercritical fluid extraction**
SFMEA	see **system failure mode and effects analysis**
SG	see **specific gravity**
SG iron	Spheroidal graphite iron; see **ductile iron**
SGIA	Smoke and gas ingress analysis; see **gas dispersion and smoke ingress analysis**
Sh	see **Sherwood number**
shaft	A rotating part of a machine that transfers energy from a motor to the process
shaft seal	A seal (qv) around a rotating machine's shaft
shaft work	The energy transferred from a machine (pump, turbine etc.) to the process fluid (a fraction of the motor work)
shakedown	A period of commissioning or running-in of a structure to stable operation; defined formally in API RP 579 - Fitness for Service
shale gas	Natural gas found in the pores of shale deposits
shale oil	An oil produced from the solid organic material in oil shale (qv)
shape factor	A fiddle factor (qv) used to correct a theorist's assumed shape of objects (usually perfect spheres) to that of real world objects (neither perfect spheres, nor exactly the same as each other)
shared control	see **shared display**

---

[93] Of which the proven effect on human health to date has been 200 cases of chloracne

shared display	*aka* shared control. A time-shared display/control/communications system. Defined formally in BS EN ISO 10209:2012 Technical product documentation - Vocabulary - Part 1: Terms relating to technical drawings: general and types of drawings and BS 1646-4:1984 Symbolic representation for process measurement control functions and instrumentation. Specification for basic symbols for process computer, interface and shared display/control functions
sharpen your pencil	*(informal)* Reduce your price
shear force	A force applied parallel to a material's cross section
shear pin	*aka* shear type coupling, shear pin device. A torque safety device in the form of a weaker pin or bolt running through a coupling (especially a rotating coupling) which shears on overtorque. Defined formally in ISO 3977 Gas turbines - Procurement - Part 3: Design requirements
shear pin device	1. A non-reclosing PRD (qv), the operation of which involves breaking a shear pin (qv). Defined formally in API Standard 521. Pressure-relieving and Depressuring Systems. Sixth Edition | January 2014 2. A shear pin (qv)
shear spool coupling	*aka* shear spool type coupling, spool type flexible coupling. A drive coupling incorporating an elastomeric link which acts as both overtorque protection and vibration damper; cf shear pin device
shear spool type coupling	see **shear spool coupling**
shear strain	Strictly, a change in angle between line elements that were originally perpendicular, due to shear stress (qv), but commonly defined as the ratio of displacement in dimensions caused by shear stress to original dimensions
shear stress	*(symbol* τ*)* A component of stress parallel to a material's cross section; cf normal stress
shear thickening	see **dilatant**
shear thinning	see **pseudoplastic**
shear type coupling	see **shear pin**. However, this alternative term is defined formally in ISO 3977 Gas turbines - Procurement - Part 3: Design requirements
shear wave	*aka* S wave, secondary wave. A transverse wave in an elastic medium subjected to oscillating shear. Defined formally in BS ISO 2041:2018 Mechanical vibration, shock and condition monitoring. Vocabulary
sheave	A pulley wheel; cf chain block, crane

sheep shank	A poor quality knot used to temporarily shorten a line of rope
sheet	1. *aka* sheet metal. A thin (<6mm thick) flat metal stock (sheet metal); cf plate
2. Less commonly, a technical drawing or segment thereof corresponding to a printed sheet; defined formally in BS EN ISO 10209:2012 Technical product documentation - Vocabulary - Part 1: Terms relating to technical drawings: general and types of drawings
3. Least commonly, a hauler's tarpaulin
shelf life	The maximum duration from manufacture to use which does not unacceptably affect performance (typically applies to product or consumables related to bioprocessing or food manufacturing). Defined formally in ASME BPE (American Society of Mechanical Engineers: Bioprocessing Equipment); cf expiration date
shell and tube heat exchanger	A heat exchanger composed of a bundle of interconnected pipes (tubes) within a pressure vessel (shell)
shell leakage	The quantity of internal fluid passing to environment during testing. Defined formally in ASME BPE (American Society of Mechanical Engineers: Bioprocessing Equipment)
shell side	A term for the contents of the shell (rather than the tubes) of a shell and tube heat exchanger (qv); cf tube side
SHEQ	Safety, health, environment, quality; see **quality, environment, safety, health**
Sherwood number	*aka* Sh. A dimensionless group, the ratio of convective mass transfer to diffusive mass transport rates
Shewhart chart	A control chart (qv) used in statistical process control to measure the variability in a process variable
shield section	*aka* shock section. In the context of fired heaters, tubes which shield other convection section tubes from direct radiation from flame. Defined formally in BS EN ISO 13705:2012/ISO 13705:2012(E) Petroleum, petrochemical and natural gas industries. Fired heaters for general refinery service, API STD 530 - Calculation of Tube Heater Thickness and API STD 560 - Fired Heaters for General Refinery Service
shielding	Material which absorbs radiation (rather than neutrons); cf absorber
shift converter	see **Z90**
shift man's persuasion	see **percussive maintenance**
shifter	*(informal)* An adjustable spanner
ship stair	see **ship's ladder**

Ship's ladder	*aka* ship stair. A fixed inclined ladder with handrail (see **Figure S3**); defined formally in OSHA 1910.29 - Fall protection systems and falling object protection-criteria and practices; cf stepladder
SHMP	Sodium hexametaphosphate
shock	A sudden change initiating transient disturbances. Defined formally in BS ISO 2041:2018 Mechanical vibration, shock and condition monitoring. Vocabulary
shock absorber	A device which dissipates energy to reduce transient mechanical disturbances. Defined formally in BS ISO 2041:2018 Mechanical vibration, shock and condition monitoring. Vocabulary
shock chilling	Rapid cooling caused by sudden contact of a surface with a colder fluid, especially for metal surfaces. Defined formally in API RP 579 - Fitness for Service
shock freezing	see **blast freezing**

Figure S3 Ship's ladders

shock isolator	A device intended to isolate a system from a number of kinds of shocks; defined formally in BS ISO 2041:2018 Mechanical vibration, shock and condition monitoring. Vocabulary; cf shock absorber
shock motion	A transient motion associated with a shock (qv). Defined formally in BS ISO 2041:2018 Mechanical vibration, shock and condition monitoring. Vocabulary
shock section	see **shield section**
shock wave	A supersonic, high energy pressure wave
shoes	see **supports**
Shokri-Beyler method	A simple method to estimate heat flux from pool fires
shop fabricated container	LPG container (qv) fabricated entirely under workshop conditions; defined formally in NFPA 59: Utility LP-Gas Plant Code, 2018 Edition
Shore durometer	see **durometer**
short imperfections	Shorter weld imperfections, defined in various ways by various standards, usually relative to overall weld length. One formal definition may be found in PD 5500:2018+A1:2018 Specification for unfired fusion welded pressure vessels
short radius	A term applied to bends, elbows and ells whose radius of curvature is equal to pipe outside diameter; cf long radius
short term exposure limit	*aka* STEL. Acceptable average concentrations of certain hazardous substances over 15 minutes in workplace air, according to UK health and safety regulation; cf long term exposure limit, maximum exposure limit
short ton	*aka* ton, US ton. A US customary units (qv) measure, 2000 lb.; cf long ton
shotcoke	A low quality grade of petcoke (qv)
shotcrete	see **gunite**
shrinking core model	A description of a reaction/dissolution model where particles decreasing in size leads to a change in rate of reaction
shut down	see **shutdown**
shutdown	*aka* shut down. 1. A period in which a plant is offline (qv) Shutdowns are not necessarily planned, unlike a turnaround (qv); cf shuttering 2. Taking a plant offline 3. Less commonly, a plant's status whilst offline
shutdown set point	The control setpoint at which shutdown (qv) is (usually automatically) initiated. Defined formally in EN ISO 10437 Petroleum, petrochemical and natural gas industries - Steam turbines - Special-purpose applications
shutoff	see **no flow**

shutoff valve	A valve whose operation guarantees isolation of plant or equipment; cf control valve
shuttering	1. see **formwork**
	2. The permanent shutdown (qv) of an activity or facility
SI units	*aka* système international d'unités. The international system of units comprising seven base units, 22 named derived units, and more or less any number of unnamed derived units (qv)
siccative	An additive which promotes drying of paints, etc.
sick process syndrome	The condition of a 'barely in control' process, which has experienced multiple failed attempts to bring it under control
side channel blower	*aka* regenerative blower, vacuum pump, vacuum compressor. A type of rotodynamic blower
side reaction	Reaction producing byproduct (qv)
side stream	A stream taken from an intermediate point in a process used to feed an ancillary process, or sometimes itself being a product
siemens	(*symbol* S) The derived SI unit of conductance; cf mho, ohm
sieve plate column	*aka* sieve tray column. A distillation column with sieve trays (qv)
sieve tray	The cheapest and therefore commonest type of contacting device used within a distillation column, used when high turndown (qv) (max ~50%) is not a concern; see **Figure S4**
sieve tray column	see **sieve plate column**
sievert	(*symbol* Sv) The derived SI unit for ionizing radiation equivalent dose; cf becquerel
sieving	The batch version of screening (qv) either in a lab or in batch production
SIF	see **safety instrumented function**

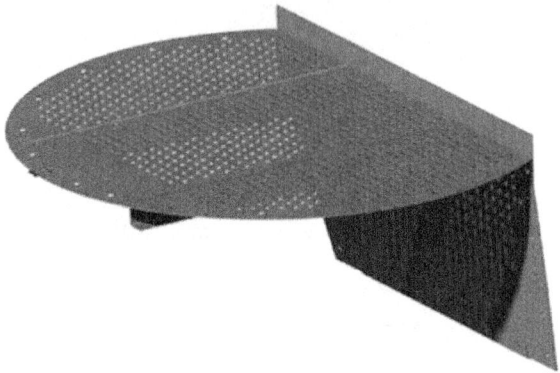

Figure S4 Sieve Tray (courtesy: Sulzer)

sight glass	1. A transparent tube used to visually monitor fluid level in a closed system 2. *aka* sight port, sight window or view port. A small transparent area used to look inside a process vessel
sight port	see **sight glass**
sight window	see **sight glass**
signal	Generically, something which contains information about a system. Most commonly in engineering contexts it will be electronic and will carry information about the status of a measured parameter
signal bandwidth	The range of possible signal frequencies. Defined formally in BS ISO 2041:2018 Mechanical vibration, shock and condition monitoring. Vocabulary
signal list	A catalogue of information about input and output signals. Defined formally in BS EN ISO 10209:2012 Technical product documentation - Vocabulary - Part 1: Terms relating to technical drawings: general and types of drawings.
signature document	A copy of a document issued for formal client approval to proceed. Defined formally in BS EN ISO 10209:2012 Technical product documentation - Vocabulary - Part 1: Terms relating to technical drawings: general and types of drawings
significant change	A change in a polymer which affects the function of something made of it. Defined formally in ASME BPE (American Society of Mechanical Engineers: Bioprocessing Equipment)
significant digits	see **significant figures**
significant figures	*aka* implied precision, implied resolution, significant digits. The number of digits rightwards from the highest to the lowest non-zero digit in a number. This should ideally be capable of being taken to imply the number of meaningful digits with respect to accuracy (qv) or precision (qv), though this is not always the case in practice, hence the word 'implied' in the alternate terms given; cf decimal places, see **Box A1**.
significant hazard	A hazard which ideally requires elimination by design. Defined formally in BS EN ISO 12100:2010 Safety of machinery. General principles for design. Risk assessment and risk reduction
SiHi	Proprietary eponym for a vacuum pump
SIL	see **safety integrity level**
silane	SiH4 or less formally, any silicon/hydrogen polymer
silica	Silicon dioxide
sill plate	Foundation plate of the wall of a building, sometimes known as a soleplate

silo	1. A structure holding bulk materials 2. *aka* organizational silo. Part of an organization which does not play well with others
Silverson mixer	Proprietary eponym for a high shear mixer
simmer	1. *aka* warn. A noticeable escape of a small amount of compressible fluid from a pressure relief valve below its set pressure. Defined formally in API RP 520 P1 7th Edition, January 2000 Sizing, Selection, and Installation of Pressure-Relieving Devices in Refineries; Part I - Sizing and Selection, API RP 576 - Inspection of Pressure Relieving Devices 2. A food processing term for holding temperature at close to boiling point
SIMOPS	see **simultaneous drilling and production**
simple harmonic vibration	Sinusoidal vibration. Defined formally in BS ISO 2041:2018 Mechanical vibration, shock and condition monitoring. Vocabulary
simplex plot	see **triangular diagram**
simulated moving bed	*aka* SMB. Making a batch process such as chromatography act like a continuous one, by connecting multiple units together and switching feed and product locations periodically
simultaneous drilling and production	*aka* SD&P, simultaneous operations, simultaneous drilling & production operations, SIMOPS, SIPROD. The (rather unsafe and therefore discouraged) approach of carrying out production and drilling operations simultaneously at one location
simultaneous drilling and production operations	see **simultaneous drilling and production**
simultaneous operations	see **simultaneous drilling and production**
simultaneous saccharification and fermentation	*aka* SSF. A biological process where the sugar metabolized by the organism is produced in the same vessel as the bioreaction is occurring
single degree of freedom system	*aka* SDoF system. A system whose state at a given instant can be defined completely by a single coordinate. Defined formally in BS ISO 2041:2018 Mechanical vibration, shock and condition monitoring. Vocabulary; qv degrees of freedom; cf multiple degrees of freedom system

single line diagram	*aka* SLD, line diagram, one line diagram. Generically, one of a number of types of diagram which show system interconnections; defined formally in BS EN ISO 10209:2012 Technical product documentation - Vocabulary - Part 1: Terms relating to technical drawings: general and types of drawings. Most commonly applied to its electrical version, the electrical engineer's equivalent of a P&ID (qv) for a three phase electrical system
single oil	A pseudocomponent (qv) sometimes used in an oil and gas context
single phase	1. Containing a single physical phase, especially if both phases are fluids; cf multiphase 2. An AC electrical supply similar in voltage and frequency to local domestic supply; cf three phase power
single stage regulator	An LPG vapor pressure regulator providing a supply at 1.0 psi (6.9 kpag) or less direct from the container. Defined formally in NFPA 58: Liquefied Petroleum Gas Code, 2017 Edition
single use bioreactor	*aka* SUB. A disposable reactor commonly used in the pharmaceutical industry where high levels of sterility are required
sink and float separation	see **dense medium separation**
sintering	*aka* frittage. 1. Generically, a process where granular materials heated below their melting point form a solid (but usually highly porous) mass 2. Specifically, a loss of catalyst activity by this mechanism, as defined formally in API RP 536 - NOx control on Fired Heaters at Refineries
SIP	1. see **State Implementation Plan** 2. see **steam in place** 3. see **sterilization in place**
siphon	see **syphon**
siphon breaker	see **syphon breaker**
SIPROD	see **simultaneous drilling and production**
SIS	see **safety instrumented system**
site	The whole area of process plant within the boundary fence, land in ownership, or bounded land within which a process plant sits
site acceptance test	*aka* SAT. The testing of an installation to confirm that the performance against specification as measured under factory conditions in the FAT (qv) is also given on site

site assembled domestic wastewater treatment plant	A stick built (qv) domestic sewage treatment plant. Defined formally in EN 12566 - Small wastewater treatment systems for up to 50 PT Part 3: Packaged and/or site assembled domestic wastewater treatment plants
site layout	Layout at a site level: the consideration of plots in relation to one other within the site as well as activities outside the site
site layout plan	see **block plan**
site plan	1. A civil engineering drawing showing location of construction works, as defined formally in BS EN ISO 10209:2012 Technical product documentation - Vocabulary - Part 1: Terms relating to technical drawings: general and types of drawings 2. A general arrangement drawing (qv) at a whole-site level; cf plot plan
site roads	Main roads within a site: roads other than plot roads (qv); see **Figure S5**
site selection	The part of project planning/plant layout design where different candidate sites are evaluated
six-sigma	*aka* 6 sigma. A suite of largely statistically-based quality management/ process improvement techniques; a trademark of Motorola when capitalized
size classification	1. see **classification** 2. Classifying the size of surface defects as per ASME BPE (American Society of Mechanical Engineers: Bioprocessing Equipment)

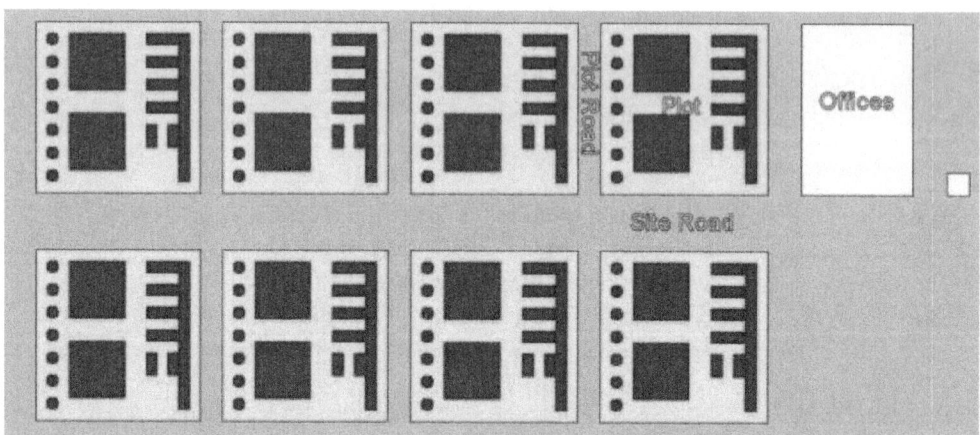

Figure S5 A process production facility and its constituent parts

size exclusion chromatography	A method of separating components based on their size relative to pores in the static medium
sizing	1. Segregation by size (qv classification) 2. Applying a non-absorbent coating to paper or textiles
sketch	A freehand or low resolution indicative drawing. Defined formally in BS EN ISO 10209:2012 Technical product documentation - Vocabulary - Part 1: Terms relating to technical drawings: general and types of drawings
skid	*aka* modular process skid. A portable frame upon which a number of items of equipment are factory preassembled or 'skid mounted' into a system capable of independent operation; or informally, such an assembly; cf module
skid tank	see **portable tank**
skilled person	An individual with sufficient education, training and experience to appropriately manage risks associated with a product. Defined formally in BS EN 82079-1:2012 Preparation of instructions for use. Structuring, content and presentation. General principles and detailed requirements; cf competent person, qualified person; see **Box Q1**.
skillet blind	A type of spade installed between flanges as a temporary barrier to flow
skud	An inclined separator (qv FWKO) for very high water cut
skyhooks	*(jocular)* Imaginary supports which an apprentice might be sent to the stores to obtain[94]
slack	see **float**
slag	A non-metallic liquid impurity (mostly silicon, aluminum, calcium and magnesium oxides) which floats on molten metal during smelting, and inclusions of the same impurity in metal; defined formally in ASME BPE (American Society of Mechanical Engineers: Bioprocessing Equipment); cf gangue, dross
slag inclusion	Slag (qv) particles in a weld. Defined formally in API STD 620 - Low Pressure Storage Tanks
Slager	Proprietary eponym for a crystallizer
slaked lime	*aka* builders' lime, cal, caustic lime, E526, hydrated lime, pickling lime. Calcium hydroxide, the product of 'slaking' quicklime (qv) with water
SLD	see **single line diagram**
slew	To move from side to side; cf luff, traverse
SLF	see **substitute liquid fuels**

---

[94] An old engineering joke related to long weights, tartan paint, copulating tools, left handed spanners etc.

slickline	In an oil industry context, a cable used to raise and lower equipment in a well shaft; cf wireline
sliding seal	A seal which moves relative to a mating surface; defined formally in ASME BPE (American Society of Mechanical Engineers: Bioprocessing Equipment)
slinger	*aka* banksman/slinger. A UK term for a rigger (qv)
slip dies	*aka* slip dyes (sic). The hardened 'teeth' on slips (qv)
slip dyes	Misspelling of slip dies (qv)
slip plate	see **spade**
slip ratio	*aka* velocity ratio. The ratio of velocities of two fluids in a two phase system; cf slip velocity
slip resistant surface	Flooring intended to improve footwear grip. Defined formally in BS EN ISO 14122 Safety of machinery. Permanent means of access to machinery. Working platforms and walkways
slip ring	A mechanism for electrically connecting a rotating component
slip tube gauge	A type of liquid level gauge for LPG tanks. Defined formally in NFPA 58: Liquefied Petroleum Gas Code, 2017 Edition
slip velocity	The difference between the velocity of two fluids in a two phase system; cf slip ratio
slippers	see **supports**
slips	*aka* dogs, casing dogs, casing slips, drill collar slips, drilling slips, drill slips, rotary slips, oilfield rotary slips. In oil and gas drilling, a term commonly used for wedge shaped retractable jaws which hold and center the drilling string (qv) in a rotary table. There are however other types of slips, though they all do the same job of gripping tubulars (qv). They are sometimes known as dogs (qv) though that term has a far wider range of meanings than slips
slogging	*aka* flogging. Using a hammer on a spanner (ideally a slogging spanner (qv), intended for the use) to tighten or loosen a nut, a popular approach in oil and gas
slogging spanner	*aka* flogging spanner. A spanner designed to be hit with a hammer to "encourage" a tight nut (qv slogging)
slop	see **slop oil**
slop oil	*aka* bad oil, slop, slops, secondary oil. Most commonly applied to crude oil contaminated with hard-to-separate solids and emulsified water, close to synonymous with rag layer (qv); also applied more generally to a wide variety of liquid oily wastes
slope	The degree of deviation from horizontal. Defined formally in ASME BPE (American Society of Mechanical Engineers: Bioprocessing Equipment)

slopover	The 'big brother' of a chip-pan fire: escape from one side of containment of a burning liquid as a result of a sudden expansion in volume, commonly as a result of the boiling of water within the containment caused by firefighting water addition; defined formally in API RP 2001 - Fire Protection at Refineries; cf frothover, boilover
slops	see **slop oil**
sloshing	Free liquid surface oscillation in a moving part-full container. Defined formally in BS ISO 2041:2018 Mechanical vibration, shock and condition monitoring. Vocabulary
slow down	Bypassing equipment and running at lower throughput in order to test it whilst still operating; cf shutdown
slow rinse	A rinse following regenerant solution in an ion exchange column
slow roll	A manufacturer's recommended warmup rotational speed of equipment. Defined formally in EN ISO 10437 Petroleum, petrochemical and natural gas industries - Steam turbines - Special-purpose applications
slow roll runout	see **electrical and mechanical run out**
slow sand filter	A type of open topped water filter with a low surface loading (relative to a rapid gravity filter (qv)) used in potable water treatment, whose effective filtration medium is a schmutzdecke (qv). It is therefore not a depth filter
SLS	Solid liquid separation
sludge	A (often viscous or semi-solid) liquid stream with a substantial solids content, usually recovered as a subnatant (qv)
sludge age	A measure of solids retention in an activated sludge plant, which is different from solids retention time (qv) and mean cell residence time (qv), even though this term is often informally used interchangeably with the other two. Sludge age in days is properly calculated as total mass of **MLSS** in aeration basin / daily mass of **TSS** in the influent
sludge bulking	Reduced settleability of activated sludge; a sign of process problems
sludge digestion	Reducing the solids content of biological sludges, usually by aerobic or anaerobic biological treatment
sludge gas	see **biogas**
sludge volume index	*aka* SVI. A measure of the settleability of activated sludge; used as a process metric
sludge yield	A dimensionless number: the mass of solids in sludge wasted from an activated sludge process for a given mass of influent BOD

sludge zone	The region of a settling tank in which settled sludge is stored
slug	1. A parcel of liquid between elongated bubbles of gas in a multiphase flow regime (also sometimes incorrectly used to describe the bubbles in such a regime) 2. Far less commonly, a now obsolete (and always rather obscure) unit of mass also known as the geepound
slug catcher	A phase separation vessel which disentrains a slug (qv) in a pipeline
slug flow	A multiphase flow regime characterized by a succession of cells comprising a slug and associated bubble; cf annular flow, wispy annular flow
sluice gate	see **penstock**
sluice valve	see **gate valve**
sluicing	1. Transporting solids in a flow of liquid 2. Performing a liquid rinse (qv) of a surface
slurry	A suspension in liquid of denser-than-liquid solids with a solids content sufficient to alter the viscosity or density of the mixture significantly from that of the liquid; or making such a suspension: in both cases, the liquid in question is most commonly water
slurry oil	The highest boiling fraction from a fluid catalytic cracking unit (qv)
small and medium sized enterprise	*aka* SME. A somewhat contested term, but commonly 'medium sized' enterprises have less than 250 employees, and 'small' less than 50, as per the EU Commission Recommendation of 6 May 2003 concerning the definition of micro, small and medium-sized enterprises
smart pig	see **intelligent pig**
SMB	see **simulated moving bed**
SMBS	Sodium metabisulfite
SME	1. see **small and medium sized enterprise** 2. see **subject matter expert**
smelting	Strictly, heating an ore in the presence of a reducing agent to extract metallic elements, though the electrolytic production of aluminum is also commonly called smelting
Smith–Brinkley shortcut method	A quick way to estimate products of multicomponent distillation; largely obsolete
smoke	A turbid dispersion of combustion products (including fine solids) in air; cf fume(s)
smoke and gas ingress analysis	*aka* SGIA; see **gas dispersion and smoke ingress analysis**

smoke and toxic gas dispersion analysis	see **gas dispersion and smoke ingress analysis**
smoke and toxic gas dispersion analysis study	see **gas dispersion and smoke ingress analysis**
smokeless capacity	In the context of flare burners, the range of flows burnable without production of smoke. Defined formally in API RP 537 Flare Details for Petroleum, Petrochemical, and Natural Gas Industries
smoldering combustion	Burning slowly with smoke, but without incandescence or flame; cf glowing combustion, flaming combustion
smooth	In a hygienic context, having a sufficiently low level of surface defects to meet operational requirements. Different formal definitions in BS EN 1672-2:2005+A1:2009 Food processing machinery. Basic concepts. Hygiene requirements and BS EN ISO 14159:2008 Safety of machinery. Hygiene requirements for the design of machinery
smoothness	The inverse of roughness (qv) (of a surface)
smother	To completely cover fire with something which excludes oxygen in order to extinguish it
SMR	see **steam methane reforming**
SMS	1. see **safety management system** 2. Short message service (text messaging)
SMT	see **steam methane reformer furnaces**
SMTE	Subject matter technical expert, see **subject matter expert**
SNAFU	Situation normal, all fucked/fouled up. Part of a spectrum (SNAFU, FUMTU, TARFU, FUBAR -qv), conveying increasing degrees of disorder/deviation from correct practice or operation and the absence of a means to resolve it
snagging	Noting construction and commissioning defects
snagging list	*aka* punch list. A list of construction and commissioning defects requiring remedy before a stage of construction or commissioning can be considered complete
snake	A tool to pass wires through things, such as a cobra (qv)
SNCR	see **selective non-catalytic reduction**
SNF	Spent nuclear fuel
SNG	see **synthetic natural gas**
Snoop	Proprietary eponym for leak detection liquid
snubber	A device which suppresses transient mechanical or electrical phenomena. The mechanical type is defined formally in BS ISO 2041:2018 Mechanical vibration, shock and condition monitoring. Vocabulary
snuffing steam	Steam used to snuff out unwanted fires in a furnace

soak	see **soaking drum**
soak process	A part of the visbreaking (qv) process where cracking reactions take place outside the furnace in a soaking drum (qv)
soakaway	An underground flow buffering chamber designed to allow collected surface water to drain away. The term is also used to describe a drainage field, which offers passive tertiary biological treatment of effluent, rather than just flow buffering. There are those who claim that a drainage field is a subtype of soakaway, but this interpretation is not supported by British Standards. It is not the same as a drainage field, though the terms are commonly confused in error
soaker	see **soaking drum**
soaker drum	see **soaking drum**
soaking cell	see **soaking drum**
soaking chamber	see **soaking drum**
soaking drum	*aka* soak, soaker, soaker drum, soaking cell, soaking chamber. A unit where cracking reactions take place outside the furnace; used for the deep conversion of vacuum bottom (qv) oils from the vacuum distillation unit (qv) in an oil refinery by visbreaking (qv)
soap	A surfactant, usually the sodium or potassium salt of a fatty acid
Soave–Redlich–Kwong equation of state	*aka* SRKEOS, SRK equation of state. A modified version of the Redlich-Kwong equation of state (qv), one of a number of empirical equations relating temperature, pressure and volume used to allow unknown properties to be estimated based on known quantities
sockolet	*aka* SOL. Proprietary eponym, an abbreviated form of SOCKet weld OutLET. A type of olet (qv)
soda	An archaic name for sodium from which soda ash, caustic soda, etc. are derived
soda lye	see **caustic soda**
sodium form cation resin	A brine regenerated cation exchange resin used for water softening
SOF	*(drawing notation)* Slip on flange
soffit	1. Most commonly, the underside of the crown (top of the internal bore) of a pipe 2. The lower surface of an architectural feature, (rather than the line of that surface); cf intrados, extrados
soffit elevation	The level above datum of the top of the internal bore of a pipe

soft acceptance criteria	A term used in commissioning to describe acceptance criteria which are not critical to safety or operation and which are acceptable as long as the limits of tolerance are within the design envelope; cf rigid acceptance criteria
soft copy	An electronic document; cf hard copy
soft foot	A defect of rotating equipment installation in which its shaft is misaligned due to one or more of the 'feet' on its casing not being in proper contact with its intended support structure
soft sensor	A proxy measurement, specifically a computer-based estimation of a hard-to-measure quantity: a calculation based on measurable quantities of an unmeasurable quantity
soft starter	A type of solid state drive starter used to manage the starting load on large motors, able to ramp up supplied frequency to a drive in a controlled manner, akin to a less sophisticated variable speed drive (qv)
softening	1. Most commonly, reducing the hardness (qv) of water 2. Far less commonly, reducing mechanical hardness (qv)
software failure mode and effects analysis	A subtype of failure mode and effects analysis (qv), as applied to a software; N.B. the acronym SFMEA is strongly deprecated to avoid confusion with system failure mode and effects analysis (qv)
SOHIC	see **stress oriented hydrogen induced cracking**
soil	1. In a hygienic context, soil is a term akin to 'weeds', being anything unwanted in the system. According to the EDEHG Glossary, it is "any undesirable/objectionable material on surfaces in the equipment or process environment". Similar definitions may be found in BS EN 1672-2:2005+A1:2009 Food processing machinery. Basic concepts. Hygiene requirements and BS EN ISO 14159:2008 Safety of machinery. Hygiene requirements for the design of machinery 2. The kind of soil which weeds grow in, if civil engineering is involved, as in BS 5930:2015 Code of practice for ground investigations 3. As a verb, to contaminate with soil (either of the above kinds)
SOL	*(drawing notation)* Sockolet (qv)
sol	A colloidal suspension of solids in a liquid continuous phase
solenoid operated pilot valve	*aka* SOPV; see **solenoid valve**
solenoid operated valve	*aka* SOV; see **solenoid valve**

solenoid valve	*aka* direct acting solenoid valve. A valve which is directly actuated by a solenoid, in which an electromagnet moves a piece of ferromagnetic material. The term is however commonly used for a valve driven by a fluid, the flow of which fluid is controlled by a solenoid valve, properly called a solenoid operated valve, or solenoid operated pilot valve
soleplate	see **baseplate**, though also sometimes used as synonym of sill plate
solid bowl centrifuge	see **bowl centrifuge**
solids retention time	*aka* SRT. A measure of solids retention in an activated sludge (qv) plant, which is different from sludge age (qv) and mean cell residence time (qv), even though this term is often informally used interchangeably with the other two. SRT in days is properly calculated as daily mass of SS to aeration basin/ daily mass of SS wasted; cf hydraulic retention time
solubility	The maximum quantity of a substance which will dissolve in a given volume of solvent at a given temperature to form a saturated solution, or to put it another way, the composition of a saturated solution, expressed as a proportion of a solute in a solvent; cf solvation
solubility coefficient	A slightly contested term: 1. The solubility (qv) of gas in a liquid expressed as volume of gas per unit volume of solvent at a given pressure and temperature, such as Bunsen coefficient (qv) or Ostwald coefficient (qv) 2. In the sugar industry, the ratio of concentration of sucrose in impure saturated solution to the concentration in a pure sucrose solution saturated at the same temperature (with concentration expressed as sucrose/water ratio). This is referred to as saturation coefficient (qv) in the beet sugar industry
solute	A component which dissolves in a solvent (qv)
solvation	*aka* dissolution. The interaction between a solvent and a solute leading to stabilization of solution; N.B. not synonymous with solubility (cf)
Solvay process	*aka* ammonia-soda process. The main process for the manufacture of sodium carbonate
solvent	A liquid in which things dissolve. Often also used informally to mean organic liquids
solvent distillation unit	*aka* solvent recovery unit. Part of a plant which recovers solvent (qv) by distillation. The NFPA definition requires that the solvent be flammable or combustible, as can be seen in NFPA 30: Flammable and Combustible Liquids Code, 2018 Edition

solvent extraction	*aka* liquid/liquid extraction. Increasing the purity of a desired material using its different solubilities in two immiscible solvents
solvent naphtha	see **mineral spirits**
solvent recovery unit	see **solvent distillation unit**
solvent-water separator	see **liquid-liquid separator**
solvolysis	The breaking down of a solute by its solvent, for example hydrolysis by water
sonic horn	An acoustic cleaning device
sonic modulus	(*symbol* E) The tensile/compressive modulus estimated by measurement of sound-wave propagation in a material, described formally in ASTM Test C 769 (Section 15.01)
sonic velocity	The speed of sound in a given medium
sonication	Using sound to agitate particles in suspension. Where ultrasound is used, it is called ultrasonication (qv)
sootblower	A device which removes residue from the outside of oil fired burner convection tubes by blowing with steam or sometimes air. Defined formally in BS EN ISO 13705:2012/ISO 13705:2012(E) Petroleum, petrochemical and natural gas industries. Fired heaters for general refinery service, API RP 536 - NOx control on Fired Heaters at Refineries, API STD 530 - Calculation of Tube Heater Thickness and API STD 560 - Fired Heaters for General Refinery Service
SOP	see **standard operating procedure**
SOR	Specification of requirements, see **requirement specification**
sorbent	*aka* sorbent material. An absorbent (qv) or an adsorbent (qv)
sorbent material	see **sorbent**
Sorel-Lewis method	see **Lewis-Sorel method**
Soret effect	*aka* (in various contexts) Ludwig–Soret effect, thermal diffusion, thermodiffusion, thermomigration, thermophoresis. Some of these terms may strictly apply to phases other than liquid, where 'Soret effect' is the strictly correct term. They all however refer to a phenomenon observed on very small scales where nanoparticles tend to move from a hotter region to a colder one
sorption	The attachment of one substance to another, conventionally differentiated into adsorption (qv) and absorption (qv); cf desorption
SOTA	see **state of the art**
Souders–Brown equation	A way to estimate maximum allowable vapor velocity in vapor-liquid contacting columns and vessels
SOUR	Specific oxygen uptake rate; see **oxygen uptake rate**

sour gas	Strictly, any petroleum gas containing significant quantities of odorous hydrogen sulfide. H_2S is also an acid gas (qv), leading to confusion between acid gas and sour gas. All sour gases are acid gases but not all acid gases are sour gases; cf sweet gas
sour products	Hydrocarbon products high in sulfur components cf sweet products
sour water	*aka* SW. A wastewater with high levels of H_2S produced from crude columns in oil refineries
source of release	In the context of loss prevention, the location at which containment is lost. Defined formally in API RP 505 2nd Edition, August 2018 Recommended Practice for Classification of Locations for Electrical Installations at Petroleum Facilities Classified as Class I, Zone 0, Zone 1, and Zone 2
sources of ignition	In the context of fire safety, things which can ignite flammable mixtures; defined formally in the context of LPG in NFPA 58: Liquefied Petroleum Gas Code, 2017 Edition
SOV	see **solenoid operated valve**
SOW	Scope of work
SOx	The generic term for sulfur oxides[95]
space envelope drawing	A drawing showing how much space is expected to be required for something prior to detailed design, and where it must connect to adjacent things. Defined formally in BS EN ISO 10209:2012 Technical product documentation - Vocabulary - Part 1: Terms relating to technical drawings: general and types of drawings
space envelope model	A physical model showing how much space is expected to be required for something prior to detailed design, and where it must connect to adjacent things. Defined formally in BS EN ISO 10209:2012 Technical product documentation - Vocabulary - Part 1: Terms relating to technical drawings: general and types of drawings
space time	(*symbol* T) The time taken to process one reactor volume of feed. Often used synonymously with residence time (qv), though it is only the same under certain circumstances
space velocity	The reciprocal of space time (qv)
spacer	see **spacer ring**
spacer ring	*aka* spacer, piping spacer. An annular insert between flanges which corrects gaps or misalignment
spade	*aka* blind, slip plate. A plate installed between flanges as a temporary barrier to flow

---

[95] There are rather more of them than the two you were thinking of!

Spalding number	One of two related dimensionless groups, BT and BM, the Spalding heat transfer number and Spalding mass transfer number respectively; useful in the study of vaporizing liquid droplets
spall	1. Flakes of material broken off from a larger body (qv spalling) 2. An oil and gas industry term for a deliberate thermal shock of heat exchanger tubes to remove fouling
spalling	Most commonly, a concrete failure mode, involving spall falling off the surface
span	The range of possible values on an instrument scale
sparger	A perforated distributor used to add gas bubbles, usually to the bottom of the liquid in a vessel. Defined formally in ASME BPE (American Society of Mechanical Engineers: Bioprocessing Equipment)
spark	A potential source of ignition: either an electrical discharge, or an incandescent moving solid particle, the latter being defined formally in NFPA 654: Standard for the Prevention of Fire and Dust Explosions from the Manufacturing, Processing, and Handling of Combustible Particulate Solids, 2017 Edition
sparkler	A type of horizontal plate filter
sparky	*(informal)* A UK term for an electrician
spatter	Small particles (commonly droplets of molten metal) expelled by the welding process which can cause weld defects. The formal definition in the context of hygienic welding in ASME BPE (American Society of Mechanical Engineers: Bioprocessing Equipment) does not include non-metallic spatter, though that might be described as spatter in other contexts
SPC	see **statistical process control**
SPE	1. see **sensitive protective equipment** 2. Society of Petroleum Engineers
Spearman's rank correlation	A non-parametric statistic used to generate correlations for ordinal data (qv)
spec	1. see **specification** 2. see **functional requirements specification**
spec sheet	1. see **datasheet** 2. see **user requirement specification**
special atmosphere	A specified gas mixture used to exclude air from within a fired heater, and also commonly to interact with the surface of furnace contents in some way. Defined formally in NFPA 86: Standard for Ovens and Furnaces, 2019 Edition

special protection	The thermal protection of LPG containers; defined formally in NFPA 58: Liquefied Petroleum Gas Code, 2017 Edition
special purpose turbine	A horizontal turbine rated for extended periods of continuous service driving critical items of equipment. Defined formally in EN ISO 10437 Petroleum, petrochemical and natural gas industries - Steam turbines - Special-purpose applications
special tool	A dedicated tool required for equipment maintenance which must be purchased from equipment vendors. Defined formally in EN ISO 10437 Petroleum, petrochemical and natural gas industries - Steam turbines - Special-purpose applications and ISO14661 Thermal turbines for industrial applications (steam turbines, gas expansion turbines) - General requirements
specialty chemicals	*aka* speciality chemicals. Fine (high value) chemicals
speciality chemicals	see **specialty chemicals**
species	1. A chemical, typically one involved in a reaction or analysis 2. A specified type of organism
specific energy consumption	*aka* SEC. Energy consumption per unit of production
specific enthalpy	(*symbol* H) Total system energy per unit mass
specific gravity	*aka* SG. A measure of density (qv) similar to relative density (qv) or vapor density (qv) in that it is the dimensionless ratio of a material's density to some reference. It has the same shortcoming as those other measures: using the wrong reference material and/or temperature will mean that the measure is incorrect. Most commonly (but by no means always) for liquids, the reference is water at a range of temperatures, and for gases, dry air at STP as per API RP 535 - Burners for Fired Heaters at Refineries
specific growth rate	(*symbol* μ) The rate of increase of biomass per unit of biomass concentration, usually estimated with the Monod equation (qv)
specific heat	A deprecated term used to mean either: 1. (*informal*) Specific heat capacity (qv) 2. The ratio of a material's specific heat capacity to that of a reference substance such as water
specific heat capacity	(*symbol* CP, CV) *aka* massic heat capacity. In SI units, the energy in joules required to raise the temperature of 1 kg of something by 1 K
specific latent heat	(*symbol* L) In SI units, the energy in joules required to change the state of 1 kg of something without raising its temperature
specific rate constant	see **rate constant**

specific speed	*aka* ns. A characteristic coefficient (qv) which allows comparison of turbomachinery with similar geometry. Defined formally in EN ISO 13709:2003 Centrifugal pumps for petroleum, petrochemical and natural gas industries
specific surface area	The surface area of a divided solid per unit mass or volume
specific volume	The reciprocal of density, i.e.; volume per unit mass
specification	*aka* spec. A definition of required product and feedstock qualities, system constraints, performance of unit operations, materials of construction, etc. Defined formally in BS EN ISO 9000 Quality management systems Fundamentals and vocabulary
specification of requirements	see **requirement specification**
specified burst pressure	The rated/marked burst pressure of a rupture disk. Defined formally in API RP 520 P1 7th Edition, January 2000 Sizing, Selection, and Installation of Pressure-Relieving Devices in Refineries; Part I - Sizing and Selection and API RP 576 - Inspection of Pressure Relieving Devices; cf marked burst pressure
specified disk temperature	The rated/marked burst temperature of a rupture disk. Defined formally in API RP 520 P1 7th Edition, January 2000 Sizing, Selection, and Installation of Pressure-Relieving Devices in Refineries; Part I - Sizing and Selection
spectacle blind	see **spectacle plate**
spectacle plate	*aka* spectacle blind. A spectacle frame shaped spade consisting of two adjacent circular plates, one with a hole, used to clearly indicate visually if a line is blanked off or open to flow
spectroscopy	The use of electromagnetic waves to infer information about the composition or state of materials
Speedsizer	Proprietary eponym for metering size presses used in the paper industry
SPFA	see **Steel Plate Fabricators Association**
spherical wave	A wave with concentric spherical wave fronts. Defined formally in BS ISO 2041:2018 Mechanical vibration, shock and condition monitoring. Vocabulary
sphericity	1. (*symbol* F) Generically, how close something is to being a perfect sphere; cf roundness 2. In ion exchange, a measure of the amount of unbroken resin beads
spheroidal graphite iron	*aka* SG iron; see **ductile iron**

spider	1. A device with a number of spokes radiating from a center, often a central support or feed point. In the context of fired heaters, an arrangement of multiple gas tip orifices of this type, as defined in API RP 535 - Burners for Fired Heaters at Refineries
	2. In the oil and gas industry, a spider is used to accommodate the slips (qv) which hold the drill pipe in a rotary table (qv)
spider coupling	Two spider (qv) devices that connect the shaft of a motor to a shaft of a pump
spider deck	A name for the moon pool (qv) area of a drilling platform; so-called due to the riser tensioner wires which give the appearance of a spider's web
spigot	1. A faucet (qv)
	2. Alternatively, the male part of a spigot and socket pipe joint
spill	see **spillage**
spillage	*aka* spill. A loss of containment
spillback	The use of a valved bypass to send some of a pump's output back to its inlet; a deprecated, inefficient way to control output
spinning	Rotation, or extrusion to fibers
spiral wound cartridge filter	*aka* string wound cartridge filter. A single use cartridge with a spirally wound filament as filter media
spiral wound membrane	The most compact configuration of a semipermeable membrane (qv), with a membrane separated from two surrounding impermeable sheets by spacing mesh wound spirally into a cylinder; see **Figure S6**

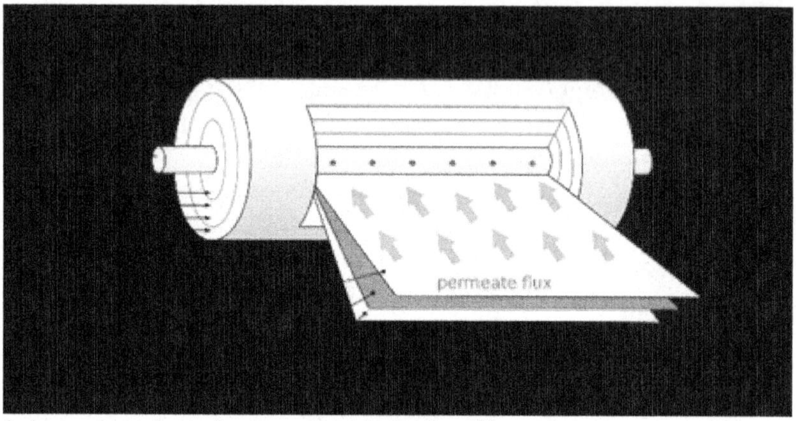

Figure S6 Spiral wound membrane (courtesy: Daniele Pugliese)[96]

---

[96] Reproduced under Creative Commons CC BY 3.0 https://creativecommons.org/licenses/by/3.0/

Spiratube	Proprietary eponym for spirally reinforced air ducting
splash area	*aka* splash contact surfaces. "Non-product contact surfaces that during normal use are subject to accumulation of soil and which require routine cleaning to avoid soil to drop or to be drawn into the main product or container", according to the EHEDG Glossary. Slightly different formal definitions for this term in BS EN 1672-2:2005+A1:2009 Food processing machinery. Basic concepts. Hygiene requirements and BS EN ISO 14159:2008 Safety of machinery. Hygiene requirements for the design of machinery; cf product contact surfaces
splash contact surfaces	see **splash area**
splitter	1. A reactor which reduces molecules to their subcomponents, especially elements 2 *(informal)* A distillation/flash column producing only one top and one bottom stream 3. A device which exists only in process simulation programs to allow theoretically possible but practically impossible separations to be modeled
SPM	see **suspended particulate matter**
spoil	see **overburden**
spoiler	see **strake**
sponge coke	A high grade of petcoke (qv)
spongy nickel	see **Raney nickel**
sponsor	1. The entity paying for a design and/or construction project (also known as the client) 2. A euphemism sometimes used to describe the 'agent' used as a proxy by those involved in bribery and corruption
spontaneous combustion	see **spontaneous ignition**
spontaneous heating	*aka* selfheating. The heating of a substance without an external heat source. Defined formally with examples in API RP 2001 - Fire Protection at Refineries
spontaneous ignition	*aka* spontaneous combustion. Ignition without external flame or heating, due to successive selfheating (qv), thermal runaway (qv) and selfignition (qv); cf autoignition
spool	A length of pipe with flanges at each end
spool piece	A removable spool (qv) kept under management control, used to positively control whether flow may pass via a potentially hazardous route
spool sheet	A list of pipe sections for fabrication taken from an isometric drawing (qv)

spool type flexible coupling	see **shear spool coupling**
spot electropolishing	A localized electrochemical polishing process often used for field repairs. Defined formally in ASME BPE (American Society of Mechanical Engineers: Bioprocessing Equipment)
spotter	see **banksman**
spray ball	*aka* sprayball. A type of spray device (qv) in the form of a hollow sphere, drilled all over with precise holes to project CIP fluid (qv) in all directions, used for automatic tank cleaning, especially in the hygienic industries
spray column	*aka* spray tower. A gas-liquid contactor with upflow of a continuous gas phase, and downflow of a discontinuous liquid phase sprayed from nozzles
spray device	*aka* spray ball, spray nozzle, nozzle. A device such as a spray ball (qv) which directs the flow of CIP fluids (qv) to a process contact surface (qv), as formally defined in ASME BPE (American Society of Mechanical Engineers: Bioprocessing Equipment)
spray dryer	A device which rapidly produces a fine, dry powder from a liquid mixture by contact with a countercurrent stream of hot, dry gas. Popular in pharmaceutical and food applications due to its suitability for heat labile (qv) materials
spray fluidizer	Proprietary eponym for a spray dryer (qv) with a fluid bed granulator
spray nozzle	see **spray device**
spray tower	see **spray column**
sprayball	see **spray ball**
spraycrete	see **gunite**
sprayed concrete	see **gunite**
spreading coefficient	A measure of firefighting foam's potential to spontaneously spread over the surface of a liquid hydrocarbon. Defined formally in NFPA 11: Standard for Low-, Medium-, and High-Expansion Foam, 2016 Edition
spring loaded pressure relief valve	see **spring loaded PRV**

spring loaded PRV	*aka* spring loaded pressure relief valve, safety valve (qv) (term usually used for gas/vapor service application), relief valve (term usually used for liquid service application), safety relief valve (qv) (term usually used for multi-phase/service applications). A pressure relief valve which is not a pilot operated pressure relief valve (qv). Defined formally in API RP 520 P1 7th Edition, January 2000 Sizing, Selection, and Installation of Pressure-Relieving Devices in Refineries; Part I - Sizing and Selection
spring return	The use of a mechanical spring to automatically drive a correcting element (qv) to a failsafe position when deenergized
springline	An imaginary horizontal line at the midpoint of the vertical axis of a pipe's cross section, dividing the lower and upper parts of the pipe
spud	1. In the context of fired heaters, a gas flow restriction device as found in a pilot (qv) burner, and as defined in API RP 535 - Burners for Fired Heaters at Refineries
2. In the context of oil and gas well drilling, using a large drill bit to get through sedimentary materials to prepare for main drilling
spud date	Usually the date of spudding in (qv); or offshore, when drilling begins on the sea floor
spud to completion	In drilling, the time from the spud date (qv) to well completion (qv)
spudding in	In drilling, adding the main drill bit to start main drilling
spurious precision	When for example someone gives us all of the figures on their calculator display without considering how many of those digits are significant, the implied precision (qv) is spurious, hence we call it 'spurious precision'; see **Box A1**.
square cut	Cut axially; defined formally in ASME BPE (American Society of Mechanical Engineers: Bioprocessing Equipment)
squareness	Commonly used in engineering as a synonym for perpendicularity; defined formally in this sense in ASME BPE (American Society of Mechanical Engineers: Bioprocessing Equipment)
squing	A vibronic level switch
SR	1. Short radius
2. see **steam rate**
Sr	see **Strouhal number**
SRB	*(drawing notation)* Short radius bend
SRK equation of state	see **Soave-Redlich-Kwong equation of state**
SRKEOS	see **Soave-Redlich-Kwong equation of state**
SRL	Short radius ell; see **short radius**

SRM	see **standard reference measure**
SRT	see **solids retention time**
SRV	see **safety relief valve**
SS	1. Stainless steel
	2. Suspended solids, see **total suspended solids**
SSC	1. Salt splitting capacity, see **salt splitting**
	2. see **sulfide stress cracking**
SSD	Safety shutdown [system], see **safety instrumented system**
SSF	see **simultaneous saccharification and fermentation**
SSHE	see **scraped surface heat exchanger**
stability	1. Generically, the ability to resist degradation or process upsets cf robustness
	2. In the context of fired heaters, the ability of a burner to handle a wide range of input flows and fuel-air mixtures. Defined formally in API RP 535 - Burners for Fired Heaters at Refineries
stabilizer	1. A chemical additive which prevents degradation
	2. Something providing mechanical stability such as a drilling stabilizer (qv)
stable liquid	A liquid which does not energetically polymerize, decompose, or otherwise self-react when shocked, pressurized, or heated; defined formally in NFPA 30: Flammable and Combustible Liquids Code, 2018 Edition
stable regime	A term used in distillation for the preferable hydrodynamic condition of aerated liquid on a sieve tray, with the aerated mixture existing as a stable froth
stack	A chimney, vertical pipe or flue which carries exhaust gases to atmosphere, especially one located downstream from a fired heater combustion chamber
stack effect	*aka* chimney effect. The draft at the bottom of a stack caused by the reduced density of heated gases
stage efficiency	1. In mass transfer, the achievement of equilibrium in a single stage (such as a tray in a column) would be 100% efficiency
	2. In steam turbines, 100% efficiency would be achieved if the work done in a stage (paired rings of static and rotating blades) were the same as the energy output
stage gate model	*aka* phase gate process, waterfall process. Proprietary eponym for a way of formalizing the natural division of process development into stages, separated by decision points; cf V model

staged air burner	A low NOx burner which mixes secondary air (qv) with the product of primary combustion. Defined formally in API RP 535 - Burners for Fired Heaters at Refineries
staged flare	A group of flares staged such that the number of flares operating is proportional to relief fluid flow. Defined formally in API Standard 521. Pressure-relieving and Depressuring Systems. Sixth Edition | January 2014 and API RP 537 Flare Details for Petroleum, Petrochemical, and Natural Gas Industries
staged fuel burner	A low NOx burner which mixes secondary fuel (qv) with the product of primary combustion. Defined formally in API RP 535 - Burners for Fired Heaters at Refineries
stages of commissioning	see **Box C3**
stages of design	see **Box S2**
stages of production processing and distribution	"Any stage, including import, from and including the primary production of a food, up to and including its storage, transport, sale or supply to the final consumer and, where relevant, the importation, production, manufacture, storage, transport, distribution, sale and supply of feed", according to Regulation (EC) No 178/2002 of the European Parliament and of the Council
staging	In-plant temporary storage, especially of flammable liquids in containers, as formally defined in NFPA 30: Flammable and Combustible Liquids Code, 2018 Edition cf break tank
stagnant film	The film model (qv) assumes that a 'stagnant film' exists at the transfer interface which offers resistance to mass transfer; the concept is also applied to heat transfer
stagnation point	The point where there is no flow; i.e.; the fluid is stagnant
stainless steel	Alloy steel which is categorized by its room temperature metallurgical structure into one of four types: austenitic, duplex, ferritic, and martensitic. These categories are further differentiated by their proportions of alloying elements into subtypes, each with different physical and chemical properties
stainless steel - ferritic - 1.4003	see **3Cr12**
stair	Fixed access, with broad steps pitched between 20° and 45°. Defined formally in BS EN ISO 14122 Safety of machinery. Permanent means of access to machinery. Working platforms and walkways
stakeholder	see **interested party**

Box S2. Stages of Design

The nature of the design process means that the minimum amount of resource is expended to get a project to the next approval point. This results in design being broken into stages corresponding to three approval points: feasibility, purchase, and construction. Opinions differ on the precise number of stages of process design, their names, and the deliverables produced at each stage but my definitions are as follows:

Conceptual design is the first stage of process plant design, in which we understand and quantify the design constraints, the sufficiency and quality of design data available, and produce a number of rough designs based on the most plausibly successful approaches. In the oil and gas industry, the conceptual stage may start from a (highly variable in scope) package of information known as basic engineering design data (BEDD), often confused with (process) basis of design, or even design philosophy. In a product design context, conceptual design is defined in BS EN ISO10209:2012.

Front end engineering design (FEED) is the second stage, also known as a preliminary or basic engineering study. This stage may be undertaken by design house, EPC company or process licensor. Whoever does it, a more accurate version of the deliverables from the previous conceptual stage will be produced, based on a more detailed design/model and, wherever possible, bespoke design items will be substituted with their closest commercially available alternatives, and the design modified to suit. Drawings at FEED stage should show the actual items proposed, as supplied by specialist suppliers and subcontractors. Even seemingly trivial items (e.g. pipework and flanges) should be included, as they will be supplied by particular manufacturers. Pricing at FEED stage should be based on firm quotes from named suppliers. Drawings should form the basis of discussions with other disciplines so that a firm pricing for civil, electrical, and software costs can be obtained. The FEED stage will also include one or more design reviews to consider layout, value engineering, safety, and robustness issues. Where necessary, the process design is modified to safely give overall best value.

Detailed design is the third stage, in which many detailed subdrawings are generated to allow control of the construction of the plant. Process engineers would normally not have in-depth involvement with the production of installation drawings, other than participation in any design reviews or Hazard and Operability Studies (HAZOPs).

Related Terms
basic engineering design data, basis of design, deliverables, feasibility study, process design, process design package, stage gate model

stamped relieving capacity	The relieving capacity on a nameplate of a PRD (qv). Defined formally in API RP 576 - Inspection of Pressure Relieving Devices
stanchion	An upright support member, especially if part of a guardrail (qv) or barrier. Defined formally in BS EN ISO 14122 Safety of machinery. Permanent means of access to machinery. Working platforms and walkways
standard	1. In ISO standards, a good practice guidance document; see BS EN ISO 10209:2012 Technical product documentation - Vocabulary - Part 1: Terms relating to technical drawings: general and types of drawings 2. In NFPA standards, a standard is however a collection of mandatory requirements; see NFPA 11: Standard for Low-, Medium-, and High-Expansion Foam, 2016 Edition 3. *aka* STD. ANSI 'Standard' pipe schedule
standard acceleration due to gravity	(*symbol* g0 or gn) *aka* standard gravity. The nominal acceleration due to gravity (on Earth). Formally defined by the General Conference on Weights and Measures, as given in BS ISO 2041:2018 Mechanical vibration, shock and condition monitoring. Vocabulary
standard ambient temperature and pressure	*aka* SATP. A 'cozier' version of STP (qv), being 25° C (298.15 K) and a standard atmosphere (101.325 kPa)
standard atmosphere	see **standard atmospheric pressure**
standard atmospheric conditions	see **standard atmospheric pressure**
standard atmospheric pressure	*aka* standard atmosphere, standard atmospheric conditions, standard sea level conditions (though these have wider meanings on other contexts). The nominal pressure of atmosphere at mean sea level on Earth; equal to around 101 kPa
standard cubic feet per minute	*aka* SCFM. Flowrate in US customary units (qv) corrected to ASME standard conditions as per standard cubic foot (qv); cf actual cubic feet per minute
standard cubic foot	*aka* SCF. Moles of gas in a cubic foot when corrected to a set of standard conditions specified by ASME
standard cubic meter	Moles of gas in a cubic meter when corrected to standard atmospheric pressure (qv) and 15°C

standard dimension ratio	*aka* SDR. The ratio of pipe OD (qv) to wall thickness, commonly used in the USA as a pressure rating or class for PVC pipe. Dimension ratio (DR) classes are standardized in ASTM D2241 Standard Specification for Poly (Vinyl Chloride) (PVC) Pressure-Rated Pipe (SDR Series)
standard error	An estimate of the approximate standard deviation of a sample distribution
standard gravity	see **standard acceleration due to gravity**
standard operating procedure	*aka* SOP. Clearly written step-by-step operator instructions intended to produce consistent routine operation
standard oxygen concentration	*aka* standard oxygen level. There isn't a standard as such, but someone using this term probably means somewhere around the average 21% v/v of the Earth's atmosphere. A formal definition may be found in API RP 536 - NOx control on Fired Heaters at Refineries
standard oxygen level	see **standard oxygen concentration**
standard pipe schedule	A schedule of wall thicknesses for NPS pipe as US ANSI/ASME standards
standard reference measure	*aka* SRM. A measure of color (qv) used in the USA brewing industry; cf European Brewing Convention
standard safety valve	*aka* high lift safety valve. Either any safety valve which 'pops' at less than 10% pressure rise, or one which opens (either with a 'pop' or a gradual linear action) on pressure rise and recloses when pressure falls, depending on the defining standard. Conflicting formal definitions, therefore, in DIN 3320-1 Safety valves; safety shut-off valves; definitions, sizing, marking and ISO 4126-1:2013 Safety devices for protection against excessive pressure — Part 1: Safety valves
standard sea level conditions	see **standard atmospheric pressure**
standard state	A baseline physical state used in thermodynamics. However, there is no standard 'standard state', so it is always necessary to determine which 'standard state' is being referred to
standard temperature and pressure	*aka* STP. Before 1982, IUPAC set this as 0° C and 101.325 kPa absolute pressure (1 atm(a)). In 1982, this was rounded up to 0° C and 100 kPa absolute pressure. IUPAC are however not the only organization offering a definition of STP. It is not the same as standard state (qv), but the same caveat applies
standard two film theory	see **Whitman two film theory**
standard tubing	A strongly deprecated term; without stating the standard, it is at best meaningless

standby	*aka* standby service, standby unit. A redundant item of equipment, available for immediate startup as in duty/standby (qv); cf idle
standby power	The standby (qv) for a primary power source. This term is defined formally in NFPA 20: Standard for the Installation of Stationary Pumps for Fire Protection, 2019 Edition
standby service	Identical in meaning to the simpler standby (qv), this term is formally defined in EN ISO 10437 Petroleum, petrochemical and natural gas industries - Steam turbines - Special-purpose applications and EN ISO 13709:2003 Centrifugal pumps for petroleum, petrochemical and natural gas industries
standby unit	see **standby**
standing storage loss	Evaporative losses of liquids in storage. Highly specific formal definition in the context of floating roof tanks in API MPMS 19.1 Manual of Petroleum Measurement Standards Chapter 19.1 Evaporative Loss from Fixed-roof Tanks
standing wave	A wave whose peak amplitude is stationary in space (but not time). Defined formally in BS ISO 2041:2018 Mechanical vibration, shock and condition monitoring. Vocabulary
Stanton number	A dimensionless group (the ratio of heat transferred to thermal capacity of fluid); used to analyze forced convection
Stanton–Pannell chart	A plot of experimentally determined friction factors (qv) versus the Reynolds number (qv)
staple fibers	Fibers cut to a defined length
star delta	UK English for a wye delta (qv) starter
starter	*aka* drive starter, drive. An electrical/electronic device which supplies the initial high current required to initiate rotation of an electrical drive
startup	The systematic, monitored commencement of plant operation following completion of commissioning (qv), though sometimes used to mean other types of operational initiation of plant or equipment
state	1. A phase of matter (vapor, liquid, solid, plasma) 2. In modeling, a variable that is differentiated
state function	A thermodynamic function dependent only on the state of a system e.g.; temperature, pressure and quantity of substance; cf path function
State Implementation Plan	*aka* SIP. In the USA, a collection of regulations and documents used by a state, territory, or local air district to implement, maintain, and enforce the National Ambient Air Quality Standards (qv), and to fulfill other requirements of the Clean Air Act

state of the art	*aka* SOTA. The set of heuristics of a designer or designers
state task network	A diagram outlining the steps in a process and how equipment interacts with changes in condition of materials
static error	The difference between the measured value of the quantity and its true value; cf dynamic error
static head	In hydraulic/pneumatic calculations, the pressure at a given point in the no-flow condition, expressed as meters of pumped fluid column height; cf static pressure, dynamic head
static liquid level	The level above a pump's horizontal centerline of its feed reservoir. Defined formally in NFPA 20: Standard for the Installation of Stationary Pumps for Fire Protection, 2019 Edition
static mixer	see **inline mixer**
static pressure	*aka* static head (though this use may be deprecated). In pneumatic/ hydraulic calculations, the pressure at a given point in the no-flow condition. Defined formally in NFPA 24: Standard for the Installation of Private Fire Service Mains and Their Appurtenances, 2019 Edition; cf impact pressure
static pressure rise	The difference in static pressure (qv) between fan outlet and inlet, according to the formal definition in BS EN ISO 13705:2012/ISO 13705:2012(E) Petroleum, petrochemical and natural gas industries. Fired heaters for general refinery service
static seal	A sealing device which does not move. Defined formally in ASME BPE (American Society of Mechanical Engineers: Bioprocessing Equipment)
static sealing pressure rating	The maximum continuous differential pressure at maximum allowable temperature which a seal can withstand with shaft stationary. Defined formally in API RP 682 - Pump Seals; cf dynamic sealing pressure rating
static spray device	A fixed spray device (qv) producing a fixed spray pattern. Defined formally in ASME BPE (American Society of Mechanical Engineers: Bioprocessing Equipment); cf dynamic spray device
stationary installation	*aka* permanent installation. A permanent, static installation of LPG containers and ancillaries; defined formally in NFPA 58: Liquefied Petroleum Gas Code, 2017 Edition
stationary phase	1. In the context of chromatography, the immobile (usually porous solid) material which materials move though or past; cf mobile phase 2. In the context of batch microbial culture, the phase where the rate of cell growth and cell death is equal; cf log phase, lag phase, death phase

stationary seat	see **mating ring**
stationary vibration	Vibration with stable statistical characteristics over time. Defined formally in BS ISO 2041:2018 Mechanical vibration, shock and condition monitoring. Vocabulary
statistical associating fluid theory	*aka* SAFT. A thermodynamics model which predicts physical properties, especially of large molecules/polymers
statistical degrees of freedom	The number of independent variables in a probability estimate. Defined formally in BS ISO 2041:2018 Mechanical vibration, shock and condition monitoring. Vocabulary
statistical process control	*aka* SPC. A quality control methodology based in statistical analysis used to minimize waste, and maximize saleable product, such as six-sigma (qv)
statutory requirement	A legal obligation. Defined formally in BS EN ISO 9000 Quality management systems Fundamentals and vocabulary
STD	see **Standard**
steady state	A simplifying assumption of constant process conditions, to be used with great caution, since all of the 'interesting' things happen away from steady state; cf unsteady state
steam	Strictly, pure water vapor, but commonly applied to various mixtures of steam and other things such as water droplets (qv wet steam), air, contaminants and deliberate additives
steam cracking	Cracking (qv) using steam (qv)
steam cycle	*aka* Rankine cycle, Rankine steam cycle, Rankine vapor cycle. A theoretical model of a closed circuit of boiler, turbine and condenser like a power plant
steam distillation	The distillation of mixtures of heat-labile water-immiscible fluids below their boiling points, using steam as heat source
steam dryness fraction	see **dryness fraction**
steam explosion	see **boilover**
steam in place	*aka* SIP. Sanitization (qv) in place or sterilization in place (qv) using steam. Defined formally in ASME BPE (American Society of Mechanical Engineers: Bioprocessing Equipment)
steam injection	Addition of live steam (qv) to a process
steam jacket	see **jacket**
steam jet ejector	An ejector (qv) with steam as motive fluid
steam lift pump	see **mist lift pump**
steam methane reformer furnaces	Essentially, heated catalytic reactors which produce hydrogen, used to turn a mix of natural gas and steam into syngas (qv), the first step in producing ammonia, hydrogen, methanol, oxy-alcohols and GTL (qv) processes, among others

steam methane reforming	*aka* SMR. Producing syngas (qv) from hydrocarbons and water, commonly used to produce gray hydrogen (qv) or blue hydrogen (qv) from natural gas
steam or gas conditions	In the context of gas turbines, conditions such as temperature, absolute pressure, or dryness fraction which define the thermodynamic state of a gas; defined formally in ISO14661 Thermal turbines for industrial applications (steam turbines, gas expansion turbines) - General requirements
steam or gas flow	In the context of gas turbines, the mass flow required for a specified power output at given operating points under specified conditions. Defined formally in ISO14661 Thermal turbines for industrial applications (steam turbines, gas expansion turbines) - General requirements; cf steam rate
steam point	The boiling point of water at 1 atmosphere[97]
steam rate	*aka* SR. The ratio of steam mass flow rate into a turbine to its power output. Defined formally in EN ISO 10437 Petroleum, petrochemical and natural gas industries - Steam turbines - Special-purpose applications and ISO14661 Thermal turbines for industrial applications (steam turbines, gas expansion turbines) - General requirements; cf steam or gas flow
steam tables	Tables of thermodynamic (and sometimes other) data for water
steam tracing	The practice of keeping process lines above a minimum temperature, or tracing (qv), with steam, or the piping used for that purpose
steam trap	A device used to collect, separate and discharge condensate and non-condensables from a steam system
steam turbine	A rotary thermal power unit driven by high pressure steam, as opposed to a gas turbine (qv). Defined formally in ISO14661 Thermal turbines for industrial applications (steam turbines, gas expansion turbines) - General requirements
steam turbine governor	A control device that adjusts steam flow rate to turbine blades, maintaining a consistent speed despite changes in load
steam wetness	A subset of gas wetness (qv). Defined formally in the case of steam in ISO14661 Thermal turbines for industrial applications (steam turbines, gas expansion turbines) - General requirements
steel	"Everyone knows what steel is", as Nowe Ateny might have said
Steel Plate Fabricators Association	*aka* SPFA. USA trade association of steel plate construction fabricators and their suppliers
Steel Tank Institute	*aka* STI. USA trade association of steel tank fabricators and their suppliers

---

[97] Around 100°C, in the unlikely event that you didn't know

steeping	*aka* maceration. Soaking in water
Stefan's law	see **Stefan–Boltzmann law**
Stefan–Boltzmann constant	(*symbol* S or Σ) A constant in the Stefan–Boltzmann law (qv), derived from several other physical constants
Stefan–Boltzmann law	Law which states that a black body's radiated energy per unit surface area per unit time is proportional to the fourth power of that body's temperature
STEL	see **short term exposure limit**
stem nut	A threaded nut used to open or close a valve by turning the rotation of an actuator into the linear motion required
stem seal	A type of rotating shaft seal
stenching	*aka* odorisation, odorization, odorising, odourising, odorizing. Adding something with an obnoxious smell to an odorless but dangerous substance (e.g.; natural gas) to facilitate leak detection
step	The horizontal foot support of a stair (qv) or stepladder (qv) (steps on ladders are more commonly known as rungs). Defined formally in BS EN ISO 14122 Safety of machinery. Permanent means of access to machinery. Working platforms and walkways
step change	In management speak (qv), a supposedly significant change in something, not necessarily sudden and sharp (cf step response), and indeed not necessarily of any real significance
step feed activated sludge	A type of biological effluent treatment plant with multiple sequential points where untreated effluent is added
step response	Most commonly, the response of a system to a sudden sharp change in input, though there are various differing and highly specific meanings in engineering science
step weir	A water flow control structure in the form of a step
step-growth polymerization	see **condensation polymerization**
stepladder	Fixed access with broad steps pitched between 45° and 75° according to the formal definition in BS EN ISO 14122 Safety of machinery. Permanent means of access to machinery. Working platforms and walkways. The term is also commonly applied to a portable, self-supporting version of this. A ladder (other than a ship's ladder (qv)) is distinguished by not being self-supporting and having narrow steps/rungs; cf stair
sterile	Free of all living organisms (or in practice having had at least a 6-log, or up to a 12-log (depending on who you ask) reduction in living organisms); defined formally in ASME BPE (American Society of Mechanical Engineers: Bioprocessing Equipment)
sterilisation	see **sterilization**

sterility	The state of being sterile (qv). Defined formally in ASME BPE (American Society of Mechanical Engineers: Bioprocessing Equipment)
sterilizability	In the context of machinery, design for sterilization (qv). Defined formally in BS EN ISO 14159:2008 Safety of machinery. Hygiene requirements for the design of machinery
sterilization	*aka* sterilisation. The process of providing at least a 6-log reduction in all types of organisms, (standards vary from 6- to 12-log) or, more formally "A process effected by chemicals, heat or other physical means, aimed at removing or killing all forms of microorganisms, including bacterial spores", according to the EHEDG Glossary. Similar (but perhaps misleadingly absolute) definition in BS EN ISO 14159:2008 Safety of machinery. Hygiene requirements for the design of machinery; cf pasteurization, ultra high temperature
sterilization in place	*aka* SIP. Sterilization without dismantling, according to the EHEDG Glossary; cf cleaning in place
sterilized	Made sterile (qv)
STFT	Standard two film theory, see **Whitman two film theory**
STHE	see **shell and tube heat exchanger**
STI	see **Steel Tank Institute**
stick built	Constructed on site; cf modular
stiction	The static friction to be overcome to get a stationary object to move; a common issue with valves and other similar devices
stiffener tripping	In the context of pressure vessels, stiffener rings twisting around a shell connection point. Defined formally in PD 5500:2018+A1:2018 Specification for unfired fusion welded pressure vessels
stiffness	1. Resistance to bending force; defined formally in BS ISO 2041:2018 Mechanical vibration, shock and condition monitoring. Vocabulary 2. In simulation, a term for a region where things change very quickly as a function of time
still	Batch distillation equipment, especially simple plant used for potable alcohol production
still gas	*aka* refinery gas. Any mixture of gases produced in refineries by distillation, cracking, reforming, and other processes
Stillies	see **Stillson**
stilling baffle	*aka* baffle block (in some contexts). A baffle which directs and smooths flow in a channel or settling tank
stills	see **Stillson**

Stillson	*aka* Stillies, stills, monkey wrench, baboon spanner, French key. The inventor of the pipe wrench (qv), and a proprietary eponym for the tool itself
stirred tank	see **agitated vessel**
Stk	see **Stokes number**
stochastic	Random
stock	see **inventory**
stocking out	Sending bulk materials to storage
stockpile	A pile of bulk material
stocktake	see **inventory**
stocktaking	see **inventory**
Stoddard solvent	see **mineral spirits**
stoichiometric air	*aka* theoretical air. The chemist's theoretical minimum amount of air required to combust a given amount of fuel. Slightly different formal definitions are given in API Standard 521. Pressure-relieving and Depressuring Systems. Sixth Edition | January 2014, API RP 537 Flare Details for Petroleum, Petrochemical, and Natural Gas Industries and API RP 535 - Burners for Fired Heaters at Refineries
stoichiometric ratio	1. Generically, the ratio of reactants in a balanced chemical equation 2. In the context of fired heaters, the minimum ratio of fuel to air giving zero oxygen in flue gases; defined formally in API RP 535 - Burners for Fired Heaters at Refineries
stoichiometry	The numerical relationship between reactants and/or products in a balanced chemical equation
stoke	(*symbol* St) A metric but non-SI unit of kinematic viscosity
Stokes diameter	The diameter of a sphere that has the same density and settling velocity as the particle
Stokes number	*aka* Stk. A dimensionless number characterizing the behavior of spherical particles or droplets suspended in a fluid flow; defined strictly as the ratio of some characteristic time of the particle or droplet to that of the flow
Stokes' law	A theoretical equation for the settling velocity of small spherical particles in a viscous fluid, balancing drag on a particle against a driving force, such as buoyancy
Stokes-Einstein equation	A theoretical equation for the diffusion coefficient of small spherical particles
stonewall	*aka* choke point. Maximum stable mass flow for a centrifugal pump or compressor operating at low discharge pressure, due to choked flow (qv) into or out of the device

stope mining	*aka* stoping. Underground mining in strong rock by means of a series of excavated voids called stopes
stoping	see **stope mining**
stoplog	A modular element of a manually erected semi-permanent flow restriction in a channel
storage occupancy	A building used as a vehicle or goods store/shelter; defined formally in NFPA 30: Flammable and Combustible Liquids Code, 2018 Edition
storage tank	A fixed vessel in a process plant which is not part of a process stream. Two definitions which differ from each other, but match this definition in essence may be found in NFPA 30: Flammable and Combustible Liquids Code, 2018 Edition and API Standard 521. Pressure-relieving and Depressuring Systems. Sixth Edition | January 2014
storage tank building	In the context of classifying buildings for fire safety, an enclosure around a storage tank which increases hazards associated with fire. Defined formally in NFPA 30: Flammable and Combustible Liquids Code, 2018 Edition
storage vessel	see **storage tank**
storm drain	*aka* storm sewer, surface water drain, stormwater drain. Drainage exclusively for surface water (qv). Some sources give surface water sewer as a synonym of storm drain, but this term is strongly deprecated for any usage. Surface water drains should only carry surface water, without sewage, so they are not sewers. Though a combined sewer (qv) carries a mixture of sewage and surface water, a sewer (qv) carries sewage.
storm overflow	A UK term for the point at which a certain amount of discharge of highly diluted, but only partially settled sewage (produced in the event that storm tanks at a municipal sewage treatment works have been completely filled by an extended storm event) is legally permitted
storm sewer	see **storm drain**
storm tank	see **storm water retention tank**
storm water	Usually means rainwater that has been collected from hard surfaces, especially if collected to a sewer (qv) or storm drain (qv), but can also be used for accumulated rainwater than has swollen a water course
storm water retention tank	Large tanks at a municipal sewage treatment plant which capture any incoming flow in excess of plant design throughput due to high storm water (qv) flows
stormwater drain	see **storm drain**

STP	1. Sewage treatment plant
	2. see **standard temperature and pressure**
strain	1. *aka* Cauchy strain. Other disciplines have other definitions of strain, but in an engineering context, strain is generically the ratio of a deformation to original size caused by an applied force (stress)
	2. A subtype of species (qv) of organism
strain gauge	An electromechanical transducer (qv), the resistance of which changes in proportion to applied strain (qv)
strain hardening	see **work hardening**
strain insulator	A cable insulator designed to operate under mechanical strain
strain limiting load	The load associated with strain (qv) limit. Defined formally in API RP 579 - Fitness for Service
strainer	A coarse filter, the main filtration mode of which is straining (qv). Defined formally in the context of fired heaters in API RP 535 - Burners for Fired Heaters at Refineries
straining	A coarse filtration mode; the capture of solid particles larger than the gaps in filtration media from a fluid cf screening
strake	*aka* helical strake, spoiler. A helical band attached around a stack to prevent wind-induced oscillations. Defined formally and identically in BS EN ISO 13705:2012/ISO 13705:2012(E) Petroleum, petrochemical and natural gas industries. Fired heaters for general refinery service, API STD 530 - Calculation of Tube Heater Thickness and API STD 560 - Fired Heaters for General Refinery Service
stranded conductors	In an electrical cable, core conductors comprised of many individual wires; cf multicore cable
strategy	A considered and resourced approach to achieving overarching, longer term aims
stratified bed	see **layered bed**
stratified flow	A two-phase flow regime in horizontal or declining pipes with liquid in the pipe bottom and gas above
stream	A slightly vague term, used to mean two related but distinct things:
	1. Most commonly, process train (qv)
	2. Less commonly, pass (qv)
stream availability	see **exergy**
stream table	A table on a process flow diagram (qv) with information on stream contents and conditions
streamline flow	see **laminar flow**

strength welded	In a shell and tube heat exchanger (qv), a tube-to-tubesheet welded joint with design strength at least equal to the axial strength of tube. Defined formally in API 660 - Shell-and-Tube Heat Exchangers and API RP 661 - Heat Exchangers
stress	Other disciplines have other definitions, but in an engineering context, stress is generically the force which produces a strain (qv)
stress concentration factor	The ratio of maximum stress to average section/bending stress; defined formally in API RP 579 - Fitness for Service
stress corrosion cracking	*aka* SCC.
	1. Properly used to only to mean crack growth (especially in a metal) caused by a synergistic combination of stress and corrosion
	2. However, the term can be used informally to describe any slow, environmentally favored cracking
stress cycle	A single instance of stress difference, going from an initial value through a maximum and a minimum and back to the initial value, as part of a repeating pattern; defined formally in API RP 579 - Fitness for Service
stress intensity factor	The theoretical stress-field intensity near the tip of an ideal crack; useful for predicting cracking; defined formally in API RP 579 - Fitness for Service
stress oriented hydrogen induced cracking	*aka* SOHIC. Hydrogen induced cracking (qv) with cracks aligned almost perpendicular to stress. Defined formally in API RP 571 - Damage Mechanism Affecting Fixed Refinery Equipment, API RP 579 - Fitness for Service and API RP 932B Corrosion Air Coolers
stress thickness	The calculation of material thickness based solely on allowable stress; defined formally in API STD 530 - Calculation of Tube Heater Thickness
stretch ratio	see **extension ratio**
string	1. *aka* stringer. The framework supporting walkway steps
	2. *aka* drilling string. A collective term used in the oil and gas industry for the bottom hole assembly (qv), transition pipe (qv) and drill pipe (qv) used to transmit fluids and torque between surface and drill bit
string wound cartridge filter	see **spiral wound cartridge filter**
stringer	1. An elongated discontinuity within bar stock (qv) caused by stretching of a non-metallic inclusion during rolling
	2. An assembly of nuclear fuel elements
	3. The framework supporting walkway steps

stringer indication	A void as a result of the loss of a stringer (qv) in bar stock. Defined formally in ASME BPE (American Society of Mechanical Engineers: Bioprocessing Equipment)
stripper	A unit operation usually involving contacting liquid and vapor phases in a column, stripping out light components to yield a stream rich in the heavy component
stripper column	*aka* stripping column; see **stripper**
stripping	The transfer of components from a liquid stream to a vapor. Where this is deliberate, the unit operation contacting the two phases is known as a stripper (qv)
stripping column	see **stripper**
stripping section	*aka* exhausting section. The distillation column section below the feed section where stripping (qv) predominates; cf rectifying section, feed section
strong acid cation resin	*aka* SAC resin, SAC exchange resin. In the context of ion exchange (qv), a resin type used in dealkalization and desalination systems; the functional group (qv) tends to be sulfonic acid
strong base anion resin	*aka* SBA resin, SBA exchange resin. In the context of ion exchange (qv), a resin type used to selectively remove a wide range of solutes, including silicates, sulfates, nitrates, and perchlorate; qv type 1 strong base anion resin, type 2 strong base anion resin
Strouhal number	*aka* Sr. A characteristic coefficient (qv)
structural damping	see **hysteresis damping**
structural design code	A purchaser-specified structural design standard. Defined formally in BS EN ISO 13705:2012/ISO 13705:2012(E) Petroleum, petrochemical and natural gas industries. Fired heaters for general refinery service, API RP 537 Flare Details for Petroleum, Petrochemical, and Natural Gas Industries, API STD 530 - Calculation of Tube Heater Thickness and API STD 560 - Fired Heaters for General Refinery Service
structural engineering drawing	A dimensioned drawing of a structure's frames and reinforcements. Defined formally in BS EN ISO 10209:2012 Technical product documentation - Vocabulary - Part 1: Terms relating to technical drawings: general and types of drawings
structural frame drawing	A drawing showing how a structure is strengthened and stabilized by its frame. Defined formally in BS EN ISO 10209:2012 Technical product documentation - Vocabulary - Part 1: Terms relating to technical drawings: general and types of drawings

structure	1. The system of organization of the parts of a system, as defined formally in BS EN ISO 10209:2012 Technical product documentation - Vocabulary - Part 1: Terms relating to technical drawings: general and types of drawings, or a system organized in accordance with such a structure 2. Something constructed of parts assembled in an organized way, as defined in NFPA 1142: Standard on Water Supplies for Suburban and Rural Fire Fighting, 2017 Edition
structure diagram	A hierarchical 'tree' type illustration of the interrelationship between objects in a system. Defined formally in BS EN ISO 10209:2012 Technical product documentation - Vocabulary - Part 1: Terms relating to technical drawings: general and types of drawings
structured packing	Vessel packing materials supplied as large prefabricated blocks, see **Figure S7**; cf random packing
struvite	Magnesium ammonium phosphate, a mineral which can cause scaling problems when it spontaneously precipitates in wastewater treatment, (but which can be a commercially useful product when formed deliberately)
stub-in tee	*aka* set-in branch. A reducing tee formed by a hole being drilled in the main pipe, a reduced bore branch pipe fitted inside the hole, and the product welded together such that the branch pipe extends to the inside of the main pipe (hence stub-in). Similar to a Weldolet (qv)

Figure S7 Structured packing (courtesy: Sulzer)

stub-on tee	*aka* set-on branch. An equal tee formed by a hole being drilled in the main pipe, and a branch pipe of the same bore being fitted over the hole, and the product welded together such that the branch pipe stays outside of the main pipe (hence stub-on)
Stubs' iron wire gauge	see **Birmingham wire gauge**
Stubs' steel wire gauge	see **Birmingham wire gauge**
Stubs' Wire Gauge	*aka* Stubs' steel wire gauge, Stubs' iron wire gauge; see **Birmingham wire gauge**
stuffing box	*aka* gland package. A compartment used to contain and compress stuffing to form a nonhygienic shaft seal. Defined formally in ASME BPE (American Society of Mechanical Engineers: Bioprocessing Equipment)
STY/DVB copolymer	A crosslinked styrene/divinylbenzene polymer
styrene butadiene rubber	*aka* SBR, Buna-S. An elastomer used in red rubber (qv) gaskets
SUB	see **single use bioreactor**
subassembly drawing	A drawing showing part of an assembly drawing (qv). Defined formally in BS EN ISO 10209:2012 Technical product documentation - Vocabulary - Part 1: Terms relating to technical drawings: general and types of drawings
subbing out	*(informal)* To subcontract (qv)
subcontract	1. A contract to carry out part of a larger contract 2. The action of engaging an entity to do something by means of a subcontract (qv subbing out)
subcooling	Cooling a liquid to below its boiling point; cf supercooling
subcritical	1. In nuclear engineering, the state in which a fission reaction is not self sustaining 2. In fluid mechanics/hydraulics, applied to a slow, stable flow regime dominated by gravitational forces, with a Froude number (qv) less than one 3. Used to describe a fluid as it approaches its liquid-vapor critical point from a lower temperature/pressure
subcritical water extraction	*aka* SWE. A technique for rapidly extracting less-polar compounds using water maintained in a liquid state under high pressure at a temperature between $100°$ C and $374°$ C; cf supercritical fluid extraction

subject matter expert	*aka* technical authority, technical expert, subject matter technical expert (SMTE). An engineer with sufficient expertise, skills and experience to be able to offer reliable solutions and oversight, especially in a design context. Defined formally in ASTM E2500-20 Standard Guide for Specification, Design, and Verification of Pharmaceutical and Biopharmaceutical Manufacturing Systems and Equipment
subject matter technical expert	see **subject matter expert**
sublaterals	Sub-sub-divisions of a main feed or drain pipe, especially a sewer. Branches feed sublaterals which feed laterals which feed a main sewer, often via a seal
sublimate	Perhaps confusingly, a term which is both synonymous with sublimation (qv), or used to mean the solid formed by desublimation (qv)
sublimation	Changing directly from solid to gas without a liquid stage; the opposite of ablimation (qv)
submersible pump	see **sump pump**; cf immersible pump
subnatant	Something lying under a supernatant (qv); cf infranatant
suboptimal	Less than ideal. A common euphemistic understatement for a solution or situation so poorly conceived or executed that negligence or malice cannot be discounted
substance	Matter, or 'stuff' (the UK PPC Regulations offer a more technical but scarcely less broad definition)
substitute liquid fuels	*aka* SLF. Organic waste used as cement kiln fuel
substitute natural gas	see **synthetic natural gas**
substitution reaction	A chemical reaction in which a functional group (qv) is substituted
substrate	1. A 'base material' to builders and printers 2. Also, by analogy, used for something subject to a chemical or biological reaction
subsurface foam injection	The introduction of firefighting foam close to the bottom of a storage tank. Defined formally in NFPA 11: Standard for Low-, Medium-, and High-Expansion Foam, 2016 Edition
subsurface safety valve	*aka* SSSV or SCSSV. See **downhole safety valve**
suburb	In firefighting, an area with population density in the range 500-1000 people/square mile, according to NFPA 1142: Standard on Water Supplies for Suburban and Rural Fire Fighting, 2017 Edition
suburban	see **suburb**

success	Achieving an objective, according to BS EN ISO 9000 Quality management systems Fundamentals and vocabulary[98]
sucker rod pump	see **nodding donkey**
suction diffuser	A type of flow straightener (qv) on a pump suction. Defined formally in NFPA 20: Standard for the Installation of Stationary Pumps for Fire Protection, 2019 Edition
suction drum	see **knockout drum**
suction gas	see **producer gas**
suction head	see **suction pressure**
suction lift	In the context of a pump, having to overcome negative liquid head in suction arrangements; cf flooded suction
suction pressure	1. The pressure at a pump suction flange, according to NFPA 20: Standard for the Installation of Stationary Pumps for Fire Protection, 2019 Edition 2. May also be used to mean the pressure at the centerline of a horizontal pump, or the pressure of gas at a compressor intake
suction pressure regulating valve	An actuated valve (qv) in pump discharge piping used to actively maintain positive suction pressure based on a pressure signal from suction piping. Defined formally in NFPA 20: Standard for the Installation of Stationary Pumps for Fire Protection, 2019 Edition
suction scrubber	see **knockout drum**
suction specific speed	A dimensionless parameter (therefore not actually a speed) that defines relationship between flow, NPSHr (qv) and speed for similar pump impellers. Defined formally in EN ISO 13709:2003 Centrifugal pumps for petroleum, petrochemical and natural gas industries
suction tank	A vessel maintained at a low pressure to serve as a passive source of vacuum. Defined in a fire protection context in NFPA 22: Standard for Water Tanks for Private Fire Protection, 2018 Edition
SuDS/SUDS	see **sustainable urban drainage systems**
sugar grist	see **grist**
sulfate process	1. A TiO_2 production process, an alternative to the chloride process (qv) 2. see **Kraft process**
sulfide stress cracking	*aka* SSC. A form of hydrogen induced cracking (qv) of metal produced by the combination of tensile stress, water and H_2S. Defined formally in API RP 932B Corrosion Air Coolers and API RP 579 - Fitness for Service

---

[98] Also something that Netherlanders - and Klingons - shout at each other!

Sulfinol process	Gas sweetening (qv) using a proprietary hybrid physical/chemical solvent called Sulfinol
summing point	In a process control context, a point at which signals are summated
sump	A low lying chamber used to collect liquids, especially effluents
sump pump	*aka* submersible pump. A waterproof (usually IP68) pump without inlet pipework used to pump from a sump (qv)
Sundyne pump	Proprietary eponym for a magnetic drive pump (though Sundyne make a lot of other kinds of pumps)
superalloy	An alloy which can operate close to its melting point, such as Hastelloy (qv); cf hyperalloy
superaustenitic stainless steel	Austenitic stainless steel with a pitting resistance equivalent number (qv) greater than 40. Defined formally in ASME BPE (American Society of Mechanical Engineers: Bioprocessing Equipment)
supercooling	Taking a liquid below its freezing point, without it freezing; cf subcooling
supercritical	1. In a thermodynamic context, a fluid at a temperature and pressure above its critical point 2. In hydraulics, a fluid velocity greater than its wave velocity
supercritical flow	Fluid flow at a velocity greater than its wave velocity
supercritical fluid extraction	*aka* SFE. Extraction with a supercritical (in a thermodynamic sense) fluid
supercritical water oxidation	*aka* SCWO. Oxidation using a supercritical (in a thermodynamic sense) aqueous liquid
superduplex stainless steel	*aka* ferralium. Duplex stainless steel with enhanced pitting corrosion resistance due to a greater chromium content than standard duplex; defined formally in ASME BPE (American Society of Mechanical Engineers: Bioprocessing Equipment)
superficial velocity	Commonly used to mean average velocity (qv), but can also mean velocity of a given phase in a multiphase system
superfluid	see **inviscid fluid**
superfractionation	Separation by fractional distillation of close-boiling liquid mixtures
Superfund	*(informal)* see **Comprehensive Environmental Response, Compensation and Liability Act**
superheat	How much hotter a vapor is than its boiling point at that pressure
superheated steam	Steam above its boiling point at that pressure; cf saturated steam

superheater	A heat exchanger where saturated steam is heated to produce superheated steam. Defined formally in the context of HRSGs (qv) in API RP 534 – Heat Recovery Steam Generators; cf desuperheater
superimposed backpressure	The outlet static pressure of a PRV (qv) at the point where it is required to lift. Defined formally in API Standard 521. Pressure-relieving and Depressuring Systems. Sixth Edition | January 2014, API RP 520 P1 7th Edition, January 2000 Sizing, Selection, and Installation of Pressure-Relieving Devices in Refineries; Part I - Sizing and Selection and API RP 576 - Inspection of Pressure Relieving Devices
supernatant	The clearer liquid above a settled lower layer (often of sludge); cf infranatant, subnatant
supernate	see **supernatant** (though a minority of sources seek to give a slightly different meaning to this term)
supersaturation	The state in which a solution contains more dissolved solute than its solubility
supersonic	A velocity greater than that of sound in a given fluid cf supercritical
supervised flame	Flame whose presence is monitored by a flame detector; defined formally in NFPA 86: Standard for Ovens and Furnaces, 2019 Edition
supervisory control	Master process control of many individual autonomous control loops on a plant
supervisory control and data acquisition	*aka* SCADA. One of the two most common forms of supervisory control, the other being DCS (qv). A high level system which can control multiple field controlled systems or PLCs (qv) and provide an easy to navigate interface for operators
supplemental gas	*aka* assist gas. Fuel gas added to a low heating value relief gas going to an endothermic flare burner to facilitate its combustion. This term defined formally in API RP 537 Flare Details for Petroleum, Petrochemical, and Natural Gas Industries
supplementary loaded safety valve	A safety valve using an additional sealing force until set pressure is reached. Defined formally in ISO 4126-1:2013 Safety devices for protection against excessive pressure – Part 1: Safety valves

supplier	*aka* provider, vendor. Generically, the commercial entity supplying a product or service. Six different formal definitions (none of which mention services) may be found in BS EN 82079-1:2012 Preparation of instructions for use. Structuring, content and presentation. General principles and detailed requirements, ISO14661 Thermal turbines for industrial applications (steam turbines, gas expansion turbines) - General requirements, IEEE 830-1998 - IEEE Recommended Practice for Software Requirements Specifications, EN ISO 10437 Petroleum, petrochemical and natural gas industries - Steam turbines - Special-purpose applications, EN ISO 13709:2003 Centrifugal pumps for petroleum, petrochemical and natural gas industries and API RP 682 - Pump Seals and BS EN ISO 9000 Quality management systems Fundamentals and vocabulary
supplier drawing	A drawing from an external part supplier. Defined formally in BS EN ISO 10209:2012 Technical product documentation - Vocabulary - Part 1: Terms relating to technical drawings: general and types of drawings
support media	Dense materials with controlled particle sizes used to support finer particles in a packed bed
supports	*aka* pipe supports. The structures which hold pipes in place during operation, which come in a variety of types such as slippers, shoes, trunnions, brackets and hangers (qv)
suppressed explosion pressure	see **vented explosion pressure**
surcharging	1. Generically, overloading 2. Specifically, hydraulically overloading a sewer (qv), especially where this causes sewage to escape from manhole (qv) covers
surface condenser	A compressor ancillary which recovers condensate by cooling
surface filter	A filter in which excluded particles are stopped at the surface of the filter; cf depth filter
surface finish	The roughness (qv) of a surface, often expressed as Ra (qv). Defined formally in ASME BPE (American Society of Mechanical Engineers: Bioprocessing Equipment)
surface inclusion	The embedding of foreign materials into the top layer of something. Defined formally in ASME BPE (American Society of Mechanical Engineers: Bioprocessing Equipment)
surface loading	*aka* rise rate. In the context of environmental engineering, the superficial velocity in a vertical direction
surface moisture	see **unbound moisture**; cf adventitious moisture

surface residual	An adherent foreign material particle. Defined formally in ASME BPE (American Society of Mechanical Engineers: Bioprocessing Equipment)
surface roughness	*aka* roughness. A variation in surface profile usually measured as the average difference between the highest and lowest points on a surface; cf Ra
surface rupture	The "breaking or tearing of a surface usually obtained through the impact of a shot- or bead blasting medium... these areas can harbor soils and microorganisms and be difficult to clean", according to the EHEDG Glossary
surface tension	(*symbol* S) A force resulting from the tendency of liquids to minimize surface area
surface treatment	"A process whereby chemical or mechanical properties of the existing surface are altered", according to the EHEDG Glossary, though this is true in all sectors
surface water	Water from rivers, streams, ditches, canals, reservoirs, ponds, lakes, sea, etc.; cf groundwater
surface water drain	see **storm drain**
surface wave	*aka* Rayleigh wave. A wave at the interface between phases; defined formally in BS ISO 2041:2018 Mechanical vibration, shock and condition monitoring. Vocabulary
surfactant	A substance which lowers interfacial tension between liquids, such as a detergent
surge	A transient increase in flow, pressure, or electrical current. Sometimes used as synonym of surging (qv)
surge analysis	*aka* transient study. A mathematical analysis of large transient high and low pressure events due to velocity changes in pipes
surge pressure	Large transient high pressure due to velocity changes in pipes
surge time	The time taken to fill a vessel from 'normal' to 'high' level by closing the outlet under normal inlet flow conditions; cf hold up time
surging	A fluid vibration in a fan or compressor caused by backpressure instability. Defined formally in BS ISO 2041:2018 Mechanical vibration, shock and condition monitoring. Vocabulary
surplus activated sludge	*aka* SAS, WAS; see **waste activated sludge**
survey	Generically, an evaluation of a site. ASME BPE (American Society of Mechanical Engineers: Bioprocessing Equipment) has its own highly specific definition
suspended growth	Having microorganisms in suspension rather than attached to a surface (cf attached growth)
suspended particulate matter	*aka* SPM; see **particulates**

suspended solids	see **total suspended solids**
suspension	An unstable mixture of particles in a fluid, which will separate on standing; cf colloid
sustainability	see **Box S3**.
sustainable aviation fuel	*aka* SAF. A non-fossil liquid fuel used to power airplanes
sustainable development	Economic development which is sustainable (qv), see **Box S3**.
sustainable drainage	see **sustainable urban drainage systems**
sustainable drainage systems	see **sustainable urban drainage systems**
sustainable urban drainage systems	*aka* SuDS/SUDS, sustainable drainage systems, sustainable drainage. Drainage systems with flow buffering capacity built in which reduce flooding risk from hard surface runoff
sustained success	Success (qv) over time, according to BS EN ISO 9000 Quality management systems Fundamentals and vocabulary
Sv	see **sievert**
SVI	see **sludge volume index**
SW	1. see **sour water** 2. Salt water
SW(F)	*(drawing notation)* Socket weld (fitting/flange)
swab valve	A valve on the top of a Christmas tree (qv) which provides vertical access to the wellbore
SWAG	see **scientific wild ass guess**
Swagelock	Proprietary eponym for a high-pressure piping system based on compression fittings
swarf	Pieces of metal, wood, or plastic debris resulting from machining
SWE	see **subcritical water extraction**
Sweco	Proprietary eponym for a type of round vibratory screen sifter
sweep velocity	*aka* sweeping velocity. The minimum velocity of a fluid required to maintain pipe surface free of accumulated material cf saltation velocity
sweeping velocity	see **sweep velocity**
Sweepolet	A proprietary eponym, an abbreviated form of SWEEPing out LET. One of various olet (qv) types
sweet gas	Petroleum gas containing insignificant quantities of odorous hydrogen sulfide; cf sour gas, acid gas
sweet products	In the context of hydrocarbon products, products low in hydrogen sulfide. Sometimes, more broadly (and many think wrongly), products low in all sulfur compounds; cf sour products

Box S3. Sustainability

Sustainability has a wide range of meanings. The general idea is that it refers to economic development which is sustainable, but it is not clear what that means.

In 2007 a report for the USA Environmental Protection Agency stated: "While much discussion and effort has gone into sustainability indicators, none of the resulting systems clearly tells us whether our society is sustainable. At best, they can tell us that we are heading in the wrong direction, or that our current activities are not sustainable. More often, they simply draw our attention to the existence of problems, doing little to tell us the origin of those problems and nothing to tell us how to solve them".

The key problem with the term sustainability is that is highly politicized. A Greenpeace member, a trade union activist, and a chemical engineer might all use the term to mean three completely different things. The IChemE have produced guidelines on what it means to chemical engineers, in the form of sustainability metrics[99], which facilitate the process of analyzing the problem and its potential solutions at least semi-quantitatively.

The IChemE's interpretation of sustainability does not support shooting for theoretical perfection in a small number of aspects of process design. Chemical engineers are concerned about the environment, but we know that the curves of process yield, energy recovery, safety, and environmental protection against cost are exponential. Both perfect processes and complete safety are infinitely costly.

Related Terms
circular economy, greenwash

swell	In the context of ion exchange (qv), resin expansion (sometimes also contraction)
swept volume	The part of a reciprocating device that performs useful work
Swg	*(drawing notation)* Swage nipple
swing bend panel	see **transfer panel**
swirl number	A dimensionless group, the ratio of angular and axial discharge momentum flux; used in analysis of combustion characteristics. Defined formally in API RP 535 - Burners for Fired Heaters at Refineries
Swiss cheese model	A model used in management of safety incidents in which the holes in layers of protection (akin to those in Swiss cheese) have to line up for an incident to occur

---

[99] Institution of Chemical Engineers (2002) *The Sustainability Metrics: Sustainable development progress metrics recommended for use in the process industries*, Rugby, UK: IChemE

switch loading	The hazardous practice of loading a fuel tanker compartment with a different fuel to that which it carried last time. Defined formally in API RP 2001 - Fire Protection at Refineries
switchboard	An electrical enclosure used to mount buses, instruments, protective devices, and switches; cf MCC, switchgear
switchgear	Enclosed, interconnected assemblies of electrical switches and their associated control, measurement, protection and regulation equipment, as well as the components of such assemblies. Defined formally in IEV ref 441-11-02; cf MCC, switchboard
SWMS	Safe working method statement, see **method statement**
swoop down	A quick planned event for a small necessary repair (along with potentially some other jobs that would fit in the down time)
SWOT analysis	Analysis of strengths, weaknesses, opportunities and threats
SWP	Safe working pressure
SWS	1. Sour water stripper, see **sour water**
	2. see **solvent-water separator**
SWUT	Shear wave ultrasonic testing
syneresis	The oozing of water from a gel, often undesirable in food
syngas	see **synthesis gas**
synthesis gas	*aka* syngas. A fuel gas mostly comprising carbon monoxide and hydrogen, produced from coal and other organic feedstocks by a number of processes; used as an intermediate in manufacture of synthetic natural gas (qv) amongst other things. When produced by coal gasification, identical to coal gas (qv)
synthetic fibers	Man-made textile fibers
synthetic foam concentrate	A concentrate of non-hydrolyzed-protein foaming agents. Defined formally in NFPA 11: Standard for Low-, Medium-, and High-Expansion Foam, 2016 Edition
synthetic hydrocarbons	Broadly, hydrocarbons produced by means other than refining of crude oil, though other usages exist
synthetic medium	see **defined medium**
synthetic natural gas	*aka* SNG, substitute natural gas. Fuel gas with a similar composition to natural gas (cf synthesis gas), mostly made currently by coal gasification
synthetic special atmosphere	A tightly specified artificial furnace atmosphere. Defined formally in NFPA 86: Standard for Ovens and Furnaces, 2019 Edition
syphon	*aka* siphon. The action of carrying liquid from high level to low under gravity via a point higher than 'high level' by virtue of liquid cohesion; also used for a pipe which syphoning flow passes through

syphon breaker	*aka* siphon breaker. Most commonly, an inverted vertical U in pipework, fitted at its apex with an inverted ball type non return valve to prevent syphoning or backsyphoning, though there are compact fittings which incorporate the working principle of this cf lute
system	An assembly of multiple interacting elements (commonly items of equipment). Four related but different formal definitions in BS ISO 2041:2018 Mechanical vibration, shock and condition monitoring. Vocabulary, BS EN ISO 9000 Quality management systems Fundamentals and vocabulary, BS EN ISO 10209:2012 Technical product documentation - Vocabulary - Part 1: Terms relating to technical drawings: general and types of drawings and NFPA 59: Utility LP-Gas Plant Code, 2018 Edition
system actuation valve	The main valve that controls the flow of water into a water spray system, such as a deluge valve (qv). Defined formally in NFPA 15 Standard for Water Spray Fixed Systems for Fire Protection
system boundary	Either an imaginary line drawn around a process on a PFD (qv), or other criteria which specify system limits, as per the rather longer formal definition in EN ISO 14040:2006 Environmental management - Life cycle assessment - Principles and framework
system curve	*aka* system head curve. A graphical representation of required pump head to achieve all possible flow rates for a hydraulic system; cf characteristic curves
system failure mode and effects analysis	*aka* SFMEA, functional failure mode and effects analysis. A subtype of failure mode and effects analysis (qv), as applied to a system
system head	The pump head required to achieve a given flow rate for a hydraulic system
system head curve	see **system curve**
system volume	1. The total system liquid volume, according to ASME BPE (American Society of Mechanical Engineers: Bioprocessing Equipment) 2. More commonly used to mean the total internal volume of a system (which may be far larger)
systematic error	see **error**
système international d'unités	see **SI units**
systems engineering	The application by management of 'systems thinking' to plant design and operation

T

T&C	1. Terms and conditions [of contract] 2. *(drawing notation)* Threaded and coupled
T&G	*(drawing notation)* Tongue and groove
TA	see **technical authority**
tack weld	A small temporary weld used to hold parts in place prior to final welding. Defined formally in ASME BPE (American Society of Mechanical Engineers: Bioprocessing Equipment)
tag number	see **reference designation**
tail	see **tails**
tail end	The last stages of a process, the opposite of front end (qv)
tail gas	1. Generically, any gaseous industrial effluent 2. Specifically applied to certain gaseous effluents in oil refining, especially those exiting the Claus process (qv) in natural gas processing 3. Applied in the chloride process (qv) to the residual gases once all economically recoverable titanium tetrachloride has been removed
tail gas treating unit	*aka* TGTU. A unit which recovers sulfur from Claus process (qv) tail gas, for recycle to process
tailing	In confectionery manufacture, a small amount of caramel/filling which gets stretched between the filling depositor and the item being filled
tailings	The gangue (qv) rich waste product from beneficiation (qv)
tailrace	A channel carrying water away from a turbine or process; cf penstock
tails	*aka* tail. The equivalent of bottoms (qv) in potable alcohol distillation: the higher boiling fractions, and the opposite of heads (qv) or foreshots (qv)
takeoff	see **material takeoff**
takeoff rate	The flowrate of distillate from a column
TAME	Tertiary amyl methyl ether
tampon	A plug installed in a catalyst dump nozzle (qv) used to keep catalyst in the reactor while the nozzle flange is removed
TAN	see **total acid number**
tan/tan	A straight vessel wall measurement (between the points at which opposite head tangents start)
tangential flow filtration	see **crossflow filtration**
Tangye pump	Proprietary eponym for manual hydraulic pump

tank	*aka* vessel. Primary containment for process material; the precise definition varies between industries. In petrochemicals, a term used only for large, atmospheric storage tanks, but in specialty chemicals it can be used to refer to smaller pressure vessels
tank blanketing	see **blanketing**
tank breather	see **vent/vac valve**
tank farm	A location with many storage tanks
tank riser	A large diameter shaft connecting an elevated or buried tank to the surface, which may be used for access or for protection of services to the tank. Defined formally in the case of fire water tanks in NFPA 22: Standard for Water Tanks for Private Fire Protection, 2018 Edition
tank wall	Plates which form the shell of a tank; defined formally in API STD 620 - Low Pressure Storage Tanks
tanker	A mobile tank for bulk liquid transport, such as a road tanker (qv) or 'oil tanker' (qv)
Tannoy	Proprietary eponym for a public address loudspeaker[100]
tantalum	A very expensive and highly chemical resistant metal, sometimes referred to as 'metallic glass'; used in the repair of glass lined vessels or very thin components (e.g.; diaphragms of pressure gauges) in highly corrosive service
tap	1. UK English for a faucet (which is why potable water is known as 'tap water' in the UK) 2. A tool used to cut a female thread
tap-o-meter	A hammer used for percussive maintenance (qv)
tapping	1. Making a connection to a pipeline or pressure vessel (hot tapping if done whilst in service, cold tapping if not) 2. Making a female thread using a tap (qv)
TAR	see **turnaround**
tar	*aka* pitch. Strictly, a dark viscous liquid produced by destructive distillation of carbonaceous materials, but often informally used to refer to any high viscosity/boiling point oil industry waste stream - particularly where chemical composition is ambiguous
tar acids	see **coal tar acids**
tar bases	see **coal tar bases**

---

[100] Taken from a company whose name was derived from tantalum (qv) alloy

TARFU	*aka* totally and royally fucked/fouled up, things are really fucked/fouled up. Part of a spectrum (SNAFU, FUMTU, TARFU, FUBAR - qv), conveying increasing degrees of disorder/deviation from correct practice or operation and the absence of a means to resolve it
target	Often used to denote the possible victims or casualties (both human and equipment) of a potential incident
target group	The intended readership of instructions for use (qv); defined formally in BS EN 82079-1:2012 Preparation of instructions for use. Structuring, content and presentation. General principles and detailed requirements
target wall	*aka* reradiating wall. The refractory wall of a fired heater upon which flames impinge directly. Defined formally in BS EN ISO 13705:2012/ISO 13705:2012(E) Petroleum, petrochemical and natural gas industries. Fired heaters for general refinery service, API STD 530 - Calculation of Tube Heater Thickness and API STD 560 - Fired Heaters for General Refinery Service
tarmac	1. *aka* rolled asphalt, asphalt, blacktop (US), pavement (US). Proprietary eponym, short for Tarmacadam: a road surfacing material comprised of aggregate and a binding material (which was originally, but is not now always tar) 2. A UK construction company
Tarmacadam	see **tarmac**
task	1. A design task involves using a well-established methodology and robust data to grind out an obvious answer (cf problem) 2. An activity performed on or near a machine, as defined formally in BS EN ISO 12100:2010 Safety of machinery. General principles for design. Risk assessment and risk reduction 3. More generally, used informally as interchangeable with 'job' or 'activity' in the context of plant maintenance or construction
Tatoray process	A proprietary process for producing mixed benzene and xylenes from toluene and heavy aromatics
Taylor bubble	Large bubbles of the lighter phase seen in slug flow (qv) and plug flow (qv) two-phase flow regimes
Taylor vortices	*aka* Taylor's vortices, toroidal vortices. A flow pattern (or patterns if we include their wavy- and modulated wavy- subtypes) which occurs in the gap between two rotating concentric cylinders
Taylor's vortices	see **Taylor vortices**
TBA	To be advised (or sometimes to be announced)

TBC	1. To be confirmed
	2. Total bacteria count
TBD	To be determined
TBE	*(drawing notation)* Thread both ends
TC	see **temperature controller**
tce	see **tonne of coal equivalent**
TCV	Temperature control valve
TDP	see **thermal death point**
TDS	1. Technical data sheet
	2. see **total dissolved solids**
	3. see **total dry solids**
TDT	see **thermal death time**
TE	*(drawing notation)* Thread end
TEA	Total exchangeable anions
TEAO motor	see **totally enclosed air over**
TEBC	Totally enclosed blower cooled, see **totally enclosed forced ventilated**
TEC	see **total exchange capacity**
technical atmosphere	*aka* ATU, ATÜ, Atmosphären Überdruck. A German unit of absolute pressure: pressure in standard atmospheres minus 1
technical authority	*aka* TA.
	1. Most commonly, a highly experienced engineer occupying a position of defined authority with respect to some technical area, (often safety critical) similar to a subject matter expert (qv)
	2. Sometimes a government body which must be consulted with respect to some technical area
technical document	A document suitable for technical purposes; defined formally in ISO 10209:2012 Technical product documentation – Vocabulary – Terms relating to technical drawings, product definition and related documentation
technical expert	The equivalent of a technical authority (qv) on a quality audit team; defined formally in BS EN ISO 9000 Quality management systems Fundamentals and vocabulary. In some companies and contexts used as equivalent to a subject matter technical expert (qv)
technical floor	see **interstitial space**
technical guidance notes	*aka* TGN. Guidance documents issued by the UK's Environment Agency on pollution control

technical manual	Any guidance document containing advice on use, operation and maintenance of a product (qv instructions for use) or plant (qv O&M manual). This particular term defined formally in BS 4884-1:1992 Technical manuals. Specification for presentation of essential information
technical product documentation	*aka* TPD. Medium of transmission of any part of a product's specification; defined formally in BS 8888:2017 Technical product documentation and specification and ISO 10209:2012 Technical product documentation — Vocabulary — Terms relating to technical drawings, product definition and related documentation
technical product specification	*aka* TPS. A collection of technical product documentation (qv) which provides a complete specification for a product. Defined formally in BS 8888:2017 Technical product documentation and specification and ISO 10209:2012 Technical product documentation — Vocabulary — Terms relating to technical drawings, product definition and related documentation
technical specification	A document which specifies requirements for a part; defined formally in ISO 10209:2012 Technical product documentation — Vocabulary — Terms relating to technical drawings, product definition and related documentation
technical tap	see **percussive maintenance**
techniques	In the context of best available techniques (qv), in EU/UK environmental legislation under the IPPC Directive, techniques are both the technology used for pollution control, and the way in which it is used
tee	see **tee fitting**
tee fitting	*aka* tee. A T-shaped pipe fitting with three connections most commonly of equal size (equal tee, see **Figure T1**) or different sizes (unequal tee), used to combine or divide flowing fluids; cf stub-in tee, stub-out tee, olets

Figure T1 Equal tee fitting

TEFC	see **totally enclosed fan cooled motor**
TEFC motor	see **totally enclosed fan cooled motor**
Teflon	Proprietary eponym for PTFE (qv), applied informally by analogy to anything (or anybody) with non-stick qualities
Teflon shoulders	*(informal, jocular)* A nickname for a co-worker who offloads work/blame onto others: a drop of the shoulder and the work/blame slides onto someone else
TEFV motor	see **totally enclosed forced ventilated**
telemetry	The collection and transmission to a remote location of field measurements
telescopic chute	see **donkey's dick**
teller ring	see **tellerette**
tellerette	*aka* teller ring. A type of random plastic packing used in adsorption towers
Temkin absorption isotherm	An empirical absorption isotherm like the more popular Freundlich adsorption isotherm (qv) and Langmuir adsorption isotherm (qv)
temperature allowance	In the context of fired heaters, a temperature increase applied to a metal design temperature to account for unknowns, especially fluid heterogeneity caused by flow distribution. Defined formally in BS EN ISO 13705:2012/ISO 13705:2012(E) Petroleum, petrochemical and natural gas industries. Fired heaters for general refinery service, API STD 530 - Calculation of Tube Heater Thickness and API STD 560 - Fired Heaters for General Refinery Service
temperature controller	*aka* TC. A device which measures and automatically controls temperature. Defined formally in the context of fired heaters in NFPA 86: Standard for Ovens and Furnaces, 2019 Edition
temperature gradient	The rate of change of temperature with respect to distance
temperature range	see **temperature window**
temperature regulating valve	*aka* TRV. A regulating valve (qv) which actually controls flow, with an intended secondary effect of controlling temperature
temperature window	*aka* temperature range. In the context of fired heater NOx control, it means the most effective temperature range, as defined formally in API RP 536 - NOx control on Fired Heaters at Refineries
temperature/entropy diagram	*aka* TS diagram. An empirically determined thermodynamic property diagram showing the relationship between temperature and entropy in a closed system; cf pressure/volume diagram
tempering	Using controlled heat treatment to modify the mechanical qualities of ferrous metals, or chocolate

temporary refuge	*aka* TR. A place designed to protect personnel from process safety hazards for a specified period between a loss of containment incident and their rescue, provided with incident monitoring and control facilities. Defined formally in ISO 13702:2015, Petroleum and natural gas industries — Control and mitigation of fires and explosions on offshore production installations — Requirements and guidelines, Second Edition, August 2015 and ISO 15544:2000, Petroleum and natural gas industries - Offshore production installations - Requirements and guidelines for emergency
tenacity	A fiber spinning metric: strength divided by titre (qv)
tender	1. Most commonly, an offer to execute works under specified conditions for a price 2. A firefighting vehicle
tensile strength	*aka* ultimate tensile strength. The breaking load per unit of cross sectional area of a material. Defined formally in API RP 579 - Fitness for Service
tension leg wellhead platform	*aka* TLWP. A type of production platform (qv)
TENV motor	see **totally enclosed nonventilated motor**
tera-	(*symbol* T-) The SI unit prefix denoting a factor of 10^{12}
teratogen	A material which causes fetal abnormality
terminal	1. Generically, a point of access 2. In the context of heat exchangers, the point at which fluid connections are made, as defined formally in BS EN ISO 13705:2012/ISO 13705:2012(E) Petroleum, petrochemical and natural gas industries. Fired heaters for general refinery service, API STD 530 - Calculation of Tube Heater Thickness and API STD 560 - Fired Heaters for General Refinery Service 3. A place where transport (road, rail, marine or pipeline) is loaded and unloaded, as defined formally in NFPA 30: Flammable and Combustible Liquids Code, 2018 Edition

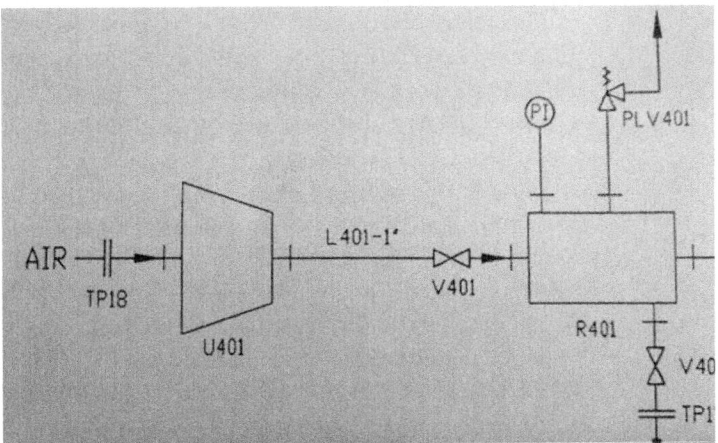
Figure T2 Termination Point (TP) on a P&ID

termination point	*aka* TP, tie-in point. One of a number of symbols on a drawing or diagram which shows the beginning or end of something. Commonly found on a P&ID (qv) to indicate the limits of the responsibilities of parties but also on logic flowcharts, (albeit with a different symbol), and more generally to show the end of a line or dimension, see **Figure T2**; cf terminator
terminator	A drawing symbol showing end of a line. Defined formally in ISO 10209:2012 Technical product documentation – Vocabulary – Terms relating to technical drawings, product definition and related documentation
ternary	Having three parts; cf tertiary
ternary graph	see **triangular diagram**
ternary phase diagram	see **triangular diagram**
ternary plot	see **triangular diagram**
Terrapin	Old UK proprietary eponym for prefabricated accommodation (superseded by Portakabin (qv) in UK)
tertiary	Third in a list; cf ternary
tertiary air	The third portion of combustion air (cf primary air, secondary air) supplied to a fired heater. Defined formally in API RP 535 - Burners for Fired Heaters at Refineries
tertiary amine	see **tertiary ammonium compound**
tertiary ammonium compound	An amine produced by the third level of substitution of ammonia with organic groups (having three such groups) cf quaternary ammonium compound
tertiary containment	In the context of loss prevention, a layer of containment around secondary containment including site drains, interceptors etc.; cf primary containment, secondary containment

tertiary effluent	Municipal wastewater which has received tertiary treatment (qv)
tertiary treatment	The third level of treatment applied to municipal wastewater, in addition to secondary treatment (qv)
Terylene	*aka* Dacron. A trade name given to polyethylene terephthalate (qv) when used in textiles
tesla	(*symbol* T) The derived SI unit of magnetic flux density
test	Determination by experiment that something meets specification; defined formally in BS EN ISO 9000 Quality management systems Fundamentals and vocabulary
test plan	A schedule setting out required resources for a program of testing in accordance with a test specification (qv). Defined formally in ISO 10209:2012 Technical product documentation — Vocabulary — Terms relating to technical drawings, product definition and related documentation
test report	Documentation of results of a test. Defined formally in ISO 10209:2012 Technical product documentation — Vocabulary — Terms relating to technical drawings, product definition and related documentation
test separator	*aka* separator, well test separator, well tester, well checker. A vessel used to test the separation under gravity of well fluids into phases
test specification	The specification of procedures to be used in a test plan (qv). Defined formally in ISO 10209:2012 Technical product documentation — Vocabulary — Terms relating to technical drawings, product definition and related documentation
Texas light sweet	see **West Texas intermediate**
TFR	see **tubular flow reactor**
TGN	see **technical guidance notes**
TGTU	see **tail gas treating unit**
Th	see **threadlike**
TH	see **total hardness**
the six Ps	see **PPPPPP**
theoretical air	see **stoichiometric air**
theoretical plate	see **theoretical stage**
theoretical stage	*aka* equilibrium stage, ideal stage, theoretical tray, theoretical plate. A stage in a multi-stage separation process at which perfect equilibrium is achieved between phases
theoretical tray	see **theoretical stage**
therm	A near-obsolete unit of heat, 100,000 BTU (qv); still used in bulk natural gas trading
thermal capacitance	see **thermal mass**
thermal capacity	see **heat capacity**

thermal conductivity	(*symbol* K or λ) Conductivity (qv) with respect to heat
thermal death point	*aka* TDP. The minimum temperature required to kill a population of a specified organism in ten minutes; cf thermal death time
thermal death time	(*symbol* tD) *aka* TDT. The time taken to kill a population of a specified organism at a given temperature
thermal diffusion	see **Soret effect**
thermal diffusivity	(*symbol* A) Thermal conductivity divided by the product of specific heat capacity and density; a property related to thermal mass (qv)
thermal efficiency	In the context of fired heaters, the ratio of heat absorption to total theoretical heat input from fuel. Defined formally in BS EN ISO 13705:2012/ISO 13705:2012(E) Petroleum, petrochemical and natural gas industries. Fired heaters for general refinery service, API STD 530 - Calculation of Tube Heater Thickness and API STD 560 - Fired Heaters for General Refinery Service
thermal expansion	The increase in volume of a material with increased temperature
thermal expansion coefficient	see **coefficient of expansion**
thermal expansion load	The loading on supports resulting from thermal expansion (qv)
thermal incinerator	see **thermal oxidizer**
thermal inertia	see **thermal mass**
thermal mass	*aka* thermal capacitance, thermal inertia. The ability of a building material to absorb, store and release heat, providing a kind of inertia with respect to temperature changes; cf thermal diffusivity
thermal oxidation process	1. In a fired heater, the controlled oxidation of the surface of silicon wafers 2. Less commonly, incineration
thermal oxidizer	*aka* thermal incinerator, incinerator. A device which removes pollutants from exhaust gases by high temperature oxidation; defined formally in NFPA 86: Standard for Ovens and Furnaces, 2019 Edition
thermal recuperative oxidizer	*aka* TRO, see **recuperative thermal oxidizer**
thermal recycling	A euphemism for incineration (qv), especially with energy recovery
thermal relief valve	*aka* TRV. See **thermal safety valve**
thermal resistance	(*symbol* R) Resistance (qv) to heat transfer
thermal runaway	see **runaway**

thermal safety valve	*aka* thermal relief valve, TRV. Usually, a spring-loaded relief valve used to protect against volume expansion associated with temperature rise; cf pressure safety valve
thermal stress	Generically, stress caused by temperature change. More specific formal definition in API RP 579 - Fitness for Service
thermal transmittance	see **U value**
thermal treatment	see **incineration**
thermal wheel	see **rotary heat exchanger**
Thermite process	A process heated by the highly exothermic reaction between iron oxide and a metal (commonly aluminum); used for the production of molybdenum and other specialty metals, occasionally for welding and less frequently still for iron production
thermocouple	A temperature sensor based on a junction between dissimilar metals which can be used at high temperatures
thermodiffusion	see **Soret effect**
thermodynamic diagram	A diagram representing relationships between thermodynamic properties such as the Mollier diagram, pressure/volume diagram, or TS diagram (qv)
thermodynamic free energy	An extensive thermodynamic state function, which may be calculated in a number of ways. Physicists usually prefer to use Helmholtz free energy (qv), chemists (and by extension chemical engineers) Gibbs free energy (qv)
thermodynamics	The engineering science of the relationships between energy, heat, temperature, and work
thermomigration	see **Soret effect**
thermophilic bacteria	Bacteria with an optimum temperature over 45° C; cf cryophilic bacteria, mesophilic bacteria
thermophoresis	see **Soret effect**
thermoplastic	*aka* plastic. A polymer which liquifies on heating, and resolidifies reversibly when cooled. Defined formally in ASME BPE (American Society of Mechanical Engineers: Bioprocessing Equipment); cf thermoset
thermoset	*aka* thermosetting polymer, plastic. Polymers that solidify irreversibly after curing with heat and/or chemical agents, and decompose rather than melt if heated. Defined formally in ASME BPE (American Society of Mechanical Engineers: Bioprocessing Equipment); cf thermoplastic
thermosetting polymer	see **thermoset**
thermosiphon	see **thermosyphon**
thermosyphon	*aka* thermosiphon. A way of circulating fluid without a pump using convection

thermosyphon reboiler	A column reboiler (qv) using a thermosyphon (qv) in place of a pump
thermowell	*aka* TW. A fitting providing a barrier between a temperature sensor and process fluids
thermutator	see **Wanson thermutator**
thickener	1. A device which thickens sludge 2. A substance added to increase viscosity (especially in the case of foodstuffs)
thickening	1. Generically, making more viscous 2. Specifically, an initial stage of enhancement of solids concentration of sludge by gravity settlement; cf drying, dewatering
Thiele-Geddes	An iterative multicomponent distillation calculation method
thin water	Used by native English speaking water specialists to describe water lacking in buffering capacity, but by native Dutch speakers to describe effluent with low pollutant levels, called 'dun water' in Dutch, (dun meaning 'thin')
thin weir	see **thincrested weir**
thincrested weir	*aka* thin weir. In hydraulics/fluid mechanics, 'thin' means having an insignificant dimension in the direction of flow. Like so much in hydraulics, it is an approximation; cf broadcrested weir
thinner	see **diluent**
third law of thermodynamics	The law which essentially states that you can't break even; qv first law of thermodynamics, second law of thermodynamics
thirsty bird pump	see **nodding donkey**
thixotropic fluid	A fluid which exhibits thixotropy (qv)
thixotropy	The opposite of rheopecty (qv) i.e.; time-dependent shear thinning, a behavior of some non-Newtonian fluids
THMs	see **trihalomethanes**
thread tape	PTFE (qv) tape used to seal threads of threaded pipe fittings; cf hemp and white
threadlike	*aka* Th. A term for an indication (qv) with measurable length, but immeasurably small width or depth; defined formally in PD 5500:2018+A1:2018 Specification for unfired fusion welded pressure vessels; cf isolated point
threadolet	A sockolet (qv) with a female threaded connection for the side branch to be screwed into

three phase	1. Most commonly, power supply with three out of phase cycles of alternating electrical current 2. Less commonly, three coexisting phases of matter (in the oil and gas industry, these usually being oil/condensate, gas and water)
three phase free water knockout	see **free water knockout**
three phase motor	The most common industrial motor type, which runs on three phase power (qv)
three phase power	The most common power transmission system, having three out of phase cycles of alternating electrical current, used to directly power large electrical appliances (motors) on process plants[101]
three phase separator	An item of process equipment designed to separate condensate/oil, gases and water in a single unit operation
three term controller	see **PID controller**
three way valve	A valve with three connection points (usually one inlet and two outlets, but sometimes vice versa), rarely larger than 100mm NB
three-position control	see **on/off control**
threshold housekeeping dust accumulations	The amount of accumulated flammable dust which triggers a requirement for clean-up. Defined formally in NFPA 654: Standard for the Prevention of Fire and Dust Explosions from the Manufacturing, Processing, and Handling of Combustible Particulate Solids, 2017 Edition
threshold limit value	*aka* TLV. A level of occupational exposure of workers to a substance which is thought to be safe. The American Conference of Governmental Industrial Hygienists sets the most commonly used levels worldwide
throat area	see **bore area**
throat bush	see **throat bushing**
throat bushing	*aka* throat bush. In the context of mechanical seals on centrifugal pumps, a closely fitted ring which separates the mechanical seal from the pumped fluid. Defined formally in EN ISO 13709:2003 Centrifugal pumps for petroleum, petrochemical and natural gas industries and API RP 682 - Pump Seals; cf throttle bushing

---

[101] For a detailed understanding of why, ask any electrical engineer when you have an hour free

throttle bushing	In the context of mechanical seals on centrifugal pumps, a closely fitted ring which separates the mechanical seal from the environment. Defined formally in EN ISO 13709:2003 Centrifugal pumps for petroleum, petrochemical and natural gas industries and API RP 682 - Pump Seals; cf **throat bushing**
throttling	Most commonly, controlling fluid flow by means of an induced pressure drop
throughput	An amount of material passing through a process per unit time (or per cycle in the case of a batch process)
throwclear	An oversize leaf in a bound publication designed to be readable with the publication open at any page. Defined formally in BS EN 82079-1:2012; cf **foldout**
thrust block	see **pipe anchor**
thrust block load	see **pipe anchor load**
TI	*(drawing notation)* Temperature indicator
ticket	*(informal)* see **CSCS**
tie compound	see **tie substance**
tie in	*aka* tie-in. As a noun: a connection; as a verb: connecting
tie line	A horizontal line drawn on a binary phase diagram (qv) used to estimate mole or mass fraction of phases using the lever rule (qv)
tie substance	*aka* tie compound. A substance known to pass unchanged through a process used in mass balance exercises
tieback	see **anchor**
tie-in	see **tie in**
tie-in point	see **termination point**
tiffie	*(informal) aka* tiffy. A UK term for an instrument technician
tiffy	see **tiffie**
TIG welding	Tungsten inert gas welding; cf **gas tungsten arc welding**
tightly threaded connection	see **national pipe thread**
tile	*aka* burner tile, burner block, muffle block, muffler block, quarl. A shaped refractory in which a burner is mounted; defined formally in API RP 535 - Burners for Fired Heaters at Refineries
Tillmann's formula	A formula setting out relationships between compounds involved in carbonate buffering, such as free CO_2 (qv)
time constant	*(symbol* T*)* 1. In control theory, the time required for response of a control system to decay to zero 2. In the analysis of complex thermal systems, the time constant is the time taken for a temperature difference to become zero

time dependent fluid	A non-Newtonian fluid (qv), the viscosity of which varies with duration of shear; may be a rheopectic fluid (qv) or thixotropic fluid (qv), dependent on whether their viscosity rises or falls with respect to time of shearing respectively
time independent fluid	A fluid whose shear rate depends only on shear stress, a category which includes, but is not limited to Newtonian fluids (qv)
time of exhaust	see **evacuation time**
time usage	The combination of standby availability and utilization for equipment
time weighted average	*aka* TWA. A calculation of the effect of exposure of a worker to hazardous substances, taking into account both concentration of the substance and time of exposure over an 8-hour shift
tinbashers	*(informal)* UK term for metal fabricators, especially of items made from sheet metal, such as storage tanks
tip	A fired equipment fuel injector
tip vortex cavitation	see **vortex cavitation**
tippler	A rotary car dumper, which holds a railcar onto its track and inverts both track and car to dump its load
TIR	1. see **total indicated runout** 2. see **total indicator reading**
title V operating permit	*aka* part 70 Federal operating permit. A permit used to control air pollution in the US under the Clean Air Act 1990
titre	Density per unit length in a fiber spinning context
TKN	Total Kjeldahl nitrogen
TLA	*(jocular)* Three letter acronym
TLE	*(drawing notation)* Thread large end
TLV	see **threshold limit value**
TLWP	see **tension leg wellhead platform**
TML	Thickness monitoring location
TNT	see **trinitrotoluene**
TNT equivalent	The energy of an explosion expressed as an equivalent detonated mass of TNT (qv)
TNT model	A relatively simple process safety consequence model; used to predict effects of an explosion of process equipment or material by approximating the equivalent energy of a TNT (qv) explosion; cf TNT equivalent
TO	1. Thermal oxidizer; cf RTO 2. *(drawing notation)* Threads only
TOC	1. *(drawing notation)* Top of concrete 2. Total organic carbon; see **organic matter**
TOE	*(drawing notation)* Thread on end

toeplate	A rigid plate affixed to the decking, guardrail or landing platform of machinery access arrangements which prevents objects at toe-height falling. Defined formally in BS EN ISO 14122 Safety of machinery. Permanent means of access to machinery. Working platforms and walkways
TOL	*(drawing notation)* Threadolet (qv)
tolerance	see **tolerance of dimension**
tolerance of dimension	*aka* tolerance, dimensional tolerance, engineering tolerance. 1. The range of acceptable deviation from a reference dimension; defined formally in ISO 10209:2012 Technical product documentation — Vocabulary — Terms relating to technical drawings, product definition and related documentation. 2. May also mean the required clearance around something
toll manufacture	*aka* tolling. The subcontracted manufacture of a product to a third party, by the IP owner of the product; common in short lifecycle products like cosmetics and biocides
tolling	see **toll manufacture**
tomography	*aka* computed tomography, CT, CAT. Imaging by slices, as in the CT scan commonly used in medicine, and less commonly in engineering contexts for non-destructive testing (qv)
ton	Either the short ton (qv) or the long ton (qv), but far more likely to be the former; see **Box B1**
TON	see **tons refrigeration**
ton refrigeration	see **tons refrigeration**
tonne	*aka* metric ton. One thousand kilograms; cf ton
tonne of coal equivalent	*aka* coal equivalent, tce. An amount of energy, 29.39 gigajoules (GJ)
tons refrigeration	*aka* ton refrigeration, refrigeration ton, ton, TON, TOR, RT, TR. A unit of cooling capacity: a rate of heat transfer equivalent to that provided by the use of a US short ton (qv) (2000 lb.) of ice per day
toolpusher	An oil and gas job role, second in command of a drilling crew under the drilling superintendent, akin to a foreman; senior to a roughneck (qv) with administration responsibilities. A toolpusher usually works 6am to 6pm, and a night toolpusher works 6pm to 6am. A tourpusher does the same job on a 12am to 12pm tour
TOP	*(drawing notation)* Top of pipe
top event	In fault tree analysis (qv), a name for ultimate cause (qv)

top management	The most senior individuals controlling and directing an organization; defined formally in BS EN ISO 9000 Quality management systems Fundamentals and vocabulary
top product	Light boiling fractions yielded from the top of a distillation column; cf bottoms
topped crude	see **atmospheric residue**
topping refinery	A basic oil refinery which only carries out atmospheric distillation
topping steam turbine	*aka* topping turbine. A high pressure, non-condensing steam turbine
topping turbine	see **topping steam turbine**
topsides	The part of a ship or oil platform above the waterline
TOR	see **tons refrigeration**
torispherical	*aka* ASME flanged and dished (ASME F&D). A common shape for the heads or caps of pressure vessels, see **Figure T3**. A torispherical shape comprises a terminal spherical segment, the radius of which is known as the crown radius (qv), tangentially intersecting the outer portion of a ring or torus, the radius of which is known as the knuckle radius (qv); cf ellipsoidal, hemispherical

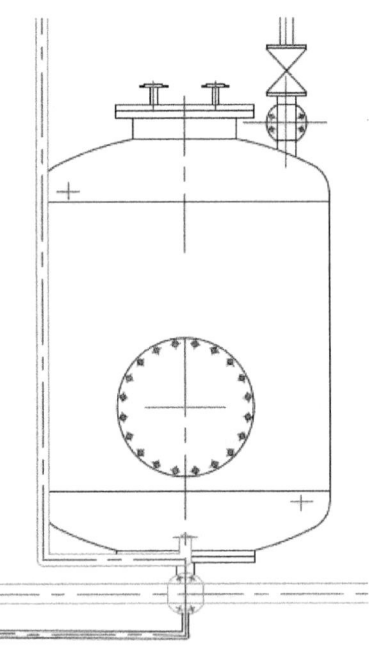

Figure T3 Torispherical heads on a pressure vessel

toroidal vortices	see **Taylor vortices**
torque	(*symbol* τ, M) *aka* moment, moment of force, moment of a force, rotational force, turning effect. The rotational equivalent of linear force, a measure of how much work something can do; though physicists and engineers may disagree slightly about the precise meaning
torr	(*symbol* Torr) A non-SI unit of absolute pressure, still sometimes used in measuring degrees of vacuum; 1 Torr is approximately equal to the head generated by a 1mm high column of mercury
torrefaction	A form of pyrolysis (qv) under less aggressive operating conditions
torsional coupling	A rotating drive coupling which avoids the transmission of damaging vibrations by damping and detuning effects. Defined formally in NFPA 20: Standard for the Installation of Stationary Pumps for Fire Protection, 2019 Edition; cf shear spool coupling
torsional vibration	Vibration in the direction of rotation around an object's axis; defined formally in BS ISO 2041:2018 Mechanical vibration, shock and condition monitoring. Vocabulary
TOS	(*drawing notation*) Top of steel
total acid number	*aka* TAN; see **acid number**. However, this particular term is most common in oil and gas
total capacity	*aka* total exchange capacity, TEC. In the case of ion exchange resins, the theoretical maximum exchange ability; cf operating capacity
total condenser	A heat exchanger which condenses all vapor which would otherwise leave the top of a distillation column, leaving no top product (qv) to move downstream to the next unit operation as a vapor; cf partial condenser
total design	According to Stuart Pugh: "Total design is the systematic activity necessary, from the identification of the market/user need, to the selling of the successful product to satisfy that need – an activity that encompasses product, process, people and organization", see **Box D1**; cf partial design
total discharge head	(*symbol* H_d) Pressure measured at a pump discharge expressed in units of head of pumped fluid, adjusted for datum and velocity head at the point of measurement. Defined formally in NFPA 20: Standard for the Installation of Stationary Pumps for Fire Protection, 2019 Edition; cf total suction head

total dissolved solids	*aka* TDS. The mass of residue left after drying a finely filtered sample of water, commonly expressed as ppm w/w or mg/l w/v of the sample; cf total suspended solids
total dry solids	*aka* TDS, total solids. Solids remaining after evaporation and extended drying at no more than 105° C
total exchange capacity	*aka* TEC; see **total capacity**
total excursion	see **excursion**
total hardness	*aka* TH. 1. In water engineering, the concentration of dissolved calcium and magnesium, often expressed as equivalent mg/l $CaCO_3$ 2. In a corrosion context, all multivalent cations may be included
total head	The total pressure (qv) which a pump generates (the sum of static and dynamic pressure), expressed as a head of fluid
total heat	A somewhat contested term: 1. Used to mean the sum of sensible heat (qv) and latent heat (qv), or these two plus any superheat (qv) 2. May also be used to mean enthalpy (qv)
total heat release	A specified fuel's calculated heat of combustion, based on its lower heating value (qv). Defined formally in BS EN ISO 13705:2012/ISO 13705:2012(E) Petroleum, petrochemical and natural gas industries. Fired heaters for general refinery service, API STD 530 - Calculation of Tube Heater Thickness and API STD 560 - Fired Heaters for General Refinery Service
total indicated runout	*aka* TIR, total runout. The difference between minimum and maximum measured heights above a reference axis of a rotating part's entire surface (not just cross section); cf runout
total indicator reading	*aka* TIR, full indicator movement (FIM); see **total indicated runout**. The formal definitions in EN ISO 13709:2003 Centrifugal pumps for petroleum, petrochemical and natural gas industries and API RP 682 - Pump Seals however use this term to mean circular runout (qv)
total isolatable inventory	The amount isolated by all emergency shutdown valves (qv) on a system
total liquid capacity	The total volume of liquid a tank can contain at its maximum design liquid level; defined formally in API STD 620 - Low Pressure Storage Tanks
total moisture	All moisture, most likely meaning both adventitious moisture (qv) and inherent moisture (qv)

total petroleum hydrocarbons	*aka* TPH. Intended as a measure of chemical pollutants ultimately derived from petroleum (in the sense of crude oil, not gasoline) in water, but may well actually measure all non-aqueous phase liquids (qv); cf fats oils and greases
total pressure	1. The sum of the partial pressures of a mixture of gases 2. In fluid mechanics, the sum of static and dynamic pressures
total rated head	The total head of a centrifugal pump at its rated capacity and speed; defined formally in NFPA 20: Standard for the Installation of Stationary Pumps for Fire Protection, 2019 Edition
total reflux	A mode of operation where all condensed vapor is returned to a distillation column
total runout	see **total indicated runout**
total solids	see **total dry solids**
total suction head	Positive head measured at a pump suction expressed in units of head of fluid, adjusted for datum and velocity head at point of measurement. Defined formally in NFPA 20: Standard for the Installation of Stationary Pumps for Fire Protection, 2019 Edition; cf total discharge head
total suction lift	(*symbol* h_s, hl) *aka* total dynamic suction lift. The negative head (required lift) measured at a pump suction expressed in units of head of fluid, adjusted for datum and velocity head at the point of measurement. Defined formally in NFPA 20: Standard for the Installation of Stationary Pumps for Fire Protection, 2019 Edition; cf total discharge head, total suction head, NPSH
total suspended solids	*aka* TSS, suspended solids (SS). Those solid particles present in a water sample which can be removed by a specified degree of settlement or filtration, sometimes specified as particles greater than two microns. Usually expressed as ppm w/w or mg/l w/v of the sample; cf total dissolved solids
total transfer velocity	see **overall mass transfer coefficient**
totally enclosed air over	*aka* TEAO motor. A totally enclosed motor (qv) used to drive a fan, (as opposed to having its own cooling fan) cooled by the fan it drives
totally enclosed blower cooled	*aka* TEBC; see **totally enclosed forced ventilated**
totally enclosed fan cooled motor	*aka* TEFC, TEFC motor. A totally enclosed motor (qv) cooled by one or more fans outside the enclosure, driven by the motor shaft
totally enclosed forced ventilated	*aka* TEFV motor, totally enclosed blower cooled (TEBC). A totally enclosed motor (qv) cooled by a blower or fans outside the enclosure, independently driven at a fixed speed

totally enclosed motor	An electric motor sufficiently enclosed to prevent the free passage of air, whilst not strictly airtight
totally enclosed naturally ventilated motor	see **totally enclosed nonventilated motor**
totally enclosed nonvented motor	see **totally enclosed nonventilated motor**
totally enclosed nonventilated motor	*aka* TENV motor, totally enclosed nonvented motor, totally enclosed naturally ventilated motor. A totally enclosed motor (qv) without a cooling fan
tote bin	Transportable containers of various sizes
toughness	The ability to deform plastically when absorbing impact energy, rather than fracturing; defined formally in API RP 579 - Fitness for Service
tourpusher	A toolpusher (qv) who works with a drill crew on a 12am to 12pm shift
tower	see **column**
town gas	Gas utility supplied to consumers via mains, often synonymous with coal gas (qv)
town mains	The municipal water supply
toxic	Poisonous
TP	1. see **termination point** 2. see **three phase**
TPD	see **technical product documentation**
TPH	see **total petroleum hydrocarbons**
TPI	*(drawing notation)* Threads per inch
TPS	see **technical product specification**
TR	see **temporary refuge**
trace heating	see **tracing**
traceability	In quality control, knowing where something and its components come from as a result of rigorous record-keeping; defined formally in BS EN ISO 9000 Quality management systems Fundamentals and vocabulary
tracing	*aka* trace heating. 1. the practice of keeping fluids in a process line above a minimum temperature with steam, or electrical heating tape (often to prevent those fluids becoming solids) 2. The steam piping or heating tape used for that purpose
train	see **process train**
tramp air	Unintended contaminant air; cf afterburn
tramp iron	Metallic tramp material (qv), usually ferrous

tramp material	Process stream contaminants from feed impurities, degradation of process equipment, items left over from maintenance/construction etc. Ranges from gases through small flakes of plastic/metal to whole animals
transfer panel	*aka* swing bend panel. A hygienic type of fluid transfer selection panel. Defined formally in ASME BPE (American Society of Mechanical Engineers: Bioprocessing Equipment)
transfer station	*aka* boiler transfer station. The arrangement of lines and safety valves associated with a boiler whose primary purpose is to protect downstream lines and equipment from overpressure, though other safety, environmental and monitoring functions may also be incorporated. Defined formally in BS EN 12952-8:2002 Water-tube boilers and auxiliary installations. Requirements for firing systems for liquid and gaseous fuels for the boiler
transient	1. A short-lived pulse of pressure, vibration or electricity, especially pulses of high or low pressure in pipework, such as water hammer or surge (qv) 2. see **unsteady state**
transient analysis	In engineering contexts, a mathematical analysis of short-lived pulses, especially of high and low pressure in pipework, such as water hammer (qv)
transient response	A shortlived excursion from a state of equilibrium in a system
transient study	see **surge analysis**
transient vibration	A short-lived vibration. Defined formally in BS ISO 2041:2018 Mechanical vibration, shock and condition monitoring. Vocabulary
transition	A process of change; or the duration of such a process
transition flow	A flow regime intermediate between laminar flow (qv) and turbulent flow (qv)
transition pipe	In the oil and gas industry, a pipe which provides a strong, flexible connection between the drill collar (qv) and drill pipe (qv)
transition point	1. The condition(s) under which there is a transition between laminar flow (qv) and turbulent flow (qv), or vice versa (cf transition point) 2. The condition(s) at which a substance changes its physical phase
transition temperature	The temperature at which the transition of a metal from ductile to brittle fracture mode occurs; defined formally in API RP 579 - Fitness for Service

transmitter	A device which transmits a signal, especially one from a transducer to a controller
transparency	The open, clear communication of information. May also imply accountability as with the anti-corruption organization 'Transparency International'
transport lag	see **distance/velocity lag**
transport phenomena	The engineering science of mass, momentum, or energy transfer
transportation lag	see **distance/velocity lag**
transshipment	Shipping to an intermediate destination prior to final delivery
transverse	In a perpendicular direction; cf axial, radial
transverse axis	The direction perpendicular to the long axis. Defined formally in the context of vibration monitoring in BS ISO 2041:2018 Mechanical vibration, shock and condition monitoring. Vocabulary
transverse dispersion	*aka* radial dispersion. Dispersion in a direction perpendicular to long axis; cf axial dispersion
transverse wave	A wave with material/ particle displacement along the transverse axis (qv). Defined formally in the context of vibration monitoring in BS ISO 2041:2018 Mechanical vibration, shock and condition monitoring. Vocabulary
trapdoor	An access hatch set into a floor. Defined formally in the context of machinery access in BS EN ISO 14122 Safety of machinery. Permanent means of access to machinery. Working platforms and walkways
trash	see **municipal solid waste**
trash rack	A very coarse screen used to exclude debris carried in water
traverse	To move backwards and forwards; cf luff, slew
trays	1. *aka* plates. Solids/liquid contacting devices used in distillation columns
	2. *aka* cable trays; see **traywork**
traywork	A collective term for the system of cable trays (qv) which contain and support power and instrument cables and sometimes flexible hoses
TRC	Total residual chlorine
tread plate	see **deckplate**
trench	1. A three-sided concrete trough, the top of which is flush with local grade. A trench or culvert (qv) often contains pipes, whilst a channel (qv) carries water or effluent without piping.
	2. A three-sided linear hole which buried underground piping is laid in, prior to backfilling to grade
trenched piping	Pipework carried in a trench (qv)

trenchless technology	Techniques used to install pipes underground without digging a trench (qv)
trench-like corrosion	see **preferential weld corrosion**
trevitesting	Proprietary eponym for a method for in-situ testing of PSVs (qv); originally developed by a company called Furmanite
trial	*aka* plant trial. A temporarily authorized deviation from an established operating envelope or method to establish at full scale whether such a change should be adopted permanently; cf performance trial
trial and error	*aka* guess and check. A problem-solving technique in which candidate solutions are generated, and then tested until a working solution is found. Sometimes characterized as not being based in insight or theory by non-practitioners, who may not appreciate the degree to which practitioners intuitively select candidates which are likely to pass the test; see **Box H2**.
trial for ignition period	*aka* flame establishing period. The time that a fuel safety shutoff valve (qv) is allowed by a safeguard device to be open, before flame detector supervision of the flame is required. Defined formally in NFPA 86: Standard for Ovens and Furnaces, 2019 Edition
triangle plot	see **triangular diagram**
triangular diagram	*aka* Gibbs triangle, simplex plot, ternary graph, ternary phase diagram, ternary plot, triangle plot. A graph in the shape of an equilateral triangle of three variables which add up to a constant
trickling filter	*aka* percolating filter. A rather old-fashioned secondary sewage treatment technology, in which sewage is distributed over the surface of a bed of carrier material upon which microorganisms grow
Triconex	Proprietary eponym used to refer to triple modular redundancy (qv) safety functions
trihalomethanes	*aka* THMs. A group of carcinogenic compounds (trichloro-, tribromo- and triodo-methane) formed when humic and fulvic acids in water react with halogens; a harmful disinfection byproduct
trim	The internal moving parts (such as stem, ball and seat) of a control valve in contact with process fluid
trinitrotoluene	*aka* TNT. A castable explosive compound
trip	The deenergization of a circuit breaker, causing rapid shut down of an item of equipment, commonly in response to a fault condition. Informally used interchangeably with interlock (qv) in some businesses/sectors, though these are not always synonymous

trip speed	(*symbol* Nt) 1. The speed of rotation at which an item of rotating machinery is tripped (qv) to automatically shut down on overspeed. This is the basis of the formal definitions in EN ISO 10437 Petroleum, petrochemical and natural gas industries - Steam turbines - Special-purpose applications, ISO14661 Thermal turbines for industrial applications (steam turbines, gas expansion turbines) - General requirements, BS EN ISO 13705:2012/ISO 13705:2012(E) Petroleum, petrochemical and natural gas industries. Fired heaters for general refinery service and EN ISO 13709:2003 Centrifugal pumps for petroleum, petrochemical and natural gas industries 2. An incompatible definition of the same term, namely speed at maximum supply frequency, may be found in EN ISO 13709:2003 Centrifugal pumps for petroleum, petrochemical and natural gas industries
triplex pump	A three-cylinder reciprocating pump popular in oil well service
tripped	Deenergized or shut down by a trip (qv)
trippers	Structures which divert bulk materials off a conveyor
triuranium octoxide	(*symbol* U_3O_8) The primary component of yellowcake (qv), the intermediate product from uranium mining before final refining to uranium dioxide or metallic uranium
TRO	see **thermal recuperative oxidizer**
troubleshooting	Fault diagnosis and correction
Trouton's rule	A heuristic which states that the entropy of vaporization of many liquids at their boiling points is around 90 J/(K·mol)
troy ounce	A unit of weight used for precious metals, approximately 31 grams (a little more than a standard (avoirdupois) ounce)
true view	1. A software program: Autodesk's free .dwg viewer 2. A drawing geometrically similar to what is represented, showing features lying on a parallel plane to the projection plane; defined formally in ISO 10209:2012 Technical product documentation — Vocabulary — Terms relating to technical drawings, product definition and related documentation
trunnions	Machinery or column supports similar to the two supports of an old-fashioned cannon
TRV	1. see **temperature regulating valve** 2. Thermal relief valve; see **thermal safety valve**
TS diagram	see **temperature/entropy diagram**
TSE	(*drawing notation*) Thread small end
TSS	see **total suspended solids**
TSV	Thermal safety valve; cf PSV, PRV

tube	*aka* tubing. A tube differs from a pipe (qv) in that it can have a non-circular cross section, and is sized by nominal outside (rather than inside) diameter. Defined formally in ASME BPE (American Society of Mechanical Engineers: Bioprocessing Equipment), though see **Box P1**.
tube bundle	An assembly of tubes inside the shell of a shell and tube heat exchanger (qv)
tube guide	A horizontal (but not axial) movement limiter for heat exchanger tubes. Defined formally in BS EN ISO 13705:2012/ISO 13705:2012(E) Petroleum, petrochemical and natural gas industries. Fired heaters for general refinery service, API STD 530 - Calculation of Tube Heater Thickness and API STD 560 - Fired Heaters for General Refinery Service
tube retainer	A horizontal radiant tube restraint feature of fired heaters. Defined formally in BS EN ISO 13705:2012/ISO 13705:2012(E) Petroleum, petrochemical and natural gas industries. Fired heaters for general refinery service, API STD 530 - Calculation of Tube Heater Thickness and API STD 560 - Fired Heaters for General Refinery Service
tube settler	A settlement tank with packs of inclined hexagonal tubes which decrease required settlement area in the same way as a lamella clarifier (qv)
tube sheet	see **tube support**
tube side	Term for the contents of the tubes of a shell and tube heat exchanger (qv); cf shell side
tube support	*aka* tube sheet. A device in fired heaters, which does just what it sounds like it does. Defined formally in BS EN ISO 13705:2012/ISO 13705:2012(E) Petroleum, petrochemical and natural gas industries. Fired heaters for general refinery service, API STD 530 - Calculation of Tube Heater Thickness and API STD 560 - Fired Heaters for General Refinery Service
tubing	*aka* tube. Usually has the same relationship to tube (qv) as piping has to pipe, also used as an abbreviation of 'flexible tubing' though see **Box P1**; cf piping
tubing hanger	An oil well component from which production tubing (amongst other things) is suspended
tubular flow reactor	*aka* TFR. A plug flow reactor in a tube
tubular heating system	A fired heater heating system with electrical heating elements enclosed in tubes, possibly filled with a special gas atmosphere. Defined formally in NFPA 86: Standard for Ovens and Furnaces, 2019 Edition

tubular membrane	A non-self-supporting semi-permeable membrane within a tube of 5-25mm bore; cf hollow fiber membrane
tubulars	An oil rig worker term for any kind of piping or tubing (qv)
tundish	A liquid-collecting device similar to a funnel
tungsten inclusion	A weld inclusion (qv) made of tungsten arising from gas tungsten arc welding (qv). Defined formally in ASME BPE (American Society of Mechanical Engineers: Bioprocessing Equipment)
turbidity	The inverse of clarity; the tendency of suspended solids, droplets or bubbles to scatter light passing through a fluid, giving a murky appearance; cf opacity
turbine	A machine which generates power from the movement of a rotor turned by means of a fast-moving flow of fluid
turbine island	see **conventional island**
turboexpander	see **gas expansion turbine**
turbomachinery	A collective name for rotating machines which impart energy to, or transfer energy from, a flowing fluid stream; includes pumps, fans, blowers, compressors, turbines etc. Broadly synonymous with 'rotating machinery'
turbulence	see **turbulent flow**
turbulent flow	*aka* turbulence. A fluid flow regime with unpredictable, highly variable patterns of local velocity and pressure; cf laminar flow, transition flow
turnaround	*aka* TAR. In the oil and gas industry, a very expensive scheduled shutdown procedure in which multiple maintenance activities requiring shutdown are carried out on a tight program. Sometimes used as synonymous with shutdown (qv) and outage (qv), but whilst all turnarounds are shutdowns/outages, not all shutdowns/outages are turnarounds
turndown	*aka* turndown ratio. The ratio of maximum to minimum controllable flow. Defined formally in the case of fired heater burners in API RP 535 - Burners for Fired Heaters at Refineries
turndown ratio	1. see **turndown** 2. A term used in distillation to denote ratio of minimum allowable to operating throughput
turning effect	see **torque**
turnkey	A product or plant which the supplier/contractor guarantees to be complete and ready for immediate operation
turpentine substitute	see **mineral spirits**
turquoise hydrogen	Hydrogen produced by methane pyrolysis; cf hydrogen colors
TW	*(drawing notation)* see **thermowell**

TWA	see **time weighted average**
twelve D process	see **botulinum cook**
two family dwelling	One building split into two dwelling units; defined formally in NFPA 30: Flammable and Combustible Liquids Code, 2018 Edition
two film theory	see **film theory**
two hand control device	A control device requiring the use of two hands simultaneously for operation. Defined formally in BS EN ISO 12100:2010 Safety of machinery. General principles for design. Risk assessment and risk reduction
two phase free water knockout	see **free water knockout**
two phase frictional multiplier	see **two phase multiplier**
two phase multiplier	*aka* two phase frictional multiplier, two phase pressure drop multiplier. A factor applied to a single phase frictional pressure drop to account for the presence of two phase flow
two phase pressure drop multiplier	see **two phase multiplier**
two property rule	A rule which states that the thermodynamic state of a pure substance can be completely defined by any two independent thermodynamic properties
two stage regulator system	An LPG vapor supply system using either separate first- and second-stage regulators, or an integral two stage regulator (qv). Defined formally in NFPA 58: Liquefied Petroleum Gas Code, 2017 Edition
two-position control	see **on/off control**
type 1 strong base anion resin	Strong base anion ion exchange media with trimethylamine functional groups
type 2 strong base anion resin	Strong base anion ion exchange media with dimethylethanolamine functional group
type A FIBC	In the context of electrostatic hazard control, a standard nonconductive flexible intermediate bulk container (qv) without protective features. Defined formally in NFPA 654: Standard for the Prevention of Fire and Dust Explosions from the Manufacturing, Processing, and Handling of Combustible Particulate Solids, 2017 Edition
type A seal	In the context of pump seals, a type of pusher seal (qv); defined formally in API RP 682 - Pump Seals

type B FIBC	In the context of electrostatic hazard control, a standard nonconductive flexible intermediate bulk container (qv), except that its materials of construction have a breakdown voltage of 6000 volts. Defined formally in NFPA 654: Standard for the Prevention of Fire and Dust Explosions from the Manufacturing, Processing, and Handling of Combustible Particulate Solids, 2017 Edition
type B seal	In the context of pump seals, a type of non-pusher (metal bellows) seal with a rotating flexible element and secondary sealing with elastomer O-rings (qv). Defined formally in API RP 682 - Pump Seals
type C FIBC	In the context of electrostatic hazard control, a flexible intermediate bulk container (qv) with a grounding tab connected through conductive material to the entire bag. Defined formally in NFPA 654: Standard for the Prevention of Fire and Dust Explosions from the Manufacturing, Processing, and Handling of Combustible Particulate Solids, 2017 Edition
type C seal	In the context of pump seals, a type of non-pusher (metal bellows) seal with a stationary flexible element and secondary sealing with graphite. Defined formally in API RP 682 - Pump Seals
type D FIBC	In the context of electrostatic hazard control, a flexible intermediate bulk container (qv) with no requirement for grounding as it is constructed of material with special properties which control electrostatic discharges. Defined formally in NFPA 654: Standard for the Prevention of Fire and Dust Explosions from the Manufacturing, Processing, and Handling of Combustible Particulate Solids, 2017 Edition
type I discharge outlet	In the context of firefighting foam, a foam outlet that delivers foam gently onto a liquid surface. Defined formally in NFPA 11: Standard for Low-, Medium-, and High-Expansion Foam, 2016 Edition
type II discharge outlet	In the context of firefighting foam, a foam outlet that does not deliver foam gently onto a liquid surface, but is designed to reduce surface agitation and foam submergence somewhat. Defined formally in NFPA 11: Standard for Low-, Medium-, and High-Expansion Foam, 2016 Edition

U

U value	*aka* thermal transmittance. Most commonly, the overall heat transfer coefficient
U&O	Utility and offsite
U/G	see **under ground**
U/S	1. Unserviceable (unusable)
	2. see **upstream station** or **utility station**
U235	see **uranium-235**
U238	see **uranium-238**
U3O8	see **triuranium octoxide**
UASB	see **upflow anaerobic sludge blanket**
UAV	see **underwater autonomous vehicle**
UBO	Ultimate beneficial owner
UEL	Upper explosive limit, see **upper flammable limit**
UF	see **ultrafiltration**
UF6	see **uranium hexafluoride**
UFD	see **utility flow diagram**
UFL	see **upper flammable limit**
UFLA	*(jocular)* Unidentified/Unnecessary four-letter acronym
UGS	see **underground gas storage**
UHHPP	see **ultra high hydrostatic pressure processing**
uHPHT	see **ultra high pressure high temperature field**
UHT	see **ultra high temperature**
UL	Underwriters Laboratories, Inc.
ULCC	see **ultra large crude carrier**
ullage	see **ullage space**
ullage space	*aka* ullage.
	1. The open volume above storage vessel contents. Defined formally in NFPA 654: Standard for the Prevention of Fire and Dust Explosions from the Manufacturing, Processing, and Handling of Combustible Particulate Solids, 2017 Edition
	2. Can however also be used, slightly confusingly, for the residual capacity of a system (available ullage = available capacity)
ULSFO	Ultra low sulfur fuel oil
ultimate analysis	1. Most commonly, the quantitative determination of the elemental composition of a fuel; cf **proximate analysis**
	2. The determination of the elemental composition of sludges, filtered solids, and scales/buildups

ultimate cause	*aka* distal cause. An event or circumstances ultimately responsible for causing something; cf proximate cause
ultimate cycle tuning method	An empirical controller tuning method; cf Ziegler-Nicols tuning method
ultimate tensile strength	see **tensile strength**
ultra high hydrostatic pressure processing	*aka* UHHPP; see **high pressure processing**
ultra high pressure high temperature field	*aka* uHPHT. An oil or gas field where the reserve is at a pressure of more than 12,500 psi and a temperature of more than 330° F (862 bar and 166° C)
ultra high pressure processing	see **high pressure processing**
ultra high speed water spray system	A rapid automatic water spray system which protects against specific deflagration hazards. Defined formally in NFPA 15 Standard for Water Spray Fixed Systems for Fire Protection
ultra high temperature	*aka* UHT, sterilization. A food preservation process using a higher temperature, and shorter duration than pasteurization (qv)
ultra large crude carrier	*aka* ULCC. An ultra large ship (even larger than a VLCC (qv)) which carries crude oil, commonly known as a supertanker
Ultra-Orthoflow process	A proprietary variant on the fluid catalytic cracker (qv)
ultraclean process	"A process using equipment disinfected before use and protected against recontamination by microorganisms that may harm the safety and suitability of the product", according to the EHEDG Glossary
ultrafiltration	*aka* UF, hyperfiltration. A somewhat contested term used for a membrane filtration process removing very fine particles, even bacteria and viruses. Sometimes used for membranes fine enough to remove macromolecules (most often by scientists) as well as particles, and at the other end of the scale, sometimes used for coarser membranes when it is more commonly known as microfiltration (qv), most commonly in US English; see **Box L1**; cf nanofiltration
Ultraforming process	A proprietary catalytic reforming process with a platinum catalyst; no longer licensed but still operating
ultrapure water	The purest grade of water, used in silicon chip production and certain other high value applications
ultrasonic flow meter	A flow meter which uses doppler effect or transit time difference to measure material velocity as the basis of a volumetric or mass flow estimate
ultrasonication	Sonication (qv) with ultrasound, commonly used to break open microorganisms

ultraviolet	Electromagnetic radiation with a wavelength of 100-400 nm. Used to disinfect surfaces, water and treated sewage; also used in hygienic industries to detect the presence of residual contamination, and in some monitoring devices
ultraviolet disinfection	*aka* UV disinfection. The disinfection of fluids by exposure to ultraviolet radiation
unacceptable leakage	The level of leakage which compromises system performance in the view of the user and the applicable regulatory body. Defined formally in ASME BPE (American Society of Mechanical Engineers: Bioprocessing Equipment)
unbound moisture	*aka* surface moisture. Moisture in excess of bound moisture (qv); cf free moisture
uncertainty analysis	1. A systematic quantification of uncertainty associated with a study such as a life cycle analysis (qv) as a result of errors in model and data used. Defined formally in an LCA context in EN ISO 14040:2006 Environmental management - Life cycle assessment - Principles and framework 2. More generally, used for many aspects of model validation. For example, uncertainty analysis can provide significant insight into the predictive ability of thermodynamic or kinetic models that can explain why real operation does not match predicted simulation results
unconfined vapor cloud explosion	*aka* UVCE. The detonation of a cloud of vapor in air: the number one process hazard, as occurred in the Flixborough disaster (qv); nowadays usually simply known as a VCE
under ground	*aka* underground (U/G). A common designation on a P&ID (qv) or piping layout drawing to indicate that a component is below the surface of the Earth; cf above ground (A/G)
undercooling	A deprecated term, as it can be used to mean both subcooling (qv) and supercooling (qv)
undercut	A welding defect in which a groove is left under the weld. Defined formally in BS EN ISO 10209:2012 Technical product documentation — Vocabulary — Terms relating to technical drawings, product definition and related documentation, API STD 620 - Low Pressure Storage Tanks, ASME BPE (American Society of Mechanical Engineers: Bioprocessing Equipment) and API RP 579 - Fitness for Service
underdrains	Pipes or channels forming a water collection and backwashing system in the base of a vessel containing a packed bed (qv)
underfill	see **concavity**
underflow	A liquid leaving the bottom of certain types of process unit, such as thickeners, settling tanks and froth cells

underground gas storage	*aka* UGS. Most commonly, storing natural gas in depleted gas reservoirs or salt caverns
undershoot	1. In the context of process control, a response falling short of target cf overshoot 2. A vibration; defined formally in BS ISO 2041:2018 Mechanical vibration, shock and condition monitoring. Vocabulary
cf overshoot	
underwater autonomous vehicle	*aka* UAV, remotely operated vehicle (ROV), remotely operated underwater vehicle (ROUV). A submarine drone
Underwood equation	More correctly, Underwood equations; used to estimate minimum reflux ratio in multicomponent distillation (qv)
unexpected startup	*aka* unintended startup. Equipment startup which endangers people because it is surprising. Defined formally in BS EN ISO 12100:2010 Safety of machinery. General principles for design. Risk assessment and risk reduction
ungeneer	see **enjuneer**
Unicracking process	A family of UOP licensed hydrocracking (qv) processes, primarily for processing vacuum gas oil (qv)
Unidak process	A proprietary catalytic hydrodealkylation (qv) process used to make naphthalene from heavier feed; no longer offered for license
Unifining process	A family of UOP licensed hydrotreating (qv) processes
Uniflex process	UOP licensed slurry-phase resid. hydrocracking (qv) processes
uniform corrosion	Consistent metal loss through corrosion across large parts of a component; defined formally in NFPA 59: Utility LP-Gas Plant Code, 2018 Edition
uniform flow	Flow with no significant change in velocity over time at any given point in space, as found for example in a long pipeline
uniform particle size	*aka* UPS. A property of a collection of particles with a high uniformity coefficient (qv)
uniformity coefficient	A measure of the uniformity of particle sizes in a granular material such as soil or sand, being the ratio of percentage passages of two filter sizes; the filter sizes and percentages used vary between applications
uniformly scattered porosity	Used in the context of welding to describe porosity scattered fairly evenly throughout a weld; defined formally in ASME BPE (American Society of Mechanical Engineers: Bioprocessing Equipment)

uninsulated	The opposite of insulated (qv). The formal definition in NFPA 15 Standard for Water Spray Fixed Systems for Fire Protection basically says this in the context of thermal insulation, albeit in a more clunky way
unintended startup	see **unexpected startup**
uninterruptible power supply	*aka* UPS. Most commonly a battery backup for electronics such as control and communication systems; defined formally in ISO 3977 Gas turbines - Procurement - Part 3: Design requirements
union	see **union coupling**
union coupling	*aka* union. A type of reusable threaded pipe coupling
Unipol PE	A proprietary gas phase polyethylene manufacturing process
Unipol PP	A proprietary gas phase polypropylene manufacturing process
Unistrut	Proprietary eponym for U-shaped metal pieces and connectors for site fabrication of relatively lightweight supports for pipework, traywork and instrumentation, see **Figure U1**
unit	A collection of equipment focused on a single operation; cf unit operation
unit area	Especially when used in the context of flux, a term whose magnitude depends on the units being used: in SI units it is 1 square meter, but in US customary units (qv) it is 1 square yard

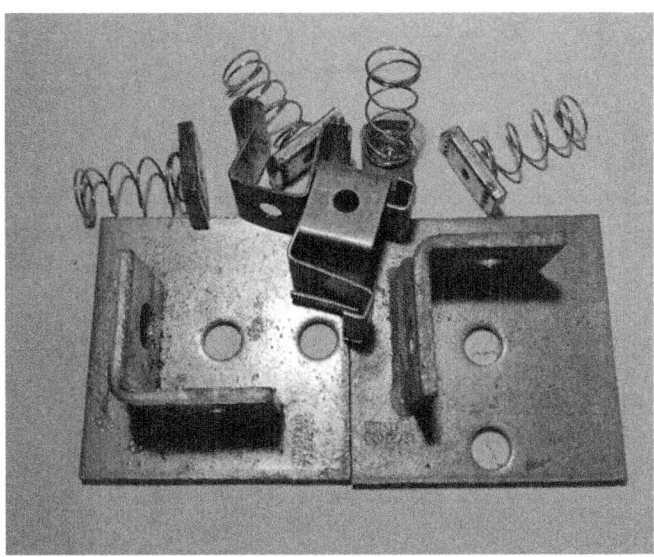

Figure U1 Some Unistrut Components

unit operation	The most basic processing step, a key chemical engineering concept. Various formal definitions can be found including those in BS EN ISO 10209:2012 Technical product documentation — Vocabulary — Terms relating to technical drawings, product definition and related documentation and NFPA 30: Flammable and Combustible Liquids Code, 2018 Edition
unit per capita loading	see **population equivalent**
unit process	1. Generally, synonymous with unit operation (qv), though a minority dispute this, claiming that unit processes differ from unit operations in that they involve chemical reaction; see the formal definition in NFPA 30: Flammable and Combustible Liquids Code, 2018 Edition 2. Also used in life cycle assessment (qv) to mean the smallest element considered, as defined in EN ISO 14040:2006 Environmental management - Life cycle assessment - Principles and framework
unit responsibility	Responsibility for a unit. Defined formally in EN ISO 10437 Petroleum, petrochemical and natural gas industries - Steam turbines - Special-purpose applications and EN ISO 13709:2003 Centrifugal pumps for petroleum, petrochemical and natural gas industries
universal cylinder	An LPG cylinder (qv) which can be installed in either a horizontal or vertical orientation; defined formally in NFPA 58: Liquefied Petroleum Gas Code, 2017 Edition
universal gas constant	see **gas constant**
universal toolkit	*(jocular)* Duck tape (qv) and WD40 (qv)[102]
unloader valve	A valve used for capacity control on a reciprocating compressor or pump. Defined formally in the context of pumps in NFPA 20: Standard for the Installation of Stationary Pumps for Fire Protection, 2019 Edition
unloading nozzle	see **catalyst dump nozzle**
UNOL	see **upper normal operating limit**
Unox process	A proprietary pure oxygen aerated activated sludge process which has a much smaller aeration system, but much larger running costs than a conventional system
unplanned maintenance	see **maintenance**
unsaturated	Used to denote compounds amenable to addition reactions, especially hydrogenation (qv)

---

[102] If it moves and it shouldn't, Duck tape it. If it doesn't and it should, WD40 it!

unscheduled maintenance	*aka* unplanned maintenance; see **maintenance**
unstable liquid	A liquid which can energetically polymerize, decompose, or otherwise self-react when shocked, pressurized, or heated. Defined formally in NFPA 30: Flammable and Combustible Liquids Code, 2018 Edition cf stable liquid
unsteady state	*aka* transient (especially in flow assurance). The condition in which variables of interest (concentration, temperature, pressure etc.) change over time: essentially the real-world situation in a process plant, though effective process control will normally keep this variability within acceptable range; cf steady state
unstructured kinetic growth model	see **unstructured model**
unstructured model	*aka* unstructured kinetic growth model. A simple model based on Monod kinetics (qv) used to approximate microorganism growth parameters
unstructured packing	*aka* random packing. Small open structures which flow to fill a space with a surface upon which reaction and/or mass transfer can take place; cf structured packing
UO_2	see **uranium dioxide**
UOP K-factor	see **K-factor**
UOPK	see **K-factor**
upcycling	*aka* creative reuse. A process which is, economically, the opposite of what is normally thought of as recycling: turning waste (qv) into a product of higher quality/value; cf downcycling
updraft	The upflow (qv) of gases; cf downdraft
upflow	Especially in a vertical column, flow upwards, from bottom to top; cf downflow
upflow anaerobic sludge blanket	*aka* UASB. A more novel compact type of anaerobic digester (qv) in which an upflow of effluent expands a bed of granular biomass
upflow velocity	Superficial upward velocity of fluid in a settlement tank etc.; cf surface loading
upgradation	A rather ugly synonym of beneficiation (qv) cf upgrading
upgrader	A plant which upgrades heavy oil (qv) or extra heavy oil (qv) into lighter, more valuable products
upgrading	see **heavy oil upgrading**
upper explosive limit	*aka* UEL; see **upper flammable limit**

upper flammable limit	*aka* UFL, upper explosive limit (UEL). The maximum concentration of a gas in air in which a flame can be propagated; identical in value to upper explosive limit. Defined formally in API RP 2510A - Fire Protection Considerations for the Design and Operation of Liquefied Petroleum Gas (LPG) Storage Facilities
upper master valve	An (often hydraulically actuated) valve on a Christmas tree (qv) used to isolate and control the flow of fluids from the wellbore in routine operations; cf lower master valve
upper normal operating limit	*aka* UNOL. The upper expected operating limit; often used as a set point for alarms to avoid reaching the upper safe operating limit (qv); frequently applied to pressure, temperature and flowrate; cf lower normal operating limit, Hx alarm
upper safe design limit	*aka* USDL, safe design limit. A term often applied to design limits of pressure, temperature and flowrate; cf lower safe design limit
upper safe operating limit	*aka* USOL. The upper acceptable operating limit, often used as a set point for trips to avoid reaching upper safe design limit (qv); often applied to pressure, temperature and flowrate
UPS	1. see **uninterruptible power supply** 2. see **uniform particle size**
upset	see **process upset**
upset gas	Gases generated by a process upset (qv), such as those sent to a flare stack (qv)
upstream	1. Generically, earlier in the process 2. Used specifically in the oil and gas industry to refer to the exploration and production steps; cf downstream, midstream
upstream station	*aka* U/S; see **utility station**
UR	User requirements; see **user requirement specification**
urania	see **uranium dioxide**
uranium dioxide	(*symbol* UO_2) *aka* urania, uranous oxide. The primary form in which uranium is handled and formed into fuel rods
uranium hexafluoride	*aka* hex (*symbol* UF_6). A compound of uranium used in isotope separation/enrichment processes, and also for fire suppression. Although it is a solid at room temperature, it sublimes at 56.5° C, and is used in gas form for both of these duties
uranium-235	(*symbol* U^{235}) The primary fissionable isotope of uranium and the rarer natural occurring isotope. Most nuclear reactors use uranium enriched in U^{235} as fuel
uranium-238	(*symbol* U^{238}) The primary natural occurring isotope of uranium; can be converted to plutonium-239 via neutron capture in a breeder reactor

uranous oxide	see **uranium dioxide**
URD	User requirements document; see **user requirement specification**
URS	see **user requirement specification**
US	User specifications; see **user requirement specification**
US customary units	*aka* British units, British-American system of units. The quasi-'British' set of units (feet and inches, usgpm, scfm, lb., psi and °F) as used in the USA and in US-influenced industries such as oil and gas. Sometimes known as imperial units (qv), particularly in former British territories, but a number of Imperial and US customary units with the same names actually differ from each other in size; see **Box B1**.
US Department of Agriculture	*aka* USDA. A federal administrative body with regulatory responsibility over many aspects of the agricultural industry; interfaces with the US Food and Drug Administration (qv) in food production matters
US Food and Drug Administration	*aka* USFDA, FDA. A federal administrative body which controls the pharmaceutical and food industries in the USA and, by extension, in all countries who wish to sell pharmaceuticals or food products in the USA
USDA	see **United States Department of Agriculture**
USDL	see **upper safe design limit**
use	There are two kinds of use to which an artifact might be put: the intended use (the purpose/activity for which an artifact is designed), and unintended uses (any other purpose/activity to which it might be put). As users can be 'inventive', the second meaning needs to be considered by designers as well as the first. Defined formally in BS EN 82079-1:2012 Preparation of instructions for use. Structuring, content and presentation. General principles and detailed requirements and IEEE 830-1998 - IEEE Recommended Practice for Software Requirements Specifications
useability	The ability to be used easily for an intended purpose. Defined formally in the context of machinery in BS EN ISO 12100:2010 Safety of machinery. General principles for design. Risk assessment and risk reduction
user	An entity which uses an artifact or service; some formal definitions also include the sense of ownership of an artifact, as in BS EN 82079-1:2012 Preparation of instructions for use. Structuring, content and presentation. General principles and detailed requirements and ASME BPE (American Society of Mechanical Engineers: Bioprocessing Equipment)

user requirement specification	*aka* URS, user requirements document (URD), requirement specification, user requirements (UR), user specifications (US), spec sheet. A document containing an early stage, high level description of the business requirements and regulatory constraints of a proposed project; cf functional requirements specification
user requirements	*aka* UR; see **user requirement specification**
user requirements document	*aka* URD; see **user requirement specification**
user specifications	*aka* US; see **user requirement specification**
USFDA	*aka* FDA; see **US Food and Drug Administration**
USGPD	US gallons [of oil] per day
USGPM	US gallons [of oil] per minute
USOL	see **upper safe operating limit**
UT	1. Ultrasonic testing, according to API RP 571 - Damage Mechanism Affecting Fixed Refinery Equipment 2. Ultrasonic examination according to API RP 579 - Fitness for Service, which appears to be an error
utilities	*aka* services. Supplies of general-purpose materials or energy to a site (qv), such as site raw water, cooling water, utility water, demineralized water, boiler feed water, condensate handling, service water, firefighting water, potable water, utility air, instrument air, steam, nitrogen, fuel gas, natural gas and electricity supplies, or the facilities providing them; cf clean utilities, black utilities
utility	The singular of utilities (qv)
utility floor	see **interstitial space**
utility flow diagram	*aka* UFD. Usually a PFD (qv) exclusively for utilities (qv), but sometimes a P&ID (qv) for utilities or a distribution drawing for utilities, depending on the site/company standard
utility gas plant	An LPG storage and vaporization plant supplying an LPG or LPG-air mixture utility (qv) to a commercial gas distribution system. Defined formally in NFPA 59: Utility LP-Gas Plant Code, 2018 Edition
utility hole	see **manhole**
utility station	*aka* U/S, upstream station. A location where operators can connect hoses for utility (qv) steam, water, air or nitrogen (as provided for in the given facility)
utility valve	*aka* UV. A designation sometimes used for an actuated isolation valve; some operator specifications would alternatively abbreviate as XV (qv)

utility water	1. In general, synonymous with process water (qv) or town mains (qv) water, though it may be raw water (qv) or filtered raw water 2. In the hygienic industries, a subset of black utilities (qv), being water which is never in direct contact with the product such as heating and cooling utilities or firefighting water. Defined formally in the context of food in EHEDG Safe Storage and Distribution of Water in Food Factories; cf clean utilities, green utilities
UV	1. Ultraviolet light 2. *(drawing notation)* Utility valve (qv)
UV disinfection	see **ultraviolet disinfection**
UVCE	see **unconfined vapor cloud explosion**
UVOx process	A form of advanced oxidation (qv) using a combination of UV radiation and ozone to treat water contaminants

V

V blender	A vee-shaped type of solid/solid mixer
V model	*aka* V-model, vee model. A stage gate model (qv) type approach linking defined final testing specification to initial equipment requirement specification, as in good manufacturing practices (qv)
v/v	A proportion of something in a mixture expressed as a volume/volume fraction
VABP	see **volume average boiling point**
vacuum	1. From an engineering point of view, a pressure so far below atmospheric that any residual gases have no effect on any process within the region (a physicist would call this a partial vacuum); less formally still, might mean any pressure below 0 bara 2. May also be used to describe a region at this pressure, as in the formal definition in NFPA 86: Standard for Ovens and Furnaces, 2019 Edition 3. A low-pressure utility (qv)
vacuum accumulator tank	see **vacuum tank**
vacuum bottom	see **vacuum residue**
vacuum bottoms	see **vacuum residue**
vacuum breaker	A valve which prevents buildup of a vacuum such as an air inlet valve (qv) or syphon breaker (qv)
vacuum compressor	see **side channel blower**
vacuum degasifier	A device which removes free and dissolved gases from water
vacuum distillation	*aka* low pressure distillation. Distillation at reduced pressure; cf atmospheric distillation
vacuum distillation unit	*aka* VDU, vacuum unit. Whilst this might in theory refer to any unit in which vacuum distillation (qv) takes place, it usually refers to a reduced pressure distillation unit which follows immediately after an atmospheric distillation (qv) unit in an oil refinery
vacuum drier	*aka* vacuum dryer. Equipment for vacuum drying (qv)
vacuum dryer	see **vacuum drier**
vacuum drying	Drying at reduced pressure. As with vacuum distillation (qv), it allows operation at reduced temperatures. This can be useful with heat labile (qv) or hygroscopic (qv) materials

vacuum evaporation	The enhancement of evaporation by reduced pressure, sometimes used in zero liquid discharge (qv) effluent treatment systems
vacuum filter	A filter in which the driving force is provided by a downstream vacuum, rather than upstream positive pressure; nowadays more of a laboratory than an engineering technique, such as a rotary vacuum filter
vacuum flask	*aka* dewar, dewar flask, dewar vessel. A vessel insulated by a silvered glass double wall containing an annulus of high vacuum; used to store relatively small quantities of liquefied gases
vacuum flotation	A form of flotation in which bubbles are generated using vacuum
vacuum fore pump	see **holding pump**
vacuum furnace	A fired heater in which a low pressure atmosphere reduces interaction with air of the surface of furnace contents in some way; cf special atmosphere
vacuum gas oil	see **light vacuum gas oil**
vacuum gauge	A pressure gauge suitable for measuring pressures below atmospheric. Defined formally in NFPA 86: Standard for Ovens and Furnaces, 2019 Edition
vacuum pan	A crystallizer (qv) operating under vacuum; used to produce sugar crystals
vacuum pressure safety valve	*aka* VPSV, low and vacuum pressure safety valve (LVPSV); see **air inlet valve**
vacuum pump	A compressor designed to remove gases from a space rather than deliver them into a space; sometimes but not always a side channel blower (qv). Defined formally in NFPA 86: Standard for Ovens and Furnaces, 2019 Edition
vacuum pumping system	In the context of a vacuum furnace (qv), the complete system which produces the required vacuum. Defined formally in NFPA 86: Standard for Ovens and Furnaces, 2019 Edition
vacuum relief valve	see **air inlet valve**
vacuum resid.	see **vacuum residue**
vacuum residue	*aka* resid., vacuum resid, vacuum bottom, vacuum bottoms, vacuum tower bottoms (VTB). Very heavy tower bottoms (qv) from a vacuum distillation unit (qv)
vacuum support rings	Stiffeners used to prevent collapse under external pressure in vessels designed for vacuum service
vacuum system	In the context of vacuum furnaces, a pressure vessel designed for vacuum service complete with a connection point to a vacuum pumping system. Defined formally in NFPA 86: Standard for Ovens and Furnaces, 2019 Edition

vacuum tank	1. *aka* vacuum vessel. A pressure vessel designed for negative internal pressure 2. *aka* vacuum accumulator tank. A tank which holds accumulated vacuum, much as an accumulator holds pressurized fluid; cf pressure tank 3. A tank which holds liquid drawn in by a vacuum created by a vacuum pump, commonly mounted on wheels to form a vacuum tanker (qv)
vacuum tanker	A vacuum tank mounted on wheels
vacuum tower bottoms	*aka* VTB; see **vacuum residue**
vacuum type insulation	Insulation like that in a vacuum flask (qv), with two reflective walls separated by a high vacuum; commonly used with cryogenic fluids. Defined formally in NFPA 86: Standard for Ovens and Furnaces, 2019 Edition
vacuum unit	see **vacuum distillation unit**
vacuum vessel	see **vacuum tank**
vain	see **nappe**
validation	A term with several meanings, see **Box V1**.
validity	1. In the context of contracts, the specified period during which an offered price can be accepted 2. In the context of statistics, the accuracy and reliability of test conclusions
valorization	The beneficial use of waste (qv)
value engineering	*aka* value management. A management technique which is intended to ensure value for money (but might simply mean cost-cutting)
value management	see **value engineering**
valve	A device to control fluid flow, most commonly through an enclosed conduit, or between such a conduit and environment
valve flow coefficient	see **flow coefficient**
valve positioner	A control system component that adjusts the position of a valve's stem or shaft, ensuring it matches the desired set point from a control system
valve proving system	A safety shutoff valve leakage detection system; defined formally in NFPA 86: Standard for Ovens and Furnaces, 2019 Edition
valve tray	A type of contacting device used within a distillation column; can handle variable loading and higher turndown than a (cheaper) sieve tray (qv)
van der Waals mixing rules	see **mixing rules**
Van Laar equation(s)	Equations which model phase equilibria of liquid mixtures

Box V1. Validation

Validation is a term with many meanings, all of which (leaving aside the more touchy-feely psychotherapy use of the term, which we need not dwell upon) share a sense of proving something to be true by comparison with objective evidence, in line with its commonplace meanings of corroboration or substantiation. So, what is the range of specific meanings of validation within chemical engineering?

In the European hygienic industries, the EHEDG give a formal definition as follows: "Obtaining evidence that the control measures managed by the HACCP plan and by the operational PRPs are capable of being effective".

In bio-pharmaceutical engineering, ASME BPE defines validation in the sense of as gathering documented evidence that a system meets specification. In a quality context, BS EN ISO 9000[103] has an essentially similar definition. The IEC 61508-4[104] definition makes specific mention of software, but there is another sense in which the term is applied to software: in computer modelling it means ensuring that modeling software matches reality by comparing outputs with real data.

Whilst the term validation has a common core meaning, it can refer to very different validation processes in different engineering contexts. It is easy to see how someone working in biochemical engineering might need to be aware that people around them could be using the same term in at least four different ways, (ignoring commonplace meanings and psychobabble). However, if we state we are using a term as defined in a named standard, we have gone a long way towards clear communication.

Related Terms
verification

Van't Hoff('s) equation	*aka* Van't Hoff('s) isochore, Vukančić-Vuković equation. An equation which relates the change in the equilibrium constant of a chemical reaction to temperature
Van't Hoff('s) isochore	see **Van't Hoff('s) equation**
vane	see **strake**
vane type mist eliminator	A device which coalesces and collects small solid particles and liquid droplets from a gas stream by means of a set of parallel vanes or louvres
vapor	*aka* vapour. Technically, a mixture of gas and liquid, but often used informally to refer to condensable gases within a system (e.g., vapor side, vapor stream)

---

[103] BS EN ISO 9000 Quality management systems Fundamentals and vocabulary
[104] IEC 61508-4 Ed. 1.0 b:1998, Functional safety of electrical/electronic/programmable electronic safety-related systems - Part 4: Definitions and abbreviations

vapor absorption cycle	A theoretical analysis of one type of refrigeration system
vapor barrier	A metallic foil barrier to flue gas flow, installed between refractory layers in fired heaters. Defined formally in BS EN ISO 13705:2012/ISO 13705:2012(E) Petroleum, petrochemical and natural gas industries. Fired heaters for general refinery service, API STD 530 - Calculation of Tube Heater Thickness and API STD 560 - Fired Heaters for General Refinery Service
vapor binding	A condition in which non condensable gases are accumulated in a reboiler or condenser, reducing heat transfer to process fluid
vapor cloud explosion	A modern term for unconfined vapor cloud explosion (qv)
vapor compression cycle	A thermodynamic description of the basis of vapor compression refrigeration
vapor density	An alternative term for specific gravity (qv) in the case of gases, as defined in API 2001 Fire Protection at Refineries
vapor density	The density of a gas relative to that of hydrogen; cf specific gravity
vapor depressuring system	The valves and piping associated with rapid vapor depressurization for safety reasons. Defined formally in API Standard 521. Pressure-relieving and Depressuring Systems. Sixth Edition | January 2014
vapor liquid equilibria	*aka* VLE. A general term for all of the engineering science around systems containing liquids and vapors where distribution between phases has reached an equilibrium, such as Dalton's law, Henry's law and Raoult's law (qv)
vapor liquid equilibrium ratio	see **equilibrium ratio**
vapor pocket	*aka* isolated vapor pocket (IVP) An unvented high point in pipework or tankage which collects vapor
vapor pressure	Generically, in an engineering context, either 1. The pressure exerted by the vapor of a substance (a formal definition along these lines is offered in API RP 2001 - Fire Protection at Refineries) 2. Shorthand for equilibrium vapor pressure (qv), a precisely defined scientific term. The NFPA has a highly formal definition based on USA standards and units, which can be found in NFPA 30: Flammable and Combustible Liquids Code, 2018 Edition. The API RP2001 definition is closely related and both are similar to the scientific definition

vapor pressure function	A dimensionless factor relating vapor pressure at average daily liquid surface temperature to atmospheric pressure; used in the estimation of evaporative losses from tankage in the oil and gas industry. Defined formally in API MPMS 19.1 Manual of Petroleum Measurement Standards Chapter 19.1 Evaporative Loss from Fixed-roof Tanks
vapor processing equipment	In the context of volatile flammable fluid transfer/filling operations, the equipment which processes displaced vapor. Defined formally in NFPA 30: Flammable and Combustible Liquids Code, 2018 Edition
vapor processing system	In the context of volatile flammable fluid transfer/filling operations, a system which captures and processes displaced vapor. Defined formally in NFPA 30: Flammable and Combustible Liquids Code, 2018 Edition
vapor recovery system	In the context of volatile flammable fluid transfer/filling operations, a system which captures but does not process displaced vapor. Defined formally in NFPA 30: Flammable and Combustible Liquids Code, 2018 Edition
vapor recovery unit	*aka* VRU. One of a number of different types of equipment which capture and/or concentrate VOCs (qv); cf vapor recovery system
vapor/liquid separator	A generic description of a unit operation which separates vapor from liquid
vaporisation	see **vaporization**
vaporization	*aka* vaporisation. A phase transition from liquid to vapor by boiling (qv) or evaporation (qv)
vaporizer	A device which vaporizes LPG by heating. Defined formally in NFPA 58: Liquefied Petroleum Gas Code, 2017 Edition
vaporizing burner	*aka* selfvaporizing liquid burner. In the context of LPG, a combined vaporizer and burner. Defined formally in NFPA 58: Liquefied Petroleum Gas Code, 2017 Edition
vaportight	In the context of volatile flammable fluids operations, the ability to contain flammable fluid under operating conditions. Defined formally in NFPA 30: Flammable and Combustible Liquids Code, 2018 Edition
vaportight barrier	In the context of hazardous area classification, a barrier to the passage of significant quantities of flammable fluid. Defined formally in API RP 505 2nd Edition, August 2018 Recommended Practice for Classification of Locations for Electrical Installations at Petroleum Facilities Classified as Class I, Zone 0, Zone 1, and Zone 2
vapour	see **vapor**

variable area flowmeter	A device, often known by the proprietary eponym Rotameter, in which flow is indicated by the rise of a weighted plug in a transparent tube of increasing cross-section
variable costs	Business costs which vary with production volume. Generally, includes raw materials and energy/utilities used in production; cf fixed costs
variable frequency drive	*aka* VFD; see **inverter**
variable liquid level gauge	A device which detects the liquid/vapor interface in a sealed, pressurized container using one of a number of dip-tube type arrangements. Defined formally in NFPA 58: Liquefied Petroleum Gas Code, 2017 Edition
variable load	*aka* live load. A cyclical load on a support slab such as that from vessels which are regularly filled and emptied, or reciprocating equipment; cf dead load, live load
variable speed drive	*aka* VSD; see **inverter**
variable speed pressure limiting control	The control of pump discharge pressure by means of a variable speed drive. Defined formally in NFPA 20: Standard for the Installation of Stationary Pumps for Fire Protection, 2019 Edition
variable speed pump	A pump with output controlled by varying driver speed, most commonly with a VSD (qv). Defined formally in the context of fire pumps in NFPA 20: Standard for the Installation of Stationary Pumps for Fire Protection, 2019 Edition
variable speed suction limiting control	Maintaining minimum pump suction pressure by means of a variable speed drive (qv). Defined formally in NFPA 20: Standard for the Installation of Stationary Pumps for Fire Protection, 2019 Edition
variables sampling	A quality sampling approach in which a parameter is measured and recorded; cf attributes sampling
variance in luster	*aka* variance in lustre. A visible difference in shine by reflected light. Defined formally in ASME BPE (American Society of Mechanical Engineers: Bioprocessing Equipment)
variance in lustre	see **variance in luster**
variation order	*aka* VO. A document issued by a contracting organization to a contractor to formally request a different amount, execution, quality, schedule or type of works to that specified in the contract
variator	*aka* mechanical variator. A mechanical power transmission device allowing a smooth variation in drive speed (as opposed to the stepwise change of a gearbox)

vault	An enclosed space whose sole purpose is containing a flammable liquid storage tank. Defined formally in NFPA 30: Flammable and Combustible Liquids Code, 2018 Edition cf storage tank building
V-belt	A type of drive belt with a 'vee' cross section
VBP	see **vehicular barrier protection**
VCE	see **vapor cloud explosion**
VCM	see **vinyl chloride monomer**
VDU	1. see **vacuum distillation unit**
	2. see **visual display unit**
vee model	see **V model**
vee notch	1. A vee-shape cut into a weir plate; see **Figure V1**
	2. Less frequently, a vee-shape cut into the ball of a ball valve
vehicle fuel dispenser	An LPG dispenser for filling vehicle fuel tanks. Defined formally in NFPA 58: Liquefied Petroleum Gas Code, 2017 Edition
vehicle security barriers	*aka* VSB, road safety barriers. Impact resistant barriers found at a site (qv) perimeter to exclude vehicles (including hostile vehicles) and maintain blast standoff; cf vehicular barrier protection
vehicular barrier protection	*aka* VBP, vehicle barriers. A physical system for excluding vehicles from an area. Defined formally in NFPA 58: Liquefied Petroleum Gas Code, 2017 Edition

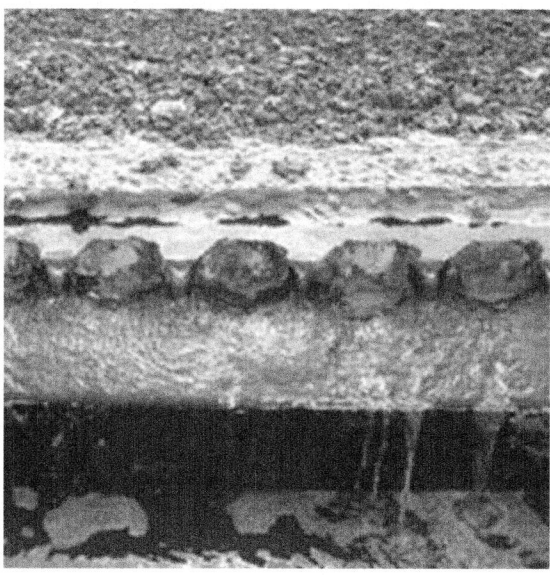

Figure V1 Vee notch

vehicular barriers	see **vehicular barrier protection**
velocity	1. Usually, speed in a given direction
	2. Relative velocity, which is still a directional speed, but possibly relative to a moving reference point, defined formally in BS ISO 2041:2018 Mechanical vibration, shock and condition monitoring
velocity head	*aka* Hv. The kinetic energy of moving fluid. Defined formally in NFPA 20: Standard for the Installation of Stationary Pumps for Fire Protection, 2019 Edition
velocity ratio	see **slip ratio**
velocity seal	*aka* orifice seal. A dry vapor seal associated with a flare. Defined formally in API Standard 521. Pressure-relieving and Depressuring Systems. Sixth Edition	January 2014 and API RP 537 Flare Details for Petroleum, Petrochemical, and Natural Gas Industries
velocity steam	*aka* coil steam. A flow of steam adjusted to control the residence time of vacuum feed in a vacuum heater; a term possibly coined by Norm Lieberman
velocity transducer	A transducer which produces an output proportional to fluid velocity
vena contracta	The point of minimum diameter and maximum speed in a fluid stream
vendor	see **supplier**
vendor list	A verified list of vendors recognized as qualified suppliers for a project
vent	1. Various meanings, all to do with fluid movement: an opening or nozzle on a contained space, a discharge point to environment of gas-based mixtures in a number of different contexts, an automatic bleed valve, and an abbreviation of ventilation
	2. Also used as a verb: to vent a system means to reduce the pressure until equalized with the system to which it is vented
vent closure	A pressure-relieving vent cover; defined formally in NFPA 654: Standard for the Prevention of Fire and Dust Explosions from the Manufacturing, Processing, and Handling of Combustible Particulate Solids, 2017 Edition
vent header	Piping which delivers relieved fluids to a vent stack; defined formally in API Standard 521. Pressure-relieving and Depressuring Systems. Sixth Edition	January 2014
vent limiter	A fixed orifice limiting vented vapor flow to atmosphere; defined formally in NFPA 86: Standard for Ovens and Furnaces, 2019 Edition

vent scrubber	see **knockout drum**
vent stack	A vertical pipe which vents vapor to atmosphere; defined formally in API Standard 521. Pressure-relieving and Depressuring Systems. Sixth Edition	January 2014
vent/vac valve	*aka* tank breather, breather valve. A combined pressure and vacuum relief valve, commonly used on a sealed atmospheric tank (qv) to vent air on filling and to admit it to prevent implosion on emptying
vented exhaust mechanical pump	see **gas ballast pump**
vented explosion pressure	*aka* suppressed explosion pressure or pressure reduced (Pred). The maximum pressure in a vented enclosure during a vented deflagration. Defined formally in NFPA 654: Standard for the Prevention of Fire and Dust Explosions from the Manufacturing, Processing, and Handling of Combustible Particulate Solids, 2017 Edition
ventilation	1. Generically, the supply and/or extraction of air, usually referring to that required for personnel comfort 2. Defined in the context of fire and explosion prevention in NFPA 30: Flammable and Combustible Liquids Code, 2018 Edition 3. The provision of air flow through a furnace, according to NFPA 86: Standard for Ovens and Furnaces, 2019 Edition 4. API RP 505 2nd Edition, August 2018 Recommended Practice for Classification of Locations for Electrical Installations at Petroleum Facilities Classified as Class I, Zone 0, Zone 1, and Zone 2 gives a generic definition, but then concerns itself largely with dilution of flammable gases by ventilation, and classification of areas based on their degree of this type of ventilation
venturi aeration	Using an eductor (qv) driven by water flow to take in, entrain and dissolve air, often in biological effluent treatment
venturi flow meter	*aka* venturi meter. A flow measurement device which uses a venturi to produce a differential pressure proportional to fluid flow, and associated pressure transducers
venturi meter	see **venturi flow meter**
venturi mixer	see **gas jet mixer**
venturi scrubber	Using an eductor (qv) driven by liquid flow to take in, entrain and preferentially dissolve readily soluble components of a waste gas stream

verification	1. In the hygienic industries, verification is carried out after a control measure, etc. has been implemented, providing "confirmation, through the provision of objective evidence, that specified requirements have been fulfilled" according to the EHEDG (cf validation). Defined similarly in the context of quality management in BS EN ISO 9000 Quality management systems Fundamentals and vocabulary, whilst IEC 61508-4: 1998 gives a similar, but far wordier definition in the context of functional safety 2. In the context of computer models, ensuring software is free of coding errors; cf validation
vermin	*aka* pests. Animals which can adversely affect food quality; defined formally in BS EN 1672-2:2005+A1:2009 Food processing machinery. Basic concepts. Hygiene requirements cf pest
vertical compressor	A compressor mounted on a vertically arranged receiver
vertical fire protection zone	A vertical fire-water supply zone in a high rise building (qv). Defined formally in NFPA 20: Standard for the Installation of Stationary Pumps for Fire Protection, 2019 Edition
vertical inline pump	A vertical axis centrifugal pump with suction and discharge connections on a common centerline. Defined formally in EN ISO 13709:2003 Centrifugal pumps for petroleum, petrochemical and natural gas industries
vertical lineshaft turbine pump	A vertical axis turbine pump with discharge coaxial with shaft. Defined formally in NFPA 20: Standard for the Installation of Stationary Pumps for Fire Protection, 2019 Edition
vertical long tube evaporator	see **rising film evaporator**
vertical shell and tube watertube heat recovery steam generator	*aka* vertical shell and tube watertube HRSG. Just as it sounds: a vertical shell and tube heat exchanger (qv) used as a heat recovery steam generator (qv) with water in the tubes, but the API still considered that a formal definition was required: see API RP 534 – Heat Recovery Steam Generators
vertical shell and tube watertube HRSG	see **vertical shell and tube watertube heat recovery steam generator**
vertical thermosyphon reboiler	A vertically oriented thermosyphon reboiler (qv); cf horizontal thermosyphon reboiler
vertical tube evaporator	see **rising film evaporator**
vertical turbine pumps total head	Usually means the sum of static and dynamic heads, as with all pump total heads (qv), but there is a slightly different definition in NFPA 20: Standard for the Installation of Stationary Pumps for Fire Protection, 2019 Edition

vertically suspended pump	A vertical-axis pump with a suspended liquid end. Defined formally in EN ISO 13709:2003 Centrifugal pumps for petroleum, petrochemical and natural gas industries
very large crude carrier	*aka* VLCC. A very large ship which carries crude oil; cf ultra large crude carrier
very tall building	Any high rise building (qv) where fire protection (qv) water demand is greater than fire department pumping capacity. Defined formally in NFPA 20: Standard for the Installation of Stationary Pumps for Fire Protection, 2019 Edition
vessel	A container used for the storage or processing of materials. Defined formally in API Standard 521. Pressure-relieving and Depressuring Systems. Sixth Edition | January 2014; cf tank
VFA	see **volatile fatty acids**
VFD	see **variable frequency drive**
VGO	see **vacuum gas oil**
VHAP	see **volatile hazardous air pollutant**
VI	see **volumetric (VI)**
vibration	Mechanical oscillations around a point. Defined formally in BS ISO 2041:2018 Mechanical vibration, shock and condition monitoring. Vocabulary
vibration isolator	A device which attenuates the transmission of a specified range of vibration frequencies. Defined formally in BS ISO 2041:2018 Mechanical vibration, shock and condition monitoring. Vocabulary
vibratory mill	A size reduction device which grinds material between two vibrating surfaces
vibromixer	A small scale, low shear mixer using vertically oscillating perforated disks
vicat hardness	see **vicat temperature**
vicat softening point	see **vicat temperature**
vicat softening temperature	see **vicat temperature**
vicat temperature	*aka* vicat hardness, vicat softening point, vicat softening temperature. The temperature at which a blunt probe under a stated load sinks to 1mm depth into a material; used as a proxy for melting point in plastics
view	*aka* projection. In drawing, the way in which a drawn object is 'seen', or the direction from which it is 'seen' such as plan, elevation, section (qv)
view factor	The fraction of radiation reflected from a surface which hits a second surface
view port	see **sight glass**

viewing copy	A replica of a document produced for visual inspection. Defined formally in ISO 10209-1:1992 Technical product documentation - Vocabulary - Part 1: Terms relating to technical drawings: general and types of drawings
Viking Johnson	*aka* V-J. Proprietary eponym for flange adaptors
vinasse	see **dunder**
vinyl chloride monomer	*aka* VCM. A halocarbon monomer produced at megatonne scale for the manufacture of polyvinyl chloride based polymers
violet hydrogen	see **purple hydrogen**
virial equation of state	A theoretical model of gas properties, unfortunately much abused in recent years
visbreaking	Reducing the viscosity of hydrocarbons by thermal cracking (qv)
viscoelasticity	A property of certain materials which initially produce an elastic response to applied stress, then produce a time-dependent viscous response
viscose	*aka* rayon. Regenerated cellulose fibers; the term can refer to the fibers themselves, or the solution they are spun from
viscosity	1. Generically, a measure of flow characteristics, either kinematic viscosity (qv) or dynamic viscosity (qv) 2. In the paper and pulp field, a measure of molar mass
viscous damping	see **linear damping**
viscous damping coefficient	see **linear damping coefficient**
viscous force	*aka* drag. The force between an object and a fluid moving past it
viscous power	Power loss due to viscous fluid friction within a pump
vision	Top management's high level aspirations for an organization; defined formally in BS EN ISO 9000 Quality management systems Fundamentals and vocabulary
visual display unit	*aka* VDU. An almost obsolete term for a computer screen, from the days of cathode ray tubes
Viton	Proprietary eponym for the highly chemically resistant FKM (qv) elastomer, used in seals and gaskets
Vitox process	A biological treatment process aerated with pure oxygen
vitreous	Glassy
vitrification	Making into glass; a process used to stabilize some kinds of nuclear waste
vitrinite	Vitreous (qv) kerogen (qv) in coal; cf coking coal
V-J	see **Viking Johnson**
VLCC	see **very large crude carrier**
VLE	see **vapor liquid equilibria**
V-model	see **V model**
VMP	Voltage at maximum power

VO	see **variation order**
VOC	see **volatile organic compound**
void	1. A space
	2. Invalidated
void fraction	see **porosity**
void volume	see **interstitial volume**
voidage	see **porosity**
volatile fatty acids	*aka* VFA. Odorous low molecular weight linear aliphatic carboxylic acids
volatile flammable liquid	Generically, just what it sounds like, but defined formally in API RP 505 2nd Edition, August 2018 Recommended Practice for Classification of Locations for Electrical Installations at Petroleum Facilities Classified as Class I, Zone 0, Zone 1, and Zone 2
volatile hazardous air pollutant	*aka* VHAP. Generically, just what it sounds like. Specifically, in the USA, a term which is defined formally in the US National Emission Standards For Hazardous Air Pollutants, Title 1, Part A, Section 112
volatile organic compound	*aka* VOC.
	1. Generally speaking, any organic compound which is volatile under ambient indoor conditions, as defined formally in BS 5930:2015 Code of practice for ground investigations
	2. The US EPA defines this term as "any compound of carbon, excluding carbon monoxide, carbon dioxide, carbonic acid, metallic carbides or carbonates, and ammonium carbonate, which participates in atmospheric photochemical reactions"
volatile solids	Those total dry solids (qv) which are lost by ignition (incineration). May be expressed as a fraction of total dry solids; used most commonly in the context of biological treatment of effluent
volatile suspended solids	*aka* VSS. Those total suspended solids (qv) which are lost by ignition (incineration). May be expressed as a fraction of total suspended solids. Used most commonly in the context of biological treatment of effluent
volatility	1. Most commonly, the tendency to evaporate
	2. Chemical, financial (or emotional) instability
vollhub safety valve	see **full lift safety valve**
volt	(*symbol* V) The derived SI unit of electrical potential
volt per meter	(*symbol* V/m) The SI unit of electric field
volume average boiling point	*aka* VABP. One of five ways used in oil refining to express the average boiling point of a mixture of hydrocarbons (the others being molal, weight, cubic and mean)

volume mean diameter	*aka* d30, d[3,0] etc. In the context of particle size distribution, a type of mean particle size, the diameter of a theoretical particle whose volume multiplied by the total number of particles in a sample will equal the total volume of the sample
volumetric (VI)	*aka* VI. A classification of a type of welding flaw defined in PD 5500:2018+A1:2018 Specification for unfired fusion welded pressure vessels; cf volumetric flaw
volumetric efficiency	(*symbol* ηv) A measure of the efficiency of gas pumping
volumetric flaw	A shape imperfection or loss of material volume; defined formally in API RP 579 - Fitness for Service
volumetric flow	A quantity of material transferred in a given time period measured by volume (e.g.; m³/hr.); cf mass flow
volumetric flux	The rate of fluid flow expressed as volume per unit time per unit of area
volumetric heat release	Net heat release per net volume of radiant section in a fired heater. Defined formally in BS EN ISO 13705:2012/ISO 13705:2012(E) Petroleum, petrochemical and natural gas industries. Fired heaters for general refinery service, API STD 530 - Calculation of Tube Heater Thickness and API STD 560 - Fired Heaters for General Refinery Service
volumetric loading	see **volumetric method filling**
volumetric method filling	*aka* volumetric loading. Filling of an LPG container to the maximum permitted liquid volume. Defined formally in NFPA 58: Liquefied Petroleum Gas Code, 2017 Edition; cf weight method filling
volumetric oxygen transfer coefficient	(*units* t⁻¹) *aka* $k_L a$. Volumetric mass transfer coefficient used to predict mass transfer of gases from bubbles to solution, as for example in aeration of effluent
volute	An expanding spiral funnel, such as the volute casing around a centrifugal pump impeller
von Babo's law	see **Babo's law**
von Kármán vortex street	see **Kármán vortex street**
vortex	A fluid rotating around an axis. May be a free vortex (qv) or a forced vortex (qv)
vortex breaker	A device which prevents formation of a free vortex (qv) in draining fluid
vortex cavitation	*aka* tip vortex cavitation. Cavitation (qv) produced by a vortex (qv) (from an impeller tip for example)
vortex flow meter	A device which uses the Kármán vortex street (qv) produced by a blunt body to measure flow

vortex separator	A type of hydrodynamic solid liquid separator, related to a hydrocyclone (qv)
vortex shedding	The process which produces a Kármán vortex street (qv) from a blunt body in a flow
VPSV	see **vacuum pressure safety valve**
VRU	see **vapor recovery unit**
VSB	see **vehicle security barriers**
VSD	see **variable speed drive**
VSS	see **volatile suspended solids**
VT	Visual testing, but given as an initialism for 'visual inspection' in API RP 571 - Damage Mechanism Affecting Fixed Refinery Equipment
VTB	vacuum tower bottoms, see **vacuum residue**
VTSR	Vertical thermosiphon reboiler
Vukančić-Vuković equation	see **Van t'Hoff('s) equation**
vulcanization	A rubber hardening process

W

w/v	A proportion of something in a mixture expressed as a weight/volume (or mass/volume) fraction
w/w	A proportion of something in a mixture expressed as a weight/weight (or mass/mass) fraction.
WABT	see **weighted average bed temperature**
WAC	Weak acid cation; see **weak acid cation resin**
Wacker process	*aka* Hoechst-Wacker process, Walker process (sic). A process for the catalytic oxidation of ethene to ethanal
Waeltz kiln	A rotary kiln used for metals recovery
Waeltz oxide	The metal-enriched product of a Waeltz kiln (qv)
Waeltz process	Recovering zinc and other metals from waste using a Waeltz kiln (qv)
Waeltz slag	The metal-lean byproduct of a Waeltz kiln (qv)
WAG	see **wild ass guess**
Walker process	A common misspelling of Wacker process (qv)
walking beam pump	see **nodding donkey**
walking line	A theoretical average user path; defined formally in BS EN ISO 14122 Safety of machinery. Permanent means of access to machinery. Working platforms and walkways
walkthrough	Usually nowadays a 3D CAD virtual tour of a plant; sometimes still the real-world equivalent
walkway	A surface used for pedestrian access to plant and equipment; defined formally in BS EN ISO 14122 Safety of machinery. Permanent means of access to machinery. Working platforms and walkways
wall shear stress	(*symbol* τ_w) The force per unit area exerted by the wall on a moving fluid, and vice-versa
wall temperature	Average temperature, often that at the wall of a pipe or vessel rather than in the bulk of the contents; cf bulk temperature
Wanson thermutator	*aka* thermutator. Proprietary eponym for a scraped surface heat exchanger
wardrobe style panel	see **Form 1 panel**
warm standby	see **idle**
warn	see **simmer**
warning message	A message to users warning of hazards, with instructions on how to avoid them. Defined formally in BS EN 82079-1:2012 Preparation of instructions for use. Structuring, content and presentation. General principles and detailed requirements

warranty	see **process guarantee**
WAS	see **waste activated sludge**
wash	A solvent used to remove impurities from a solid or liquid product, or the process of using a solvent for this purpose
wash in place	*aka* WIP. A semi-automatic cleaning process intermediate between cleaning in place (qv) and cleaning out of place (qv)
washback	A brewing container in a whisky distillery
washer	A thin plate (typically disk-shaped, but sometimes square) with a hole (typically in the middle), normally used to distribute the load of a threaded fastener, such as a bolt or nut
washings	A liquid (commonly aqueous) waste stream which has been used to wash a product
Washoe process	A variation of the pan amalgamation process (qv) once used for silver recovery
waste	A term with a broad range of possible meanings, see **Box W1**.
waste activated sludge	*aka* WAS, surplus activated sludge (SAS). Sludge produced in excess of that required for recycle (qv return activated sludge) in an activated sludge process
waste disposal	A somewhat contested term, commonly defined as collection, handling and deposition of waste, but disposal is strictly the last part; the rest is arguably waste management (qv)
waste gas	see **relief gas**
waste heat	*aka* low grade heat. Heat considered uneconomic to recover; commonly dumped to environment
waste heat boiler	An oxymoronic term for a boiler used to recover heat exhausted from another process (see **Box W1**.)
waste heat recovery unit	*aka* WHRU. An oxymoronic term for a heat exchanger (such as a rotary heat exchanger (qv) or economizer (qv)) used to recover heat exhausted from another process (see **Box W1**.)
waste heat system	An oxymoronic term for a system which recovers heat from hot turbine exhaust to produce steam or hot oil (see **Box W1**.)
waste hierarchy	*aka* the '3 Rs' of waste management: reduce, reuse and recycle. Sometimes there are 4 Rs: reduce, reuse, recycle, rethink
waste management	A contested term; most definitions make this term a synonym of waste disposal, i.e., getting rid of waste; others emphasize the waste hierarchy (qv) as part of waste management
waste minimization	see **resource efficiency**
waste vapor	see **relief gas**
wastewater	*aka* effluent, sewage. Water considered uneconomic to recover, commonly dumped to environment (usually after some treatment)

Box W1. Waste/Circular Economy

Waste is a contentious concept. We might broadly define it as 'substances or objects which the holder intends or is required to dispose of', but the various terms in the dictionary which contain the word waste show that waste is in the eye of the disposer. How can there be such a thing as a waste heat boiler? If heat can economically be used to drive a boiler, how could it have been waste?

What if there was no such thing as waste? Should we institute a closed loop or circular economy, with zero waste targets? There is now a British Standard[105] on implementation. Setting aside the economic assumption behind the idea, (whether closing material and product loops would, in fact, prevent primary production), the key issue with the circular economy is that each increment in process efficiency costs a little more than the last. Process engineers should also be aware that past a certain efficiency, (even if all we cared about was societal benefit) any additional resources devoted to an increase in efficiency may have greater societal benefit used elsewhere.

Engineers are not tasked with defining or indeed maximizing societal benefit – our role is to provide commercially sound solutions, within the constraints of laws and codes which set minimum standards of health, safety, and environmental performance. Society sets the constraints and engineers then provide societally acceptable process outcomes.

The idea of designing for maximum environmental benefit irrespective of cost is like the similar concept of maximum safety irrespective of price, and carries the same critique of myopia. Perversely, we are wasting resources if we use the minimum possible amount of raw materials with no thought of the incremental cost of each advance in 'efficiency'.

All of this said, in my own work in waste minimization, I visited hundreds of production facilities, none of which were even close to the next incremental improvement being uneconomic. There was always room for greater and economically viable resource efficiency (other than in a single case, where greenwash measures were already rife, serving a valuable marketing - if not technical - purpose). It is always worth checking periodically for resource minimization opportunities, as changes in technology, the cost of feedstocks, energy and 'waste' disposal may alter the threshold for economically viable recovery.

This is the only form of 'working towards the circular economy' that makes any rational sense, but it remains something of a misnomer, because we are not actually working towards the circular economy, any more than Zeno's tortoise was coming to the end of its race. Rather, it is a rolling programme of improvement, balancing cost, health, safety and environment; something engineers have been doing for decades.

Related Terms
downcycling, upcycling, valorization, waste disposal, waste heat recovery unit, waste management

---

[105] BS8001:2017 Framework for implementing the principles of the circular economy in organizations. Guide

water	Tempting as it is to wax lyrical about the author's specialist subject, water is the most commonly used chemical in every process plant, supplied at very high purity for very low cost almost everywhere
water activity	(*symbol most commonly* aw, *less frequently* aw, *sometimes (and perhaps wrongly)* Aw). The partial vapor pressure of water in a substance, divided by the standard state partial vapor pressure of water. In the field of food science, the standard state is most often defined as the partial vapor pressure of pure water at the same temperature; used as a measure of water levels in food
water balance	In chemical engineering contexts, a mass balance (qv) for water alone
water capacity	The capacity of an LPG container expressed as the amount of water at 16° C it can contain; defined formally in NFPA 58: Liquefied Petroleum Gas Code, 2017 Edition
water compatible	Whilst other meanings are clearly possible, in the context of fire safety, a term for the property of having water extinguishable fires. Defined formally in NFPA 654: Standard for the Prevention of Fire and Dust Explosions from the Manufacturing, Processing, and Handling of Combustible Particulate Solids, 2017 Edition; cf water incompatible
water delivery rate	The minimum volumetric flow of water required by a standard such as NFPA 1142 to be delivered to a fire scene. Defined formally in NFPA 1142: Standard on Water Supplies for Suburban and Rural Fire Fighting, 2017 Edition
water depth load	see **environment load**
water for injection	*aka* WFI. A high quality grade of pharmaceutical water. Formal definition in ASME BPE (American Society of Mechanical Engineers: Bioprocessing Equipment)
water gas	A hydrogen enhanced CO/H_2 mixture produced nowadays from synthesis gas (qv), or historically from steam and white hot coke (coal gas (qv))
water gas shift reaction	*aka* WGSR. Carbon monoxide and water vapor reacting to form carbon dioxide and hydrogen
water hammer	*aka* hydraulic shock, surge, transient. Potentially destructive pressure waves (can be positive and/or negative) caused by a momentum change in water; cf fluid hammer

water incompatible	Though other meanings are clearly possible, in the context of fire safety, a term for the property of not having water extinguishable fires. Defined formally in NFPA 654: Standard for the Prevention of Fire and Dust Explosions from the Manufacturing, Processing, and Handling of Combustible Particulate Solids, 2017 Edition cf water compatible, water reactive
water main	A large, commonly underground, pipe used to supply water (especially potable water) to a site
water miscible liquid	A liquid that can be mixed with water in all proportions without the need for additives
water mist	Broadly, a water spray (qv) having droplet size less than 1 mm at minimum nozzle design pressure. Both API RP 2001 - Fire Protection at Refineries and API RP 2001 - Fire Protection at Refineries refer to the formal definition in NFPA 750 Standard on Water Mist Fire Protection Systems, which spells out the droplet size specification a little more scientifically than I have here.
water mist positive displacement pumping unit	In the context of fire protection, multiple parallel positive displacement pumps (qv) supplying a single water mist distribution system. Defined formally in NFPA 20: Standard for the Installation of Stationary Pumps for Fire Protection, 2019 Edition
water pollution	The pollution (qv) of water (qv)
water preheater	Part of a heat recovery steam generator (qv), amongst other things
water quality based effluent limit	*aka* WQBEL. Effluent environmental discharge permit conditions set in the USA, similar to water quality criteria (qv)
water quality criteria	A term with slightly different meanings in different jurisdictions; generically, concentrations of pollutants in an effluent which are set as legally binding maxima by an environmental regulator, on the basis that they would be expected to have no adverse effect on aquatic life in the body of water they are discharged to. Less commonly the criteria include other parameters such as DO (qv) or temperature
water reactive	1. Generically, the property of reacting chemically with water 2. Used in the context of fire safety when the reaction product creates a new fire safety concern. Defined formally in NFPA 654: Standard for the Prevention of Fire and Dust Explosions from the Manufacturing, Processing, and Handling of Combustible Particulate Solids, 2017 Edition; cf water incompatible

water retention	In the context of ion exchange (qv), the percentage of the weight of wetted resin which is water
water seal	see **liquid seal**
water softening	Removal of hardness (qv) from water
water spray	In the context of fire safety, the tightly controlled pattern of water droplets produced by spray nozzles used for fire protection. Defined formally in NFPA 15 Standard for Water Spray Fixed Systems for Fire Protection; cf water mist
water spray system	A fixed pipe water supply system bearing water spray nozzles used for fire protection; defined formally in NFPA 15 Standard for Water Spray Fixed Systems for Fire Protection
water supply and sewerage drawing	A drawing of water utilities. Defined formally in BS EN ISO 10209:2012 Technical product documentation – Vocabulary – Terms relating to technical drawings, product definition and related documentation
water supply officer	*aka* WSO. An individual designated by a fire department as responsible for the supply of firefighting water. Defined formally in NFPA 1142: Standard on Water Supplies for Suburban and Rural Fire Fighting, 2017 Edition
water table	The level below which the ground (qv) is saturated with water
water treatment	The process of preparation of water for a stated purpose, usually by purification
water trials	Commissioning activities conducted using water in place of more hazardous process fluids to allow run-in and operational testing of the equipment as well as to provide operator training and familiarization
water tube boiler	see **watertube boiler**
water vapor	Water (qv) in its vapor phase (also used for water in its gas phase)
water wastage	In the context of firefighting, a water spray (qv) which does not hit a protected surface. Defined formally in NFPA 15 Standard for Water Spray Fixed Systems for Fire Protection
waterbath vaporizer	*aka* immersion type vaporizer. An LPG vaporizer in which a temperature-controlled waterbath heated by an immersion heater is the heat source; defined formally in NFPA 58: Liquefied Petroleum Gas Code, 2017 Edition
waterfall plot	see **cascade plot**
waterfall process	see **stage gate model**
watertight	Impervious to water; functionally identical to liquidtight (qv) under most circumstances
watertube boiler	*aka* water tube boiler. A boiler in which water is circulated through the tubes of a fired heat exchanger; cf firetube boiler

watertube low pressure casing HRSG	A heat recovery steam generator (qv) with multiple water-containing tubes in a gas-filled casing; defined formally in API RP 534 – Heat Recovery Steam Generators
Watson characterization factor	*aka* Watson factor, characterization factor. The ratio of mean boiling point and specific gravity of hydrocarbon fractions; a factor of at least 12.5 suggests predominantly paraffinic (qv), and lower values suggest predominantly naphthenic (qv) or aromatic (qv) hydrocarbons
Watson factor	see **Watson characterization factor**
watt	(*symbol* W) The SI unit of power
watt hour	(*units* Wh, kWh, MWh) Units of power consumed or generated over unit time; kWh and MWh are the more commonly used multiples in professional practice
wave load	see **environment load**
wavelength	The distance between successive peaks in a sinusoidal wave in the direction of propagation. Defined formally in BS ISO 2041:2018 Mechanical vibration, shock and condition monitoring. Vocabulary
waviness	The degree of surface rippling; defined formally in ASME BPE (American Society of Mechanical Engineers: Bioprocessing Equipment)
wavy flow	One of the various possible two phase flow regimes in a horizontal pipe, where waves form at the phase interface
wax	1. Generically, a hydrocarbon or lipid, more solid than grease (qv) 2. In the oil and gas industry, material comprising large hydrocarbon chains, solid at room temperature and soluble in n-heptane (cf asphaltenes, which are n-heptane insoluble)
WBA	Weak base anion, see **weak base anion resin**
WBB	see **wet bottom boiler**
WBC	see **weak base capacity**
WBT	see **wet bulb temperature**
WCB	see **working cell banks**
WD40	Proprietary eponym for spray lubricants[106]
weak acid cation resin	In the context of ion exchange, a resin type used In dealkalization and desalination systems; the functional group tends to be carboxylic acid
weak base anion resin	In the context of ion exchange, a resin type used to selectively remove a wide range of solutes, including mineral acids and total organic carbon; the functional group tends to be amine

---

[106] Half of the universal toolkit (qv)

weak base capacity	*aka* WBC. A measure of reduction in ion exchange resin salt splitting capacity as the result of the degradation of quaternary ammonium groups to tertiary ammonium groups
wearing part	see **normal wear part**
weber	(*symbol* Wb) SI derived unit of magnetic flux
Weber number	(*symbol* We) A dimensionless group relating fluid inertia and surface tension; used in certain specialist applications of fluid mechanics
WEEL	see **workplace environmental exposure levels**
weep hole	1. Generically, a small drain hole at a low point in a system 2. Specifically, in oil and gas, a small drain hole which allows liquids to drain from decks, trays and other internals after a vessel is removed from operation; defined formally in API RP 2510A - Fire Protection Considerations for the Design and Operation of Liquefied Petroleum Gas (LPG) Storage Facilities
weeping	The leaking of liquid through tray perforations in a distillation column; occurs if there is insufficient vapor flow and pressure drops below that required to hold up the liquid. If uncontrolled, can lead to dumping (qv)
weighbridge	A purpose built machine dedicated to weighing large industrial vehicles and their contents, usually for commercial purposes
weighments	Acts of weighing
weight	A force acting on a mass due to gravity
weight average bed temperature	see **weighted average bed temperature**
weight average boiling point	One of five ways used in oil refining to express the average boiling point of a mixture of hydrocarbons (the others being molal, volume, cubic and mean)
weight hourly space velocity	*aka* WHSV. A type of space velocity (qv) used in catalytic reactors, being the weight of feed flow per unit catalyst weight per hour
weight method filling	Filling LPG containers by control of LPG weight; defined formally in NFPA 58: Liquefied Petroleum Gas Code, 2017 Edition; cf volumetric method filling
weight ton	see **long ton**
weighted average bed temperature	*aka* WABT, weight average bed temperature. A key parameter in hydroprocessing: an estimate of catalyst temperature, whose maximization enhances desirable reactions, especially those which remove sulfur and nitrogen from the product
weightometer	*aka* belt weigher. An instrument which measures mass flow of bulk materials on a conveyor belt

weir	A linear obstruction to open channel flow used for flow control or measurement
Weir pump	Proprietary eponym referring to a direct-impingement piston-type reciprocating pump
Weizmann process	see **ABE fermentation**
WEL	see **workplace exposure limits**
weld	Most commonly, attaching two objects made from fusible material (usually metals, but can be plastics) together by localized melting (using heat, or sometimes solvents in the case of plastics), or the localized coalescence of material that this process produces. Definitions based on metal welding may be found in API RP 579 - Fitness for Service and API RP 945 - Avoiding Stress Cracking in Amine Units
weld bead	*aka* bead. The deposit of filler material visible on the surface of a weld
weld joint design	The intended size and shape of a weld; defined formally in ASME BPE (American Society of Mechanical Engineers: Bioprocessing Equipment)
weld metal	The part of a weld which is melted by welding; defined formally in API STD 620 - Low Pressure Storage Tanks
weld whitening	A grain structure difference between weld and base metal, which is visible after electropolishing. Defined formally in ASME BPE (American Society of Mechanical Engineers: Bioprocessing Equipment)
welded joint	The point of union of members by welding; defined formally in API STD 620 - Low Pressure Storage Tanks
welder	Usually refers to someone who welds metal manually or semiautomatically; defined formally in API STD 620 - Low Pressure Storage Tanks, though may also include the less skilled operators of automatic welding equipment. See **welding operator**
welder's cap	A brightly decorated cap with a bill worn by a welder (qv). The bill protects the ears from hot metal and sparks; the customary bright decoration is more of a mystery

welding operator	Considering how obvious this term appears, it is perhaps surprising that there is a difference between the formal definitions in ASME BPE (American Society of Mechanical Engineers: Bioprocessing Equipment) and API STD 620 - Low Pressure Storage Tanks. The ASME BPE definition includes operators of automatic welding equipment, whilst API does not[107]
weldment	1. Generically, a number of parts joined by welding 2. However, API RP 945 - Avoiding Stress Cracking in Amine Units, defines the term as those parts of the weld and base metal affected by heat cf welded joint
Weldolet	*aka* WOL. Proprietary eponym, an abbreviated form of butt WELDed pipe branch OutLET connection. A type of olet (qv)
well	1. Generically, an excavation in the ground from which a fluid is extracted, especially groundwater, often as opposed to a drilled borehole (qv) 2. In the oil and gas industry, an abbreviation of 'oil well', or 'gas well': a borehole used for hydrocarbon extraction 3. An enclosure around a fixed ladder as per OSHA 1910.29 - Fall protection systems and falling object protection-criteria and practices
well checker	see **test separator**
well completion	*aka* completion. The process of making a fully drilled well-bore ready for production operations
well control	*aka* blowout prevention. Preventing a sudden and highly hazardous release of well fluids (qv) from an oil well caused by sudden entry of high pressure gas, known as a blowout
well control event	*aka* well control incident. An incident in oil and gas drilling and completion with the possible consequence of a blowout
well control incident	see **well control event**
well fluids	*aka* produced fluids. The mixture of oil, gas, water, sand, etc. arising from an oil or gas well
well fracturing pump control	see **frac pump control**

---

[107] This may be because the oil and gas industry does not use such equipment (used to make high integrity welds in the hygienic industries). Welding operator does however tend to sound to an engineer less skilled than 'welder', in the same way as CAD operator (qv) sounds less skilled than draughtsman (qv)

well interventions	Operations performed on an oil and gas well to optimize its performance, address issues, and maintain its integrity without requiring a full workover. These interventions can be "light", involving tools and sensors lowered into a live well, or "heavy", which may require removing wellhead equipment and killing the well
well test separator	see **test separator**
well tester	see **test separator**
well trained user	In the context of safe working at height, a person trained and experienced in using a fall arrester. Defined formally in BS EN ISO 14122 Safety of machinery. Permanent means of access to machinery. Working platforms and walkways
well water	Groundwater obtained from a well (qv); often as opposed to a drilled borehole
well-bore	An oil and gas term for a borehole (qv)
Wellman-Lord process	A type of wet flue gas desulfurization process
West Texas intermediate	*aka* WTI, Texas light sweet. A benchmark classification of sweet light crude oil; cf Brent crude, Dubai crude
wet air oxidation	see **Zimmermann process**
wet barrel hydrant	A fire hydrant type used where there is no risk of freezing weather; defined formally in NFPA 24: Standard for the Installation of Private Fire Service Mains and Their Appurtenances, 2019 Edition; cf dry barrel hydrant
wet basis	*aka* wet basis moisture content, moisture content. The weight of water in a wet solid as a percentage of its total mass
wet basis moisture content	see **wet basis**
wet bottom boiler	*aka* WBB. A boiler whose bottom temperature is above ash melting point, from which ash is removed in a molten form
wet bulb temperature	*aka* WBT. The temperature shown by a thermometer with a sensor wrapped in a wet wick; cf dry bulb temperature
wet cleaning	*aka* cleaning out of place. "Cleaning (and disinfection if necessary) of equipment or processing environment with aqueous solutions of detergent (and disinfectant if necessary) followed by rinsing." according to the EHEDG Glossary
wet commissioning	The later stages of cold commissioning (qv), undertaken with less-hazardous fluids (mainly air, water and solvents) in the system, such as water trials (qv)
wet critical speed	The critical speed of a pump when filled with pumped fluid. Defined formally in EN ISO 13709:2003 Centrifugal pumps for petroleum, petrochemical and natural gas industries; cf dry critical speed

wet gas	1. Usually, gas containing a small amount of entrained liquid 2. In oil and gas, it can mean either (1.), or alternatively, natural gas containing significant quantities of condensables cf dry gas
wet pit	1. Generically, a pump sump or wet well 2. In the context of fire safety, a sump filled passively from an open body of freshwater via a screening arrangement. Defined formally in NFPA 20: Standard for the Installation of Stationary Pumps for Fire Protection, 2019 Edition
wet porch	see **moon pool**
wet process	1. One of a large number of processes involving mixing solids with water, often to produce a slurry for subsequent processing 2. A sulfuric acid production process
wet scrubbing	A unit operation which removes acid or odorous gases from exhaust gases by scrubbing with an aqueous liquid
wet steam	Steam with suspended water particles, usually defined as more than 0.5% moisture
wet well	A sewage pumping station consisting of a single chamber which acts both a sump and a pump mounting chamber, necessitating the use of submersible pumps; cf drywell
wet volume capacity	In the context of ion exchange (qv), exchange capacity per unit hydrated resin volume
wetlands	A geographical area constantly exposed to surface or groundwater
wetness	see **gas wetness**
wetted parts	Parts of equipment in contact with liquid, usually process fluid, especially aqueous process fluid
wetted perimeter	The length of an open channel perimeter wetted by flowing fluid
wetting agent	Surfactant
WFI	see **water for injection**
WFMT	Wet fluorescent magnetic particle testing
WGSR	see **water gas shift reaction**
wharf	A platform alongside a body of water used for vessel docking, loading and unloading; defined formally in NFPA 30: Flammable and Combustible Liquids Code, 2018 Edition
what-if analysis	1. A structured brainstorming method for determining what things can go wrong and judging the likelihood and consequences of those situations occurring, such as HAZOP (qv) 2. A set of tools in MS Excel which allow iteration
Whessoe	Proprietary eponym for a vent/vac valve (qv)
whirlybird	*(informal)* A helicopter

white hydrogen	Naturally occurring hydrogen; cf hydrogen colors
white liquor	Turbid, white, alkaline solution/suspension used in the Kraft papermaking process, which becomes black liquor (qv) when spent
white oil	see **mineral oil**
white rust	Zinc oxide
white spirit	see **mineral spirits**
Whitman two film theory	*aka* Lewis-Whitman two film theory, standard two film theory, STFT. A theoretical model of steady state mass transfer
wholesomeness	A deprecated term for food suitability (qv)
WHRU	see **waste heat recovery unit**
WHSV	see **weight hourly space velocity**
WI	1. see **Wobbe index** 2. see **wrought iron**
wild ass guess	*aka* WAG. Less scientific than a SWAG (qv)
wildcat	An exploratory oil well in unproven territory; cf development well
wildcat connection	A delta (qv) pattern three phase connection
Wilden pump	*aka* Wildon pump (in error). Proprietary eponym for air operated pumps; cf Blagdon pump, Sandpiper pump
Wildon pump	Common misspelling of Wilden pump
Wilke-Chang correlation	A way to estimate unknown diffusion coefficients in dilute solutions
wind fence	A ground level structure protecting a flare from crosswinds; defined formally in API RP 537 Flare Details for Petroleum, Petrochemical, and Natural Gas Industries
wind load	see **environment load**
wind rose	A graphic tool used by meteorologists to give a succinct view of how wind speed and direction are typically distributed at a particular location; used when selecting stack position
windage (loss)	see **drift loss**
windbox	*aka* plenum. A burner air distribution chamber; defined formally in API STD 530 - Calculation of Tube Heater Thickness
windmilling	A situation when an idle fan in a cooling tower rotates backwards
Windscale fire	The UK's worst nuclear accident to date, level 5 out of a possible 7 on the International Nuclear Event Scale, involving the ignition of several tonnes of uranium

windshield	A device shielding downwind flare burner components from flame impingement. Very similar formal definitions may be found in API Standard 521. Pressure-relieving and Depressuring Systems. Sixth Edition	January 2014 and API RP 537 Flare Details for Petroleum, Petrochemical, and Natural Gas Industries
winterizing	see **dewaxing**
WIP	see **wash in place**
wiped film evaporator	A type of evaporator (qv) used to concentrate heat-sensitive liquid mixtures
wire mesh type mist eliminator	A device which coalesces and collects small solid particles and liquid droplets from a gas stream by means of a wire mesh pad
wireline	An oil industry term for a number of types of cable which go down the wellbore, in summary: 1. A thin loadbearing cable without an electrical conductor used to raise and lower tools in a well shaft called a slickline 2. An electric cable connecting sensors in a well to instruments aboveground called an e-line with three subtypes (multi-conductor, single conductor and braided)
wireline operator	A worker who sets up and uses a wireline (qv)
wireline work areas	Areas where wirelines (qv) are being deployed. Defined formally in API RP 505 2nd Edition, August 2018 Recommended Practice for Classification of Locations for Electrical Installations at Petroleum Facilities Classified as Class I, Zone 0, Zone 1, and Zone 2
wiring	An electrical installation of cabling and associated devices such as switches, distribution boards, sockets, and light fittings in a structure
wiring schematic	A schematic diagram of equipment or plant wiring (qv)
wispy annular flow	One of the various possible two-phase flow regimes in a pipe, with a rapidly moving core of gas and liquid, and a slower moving continuous film of liquid at the wall; cf annular flow, slug flow
witch's hat	*aka* witch's tit. A temporary conical strainer inserted into a flange set during commissioning
witch's tit	*(potentially offensive)* Oil and gas industry version of witch's hat (qv)
withdrawal loss	In the context of petrochemical industry storage tanks, loss by liquid evaporation. Defined formally in API MPMS 19.1 Manual of Petroleum Measurement Standards Chapter 19.1 Evaporative Loss from Fixed-roof Tanks
within sight	see **in sight from**

within sight from	see **in sight from**
witnessed	A contractual inspection or test which the purchaser is notified of, and continuation of production is conditional on attendance of their representative. Defined formally in EN ISO 10437 Petroleum, petrochemical and natural gas industries - Steam turbines - Special-purpose applications and EN ISO 13709:2003 Centrifugal pumps for petroleum, petrochemical and natural gas industries cf observed
witnessed inspection	*aka* witnessed inspection test, witnessed test. An inspection which is witnessed (qv). A formal definition of this term compatible with 'witnessed' may be found in ISO14661 Thermal turbines for industrial applications (steam turbines, gas expansion turbines) - General requirements. A slightly different definition can be found in API RP 682 - Pump Seals cf observed inspection
witnessed inspection test	see **witnessed inspection**
witnessed test	see **witnessed inspection**
WNF	*(drawing notation)* Weld neck flange
WO	see **work order**
Wobbe Index	*aka* WI. The gross heating value of a gaseous fuel divided by the square root of its specific gravity (qv). Defined formally in API RP 535 - Burners for Fired Heaters at Refineries and ISO 3977 Gas turbines - Procurement - Part 3: Design requirements
wobble pump	see **progressing cavity pump**
WOD	see **write only documents**
WOL	see **Weldolet**
wood gas	see **producer gas**
work environment	Conditions in the workplace. Defined formally in BS EN ISO 9000 Quality management systems Fundamentals and vocabulary
work hardening	*aka* strain hardening. Increasing material strength by plastic deformation
work method statement	see **method statement**
work order	*aka* WO. A detailed description of a work process, often with assignment of responsibilities to named individuals, commonly used in management of maintenance
work permit	1. Most commonly, a document required by a foreign national to be able to work in a country 2. Rarely, a synonym for permit to work (qv)
working capital	*aka* net working capital (NWC). In accounting, the difference between current assets and current liabilities

working cell banks	*aka* WCB. Cell banks derived from a master cell bank (qv) under GMP (qv); used as source of cells in bioprocess engineering
working drawing	A drawing with dimensional information for field use by construction or fabrication operatives; cf construction drawing, production drawing
working fluid	see **pump fluid**
working medium	see **pump fluid**
working platform	A horizontal platform used whilst working on equipment; defined formally in BS EN ISO 14122 Safety of machinery. Permanent means of access to machinery. Working platforms and walkways; cf walkway
working pressure	*aka* WP; see **maximum allowable working pressure**
workover	The necessary removal and replacement of oil well wireline (qv) or casing to extend its life
workplace environmental exposure levels	*aka* WEEL. Occupational exposure limits in the form of time weighted average concentrations for air contaminants used in the USA, similar in nature but broader in scope than threshold limit value (qv), approximately equivalent to the UK's workplace exposure limits (qv)
workplace exposure limits	*aka* WEL. Time weighted average concentrations of certain hazardous substances in workplace air; used in UK health and safety regulation; cf STEL, LTEL, workplace environmental exposure levels
works	1. Process plant; defined formally in BS EN ISO 10209:2012 Technical product documentation – Vocabulary – Terms relating to technical drawings, product definition and related documentation 2. Construction activity 3. see **gubbins**
worm	1. see **screw** 2. *(informal)* A beginner in oilfield work
WP	see **working pressure**
WQBEL	see **water quality based effluent limit**
WRC	In the context of ion exchange (qv), water retention (qv) capacity
WrC	Water Research Council; a UK research organization
wrinkles	Pipeline anomalies caused by bending which may require repair; cf buckle
write only documents	*aka* WOD, write only documentation. Documents which must be written and filed somewhere for some reason (usually to do with paying lip service to something), but are never actually read

wrought iron	*aka* WI. A term properly used for a low carbon steel with fibrous slag inclusions which can be worked at higher temperatures than mild steel (qv), and weathers far better. Commonly however used to cover any low carbon steel, especially mild steel; cf gray iron
WSO	see **water supply officer**
WT	*(drawing notation)* Wall thickness
WTI	see **West Texas intermediate**
WUB	Whole uncracked beads, in the context of ion exchange (qv)
Wulff process	A process for production of acetylene (ethyne) from hydrocarbon gas and steam in a regenerative furnace
WVC	Wet volume capacity, in the context of ion exchange (qv)
WWT	Wastewater treatment, or water and waste treatment
wye	1. Abbreviation of wye fitting (qv)
	2. A star- or Y-shaped pattern of three-phase electrical connection, e.g.; wye delta (qv)
wye delta	*aka* star delta. The commonest kind of three phase motor starter, allowing a reduced motor starting current compared with a direct on line (DOL) starter. The connections are in two configurations, 'delta' and 'wye' (*aka* 'star')
wye fitting	*aka* wye, Y fitting. A pipe fitting with three openings used to create branch lines which is more Y shaped than tee (qv) shaped

X

X ray view	A perspective drawing showing an object as if partially transparent; defined formally in BS EN ISO 10209:2012 Technical product documentation — Vocabulary — Terms relating to technical drawings, product definition and related documentation
XCAT	A proprietary catalytic propylene production process
XFMR	*(drawing notation)* Transformer
XH	Extra heavy, see **extra strong**
XIS	A proprietary p-xylene isomerization (qv) process
Xmas tree	see **Christmas tree**
XR	*(drawing notation)* X-ray [at pipe welds]
XS	1. Excess 2. see **extra strong**
XTLA	see ETLA
XV	*(drawing notation)* see **utility valve**
XXH	Extra extra heavy; see **double extra strong**
XXS	Extra extra strong; see **double extra strong**
xylene	see **xylol**
Xylenes-plus	*aka* xylol refining process. A proprietary catalytic isomerization (qv) process turning toluene into benzene and xylene
Xylofining process	*aka* xylol refining process. A proprietary catalytic isomerization (qv) process turning a mix of ethyl- and methyl- benzene to xylene, ethylene and benzene
xylol	*aka* xylene. A mixture of the o-, m- and p- isomers of dimethylbenzene
xylol refining process	see **Xylenes-plus** or **Xylofining process**

Y

Y fitting	see **wye fitting**
Y strainer	A Y-shaped line strainer
yard	An Imperial unit of length, slightly less than a meter, obsolete everywhere except as part of US customary units (qv)
Yarway valve	Proprietary eponym for an automatic recirculation valve
yeast	A single celled fungus used in brewing and more modern biotechnology
yellow hydrogen	Hydrogen produced by solar powered electrolysis; cf hydrogen colors
yellowcake	The concentrate produced by initial uranium ore beneficiation (qv)
yield	1. In mechanical engineering, bend(ing) or break(ing) under pressure 2. In reaction engineering, usually means a percentage conversion (strictly percentage yield), but used in different senses such as sludge yield (qv), and as a simple amount of product
yield per pass	The yield (qv) per cycle in a reaction with recycle
yield point	Stress at the point where a material becomes plastic
yield strength	Stress at the point where a defined amount of permanent deformation has occurred; defined formally in API RP 579 - Fitness for Service
yield stress	see **yield strength**
yocto-	(*units* y-) SI unit prefix denoting a factor of 10^{-24}
yotta-	(*units* Y-) SI unit prefix denoting a factor of 10^{24}
Young–Laplace equation	An equation relating surface tension and weight; used to determine liquid drop contact angle on a surface
Young's modulus	*aka* elastic modulus. A measure of the stiffness of a material (rather than its hardness, strength, or toughness)

Z

Z value	The temperature increase, in degrees, required to obtain a 1 log reduction in D value (qv); it applies only at the temperature/D-value at which it is measured
Z90	The length of catalyst bed needed to achieve a 90% temperature change in an adiabatic reactor vessel; used for quick estimation of catalyst activity in ammonia or HyCO (qv) plant reactors such as pre-reformers and shift convertors
ZDL valve	see **zero deadleg valve**
ZEBRA	see **zero energy best rate account**
zebra stripe display	Zebra stripes allow the detection of small surface defects which are otherwise hard to see by simulating the reflection of long strips of light on a very shiny surface; they also allow an observer to verify that two adjacent faces are in contact or tangential, or have continuous curvature
Zenith process	An unconventional semicontinuous edible oil refining process, comprising treatment with phosphoric acid to remove non-fatty substances, centrifugation to remove solids, neutralization and bleaching
zeolite	*aka* molecular sieve. A mineral composed of hydrated aluminosilicates; used as an adsorbent, catalyst etc.
zepto-	(*units* z-) The SI unit prefix denoting a factor of 10^{-21}
zero dead leg valve	see **zero deadleg valve**
zero deadleg valve	*aka* zero dead leg valve, ZDL valve. A type of diaphragm valve favored in pharma; cf mixproof valve
zero energy best rate account	*aka* ZEBRA. A 'Mountweasel', supposedly a financial product formerly offered by the Lirpa Loof investment bank[108]
zero liquid discharge	*aka* ZLD. In short, rarely what it sounds like: usually greenwash (qv); really means a reduced liquid discharge, diversion of liquid discharge to irrigation etc.
zero order reaction	A reaction whose rate is independent of concentration of reactants
zeroth law of thermodynamics	The law which states that two thermodynamic systems in thermal equilibrium with a third are in thermal equilibrium with each other
zeta potential	see **electrokinetic potential**

---

[108] The clue is in the name of the bank, though it would have been more obvious on the date of issue that it was less than legitimate

zetta-	(*units* Z-) The SI unit prefix denoting a factor of 10^{21}
Ziegler process	*aka* Ziegler-Alfol synthesis. The production of fatty alcohols from ethylene with triethylaluminium
Ziegler-Alfol synthesis	see **Ziegler process**
Ziegler–Natta catalysts	Mostly titanium based compounds used in the synthesis of many commercially important polymers of alkenes (olefins), such as polyethylene
Ziegler–Nichols tuning method	A heuristic PID controller (qv) tuning method with more limited applicability than may be implied in academia[109]
Zimmerman process	*aka* wet air oxidation, Zimpro. A wastewater treatment process using air or oxygen and heat in a pressurized reactor
Zimpro	see **Zimmermann process**
Zincex process	A zinc recovery process based on solvent extraction; rarely used unmodified nowadays due to the complexity of the original process
zirconium	A reactive metal used in the fabrication of extremely corrosion-resistant process equipment in highly demanding applications, e.g.; acid service heat exchanger tubes
Zirpro process	A textile flameproofing process
ZLD	see **zero liquid discharge**
zone 0	An area where an explosive atmosphere of a flammable gas/air mixture is commonly present; defined formally in ATEX 137 (1999/92/EC), IEC 60079-10-1 Area Classification – Gases and Vapors, IEC 60079-10-2 Area Classification – Hazardous Dusts, BS EN 60079 - Explosive atmospheres and API RP 505 2nd Edition, August 2018 Recommended Practice for Classification of Locations for Electrical Installations at Petroleum Facilities Classified as Class I, Zone 0, Zone 1, and Zone 2; qv hazardous (classified) location, hazardous area classification study; cf zone 1, zone 2

---

[109] Rather than summarize the long list of circumstances under which this method should not be applied, it is simpler to quote process control expert Myke King: "*Don't apply the Ziegler-Nichols tuning method*"

zone 1	An area where an explosive atmosphere of a flammable gas/air mixture is occasionally present; defined formally in ATEX 137 (1999/92/EC), IEC 60079-10-1 Area Classification – Gases and Vapors, IEC 60079-10-2 Area Classification – Hazardous Dusts, BS EN 60079 - Explosive atmospheres and API RP 505 2nd Edition, August 2018 Recommended Practice for Classification of Locations for Electrical Installations at Petroleum Facilities Classified as Class I, Zone 0, Zone 1, and Zone 2 qv hazardous (classified) location, hazardous area classification study; cf zone 0, zone 2
zone I/div I	An area with a high risk of explosive atmosphere in normal operation; defined formally in ISO 3977 Gas turbines - Procurement - Part 3: Design requirements
zone 2	An area where an explosive atmosphere of a flammable gas/air mixture is unlikely to be present; defined formally in ATEX 137 (1999/92/EC), IEC 60079-10-1 Area Classification – Gases and Vapors, IEC 60079-10-2 Area Classification – Hazardous Dusts, BS EN 60079 - Explosive atmospheres and API RP 505 2nd Edition, August 2018 Recommended Practice for Classification of Locations for Electrical Installations at Petroleum Facilities Classified as Class I, Zone 0, Zone 1, and Zone 2 qv hazardous (classified) location, hazardous area classification study; cf zone 0, zone 1
zone 20	An area where an explosive atmosphere of a dust/air mixture is commonly present; defined formally in IEC 60079-10-2 Area Classification – Hazardous Dusts and BS EN 60079 - Explosive atmospheres qv hazardous (classified) location, hazardous area classification study; cf zone 21, zone 22
zone 21	An area where an explosive atmosphere of a dust/air mixture is occasionally present; defined formally in IEC 60079-10-2 Area Classification – Hazardous Dusts and BS EN 60079 - Explosive atmospheres qv hazardous (classified) location, hazardous area classification study; cf zone 20, zone 22
zone 22	An area where an explosive atmosphere of a dust/air mixture is unlikely to be present; defined formally in IEC 60079-10-2 Area Classification – Hazardous Dusts and BS EN 60079 - Explosive atmospheres qv hazardous (classified) location, hazardous area classification study; cf zone 20, zone 21
zone II/div I	An area with a low risk of explosive atmosphere in normal operation; defined formally in ISO 3977 Gas turbines - Procurement - Part 3: Design requirements

zone refining	Refining by repeated zone melting (passing a molten zone down a bar or rod of material)
zoning	1. Most commonly used in the sense of zoning study (qv); cf high care areas, low care areas, medium care areas 2. Less commonly, in a hygienic context, "the physical or visual division of the plant into sub-areas, leading to the segregation of different activities with different hygiene levels" according to the EHEDG Glossary
zoning study	see **hazardous area classification study**

Printed in Dunstable, United Kingdom